Primate Locomotion
Recent Advances

Primate Locomotion
Recent Advances

Edited by

Elizabeth Strasser
California State University, Sacramento
Sacramento, California

John Fleagle
State University of New York at Stony Brook
Stony Brook, New York

Alfred Rosenberger
National Zoological Park
Washington, D.C.

and

Henry McHenry
University of California, Davis
Davis, California

Plenum Press • New York and London

Library of Congress Cataloging-in-Publication Data

Primate locomotion : recent advances / edited by Elizabeth Strasser
... [et al.]
         p.   cm.
      "Proceedings of a symposium on Primate Locomotion, held March
27-28, 1995, in Davis, California"--T.p. verso.
      Includes bibliographical references and index.
      ISBN 0-306-46022-X
       1. Primates--Locomotion--Congresses. 2. Primates--Behavior-
-Congresses. 3. Primates--Evolution--Congresses. I. Strasser,
Elizabeth.   II. Symposium on Primate Locomotion (1995 : Davis,
Calif.)
QL737.P9P7256 1998
573.7'9198--dc21   98-38702
                                                            CIP

Proceedings of a symposium on Primate Locomotion, held March 27 – 28, 1995,
in Davis, California

ISBN 0-306-46022-X

© 1998 Plenum Press, New York
A Division of Plenum Publishing Corporation
233 Spring Street, New York, N.Y. 10013

http://www.plenum.com

10 9 8 7 6 5 4 3 2 1

Printed in the United States of America

PREFACE

The study of primate locomotion is a unique discipline that by its nature is interdisciplinary, drawing on and integrating research from ethology, ecology, comparative anatomy, physiology, biomechanics, paleontology, etc. When combined and focused on particular problems this diversity of approaches permits unparalleled insight into critical aspects of our evolutionary past and into a major component of the behavioral repertoire of all animals. Unfortunately, because of the structure of academia, integration of these different approaches is a rare phenomenon. For instance, papers on primate behavior tend to be published in separate specialist journals and read by subgroups of anthropologists and zoologists, thus precluding critical syntheses.

In the spring of 1995 we overcame this compartmentalization by organizing a conference that brought together experts with many different perspectives on primate locomotion to address the current state of the field and to consider where we go from here. The conference, *Primate Locomotion-1995*, took place thirty years after the pioneering conference on the same topic that was convened by the late Warren G. Kinzey at Davis in 1965. The 1965 conference (published in 1967) brought together students of primate behavior (Ripley), morphology (Tuttle, Napier, Oxnard, Prost), paleontology (Simons), and others to address a common theme and, for many, to define the study of primate locomotion as a distinct area of research. As the papers in the current volume demonstrate, we have come a long way in thirty years with naturalistic studies on dozens of species, new skeletal remains of numerous fossil species, and an expanding battery of morphometric and experimental techniques. We hope that the excitement and spirit of interdisciplinary collaboration that characterized the 1965 conference is still evident in the field these three decades later.

At the 1995 conference, the papers were presented in six sessions. Upon hearing the presentations, we subsequently decided that a book divided into four sections would accommodate the material better. The first section of this book (Part I: *Naturalistic Behavior*) contains six papers describing the positional behavior of primates from all the major extant groups. Five papers are grouped in Part II (*Morphology and Behavior*). These papers integrate functional studies of primate morphology with experimental studies on a wide range of primate behaviors. They provide the critical link that enables us to reconstruct the behavior of extinct species known only from their bones. The six papers in Part III (*Data Acquisition and Analytic Techniques*) demonstrate the potential for new techniques as well as the promises and problems inherent in currently used techniques. In the final section of this book (Part IV) six papers are grouped together under the section title

Fossils and Reconstructing the Origins and Evolution of Taxa. While the origin and radiation of major groups of primates have been linked to the acquisition of novel locomotor strategies, it appears that the impetus for the development of novel locomotor behaviors is first of all related to diet. The six papers in this section address this issue.

Given the breadth of approaches used in this book, we hope that its promise to be an exciting update to the 1965 conference on primate locomotion holds true.

The conference *Primate Locomotion-1995* was generously supported by the National Science Foundation (SBR 9507711), The Wenner-Gren Foundation for Anthropological Research (CONF-167), and the California State University, Sacramento Foundation (#120025). In an effort to have the highest quality papers in this book, each manuscript was reviewed by a combination of the other contributors, the coeditors, and an assortment of colleagues to whom we are truly grateful for their efforts. We thank Geoffrey Kushnick for his assistance in compiling the index for this volume, Diana Norman for the cover artwork, and Eileen Bermingham, Donna Carty, MaryAnn McCarra, and Susan Safren for their patience in working with us on this book.

Elizabeth Strasser, Sacramento, California
John G. Fleagle, Stony Brook, New York
Alfred L. Rosenberger, Washington, D.C.
Henry M. McHenry, Davis, California

REFERENCES

Kinzey WG, editor (1967) Symposium on Primate Locomotion. Am. J. Phys. Anthropol., *26(2)*.

CONTENTS

Part III: Data Acquisition and Analytic Techniques

Part IV: Fossils and Reconstructing the Origins and Evolution of Taxa

I

NATURALISTIC BEHAVIOR

INTRODUCTION TO PART I

Alfred L. Rosenberger

This first section of the book has a variety of interesting papers that deal with the study of positional behavior in the wild and that sample all the major taxa of primates. The first three papers explore topics ranging from methodological issues to the first, surprisingly, comparative study of the positional behavior of prehensile-tailed primates. The last three papers of this section examine the effect of habitat characteristics on positional behavior and come to some surprising conclusions.

In Chapter 1, as our reviewers noted, Marian Dagosto and Daniel Gebo provide a thoughtful, balanced, and judicious treatment of all the major problems and points of importance concerned with the study of primate positional behavior. Their paper is both a cautionary tale about incommensurate research designs and a sobering summary of the state-of-the-art data base of field-workers. Not only will behaviorists find this contribution useful, but functional morphologists interested in the design and performance of the primate postcranial skeleton also will find it to be valuable and thought-provoking on many levels. Dagosto and Gebo's paper is appropriate as the first in this volume, since it is such a fitting tribute to Warren G. Kinzey as well as his colleague Suzanne Ripley.

Suzanne Walker's paper (Chapter 2) addresses some current deficiencies in the definition and use of gross categories of positional behavior. She illustrates the deficiencies with examples of positional behaviors that require finer-grained descriptions than the gross categories of "sit" or "stand". In addition, she examines some associations between the newly identified positional behaviors and characteristics of the supports that would be missed using gross characterizations.

David Bergeson (Chapter 3) uses the results of a naturalistic study of tail use in three Costa Rican monkeys to examine a series of hypotheses about the ecological significance of fore- and hindlimb suspensory behavior in primates with prehensile tails. In general, he found that hypotheses put forth to explain suspensory behavior in other groups of primates do not seem to accord with patterns of suspension using prehensile tails. Moreover, the three platyrrhine species that he studied show different patterns of tail suspension.

In Chapter 4, Paul Garber examines both between- and within-site variation in the positional behavior of Peruvian moustached tamarins. He found only limited between-site and within-site variation in positional patterns: rare modes were always rare and common modes were always common. Support characteristics were found to vary more than positional behaviors. As a consequence, as one reviewer noted, in this paper Garber reinforces the growing awareness that support choice is less constrained by morphology or ecology than are positional behaviors.

In Chapter 5, our reviewers note that Scott McGraw's is a timely paper on the debate of the relationship between habitat structure and locomotor behavior. McGraw examines within-species variability among three sympatric cercopithecids to determine if the monkeys change their locomotion and support use during the two maintenance activities of traveling and foraging. Like Garber (Chapter 4) he found that the pattern of support use is more variable than are the accompanying locomotor behaviors. Furthermore, the effect of maintenance activities on locomotor profiles is nil. As McGraw points out, the fact that the focal species moved in the same general manner independent of differences in their respective habitats has implications for the study of fossils, in that it strengthens our ability to predict locomotor behavior from morphology.

In Chapter 6, Melissa Remis reports on the positional behavior and substrate use of lowland gorillas and compares them to similar data for mountain gorillas. Her main argument is that the differences in positional behavior between lowland and mountain gorillas are related to habitat differences. In contrast to lowland gorillas, mountain gorillas are constrained from engaging in more arboreal behaviors by the distribution of food items. As noted by Remis, her results do not correspond with the generalizations given by Garber (Chapter 4) and McGraw (Chapter 5).

METHODOLOGICAL ISSUES IN STUDYING POSITIONAL BEHAVIOR

Meeting Ripley's Challenge

Marian Dagosto[1] and Daniel L. Gebo[2]

[1]Department of Cell and Molecular Biology
Northwestern University School of Medicine
Chicago, Illinois 60611
[2]Department of Anthropology
Northern Illinois University
DeKalb, Illinois 60115

1. INTRODUCTION

A major goal of primate evolutionary morphology is to relate limb anatomy to aspects of locomotor or postural behavior or to an entire regime of positional behavior. If well established in living forms these relationships provide the necessary framework for reconstruction of the behavior of extinct species and to answer questions concerning the evolution of primate locomotor systems that were the impetus for the 1965 conference (Kinzey, 1967). Studies of this sort usually proceed by noting salient differences in behavior and morphology in two or more species, establishing correlations between the two, and attempting to explain the correlations using causal arguments derived from biomechanical principles (e.g., Ashton and Oxnard, 1964; Fleagle, 1977a,b; Fleagle and Meldrum, 1988). Obviously, success in such an endeavor depends on the quality of both morphological and behavioral data sets (Ripley, 1967; Fleagle, 1979). In her contribution to the 1965 symposium, Ripley pointed out how the lack of detailed information on behavior hampered efforts to assess the evolutionary importance of behaviors and to construct realistic locomotor groupings or classifications. Morphologists tend to "underestimate the rich complexity of locomotor behavior and to underestimate the difficulties involved in relating morphology and habitual behavior" (Ripley, 1965:167). She especially criticized attempts to define locomotor types in primates on the basis of insufficient information. At the 1965 conference, both Ripley (1967) and Kinzey (1967) challenged primatologists to collect more data to alleviate these problems. Ripley (1967:149) outlined the elements of

Primate Locomotion, edited by Strasser *et al.*
Plenum Press, New York, 1998

"an exhaustive study of the total locomotor pattern" necessary before such categorizations would be meaningful.

Although there has been progress since 1965 we believe the study of positional behavior still falls short of the goals set by Ripley and Kinzey to the point where integration of behavioral and morphological data is still compromised. There are several aspects of positional behavior studies that contribute to this deficiency. (There are problems with morphological studies as well, but it is not our intent to address these here.) First, long term, detailed studies of positional behavior as outlined by Ripley (1967) are still rare compared to those of diet or social structure, and thus the base for studies built around this information is weak. Secondly, there have been very few discussions concerning the methods of studying positional behavior. Several techniques of data collection and analysis exist, and whether results are comparable across studies is debatable and has never been rigorously tested. Comparability of data is a serious issue for the type of broadly interspecific quantitative studies that are currently being attempted (e.g., Crompton et al., 1987; Oxnard et al., 1990). Thirdly, sources of intraspecific variation in behavior have only begun to be looked at, and there is disagreement about the existence and importance of these factors.

2. RARITY OF POSITIONAL BEHAVIOR STUDIES

Previous to the 1965 symposium, assessments of positional behavior were based largely on qualitative anecdotal reports or on observations in a captive setting (e.g., Avis, 1962; Ashton and Oxnard, 1964). To assess improvement since that time we counted up studies that were directed at documenting positional behavior in the wild in a quantitative fashion, or that at least contained usable quantitative accounts of positional behavior or substrate use (Table 1). Other notable studies of positional behavior that are non-quantitative or were conducted in a captive setting are listed in Table 2. Since the 1965 symposium, there have been approximately 50 field based studies (about 1.3 studies per year) but almost half of these were undertaken within the last 10 years. There are about 50 species on which there is some such data, although still barely a handful (maybe none) that meet Ripley's criteria for an exhaustive study of total locomotor pattern. All major taxonomic groups are represented, but there is a heavy emphasis on New World monkeys (*Alouatta* has been the subject of 8 studies, *Saguinus* 6, and *Ateles* 6). By 1985 quantitative data on positional behavior were known for only 29 species of primates. In contrast, by 1985 quantitative data on group size had been collected for 111 species (Jolly, 1985), and quantitative data on diet for 56 species (Richard, 1985).

The addition of these data has led to more realistic categorization of positional behavior, a subject of some concern at the 1965 conference. For example, careful field studies of *Ateles, Colobus*, and *Presbytis* by Mittermeier and Fleagle (1976), Morbeck (1974), and Ripley (1967) resulted in a successful challenge of the usefulness of the "semibrachiator" category (Mittermeier and Fleagle, 1976). At the same time the questions have changed; there is now less emphasis on broad categorization, which may explain major differences in morphology, but leaves us with a lot of unexplained variation in both behavior and morphology within groups. We also have skeletal remains from many more fossil primates than Simons was able to review at the 1965 conference (Simons, 1967). These extinct species differ in more subtle aspects of morphology than those associated with major locomotor categories, but the implication of these differences for behavior is also something that we strive to understand. There is also renewed interest in ecomorphology (Bock, 1990; Wainwright and Reilly, 1994); in particular relationships among habitat,

Table 1. Studies of positional behavior in primates since 1965. Only studies that were conducted in the wild and that are quantitative in nature are included. Studies using the same data are grouped together. See Table 2 for other studies. The last column indicates if data on locomotion (L), posture (P), substrate use (S; either size or orientation), or habitat (H; usually height data) are included

Years	Publication	Species	Data
1965-1970	Richard, 1970	*Ateles geoffroyi*	L
		Alouatta palliata	
1971-1975	Rose, 1974, 1978, 1979	*Colobus guereza*	L,P,S
		Cercopithecus aethiops	
	Morbeck, 1974, 1977a,b, 1979	*Colobus guereza*	L,P,S,H
1976-1980	Fleagle, 1976, 1980	*Hylobates syndactylus*	L,P,S
		Hylobates lar	
	Mendel, 1976	*Alouatta palliata*	L,P,S,H
	Rose, 1976, 1977	*Papio anubis*	L,P
		Cercopithecus ascanius	
		Cercopithecus mitis	
	Fleagle, 1978	*Presbytis melalophos*	L,P,H
		Presbytis obscura	
	Mittermeier, 1978	*Ateles geoffroyi*	L,P,S,H
		Ateles paniscus	
	Richard, 1978	*Propithecus verreauxi*	P,S,H
	Fleagle and Mittermeier, 1980; Fleagle et al., 1981	*Ateles paniscus*	L,S,H
		Alouatta seniculus	
		Pithecia pithecia	
		Saguinus midas	
		Saimiri sciureus	
		Chiropotes satanas	
		Cebus apella	
	Garber, 1980, 1984	*Saguinus oedipus*	L, P, S
	MacKinnon and MacKinnon, 1980	*Tarsius spectrum*	L,S,H
	Susman et al., 1980	*Pan paniscus*	L
1981-1985	Happel, 1982	*Pithecia hirsuta*	L,H
	Crompton, 1983, 1984	*Galago moholi*	L,P,S,H
		Otolemur crassicaudatus	
	Gittins, 1983	*Hylobates agilis*	L,P,S,H
	Schön-Ybarra, 1984	*Alouatta seniculus*	L,P
	Srikosamatara, 1984	*Hylobates pileatus*	L,S,H
	Susman, 1984	*Pan paniscus*	L,H
	Oliveira et al., 1985	*Pithecia pithecia*	L,H
	Tuttle and Watts, 1985	*Gorilla gorilla*	L,P
1986-1990	Cant, 1986	*Alouatta pigra*	L,P,S
		Ateles geoffroyi	
	Crompton and Andau, 1986	*Tarsius bancanus*	L,P,S,H
	Sugardjito and van Hooff, 1986	*Pongo pygmaeus*	L,P,H
	Cant, 1987a,b	*Pongo pygmaeus*	L,P,S
	Schon-Ybarra and Schon, 1987	*Alouatta seniculus*	L,P,S
	Cant, 1988	*Macaca fascicularis*	L,P,S
	Boinski, 1989	*Saimiri oerstedii*	L,P,S,H
	Wilson et al.,1989	*Eulemur coronatus*	L,H
	Fontaine, 1990	*Ateles geoffroyi*	L,S
1991-1996	Garber, 1991	*Saguinus fuscicollis*	L,S
		Saguinus mystax	
		Saguinus geoffroyi	
	Hunt, 1991, 1992, 1994	*Papio anubis*	L,P,S,H
		Pan troglodytes	
	Doran, 1992a,b 1993a,b, 1996; Doran and Hunt, 1994	*Pan troglodytes*	L,P,S,H
		Pan paniscus	

(*continued*)

Table 1. (*continued*)

Years	Publication	Species	Data
	Gebo, 1992	*Alouatta palliata*	L,P,S,H
		Cebus capucinus	
	Bicca-Marques and Calegano-Marques, 1993, 1995	*Alouatta caraya*	L,P
	DaSilva, 1993	*Colobus polykomos*	P
	Tremble et al., 1993	*Tarsius dianae*	L,P,S,H
	Youlatos, 1993	*Alouatta seniculus*	L,S
	Walker, 1994, 1996	*Chiropotes satanas*	L,P
		Pithecia pithecia	
	Cannon and Leighton, 1994	*Hylobates agilis*	L,S,H
		Macaca fascicularis	
	Dagosto, 1994, 1995	*Varecia variegata*	L,P,S,H
		Eulemur fulvus	
		Eulemur rubriventer	
		Propithecus diadema	
	Gebo et al., 1994; Gebo and Chapman, 1995a,b	*Colobus badius*	L,P,S,H
		Colobus guereza	
		Cercopithecus mitis	
		Cercopithecus ascanius	
		Cercocebus albigena	
	Garber and Preutz, 1995	*Saguinus mystax*	L,S,H
	Remis, 1995	*Gorilla gorilla*	L,P,S,H
	McGraw, 1996	*Colobus polykomos*	L,P,S,H
		Colobus badius	
		Colobus verus	
		Cercopithecus diana	
		Cercopithecus campbelli	
	Doran, 1996	*Gorilla gorilla*	L,P,S
	Walker and Ayres, 1996	*Cacajao calvus*	L,P,S

body size, diet, and positional behavior (e.g., Fleagle and Mittermeier, 1980; the niche metrics of Crompton et al., 1987 and Oxnard et al., 1990). Addressing these questions in a broad comparative perspective often relies heavily on quantitative approaches, and therefore quantitative assessments of behavior.

3. METHODS OF COLLECTION AND ANALYSIS OF POSITIONAL BEHAVIOR

Despite the increase in the number of studies since 1965, there has been surprisingly little review and critique of methods. With the exception of Ripley's (1967) paper, very little has been written about how to conduct such studies. What sort of data should be collected, how should they be collected, and how should they be analyzed? The answers to these questions depend of course on what we want to ask of our data, and this varies among studies (Rose, 1979). Given the rarity of field studies, the majority of them are still directed towards simply documenting the behavior of species, almost always with the notion that these data will be useful for explaining anatomical form. Thus, we will concentrate our remarks on these types of studies, and on locomotor behavior in particular (although many of the same arguments will apply to the study of posture or substrate use).

Table 2. Studies of positional behavior that are non-quantitative or that were conducted on animals in captivity or semi-captivity

Publication	Species	Data
Ripley, 1967, 1977, 1979	*Presbytis entellus*	L,P,S,H
Walker, 1969	*Perodicticus potto*	L,P,S
Mittermeier and Fleagle, 1976	*Ateles geoffroyi*	L,P
Kinzey et al., 1975	*Cebuella pygmaeus*	S
Charles-Dominique, 1977	*Arctocebus calabarensis*	S,H
	Perodicticus potto	
	Euoticus elegantulus	
	Galago alleni	
	Galago demidovii	
Kinzey, 1977	*Callicebus torquatus*	S,H
Tattersall, 1977	*Eulemur fulvus*	S,H
Rodman, 1979, 1991	*Macaca fascicularis*	H
	Macaca nemestrina	
Walker, 1979	*Galago demidovii*	L,S
Dykyj, 1980	*Nycticebus coucang*	S
Rollinson and Martin, 1981	*Cercocebus galeritus*	L,S
	Cercocebus albigena	
	Cercopithecus neglectus	
	Cercopithecus nictitans	
	Cercopithecus pogonias	
	Miopithecus talapoin	
Sugardjito, 1982	*Pongo pygmaeus*	L,P
Glassman and Wells,1984	*Nycticebus coucang*	L,P,S
Niemitz, 1984a,b	*Tarsius bancanus*	L,P,H
Roberts and Cunningham, 1986;	*Tarsius bancanus*	S,H
Roberts and Kohn, 1993		
Gebo, 1987	*Lemur, Eulemur* (7 species*), Hapalemur, Propithecus, Cheirogaleus* (2 species*), Mirza, Microcebus, Perodicticus, Nycticebus, Loris, Tarsius* (2 species*)*	L
Tilden, 1990	*Eulemur rubriventer*	L,S
Fontaine, 1990	*Saimiri sciureus*	L,P,S
Curtis, 1992; Curtis and Feistner, 1994	*Daubentonia madagascariensis*	L,P,S
Rosenberger and Stafford, 1994	*Leontopithecus rosalia*	L
	Callimico goeldii	

It is also desirable to develop more specific questions (e.g., Cant, 1992), which might require other techniques and methods not addressed in this paper.

Some aspects of methodology, such as sampling strategies, data analysis, and definitions of behavioral categories directly affect comparability among studies, an important issue for both broad, interspecific comparisons, and even more narrowly defined contrasts if the behavioral data are derived from different sources. Even though both can be described numerically, morphological and behavioral data sets are different. Most aspects of morphology that we assess are relatively static over the adult life of an individual, they occur in discrete states or are easily quantified, each individual is represented by a single score or measurement that is generally easily repeatable, samples are characterized by well understood summary statistics, and differences among species can be tested by traditional univariate and multivariate techniques. In other words, in morphological studies we generally have a pretty good idea what the numbers mean.

Positional behavior is a more complicated matter. For perfectly understandable practical reasons, behavior is sampled during a very brief time frame (weeks, months) so we have a very poor idea of how plastic behavior may be during an individual's or species' life span. Studies are also limited in space, so behavior from only a small part of the species' range is ever sampled. The potential for behavior, substrate use, and habitat selection to be heavily influenced by immediate circumstances, while morphological measurements are less labile is, in fact, the impetus for ecomorphologists to discover morphological correlates that can be substituted for the more intractable environmental variables (Bock, 1990; Rickelfs and Miles, 1994).

Behavioral attributes studied during these brief interludes are collected by sequential observations on individuals, these are tallied and transformed into proportions. The nature of the data makes it difficult to analyze statistically, and the appropriate summary statistic or comparative test is a matter of contention (Dagosto, 1994). Even though these studies result in hundreds or thousands of observations, should we be satisfied? Do these impressive numbers reflect a deep understanding and adequate characterization of behavior, or are they just an artifact of method (pseudoprecision). In other words, we are not sure we understand what the numbers mean: do these data adequately epitomize the positional behavior of a species in the same way that osteological measurements characterize morphology? Several aspects of data collection and analysis can affect the values published in any single study and must be taken into account when comparing studies.

3.1. Time Samples versus Bouts

Typically, studies of positional behavior follow one of two methods, continuous sampling (bouts; e.g., Fleagle, 1976) or instantaneous time sampling. It is usually assumed that results (in terms of proportions of behaviors) will be comparable: behaviors that occur more frequently than others should take up more of an animal's time. But this depends on the duration of the behavior. Behaviors that occur infrequently but have a long duration may take up more time than behaviors that occur more frequently but have a short duration. For example, *Propithecus diadema* leaps much more often than it sits (leaping = 57% of all bouts; sitting = 15% of all bouts), but bouts of leaping last much shorter than bouts of sitting (sifakas spend only 3.5% of their time leaping and 64% of their time sitting). Obviously one must be careful when comparing such numbers.

The comparability of results using bouts and time samples has rarely been tested. In one such test, Doran (1992a) found that the resultant proportions were quite different, but if bouts were corrected by distance traveled, the two methods yielded similar results. In Dagosto's (1994, 1995) study of lemurs, data were also collected with both methods simultaneously. As in Doran's study, there were significant differences in the proportions of behaviors measured with data from bouts and time samples; in *Eulemur rubriventer* and *Propithecus diadema*, the differences are statistically significant (Table 3). In this case, however, correcting bouts for distance makes the results more comparable only for *E. fulvus*. In *E. rubriventer*, there is no great disparity in the average distance traveled per bout for each type of locomotor behavior (Table 4), so correcting bouts for distance changes the proportions very little, and not enough to make them equivalent to the proportions calculated for time samples. Distance traveled during an event of leaping in *P. diadema* is greater than for quadrupedalism or climbing, so that if bouts are "corrected" for distance, the proportion of leaping increases relative to quadrupedalism and climbing, making the difference between bouts and time sampling even greater (Table 3)!

Table 3. Proportions of locomotor behavior in Malagasy lemurs calculated by several methods. In IND, proportions are calculated for each individual and the results averaged. The observed range of proportions is given below. In LAO, all observations are lumped before proportions are calculated. With the bout method (Bout), proportions are based on the number of occurrences of the behavior divided by the number of occurrences of all locomotor behaviors (N). With time samples (TS) the proportions are based on the number of time samples during which the behavior was being performed divided by the total number of time samples of locomotor behavior (N). In Bout with distance (Bout(wd)), the proportions in the Bout column are "corrected" by the average distance traveled using the mode of displacement (Table 4) following the method of Doran (1992a). P gives the statistical significance of a test of the differences in calculated proportions between the Bout and TS methods, using the two-sample test of Manly (1991); *, P=1.0–0.05, **P<0.05, ns = not significant.

	Eulemur fulvus				Eulemur rubriventer				Propithecus diadema			
	Bout	TS	Bout (wd)	P	Bout	TS	Bout (wd)	P	Bout	TS	Bout (wd)	P
IND												
N	3987	267			5285	291			5696	199		
Leap	55.7	50.6	50.9	ns	62.2	49.2	60.0	**	88.2	75.6	92.9	**
(range)	34-72	24-73		ns	52-74	38-60			85-91	63-96		
Quad.	28.9	35.1	35.6	ns	23.1	30.3	26.3	**	1.0	2.8	0.7	ns
(range)	17-46	11-61			13-33	14-44			0-3	0-13		
Climb	12.8	13.7	12.9	ns	13.4	19.1	13.4	**	8.9	18.2	7.1	**
(range)	7-21	0-33			10-18	11-27			6-11	4-33		
Other	2.3	.5		ns	1.3	1.4		ns	1.9	3.5		ns
(range)	0-6	0-3			0-5	0-7			0-5	0-17		
LAO												
Leap	61.8	48.3	56.4		62.2	49.0	60.3		88.4	76.4	93.1	
Quad.	25.3	39.0	31.1		23.2	32.0	26.5		0.9	2.5	0.6	
Climb	11.1	11.6	11.2		12.8	17.7	12.8		9.0	18.1	7.1	
Other	1.9	1.1			1.9	1.3			1.7	3.0		

Table 4. Average distance traveled (in meters) with each type of locomotor behavior in Malagasy primates

	E. fulvus	E. rubriventer	P. diadema
Leap	1.4	1.5	1.9
Quad.	1.9	1.8	1.3
Climb	1.5	1.6	1.5

Regardless of whether or not results can be made comparable, it must be remembered that these data collection techniques are designed to answer different questions, so there is no reason to expect that they will converge on the same result. Continuous recording is designed to measure the *frequency of occurrence* of discrete events, it answers the question, "How often does this event occur?" Time sampling is designed to measure the *percent of time devoted* to states (Altmann, 1974; Martin and Bateson, 1986), it answers the question "How much time is spent in this activity?" "How much does each type of locomotor activity contribute to total distance traveled?" is yet another important question. Investigators need to decide which aspect(s) of behavior they want to measure before choosing a data collection technique. Using these definitions, we would modify Doran's (1992a) initial assessment that bout (without distance) overestimates the *frequency* of locomotor activities used often, but for short distances; and underestimates the frequency of locomotor activities used rarely, but for long distances. Bout (without distance) estimates the *frequencies* (i.e., how often behaviors occur) perfectly well; what it may not reflect well is the relative amount of time spent performing the activity, nor its contribution to distance traveled. The former requires time sampling or some estimate of duration and the latter requires information on average distance per bout, especially if all types of locomotor behaviors do not have similar average distances, as was the case in the chimpanzees studied by Doran. In that case, the distance traveled during a bout was probably highly correlated with the duration of the event, so that correcting bouts for distance also corrected them for duration, and thus made the resultant proportions more comparable to those estimated with time samples. In the sifaka example, distance traveled does not correlate well with bout duration (i.e., an event of leaping takes less time, yet covers a greater distance than an event of climbing), so that correcting bouts by distance does not make them more comparable to time samples; it does, however, yield a better picture of the contribution of each mode of locomotion to total distance traveled. It might be possible to translate time samples to bouts with a measure of time duration, (this will be necessary for postures, as they do not have associated distances), but since we did not collect this type of data, we cannot explore this further.

These two studies show that even data collected by the same investigator at the same time will yield different results if different methods are employed, and means that direct comparisons of proportions among studies using different techniques is potentially hazardous. The rank order of behaviors as estimated by the two methods is comparable in the lemur examples, but not in Doran's study, so even this less discriminating comparison may not always be reliable.

These examples, however, do not answer the question of which technique (if either) gives the "right" answer, i.e., which set of proportions best reflects what actually happened during the time of the study. To answer this, one needs some way of determining what actually happened, something which will rarely be available for a field positional behavior study unless all observation time is videotaped (e.g., Rosenberger and Stafford, 1994) We did, however, conduct an "experiment" in event sampling using Chicago Bulls' basketball games as the data source. For 10 games during the regular 1995–96 season, we scored shot attempts using both bout and time sampling (at 1 minute and 2 minute intervals) and calculated the proportion of shot attempts attributable to each player. In this case, the results of bouts and time sampling should be directly comparable since there are no differences in event duration between players. Because sitting on the couch and watching a basketball game is much easier than running around the forest chasing primates (our respective departmental chairs were too cheap to spring for tickets for this very important scientific endeavor so we had to watch at home) a random sample of half of the bouts observed in a game was extracted for analysis. We compared the results of sampling with what actually happened (at least the official scorer's version), since game results are printed in the newspaper the next day. The results of this experiment are presented in Table 5. Bout scoring differed less from what actually happened than did either 1 minute or 2 minute time samples in every single game and even when the events from all 10 games are combined. Bout scoring never differed by more than more than 3% from the actual proportions, while time samples differed by as much as 15%. The rank order of player shooting is the same for the top 3 positions for all sampling techniques except 2 minute time sampling, but varies a lot for any player contributing less than 10% of shot attempts (Spearman rank correlation does not, however, indicate any differences between the methods). The same results pertain if other events or statistics are used (scoring, 2 point versus 3 point shots, field goal percentage, etc.). The relatively poor performance of time sampling is no doubt due to the small number of events scored with this method. Shot attempts are frequent (more than 80 per 48 minute game for each team), but because they have a very short duration (Mean=0.96 sec, SD=0.3 sec, N=45), they take up only 5% of total game time and thus are rarely scored with a time sampling method. The same problem occurs in the study of locomotor behavior because primates spend very little time (usually less than 10% of a day) moving. Very few events of locomotion are scored even in many hours of field time. For example, in Dagosto's (1994) study 234 hours of observation of *E. rubriventer* resulted in only 291 time samples of locomotor behaviors (compared to over 5000 bouts); in *E. fulvus* 191 hours of observation yielded 267 time samples but 4000 bouts, and in *P. diadema* 250 hours produced 199 time samples and 5700 bouts. Therefore, one might very well ask whether time sampling is a very efficient method for collecting data on locomotor behavior even if it will ultimately converge on the correct result (Doran, 1992a; Dagosto, 1994). Those researchers particularly interested in locomotion will have to ask themselves if the perceived advantages of time sampling are worth the extra time commitment necessary to get a reasonable sample of locomotor events.

What constitutes a reasonable sample of events, or a reasonable amount of observation time, is also a question that has gone largely unaddressed. Gebo (1992; Gebo and Chapman, 1995a) demonstrates that frequencies are labile when the sample consists of less than 1000 bouts, but settles into a pattern of little change after 3000 observations (about 2/3 of which are locomotor rather than postural events). (He did not, however, test for statistical differences in behavior.) Based on our experience with lemurs and monkeys, one can obtain 2000 locomotor bouts in about 50–100 hours of observation time; with 2 minute time sampling it might take as long as 1000 hours to get 2000 observations of locomotor behavior.

On the other hand, Kawanaka (1996) calculated that 25 hours of observation is necessary to accurately estimate the amount of time spent in five general behavioral categories for an individual, suggesting that less time might be adequate. This result pertains, however, only when the whole suite of behaviors is compared using a correlation; direct

Table 5. *Top*. Proportions of shot attempts by different players on the Chicago Bulls basketball team, sampled during ten games scored by the bout method (Bout; a random sample of half the bouts scored were used), one-minute time samples (1TS), and two minute time samples (2TS). These are compared to the actual proportions determined by the official scorer (Actual). N, number of observations recorded. The difference in the proportions between each method and the actual record is calculated as the absolute value of the actual proportion minus the calculated proportion. The proportions calculated with the bout method are less different from the actual proportions than is either time sampling method. *Bottom*. Comparison of the proportions of shot attempts during the whole season compared with those from the sampling period

Player	Actual	Bout	ABS (Actual-Bout)	1TS	ABS (Actual-1TS)	2TS	ABS (Actual-2TS)
N	867	415		41		16	
Brown	1.4	1.7	0.3	0.0	1.4	0.0	1.4
Buechler	2.4	2.9	0.5	2.4	0.0	6.3	3.8
Caffey	0.6	0.7	0.1	0.0	0.6	0.0	0.6
Edwards	0.5	1.2	0.7	0.0	0.5	0.0	0.5
Harper	7.3	6.5	0.8	4.9	2.4	0.0	7.3
Jordan	29.8	27.0	2.8	41.5	11.7	31.3	1.5
Kerr	6.1	7.7	1.6	4.9	1.2	6.3	0.1
Kukoc	10.7	12.5	1.8	14.6	3.9	6.3	4.5
Longley	9.5	8.2	1.3	4.9	4.6	6.3	3.2
Pippen	22.4	22.4	0.0	24.4	2.0	37.5	15.1
Rodman	4.6	4.1	0.5	0.0	4.6	0.0	4.6
Simpkins	2.0	2.7	0.7	0.0	2.0	0.0	2.0
Wennington	2.9	2.4	0.5	2.4	0.4	6.3	3.4
Sum ABS(actual-sample)			*11.6*		*35.3*		*47.9*
Mean diff.			*0.89*		*2.71*		*3.68*
Range			*0-2.8*		*0-11.7*		*0.1-15.1*

Player	Season	ABS (Season-Actual)	ABS (Season-Bout)	ABS (Season-1TS)	ABS (Season-2TS)
N	6997				
Brown	2.7	1.4	1.1	2.7	2.7
Buechler	3.5	1.0	0.6	1.0	2.8
Caffey	2.3	1.7	1.6	2.3	2.3
Edwards	1.6	1.1	0.4	1.6	1.6
Harper	7.2	0.1	0.7	2.3	7.2
Jordan	26.4	3.3	0.5	15.0	4.8
Kerr	6.9	0.8	0.8	2.0	0.6
Kukoc	11.2	0.5	1.3	3.4	5.0
Longley	7.2	2.3	1.0	2.3	0.9
Pippen	17.4	5.0	5.0	7.0	20.1
Rodman	4.3	0.3	0.2	4.3	4.3
Simpkins	2.3	0.3	0.4	2.3	2.3
Wennington	4.9	2.0	2.5	2.5	1.3
Other	2.1	2.1	2.1	2.1	2.1
Sum ABS(actual-sample)		*22.0*	*18.1*	*50.8*	*58.1*
Mean diff.		*1.57*	*1.3*	*3.63*	*4.15*
Range		*0.1-5.0*	*0.2-5.0*	*1.0-15.0*	*0.9-20.1*

comparisons of proportions of time spent were not made, nor was the variance in proportions among the trials reported.

In any case, these are probably minimum estimates. If the data are going to be subcategorized by age-sex groups, habitat, season, substrate context, activity context, etc., even more observations are necessary to ensure that each subcategory is adequately sampled. The number of events recorded is, however, not the only important factor. Researchers also need to consider the number of behavioral categories being scored, number of days over which the data are collected (since daily variation in behavior can be enormous--[Garber and Preutz, 1995]), and also the number of different individuals on which observations are made.

It is distressingly difficult to extract information on number of observations or contact hours from publications because of the variety of ways in which it is reported. Of the studies from which we were able to determine this, only 55% are based on more than 1000 observations of locomotor behavior, and only 30% on more than 2000. Of the studies that report the number of contact hours, only 33% are based on more than 250 hours, and 27% are based on less than 100 hours of observation time.

3.2. Data Analysis and Statistics

In most studies all observations are lumped because they cannot be attributed to particular individuals. We see several problems with this typical approach. Sequential observations of individuals can lead to a data set in which temporal autocorrelation (when the probability of occurrence of one event is affected by the previous event) is a problem (e.g., Mendel, 1976). One approach to dealing with this is to have a long time interval between observations; this is the usual rationale for time sampling (but the longer the interval, the longer it will take to get a reasonable sample of locomotor behaviors). Rarely, though, has anyone tested their data to see if significant temporal autocorrelation still exists. There are statistical approaches to account for temporal autocorrelation (e.g., Altham, 1979), but these techniques have not been applied to positional behavior.

Even if several hundred or thousand temporally non-correlated observations are collected, in reality these are attributable to only a small number of individuals, and thus are still autocorrelated. We maintain that the proper unit of analysis in studies of positional behavior is the individual animal; the multiple observations (bouts or time samples) are crucial in that they contribute to the accuracy and precision with which the behavior of each individual is measured, but they do *not* increase the number of degrees of freedom for a statistical test. The correct sample size for determining the P value of such a data set is thus the number of individuals studied, not the number of observations. An incorrectly inflated sample size will result in much lower P values, more instances of statistical significance, and therefore greater possibility of Type 1 errors (finding significance where none exists) (Hurlbert, 1984; Machlis et al., 1985). A comparable problem in morphological studies has been the topic of much discussion (e.g., Felsenstein, 1985; Cheverud et al., 1989; Harvey and Pagel, 1991; Smith, 1994).

To demonstrate this problem, Dagosto (1994) analyzed data in two ways: (1) lumping all observations (LAO) to calculate total species proportions and testing between species differences with a Chi-square test versus (2) lumping the observations and calculating proportions for each individual (IND) and testing between species differences with ordinary parametric and randomization tests. For the same data LAO (N= number of observations, thousands) almost always gave a significant result while IND (N= number of individuals studied, 12–20) almost always gave a nonsignificant result. Although there is little difference in species' "central tendency" statistics calculated by either method (calculating species proportions after lumping all observations versus calculating proportions for each individual and taking the average; see Table 3), the results of statistical tests are greatly affected by method.

The species-wide lumping approach (LAO) has additional statistical problems. This procedure yields a suite of categories that must be tested as a whole with a Chi-square test, G test, or a rank order correlation; these tests do not generally allow comparison of each category (e.g., leaping, climbing) separately. If the test is significant, most researchers argue that one category or another contributes most to the total difference by looking at raw differences in frequency or the contribution of the deviation in that category to the Chi-square statistic, but the test itself does not bear directly on this question. This could be accomplished by appropriately compressing categories (e.g., leaping versus not leaping).

Assuming that individual animals are the proper units of analysis, the Chi-square and G tests are invalid for reasons outlined by Hurlbert (1984). The pooling of data from different individuals, which is necessary for the tests, results in: (1) loss of independence, thus violating the underlying assumption of the tests; (2) loss of information on variability among the individuals in each group; and, (3) weighted averages being compared if the number of observations for each individual is different. In LAO, species are characterized with a single number (presumably some sort of measure of central tendency), but there is no way to estimate dispersion around that number (variance). We seem to treat behavioral repertoires as properties of a species, rather than as an epiphenomenon of the behavior of the individuals of the species. What the LAO approach actually measures is something like "of all locomotor events taking place in this species, X% are leaping", when what we really ought to measure is "how often does each individual leap", and construct an estimate of species' central tendency and dispersion from a sample of individuals. The two approaches will converge only if (1) there is no intraspecific variation in behavior or (2) if there is intraspecific variation, no individual contributes more observations to the sample than any other. In any case, the lack of any way to measure dispersion in LAO makes it difficult to make interspecific (or any intergroup) comparisons as in quantitative morphology, where interspecific variation is interpreted in the context of intraspecific variation.

If observations can be associated with individuals (IND), then proportions of behaviors can be calculated for individuals thus solving these problems. Individual values for categories (e.g., leaping, climbing) can be tested independently of other categories and a measure of central tendency and the variation around it can be calculated with standard statistics (Boinski, 1989; Fontaine, 1990; Dagosto, 1994). Interspecific comparisons of behavior can proceed much like morphological comparisons, using standard parametric, nonparametric, or randomization approaches for continuous quantitative variables. We appreciate that access to individually identifiable study subjects is often practically difficult. Other approaches to lumping data are suggested in Dagosto (1994) and Gebo and Chapman (1995a).

An emphasis on individual behavior rather than species-wide measures is important in other areas of endeavor within evolutionary morphology. The research strategy for studying selection and adaptation outlined by Arnold (1983), for example, relies critically on determining the relationship between performance and morphology of individuals.

3.3. Categories

Because data on positional behavior are always presented as proportions, the numbers arrived at in any study depend entirely on the number of categories used and how different subcategories are lumped. This greatly affects comparability among studies. There are two important questions regarding categorization of positional behavior. The first concerns definitions: the names given to different kinds of behavior. We will not deal with this here, suffice it to say that a common language for modes of behavior would be extremely useful (Hunt et al., 1996). There is simply no point in comparing frequencies of climbing among studies if climbing is defined in a different way by each investigator. Similar confusion would arise in morphology if each researcher called the bones by different names or defined the same measurements using different landmarks.

What constitutes a useful categorization of behavior is another important question. For morphology-behavior associations, we presumably desire categories that capture kinematically meaningful differences in behavior having direct relationships with underlying anatomical structures. This is another area where the ideal (using enough categories to capture as detailed depictions of the kinematic differences as possible) runs into practical difficulties (what can actually be seen in the field; Rose, 1979; Rosenberger and Stafford, 1994). For example, we have no doubt that what we call "leaping" in *Propithecus* and *Eulemur* is kinematically different, or that within each species there are several kinds of kinematically different behaviors we call leaping (some of these are discussed in Oxnard, 1984). Although we recognize that several types of leaping occur, we are forced to lump these behaviors when calculating frequency, because in our experience it is tremendously difficult to reliably, regularly recognize these subtly (or even not so subtly) different modes, especially for quickly occurring locomotor behaviors (e.g., did the animal land feet first, hands first, or with all limbs simultaneously; was there an aerial phase in a quadrupedal sequence, etc.). Thus, there will always be a certain level of unavoidable generality to an observational study, which may compromise its usefulness to morphologists. If such subtle, yet important kinematic characterizations are the question of interest, a significant amount of videotaping is necessary (Rosenberger and Stafford, 1994; Terranova, 1995; Demes et al., 1995, 1996; but see Fontaine, 1990). Obtaining frequencies of occurrence of such categories based on large numbers of videotaped field observations is likely to be a very demanding enterprise. The ability to accurately apprehend subtle differences in body position is not as much of a problem for behaviors of long duration, like most postures.

4. INTRASPECIFIC VARIATION

A comparison in which the data consists of observations of one species collected at one time in one locality and of another species collected at a different time at a different locality by another observer is a poor experimental design. One may use inferential statistics and believe that one is testing for a species' effect, but in fact a number of other effects, including those from habitat structure, seasonal differences in behavior, interobserver 'error', degree of habituation, presence and density of predators, age/sex composition of the groups, etc. could influence the result. Such comparisons do not control for either these or for stochastic effects, and thus must be interpreted cautiously (James, 1982; Simberloff, 1982; Hurlbert, 1984; Garland and Adolph, 1994). In accepting the results of such "tests" or "natural experiments" we are making the assumption that these other effects are negligible compared to the species' effect. Depending on the scale of the question being addressed or the magnitude of the observed difference, such an assumption may be a perfectly valid course of action. Nevertheless, the rather large assumption of negligibility has not yet been adequately tested. It likely will be impossible to control for all of these variables in any field study, however, we can at least attempt to assess the effect of these factors in a few cases.

In the section above it was argued that interspecific comparisons are stronger when they can be made in the light of intraspecific variation. Unfortunately, we have as yet only a poor idea of the sources of variation in positional behavior within species and the extent of their effects. Intraspecific variation due to age and sex have been documented in many primates (Ripley 1967; Sugardjito and van Hooff, 1982; Crompton 1983; Cant, 1987a,b; Boinski, 1989; Doran 1992b, 1993b; Hunt, 1994; Remis, 1995). Simple variation in behavior among adult individuals of the same species has rarely been assessed, however, largely because observations are not associated with particular individuals. Positional behavior is known to alter with activity context (e.g., feeding, traveling, resting); studies vary in what contexts behavior is measured and reported. In addition, variation in positional behavior that is associated with differences in habitat or season has only begun to be documented. Understanding the causes and extent of intraspecific variation is critical for assessing the meaning (and statistical significance) of interspecific variation. If significant intraspecific variation exists, research designs must be constructed carefully in order to arrive at accurate summary values for species. In addition, intraspecific variation is often the key to understanding interspecific variation: "the analysis of population variation should become a powerful weapon in the morphologist's arsenal" (Arnold, 1983: 348; see also James, 1982; Bradshaw, 1987).

In a field study we can only measure performance, not potential—we can only see what animals do, not necessarily all they are capable of doing (Rose, 1979; Morbeck, 1979; Gomberg et al., 1979). An important question for the study of positional behavior is "Does performance change due to immediate circumstances of habitat or season?" By necessity most studies of positional behavior are limited in time and space. We do not question here whether a researcher's data accurately measures what happened in that short period of time, but rather ask if what happened in that short period of time or in that single place accurately characterizes long term behavior or species-wide potential.

Using our Bulls example, we compared how well our 10 game sample (collected during December and January) reflected what happened during the total 82 game season (October-April). Even though the Bulls had a remarkably consistent season, winning a record 72 of 82 games and with no significant player injuries, Table 5 shows that the sampling methods did a better job predicting what happened in the sample games than in the overall season. Similarly, in our studies of positional behavior, we have both found significant intraspecific variation in frequencies of behavior associated with differences in season and habitat. Other studies have also discussed seasonal or habitat related differences in positional behavior and substrate use, but have reached conflicting conclusions about its presence and importance (Boinski, 1989; Crompton, 1984; Doran and Hunt, 1994; Garber and Preutz, 1995; McGraw, 1996; Remis, this volume).

4.1. Habitat

That habitat architecture can impose constraints on the positional behavior of primates has been argued by several workers, most notably Ripley (1967, 1977, 1979). Cant (1992, p. 277) has also voiced the caution that field studies of positional behavior should aim to control for habitat structure "in order to avoid the possibility that behavioral differences between species are artifacts of observing them in different structural contexts" (see also Pounds, 1988, 1991).

We approached this question by studying the same species in different habitats in two different primate groups, which minimizes problems of interobserver differences in data collection techniques. The following comparisons are based on overall frequencies of locomotor behavior (all activity contexts included). Gebo and Chapman (1995b) documented the behavior of the red colobus monkey (*Colobus badius*) in the Kibale Forest, Uganda. Dagosto (1995; Dagosto and Yamashita, in press) studied four species of Malagasy primates, *Propithecus diadema, Eulemur rubriventer, Eulemur fulvus,* and *Varecia variegata* at Ranomafana National Park, Madagascar.

Colobus badius was observed in primary, secondary, and pine forests. These forests differ in two structural attributes: the size of gaps to the nearest tree, which was largest in the secondary forest and least in the pine forest, and continuity, which measures the percent of trees for which there is no gap to the nearest tree. This is largest in the pine forest and lowest in the secondary forest. The locomotor behavior of the red colobus monkey is significantly different in these habitats (Table 6). Leaping and quadrupedalism are least frequent in the primary forest and most frequent in the secondary forest. Climbing was least frequent in the secondary forest.

At Ranomafana National Park, *Eulemur* and *Propithecus* were observed in two different areas, Vatoharanana and Talatakely. The Vatoharanana forest has a higher percentage of large trees (as measured by height, diameter at breast height, or crown diameter), and a higher percentage of large gaps between the crowns of trees (Dagosto and Yamashita, in press; White et al., 1995) compared to Talatakely. All three species showed a similar response to this habitat difference; leaping is less frequent and quadrupedalism and climbing more frequent at Vatoharanana. The degree of response differed, being least in *Propithecus* and most in *Eulemur fulvus*. The difference in the species behaviors between forests may be mediated through activity budget. All species spent more time feeding/foraging and less time traveling at Vatoharanana, presumably because of the larger food patch size available there. Since leaping is more common in travel than during feeding in most species of primates so far studied (Fleagle and Mittermeier, 1980; Gebo and Chapman,1995a; Dagosto and Yamashita, in press), increasing the amount of time spent traveling, as at Talatakely, increases the overall frequency of leaping at this site.

4.2. Season

One might expect that as the kinds of foods exploited change during the year, microhabitat selection or activity pattern might also change, possibly resulting in differences in positional behavior. For the red colobus monkeys at Kibale Forest, the seasonal comparison was made in the primary forest only. In the dry season quadrupedalism is more frequent, and leaping and climbing less frequent (Table 6). This is associated with increased use of the middle canopy in the wet season (Gebo and Chapman, 1995b).

Among the lemurs of Ranomafana, *Propithecus* is the only species that shows no significant difference in positional behavior by season. The three lemur species (*E. fulvus, E. rubriventer,* and *V. variegata*) show a similar pattern: leaping is more frequent and quadrupedalism less frequent in the dry season than in the wet season. In this case, we were unable to identify a causal factor. Activity pattern does not differ much between seasons except in *Varecia*, and in this species it did not predict the observed difference in locomotor behavior, since leaping is more frequent in the dry season, when travel is actually less frequent. Although there are seasonal changes in food species used in all the lemurs, this is only sometimes correlated with changes in microhabitat as estimated by the struc-

Table 6. Habitat and seasonal variation in proportions of locomotor behavior during all activities (using bouts) in *Colobus badius* and Malagasy lemurs

	Colobus badius				*Eulemur fulvus*			*Eulemur rubriventer*			*Propithecus diadema*			*Varecia variegata*		
Habitat	Primary	Secondary	Pine	P	Vato	Talata	P	Vato	Talata	P	Vato	Talata	P	Vato	Talata	P
Leap	28.8	33.8	34.4	**	44.2	67.7	**	58.6	64.3	ns	85.2	88.8	**			
Quad.	27.3	38.2	31.3	**	35.6	22.2	ns	23.8	22.5	ns	1.9	0.6	ns			
Climb	36.4	22.0	26.6	**	17.1	8.6	**	15.1	11.7	ns	9.4	8.8	**			
Other	7.6	5.9	7.8	ns	3.1	3.0	ns	3.7	1.8	ns	2.3	1.8	ns			
Season	Wet	Dry		P	Wet	Dry	P	Wet	Dry	P	Wet	Dry	P	Wet	Dry	P
Leap	31.2	27.0		ns	62.7	73.3	**	59.7	69.3	*	89.1	88.6	ns	48.0	55.5	ns
Quad.	26.6	33.7		**	25.8	18.9	**	27.3	18.0	*	0.5	0.8	ns	44.0	29.0	**
Climb	36.0	33.6		ns	8.5	7.1	ns	11.3	12.7	ns	8.6	8.9	ns	6.5	12.5	**
Other	6.2	5.7		ns	1.9	0.7	*	2.0	0.7	ns	1.7	1.7	ns	0.0	0.0	ns

P gives the statistical significance of a test of the differences in calculated proportions between the habitats or seasons; * P=1.0-.05, **P<0.05, ns = not significant. The *Colobus* data were tested using ANOVA, and the lemur data were tested using the two-sample test of Manly (1991). Data from Gebo and Chapman, 1995b; Dagosto, 1995; and Dagosto and Yamashita (in press).

ture of the food trees; variation in structural attributes of resource trees does not correlate in a regular manner with variation in positional behavior (Dagosto, 1995).

4.3. Comparison of Studies

All five species that we studied show some statistically significant variation in the frequency of use of locomotor behaviors associated with different habitat, season or both. Variation due to season is generally less often significant and less in magnitude than that due to habitat. In both the *Colobus* and lemur studies, locomotion during travel is more conservative (less likely to show seasonal or habitat differences) than locomotion during feeding.

These examples indicate that studies limited in time or place may not fully capture the expressed range of behavior of a species and could compromise comparisons between species. For example, whether *Eulemur fulvus* or *E. rubriventer* leaps most does depend on the site at which observations are made: at Vatoharanana *E. rubriventer* leaps significantly ($P<0.05$) more than *E. fulvus*; at Talatakely, *E. fulvus* leaps more than *E. rubriventer*; if data from both sites are lumped, they are indistinguishable in leaping frequency (Dagosto, 1994). *Propithecus* leaps more ($P<0.001$) than either *Eulemur* species regardless of site or season although the magnitude of the difference is much less at some times and in some places than in others (Table 6). The scale of the expected or observed differences in behavior is important. Comparisons of closely related or behaviorally similar species will have to be designed very carefully and be fairly exhaustive before secure conclusions can be attained; we cannot immediately accept that the 'species effect' outweighs other factors. In comparisons of behaviorally quite different species we may be able to relax our skepticism, although we still might not be able to determine how much of the difference is the result of ultimate (i.e., morphology) or proximate (habitat, season) causes.

Our results may appear to be contradictory to the conclusions reached by Garber and Preutz (1995) in their study of *Saguinus mystax* in which they found no differences in the rank order of positional behaviors despite marked structural differences in the two habitats in which the behaviors were observed. They concluded from these data and from other studies that positional behavior is conservative to differences in habitat. In fact, our results are quite similar. In *Colobus badius* and in the Malagasy lemurs, it is also evident that there are no radical changes in behavior; the most common behaviors are still the most common, and rare behaviors are still rare—each species is constrained to some extent by anatomical design. Among lemurs, as in *Saguinus*, the rank order of behaviors does not differ by habitat or season, but in *Colobus badius,* there is a significant difference. The latter case is not surprising, nor particularly meaningful, since the three most common behaviors each contribute about 1/3 to the total behavioral repertoire.

Demonstrating a lack of difference in the rank order of proportions of an entire suite of behaviors answers a different question than the one we asked, which is whether there are meaningful contrasts in the frequency of occurrence of specific types of behaviors between habitats or seasons. We are able to demonstrate statistically significant differences among behavioral differences and habitat, microhabitat, activity pattern, and diet and believe that these associations are crucial to understanding the transformations in behavior and morphology among species. Of the 9 studies cited by Garber and Preutz (1995: Table 6) to indicate consistent positional repertoires in different habitats, we would interpret at least 6 (studies on *S. sciureus, C. capucinus, A. palliata, A. geoffroyi, C. guereza,* and *S. mystax*) to show potentially important differences in frequencies of specific types of behavior (although differences in methods makes direct comparisons among studies quite difficult). For example, in the Rio Blanco habitat, the frequency of quadrupedal walking in *S. mystax* is 23.1% and at Padre Isla it is 35.4%, a difference that exceeds in magnitude many of the statistically significant differences in frequency we found in our studies. In addition, Garber and Preutz's analysis is based on positional behavior during travel only, the context in which we found behavior to be the most conservative.

The primary difference between studies that have found no habitat differences and those that have is a matter of the scale at which the question is asked. On a large scale (rank order of behaviors, whole suites of behaviors compared with the G test) few differences are apparent (Garber and Preutz, 1995; McGraw, 1996; Doran and Hunt, 1994). At a smaller scale (the frequency of occurrence of particular behaviors) many distinctions are evident (Gebo and Chapman, 1995b; Dagosto, 1995). Morphologists must be careful to use behavioral data and statistical tests that are appropriate to the scale of the anatomical question being investigated. For example, Doran and Hunt (1994) find no large scale difference in the repertoire of positional behavior of chimpanzees in different habitats. There is, however, a significant difference in the time spent in arm hanging and climbing between rainforest and woodland chimpanzees, which influences explanations for the evolution of chimp shoulder morphology (Doran, 1996).

McGraw (1996) cites other possible explanations for the lack of consensus on this issue, including the likelihood that studies differ in their degree of habitat contrast. As with positional behavior, there is no standard way of measuring or testing habitat differences. All investigators do not measure the same aspects of habitat structure and, since we do not as yet understand the causal relationships among properties of habitat structure and positional behavior, we do not know which of the variables are most important to the primates. Nonstructural variables, such as resource spacing, timing, and diversity, may also be important, yet are rarely measured (Remis, 1995; Dagosto and Yamashita, in press). Finally, species may vary in their ability to respond to habitat or seasonal differences, thus there may well be no generalizations about the effect of such differences.

All of these studies, including our own, suffer from the problem of 'pseudoreplication' (Hurlbert, 1994), since they consist of data collected during only one or two replicates of the contrasted seasons or habitats, and thus do not provide an estimate of variability within each season/habitat with which to assess the differences between them. In such contrasts, stochastic effects are not controlled for, and statistical inferences are not warranted, or are at least subject to the same sorts of assumptions outlined at the beginning of this section. Better designs, perhaps including controlled experiments (e.g., Pounds, 1988), are needed to adequately address this issue.

5. CONCLUSION

Many studies of positional behavior are taken up with the aim of providing data useful for integration with studies of morphology. At the 1965 conference, Ripley pointed out problems with how the behavioral data then available was being used and outlined an approach to the study of positional behavior that would provide a sounder basis for these studies. In 1965 the fundamental problem was lack of data. This has been somewhat alleviated (although we would argue that we still have a long way to go), but the increase in the number of studies of positional behavior has produced a new set of problems related to comparability of results. This lack of comparability is likely to frustrate anyone attempting a broadly comparative synthesis of primate locomotion. Perhaps this is why the increase in the amount of data in the last thirty years has not resulted in any increase in synthetic studies. The ones that exist are limited in taxonomic scope (e.g., Fleagle and Mittermeier, 1980; Gebo and Chapman, 1995a: in both cases all the behavioral data were collected by the same investigators) or are forced to translate qualitative and quantitative data into relative scales (Crompton et al., 1987; Oxnard et al., 1990). Two-species comparisons have been successfully used to elucidate skeletal adaptations to positional behavior, in part because the behavioral data are again internally consistent since they are often collected by the same investigator (e.g., Fleagle, 1977a,b). However, these types of studies have their limitations (Fleagle, 1979; Oxnard, 1979; Bradshaw, 1987; Garland and Adolph, 1994). Broader based comparative studies are thought to provide stronger evidence for adaptation by demonstrating multiple occurrences of evolutionary convergence (Harvey and Pagel, 1991). Since it is very unlikely that many species will be studied by the same investigator, the issue of comparability of both the behavioral and anatomical data is an important concern.

As morphologists, there are several kinds of information we would like to know about the positional behavior of any species in order to explain form and differences in form:

1. *Repertoire*: What kinds of behaviors/substrates are used? It may be possible to obtain adequate information from captive animals. Quantification is not required, but how behaviors are defined (categorization) is critical.

2. *Frequency/Time Spent*: How often do kinds of events occur/how much time is spent performing them? Answering these questions requires quantitative data. The results may be sensitive to substrate/habitat context, so it is preferable to conduct a field, rather than captive, study (e.g., compare results of Gebo, 1987 with Dagosto, 1994).

3. *Context*: To understand the impetus for the origin of different behaviors it is important to document the associations of positional behaviors with habitat types, substrate use, season, diet, age, gender: these are the elements of Ripley's study of total locomotor pattern. Like frequency, these aspects of positional behavior are also best examined with quantitative data, and should be studied in the field rather than in captivity if biological role (actual use, Bock and von Wahlert, 1965) rather than function (potential use) is the object of investigation.

4. *Kinematics/ Biomechanics*: How are the behaviors performed; what are the joint angles, excursions, forces acting on the bones and joints? Some aspects of these can be studied in the field if fairly broad categories are used, but filming (in the field or in captivity) is necessary if more subtle distinctions are required, or if events occur too quickly to be reliably apprehended by eye.

5. *Energetics*: How much energy is required to move a certain distance using a locomotor behavior, or to maintain a particular posture? These can only be addressed in a laboratory setting using experimental techniques.

Field studies are only valuable for 1, 2 and 3; 4 and 5 are best studied in a captive or laboratory setting. Frequency and context, however, can *only* be reliably assessed from field data, and to be most useful, they must involve the collection of quantitative data. Morphologists could articulate more clearly which of these types of data would be most useful for answering the particular questions they are most interested in.

The scale of the question being asked must also be considered. General categorizations of behavior are largely based simply on repertoire, frequency is only involved at a very low level of specificity—i.e., which kind of behavior is used the most. If relationships between such general differences in behavior and morphology are all that is desired or all that we can reasonably hope to demonstrate given problems with executing adequate experimental designs in a field situation, then one might very well ask if quantitative data are necessary at all, and if they are not, then most of our concerns about comparability among studies or intraspecific variation are largely irrelevant. For example, despite the large number of quantitative studies since 1965, the basic categorization of primate loco-motor behavior outlined by Napier and Napier (1967) and Napier and Walker (1967) has generally survived intact, and is the basis for most anatomical and paleontological studies. Most investigators (e.g., Oxnard, 1974) have, however, expressed some dissatisfaction with the generality of this scheme. At such general levels we run the risk of reducing the complexity of behavior to overly simple variables, thus missing the opportunity to investigate differences in behavior and anatomy that exist within broadly defined locomotor groups. Bock (1990) gives examples in which the oversimplification of avian morphology by ecologists compromised the results of their ecomorphological analyses. We do not want to similarly misrepresent behavior with the same unfortunate result.

If the establishment of more specific relationships between morphology and behavior are desired, then questions about differences in performance other than repertoire become more important. One might suspect that differences in frequency/time spent on behaviors or in substrate milieu are responsible for morphological differentiation (e.g., Fleagle 1977a,b; Fleagle and Meldrum, 1988; Rodman, 1979; Gebo and Sargis, 1994). Difference in frequency/time spent is the hypothesis most often proposed to explain anatomical differences among fossil species, as well. Perhaps one believes that the kinematics of the 'same' behavior are different (e.g., Demes et al., 1996), or that the energetic cost of the same behavior is different. These more specific questions (especially those dealing with frequency/time spent) generally require quantitative data and statistical approaches, and thus the issue of how well the numbers are estimated becomes crucial. Simply put, if the actual numbers are important for the question one is asking, then it is important that we obtain good estimates of the actual numbers. Quantification without the accuracy, precision, and applicability of statistical inference gained by good experimental design is of no more value than a qualitative description and may potentially be more misleading.

Our point is that most current data on positional behavior are collected under research designs that are severely limited by the practicalities of the field situation. This needs to be taken into account when interpreting the numbers obtained. All estimates have associated error. Our concerns are that (1) the most commonly used method of summarizing positional data does not lend itself to the calculation of error, and (2) because of the inadequacies of experimental design the amount of error may be larger than we realize; and, depending on the scale of the question, could have consequences for derivative studies based on these data.

Quantitative studies raise the issue of comparability of numbers (but even qualitative descriptions are often difficult to compare). We have identified several factors that affect interstudy comparability: (1) the definition of behaviors and the categorization scheme used; (2) the method of data collection (bouts/time samples); (3) how data are summarized and tested (IND versus LAO; comparisons of whole repertoires (G test, rank correlations) versus specific behaviors); and, (4) sampling composition such as the age/sex composition of the sample, the number of different seasons and habitats sampled, and the behavioral contexts (e.g., traveling/feeding/resting; arboreal/terrestrial) sampled.

We are not implying that there is only one way of doing a positional behavior study. The appropriate way to do any study depends entirely on the questions one wishes to ask. Field studies are time consuming and expensive, therefore most researchers desire that the results of their hard work will be useful beyond their own immediate goals. We are only pointing out that for this purpose, some concern for comparability of results among studies should enter into decisions about data collection techniques, so that we can someday achieve the goals set by Suzanne Ripley and Warren Kinzey in 1965.

ACKNOWLEDGMENTS

We would like to thank the organizers of the symposium (E. Strasser, A. Rosenberger, J. Fleagle, and H. McHenry) for inviting us to contribute to this volume. Funding for our positional behavior projects has come from The LSB Leakey Foundation, Northern Illinois University (DLG), the National Science Foundation and Northwestern University (MD). We especially thank W.G. Kinzey for his advice and support of our projects. We thank the reviewers for their helpful comments. We would also like to express our appreciation to all of the contributors to the original conference for producing the stimulating papers that have so heavily influenced the field of primate locomotion.

REFERENCES

Altham PME (1979) Detecting relationships between categorical variables observe over time: a problem of deflating a Chi-squared statistic. Appl. Stat. 28:115–25.
Altmann J (1974) Observational study of behavior: Sampling Methods. Behaviour 49:227–67.
Arnold SJ (1983) Morphology, performance, fitness. Amer. Zool. 23:347–61.
Ashton EH, and Oxnard CE (1964) Locomotor patterns in primates. Proc. Zool. Soc. Lond. 142:1–28.
Avis V (1962) Brachiation: the crucial issue for man's ancestry. Southwestern J. Anthro. 18:119–48.
Bicca-Marques JC, and Calegaro-Marques C (1993) Feeding postures in the black howler monkey, Alouatta caraya. Folia Primatol. 60:169–72.
Bicca-Marques JC, and Calegaro-Marques C (1995) Locomotion of black howlers in a habitat with discontinuous canopy. Folia Primatol. 64:55–61.
Bock WJ (1990) From Biologische Anatomie to ecomorphology. Neth. J. Zool. 40:254–77.
Bock WJ, and von Wahlert G (1965) Adaptation and the form-function complex. Evol. 19:269–99.
Boinski S (1989) The positional behavior and substrate use of squirrel monkeys: ecological implications. J. Hum. Evol. 18:659–77.
Bradshaw AD (1987) Comparison--its scope and limits. New Phytol. 106:3–21.
Cannon CH, and Leighton M (1994) Comparative locomotor ecology of gibbons and macaques: Selection of canopy elements for crossing gaps. Am. J. Phys. Anthropol. 93:505–24.
Cant JGH (1986) Locomotion and feeding postures of spider and howling monkeys: Field study and evolutionary interpretation. Folia Primatol. 46:1–14.
Cant JGH (1987a) Effects of sexual dimorphism in body size on feeding postural behavior of Sumatran orangutans (Pongo pygmaeus). Am. J. Phys. Anthropol. 74:143–8.
Cant JGH (1987b) Positional behavior of female Bornean orangutans (Pongo pygmaeus). Am. J. Primatol. 12:71–90.
Cant JGH (1988) Positional behavior of long-tailed macaques (Macaca fascicularis) in Northern Sumatra. Am. J. Phys. Anthropol. 76:29–37.
Cant JGH (1992) Positional behavior and body size of arboreal primates: A theoretical framework for field studies and an illustration of its application. Am. J. Phys. Anthropol. 88:273–83.
Charles-Dominique P (1977) Ecology and behavior of nocturnal primates. New York: Columbia University Press.
Cheverud JM, Wagner GP, and Dow MM (1989) Methods for the comparative analysis of variation patterns. Syst. Zool. 38:201–13.
Crompton RH (1983) Age differences in locomotion in the subtropical Galaginae. Primates 24:24–59.
Crompton RH (1984) Foraging, habitat structure, and locomotion in two species of Galago. In P Rodman and J Cant (eds.): Adaptations for Foraging in Nonhuman Primates. New York: Columbia University Press, pp. 73–111.
Crompton RH, and Andau PM (1986) Locomotion and habitat utilization in free-ranging Tarsius bancanus: a preliminary report. Primates 27:337–55.

Crompton RH, Lieberman SS, and Oxnard CE (1987) Morphometrics and niche metrics in prosimian locomotion: an approach to measuring locomotion, habitat, and diet. Am. J. Phys. Anthropol. *73*:149–78.

Curtis D (1992) Substrate use in captive aye-ayes *Daubentonia madagascariensis*. J. Wildl. Pres. Trust *28*:30–44.

Curtis DJ, and Feistner ATC (1994) Positional behavior in captive aye-ayes (*Daubentonia madagascariensis*). Folia Primatol. *62*:155–9.

Dagosto M (1994) Testing positional behavior of Malagasy lemurs. A randomization approach. Am. J. Phys. Anthropol. *94*:189–202.

Dagosto M (1995) Seasonal variation in positional behavior of Malagasy lemurs. Int. J. Primatol. *16*:807–33.

Dagosto M, and Yamashita N (in press) Effect of habitat structure on positional behavior and support use in three species of lemur. Primates.

DaSilva GL (1993) Postural changes and behavioral thermoregulation in *Colobus polykomos*: the effect of climate and diet. Afr. J. Ecol. *31*:226–41.

Demes B, Jungers WL, Gross TS, and Fleagle JG (1995) Kinetics of leaping primates: Influence of substrate orientation and compliance. Am. J. Phys. Anthropol. *96*:419–30.

Demes B, Jungers W, Fleagle J, Wunderlich R, Richmond B, and Lemelin P (1996) Body size and leaping kinematics in Malagasy vertical clingers and leapers. J. Hum. Evol. *31*:366–388.

Doran DM (1992a) Comparison of instantaneous and locomotor bout sampling methods: A case study of adult male chimpanzee locomotor behavior and substrate use. Am. J. Phys. Anthropol. *89*:85–99.

Doran DM (1992b) The ontogeny of chimpanzee and pygmy chimpanzee locomotor behavior: a case study of paedomorphism and its behavioral correlates. J. Hum. Evol. *23*:139–57.

Doran DM (1993a) Comparative locomotor behavior of chimpanzees and bonobos: The influence of morphology on locomotion. Am. J. Phys. Anthropol. *91*:83–98.

Doran DM (1993b) Sex differences in adult chimpanzee positional behavior: The influence of body size on locomotion and posture. Am. J. Phys. Anthropol. *91*:99–116.

Doran DM (1996) Comparative positional behavior of the African apes. In W McGrew, L Marchant and T Nishida (eds.): Great Ape Societies. Cambridge: Cambridge University Press, pp. 213–224.

Doran DM, and Hunt K (1994a) The comparative locomotor behavior of chimpanzees and bonobos: species and habitat differences. In R Wrangham, W McGrew, F de Waal, and P Heltne (eds.): Chimpanzee Cultures. Cambridge, MA: Harvard University Press, pp. 93–108.

Dykyj D (1980) Locomotion of the slow loris in a designed substrate context. Am. J. Phys. Anthropol. *52*:577–86.

Felsenstein J (1985) Phylogenies and the comparative method. Am. Nat. *125*:1–15.

Fleagle JG (1976) Locomotion and posture of the Malayan siamang and implications for hominoid evolution. Folia Primatol. *26*:247–69.

Fleagle JG (1977a) Locomotor behavior and muscular anatomy of sympatric Malaysian leaf monkeys *Presbytis obscura* and *Presbytis melalophos*. Am. J. Phys. Anthropol. *46*:297–308.

Fleagle JG (1977b) Locomotor behavior and skeletal anatomy of sympatric Malaysian leaf monkeys *Presbytis obscura* and *Presbytis melalophos*. Yrbk. Phys. Anthro. *20*:440–53.

Fleagle JG (1978) Locomotion, posture, and habitat use of two sympatric leaf-monkeys. In G Montgomery (ed.): Ecology of Arboreal Folivores. Washington: Smithsonian Institution Press, pp. 243–51.

Fleagle JG (1979) Primate positional behavior and anatomy: naturalistic and experimental approaches. In M Morbeck, H Preuschoft, and N Gomberg (eds.): Environment, Behavior and Morphology: Dynamic Interactions in Primates. New York: G. Fischer, pp. 313–25.

Fleagle JG (1980) Locomotion and posture. In DJ Chivers (ed.): Malayan Forest Primates. New York: Plenum Press, pp. 191–207.

Fleagle JG, and Meldrum DJ (1988) Locomotor behavior and skeletal morphology of two sympatric pitheciine monkeys, *Pithecia pithecia* and *Chiropotes satanas*. Am. J. Primatol. *16*:227–49.

Fleagle JG, and Mittermeier RA (1980) Locomotor behavior, body size, and comparative ecology of seven Surinam monkeys. Am. J. Phys. Anthropol. *52*:301–14.

Fleagle JG, Mittermeier RA, and Skopec AL (1981) Differential habitat use by *Cebus apella* and *Saimiri sciureus* in central Surinam. Primates *22*:361–7.

Fontaine R (1990) Positional behavior in *Saimiri bolivensis* and *Ateles geoffroyi*. Am. J. Phys. Anthropol. *82*:485–508.

Garber PA (1980) Locomotor behavior and feeding ecology of the Panamanian tamarin (*Saguinus oedipus geoffroyi*, Callitrichidae, Primates). Int. J. Primatol. *1*:185–230.

Garber PA (1984) Use of habitat and positional behavior in a Neotropical primate *Saguinus oedipus*. In P Rodman and J Cant (eds.): Adaptations for Foraging in Nonhuman Primates. New York: Columbia University Press, pp. 112–33.

Garber PA (1991) A comparative study of positional behavior in three species of tamarin monkeys. Primates *32*:219–30.

Garber PA, and Preutz JD (1995) Positional behavior in moustached tamarin monkeys: Effects of habitat on locomotor variability and locomotor stability. J. Hum. Evol. *28:*411–26.

Garland T, and Adolph SC (1994) Why not to do two-species comparative studies: Limitations on inferring adaptation. Physio. Zool. *67:*797–828.

Gebo DL (1987) Locomotor diversity in prosimian primates. Am. J. Primatol. *13:*271–81.

Gebo DL (1992) Locomotor and postural behavior in *Alouatta palliata* and *Cebus capucinus.* Am. J. Primatol. *26:*277–90.

Gebo DL, and Chapman CA (1995a) Positional behavior in five sympatric Old World monkeys. Am. J. Phys. Anthropol. *97:*49–76.

Gebo DL, and Chapman CA (1995b) Habitat, annual and seasonal effects on positional behavior in red colobus monkeys. Am. J. Phys. Anthropol. *96:*73–82.

Gebo DL, Chapman CA, Chapman LJ, and Lambert J (1994) Locomotor response to predator threat in red colobus monkeys. Primates *35:*219–23.

Gebo DL, and Sargis EJ (1993) Terrestrial adaptations in the postcranial skeletons of guenons. Am. J. Phys. Anthropol. *93:*341–72.

Gittins SP (1983) Use of the forest canopy by the agile gibbon. Folia Primatol. *40:*134–44.

Glassman DM, and Wells JP (1984) Positional and activity behavior in a captive slow loris: A quantitative assessment. Am. J. Primatol. *7:*121–32.

Gomberg N, Morbeck M, and Preuschoft H (1979) Multidisciplinary research in the analysis of primate morphology and behavior. In M Morbeck, H Preuschoft, and N Gomberg (eds.): Environment, Behavior, and Morphology: Dynamic Interactions in Primates. New York: Gustav Fischer, pp. 5–21.

Happel R (1982) Ecology of *Pithecia hirsuta* in Peru. J. Hum. Evol. *11:*581–90.

Harvey P, and Pagel MD (1991) The Comparative Method in Evolutionary Biology. New York: Oxford University Press.

Hunt KD (1991) Positional behavior in the Hominoidea. Int. J. Primatol. *12:*95–118.

Hunt KD (1992) Positional behavior of *Pan troglodytes* in the Mahale mountains and Gombe Stream National Parks, Tanzania. Am. J. Phys. Anthropol. *87:*83–106.

Hunt KD (1994) Body size effects on vertical climbing among chimpanzees. Int. J. Primatol. *15:*855–66.

Hunt K, Cant J, Gebo D, Rose M, Walker S, and Youlatos D (1996) Standardized descriptions of primate locomotor and postural modes. Primates *37:*363–387.

Hurlbert S (1984) Pseudoreplication and the design of ecological field experiments. Ecological Monographs *54:*187–211.

James FC (1982) The ecological morphology of birds: A review. Ann. Zool. Fennici *19:*265–275.

Jolly A (1985) The Evolution of Primate Behavior. New York: MacMillan.

Kawanaka K (1996) Observation time and sampling intervals for measuring behavior and interactions of chimpanzees in the wild. Primates *37:*185–196.

Kinzey WG (1967) Preface. Am. J. Phys. Anthropol. *26:*115–8.

Kinzey WG (1977) Positional behavior and ecology in *Callicebus torquatus.* Yrbk. Phys. Anthropol. 20:468–80.

Kinzey WG, Rosenberger AL, and Ramirez M (1975) Vertical clinging and leaping in a neotropical anthropoid. Nature *255:*327–8.

Machlis L, Dodd PWD, and Fentress JC (1985) The pooling fallacy: Problems arising when individuals contribute more than one observation to the data set. Z. Tierpsychol. *68:*201–14.

MacKinnon J, and MacKinnon K (1980) The behavior of wild spectral tarsiers. Int. J. Primatol. *1:*361–79.

Manly B (1991) Randomization and Monte Carlo Methods in Biology. New York: Chapman and Hall.

Martin P, and Bateson P (1986) Measuring Behaviour. Cambridge: Cambridge University Press.

McGraw WS (1996) Cercopithecid locomotion, support use, and support availability in the Tai Forest, Ivory Coast. Am. J. Phys. Anthropol. *100:*507–22.

Mendel F (1976) Postural and locomotor behavior of *Alouatta palliata* on various substrates. Folia Primatol. *26:*36–53.

Mittermeier RA (1978) Locomotion and posture in *Ateles geoffroyi* and *Ateles paniscus.* Folia Primatol. *30:*161–93.

Mittermeier RA, and Fleagle JG (1976) The locomotor and postural repertoires of *Ateles geoffroyi* and *Colobus guereza,* and a reevaluation of the locomotor category semibrachiation. Am. J. Phys. Anthropol. *45:*235–56.

Morbeck ME (1974) Positional behaviour in *Colobus guereza*: A preliminary analysis. In S Kondo (ed.): Symp 5th Cong. International Primatological Soc. Tokyo: Japan Science Press, pp. 331–43.

Morbeck ME (1977a) Leaping, bounding and bipedalism in *Colobus guereza*: A spectrum of positional behavior. Yrbk. Phys. Anthropol. *20:*408–20.

Morbeck ME (1977b) Positional behavior, selective use of habitat substrate and associated non-positional behavior in free-ranging *Colobus guereza* (Ruppel, 1835). Primates *18:*35–58.

Morbeck ME (1979) Forelimb use and positional adaptation in *Colobus guereza*: Integration of behavioral, ecological, and anatomical data. In M Morbeck, H Preuschoft, and N Gomberg (eds.): Environment, Behavior, Morphology: Dynamic Interactions in Primates. New York: Gustav Fischer, pp. 95–117.

Napier J, and Napier PH (1967) Handbook of Living Primates. New York: Academic Press.

Napier JR, and Walker A (1967) Vertical clinging and leaping--a newly recognised category of locomotor behaviour of primates. Folia Primatol. *6:*204–19.

Niemitz C (1984a) Activity rhythms and use of space in semi-wild Bornean tarsiers, with remarks on wild spectral tarsiers. In C Niemitz (ed.): Biology of Tarsiers. Stuttgart: Gustav Fischer, pp. 85–116.

Niemitz C (1984b) Locomotion and posture of *Tarsius bancanus*. In C Niemitz (ed.): Biology of Tarsiers. New York: Gustav Fischer, pp. 191–226.

Oliveira JMS, Lima MG, Bonvincino C, Ayres JM, and Fleagle JG (1985) Preliminary notes on the ecology and behavior of the Guianan saki (*Pithecia pithecia*, Linnaeus 1766; Cebidae, Primate). Acta Amazonica *15:*249–63.

Oxnard CE (1974) Primate Locomotor Classifications for evaluating fossils: Their inutility and an alternative. In S Kondo (ed.): Symp. 5th Cong. Int'l. Primat. Soc. Tokyo: Japan Science Press, pp. 269–84.

Oxnard CE (1979) The morphological behavioral interface in extant primates: Some implications for systematics and evolution. In M Morbeck, H Preuschoft, and N Gomberg (eds.): Environment, Behavior, Morphology: Dynamic Interactions in Primates. New York: Gustav Fischer, pp. 209–28.

Oxnard CE(1984) The Order of Man. New Haven: Yale University Press.

Oxnard CE, Crompton RH, and Lieberman SS (1990) Animal Lifestyles and Anatomies. Seattle: University of Washington Press.

Pounds J (1988) Ecomorphology, locomotion, and microhabitat structure: Patterns in a tropical mainland *Anolis* community. Ecological Monographs *58:*299–320.

Pounds J (1991) Habitat structure and morphological patterns in arboreal vertebrates. In S Bell, E McCoy, and H Mushinsky (eds.): Habitat Structure: The Physical Arrangement of Objects in Space. New York: Chapman-Hall, pp. 109–117.

Remis M (1995) Effects of body size and social context on the arboreal activities of lowland gorillas in the Central African Republic. Am. J. Phys. Anthropol. *97:*413–33.

Richard A (1970) A comparative study of the activity patterns and behavior of *Alouatta villosa* and *Ateles geoffroyi*. Folia Primatol. *12:*241–63.

Richard A (1978) Behavioral variation: Case study of a Malagasy lemur. Lewisberg, Pa: Bucknell University Press.

Richard A (1985) Primates in Nature. New York: WH Freeman.

Rickelfs RE, and Miles DE (1994) Ecological and evolutionary inferences from morphology: An ecological perspective. In P Wainwright and S Reilly (eds.): Ecological Morphology. Chicago: University of Chicago Press, pp. 13–41.

Ripley S (1967) The leaping of langurs: A problem in the study of locomotor adaptation. Am. J. Phys. Anthropol. *26:*149–70.

Ripley S (1977) Gray zones and gray langurs: Is the "semi-" concept seminal? Yrbk. Phys. Anthropol. *20:*376–94.

Ripley S (1979) Environmental grain, niche diversification, and positional behavior in Neogene primates: An evolutionary hypothesis. In M Morbeck, H Preuschoft, and N Gomberg (eds.): Environment, Behavior, and Morphology: Dynamic Interactions in Primates. New York: Gustav Fischer, pp. 37–74.

Roberts M, and Cunningham B (1986) Space and substrate use in captive western tarsiers, *Tarsius bancanus*. Int. J. Primatol. *7:*113–30.

Roberts M, and Kohn F (1993) Habitat use, foraging behavior and activity patterns in reproducing Western tarsiers, *Tarsius bancanus* in captivity. Zoo Biology *12:*217–32.

Rodman PS (1979) Skeletal differentiation of *Macaca fascicularis* and *Macaca nemestrina* in relation to arboreal and terrestrial quadrupedalism. Am. J. Phys. Anthropol. *51:*51–62.

Rodman PS (1991) Structural differentiation of microhabitats of sympatric *Macaca fascicularis* and *M. nemestrina* in East Kalimantan, Indonesia. Int. J. Primatol. *12:*357–75.

Rollinson J, and Martin RD (1981) Comparative aspects of primate locomotion, with special reference to arboreal cercopithecines. Symp. Zool. Soc. Lond. *48:*377–427.

Rose MD (1974) Postural adaptations in New and Old World monkeys. In F Jenkins (ed.): Primate Locomotion. New York: Academic Press, pp. 201–222.

Rose MD (1976) Bipedal behavior of olive baboons (*Papio anubis*) and its relevance to an understanding of the evolution of human bipedalism. Am. J. Phys. Anthropol. *44:*247–262.

Rose MD (1977) Positional behavior of olive baboons *Papio anubis* and its relationship to maintenance and social activities. Primates *18:*59–116.

Rose MD (1978) Feeding and associated positional behavior of black and white colobus monkeys (*Colobus guereza*). In G Montgomery (ed.): The Ecology of Arboreal Folivores. Washington, D.C.: Smithsonian Press, pp. 253–62.

Rose MD (1979) Positional behavior of natural populations: Some quantitative results of a field study of *Colobus guereza* and *Cercopithecus aethiops*. In M Morbeck, H Preuschoft and N Gomberg (eds.): Environment, Behavior, Morphology: Dynamic Interactions in Primates. New York: Gustav Fischer, pp. 75–93.

Rosenberger AL, and Stafford BJ (1994) Locomotion in captive *Leontopithecus* and *Callimico*: A multimedia study. Am. J. Phys. Anthropol. *94:*379–94.

Schön Ybarra MA (1984) Locomotion and postures of red howlers in a deciduous forest-savanna interface. Am. J. Phys. Anthropol. *63:*65–76.

Schön Ybarra MA, and Schön III MA (1987) Positional behavior and limb bone adaptations in red howling monkeys (*Alouatta seniculus*). Folia Primatol. *49:*70–89.

Simberloff D (1982) The status of competition theories in ecology. Ann. Zool. Fennici *19:*241–253.

Simons EL (1967) Fossil primates and the evolution of some primate locomotor systems. Am. J. Phys. Anthropol. *26:*241–54.

Smith RJ (1994) Degrees of freedom in interspecific allometry: An adjustment for the effects of phylogenetic constraint. Am. J. Phys. Anthropol. *93:*95–108.

Srikosamatara S (1984) Ecology of pileated gibbons in South-East Thailand. In H Preuschoft, D Chivers, W Brockelman, and N Creel (eds.): The Lesser Apes. Edinburgh: Edinburgh University Press, pp. 242–57.

Sugardjito J (1982) Locomotor behaviour of the Sumatran Orang Utan (*Pongo pygmaeus abelii*) at Ketambe, Gunung Leuser National Park. Malay Nat. J. *35:*57–64.

Sugardjito J, and van Hooff JARAM (1986) Age-sex class differences in the positional behavior of the Sumatran Orang-utan (*Pongo pygmaeus abelii*) in the Gunung Leuser National Park, Indonesia. Folia Primatol. *47:*14–25.

Susman RL (1984) The locomotor behavior of *Pan paniscus* in the Lomako forest. In R Susman (ed.): The Pygmy Chimpanzee. New York: Plenum Press, pp. 369–93.

Susman RL, Badrian NL, and Badrian AJ (1980) Locomotor behavior of *Pan paniscus* in Zaire. Am. J. Phys. Anthropol. *53:*69–80.

Tattersall I (1977) Ecology and behavior of *Lemur fulvus mayottensis* (Primates, Lemuriformes). Anth. Pap. Am. Mus. Nat. Hist. *54:*421–82.

Terranova C (1995) Functional morphology of leaping behaviors in galagids: Associations between landing limb use and diaphyseal geometry. In L Alterman, G Doyle, and M Izard (eds.): Creatures of the Dark. New York: Plenum Press, pp. 473–494.

Tilden CD (1990) A study of locomotor behavior in a captive colony of red-bellied lemurs (*Eulemur rubriventer*). Am. J. Primatol. *22:*87–100.

Tremble M, Muskita Y, and Supriatna J (1993) Field observations of *Tarsius dianae* at Lore Lindu National Park, Central Sulawesi, Indonesia. Tropical Biodiversity *1:*67–76.

Tuttle RH, and Watts DP (1985) The positional behavior and adaptive complexes of *Pan gorilla*. In S Kondo (ed.): Primate Morphophysiology, Locomotor Analyses and Human Bipedalism. Tokyo: University of Tokyo Press, pp. 261–88.

Wainwright P, and Reilly SM, eds. (1994) Ecological Morphology. Chicago: Chicago University Press.

Walker A (1969) The locomotion of the lorises, with special reference to the potto. E. Afr. Wildl. J. *7:*1–5.

Walker AC (1979) Prosimian locomotor behavior. In G Doyle and R Martin (eds.): The Study of Prosimian Behavior. New York: Academic Press, pp. 543–65.

Walker SE (1994) Positional behavior and habitat use in *Chiropotes satanas* and *Pithecia pithecia*. In B Thierry, J Anderson, J Roseder and N Herrenschmidt (eds.): Current Primatology, Vol. 1: Ecology and evolution. Strasbourg: International Primatological Society, pp. 195–202.

Walker SE (1996) The evolution of positional behavior in the Saki/Uakaris *(Pithecia, Chiropotes, and Cacajao)*. In M Norconk, A Rosenberger, and P Garber (eds.): Adaptive Radiations of Neotropical Primates. New York: Plenum Press, pp. 335–368.

Walker SE, and Ayres JM (1996) Positional behavior of the white uakari (*Cacajao calvus calvus*). Am. J. Phys. Anthropol. *101:*161–172.

White FJ, Overdorff DJ, Balko EA, and Wright PC (1995) Distribution of ruffed lemurs (*Varecia variegata*) in Ranomafana National Park, Madagascar. Folia Primatol. *64:*124–31.

Wilson JM, Stewart PD, Ramangason G-S, Denning AM, and Hutchings MS (1989) Ecology and conservation of the crowned lemur, *Lemur coronatus*, at Ankarana, N. Madagascar. Folia Primatol. *52:*1–26.

Youlatos D (1993) Passage within a discontinuous canopy: Bridging in the red howler monkey (*Alouatta seniculus*). Folia Primatol. *61:*144–147.

2

FINE-GRAINED DIFFERENCES WITHIN POSITIONAL CATEGORIES

A Case Study of *Pithecia* and *Chiropotes*

Suzanne E. Walker

Department of Anthropology
Humboldt State University
Arcata, California 95521

1. INTRODUCTION

The study of primate locomotion and posture has advanced considerably since its beginnings, approximately thirty years ago. Early studies were primarily concerned with inferring the locomotor behavior of the earliest hominids, using the living great apes as analogues (Ashton and Oxnard, 1963; Napier, 1963). A turning point for studies on locomotion and posture was the 1965 Primate Locomotion Symposium, organized by Warren Kinzey. In this symposium, the importance of field studies was first emphasized, as was the importance of distinguishing between locomotor categories based upon natural behavior rather than upon skeletal anatomy or observations of zoo animals (Kinzey, 1967). At the same time, Prost (1965) stressed the importance of a standardized system to classify positional (i.e., locomotor and postural) behavior, which would allow for more precise comparison between studies, and emphasized the importance of postural (in addition to locomotor) behaviors in shaping the postcranium.

Many of the useful ideas and terms that are now frequently utilized in the literature were introduced in the 60s and 70s as a result of field studies of primate positional behavior (e.g., Ripley, 1967; Rose, 1974; Morbeck, 1976). Studies undertaken in the field are essential in order to provide more realistic views of how primates move in their natural habitats; they provide the basis for making inferences about environmental variables that may have provided the selective pressures that shaped primate anatomies. Ripley (1977) was instrumental in replacing the "semi" locomotor categories with others more descriptive of actual behavioral patterns (e.g., an animal cannot "semi-brachiate"). Others (e.g., Fleagle and Mittermeier, 1980) tested the predictions of Napier (1962) and Cartmill and Milton (1977) about the influence of forest structure and body size on positional behavior. Further research resulted in ex-

Primate Locomotion, edited by Strasser *et al.*
Plenum Press, New York, 1998

amination of more fine-grained environmental influences: that of the structure of individual trees (Ripley, 1977), use of various tree portions (Mendel, 1976), and the location of various classes of primate food in the forests (Crompton, 1984). Some of these important field studies have emphasized a problem-oriented approach (e.g., Cant and Temerin, 1984), focusing on the ways in which primates deal with problems presented to them by their environment and the location of foods within the forest matrix. These and other studies have paved the way for more sophisticated and detailed research (see Dagosto and Gebo, this volume).

Research techniques in the study of primate positional behavior are becoming increasingly "fine-tuned," with researchers using new data collection techniques and more detailed behavioral categories (e.g., Fontaine, 1990; Hunt et al., 1996) that make for more feasible cross-study comparisons. With ever-improving video capabilities and interfacing computer programs (e.g., Rosenberger and Stafford, 1994), we can now attempt some of the analyses of positional behavior of wild primates previously possible only for those in captivity. Only recently have a rigorous set of statistical tests been applied to results of studies of positional behavior (e.g., Boinski, 1989; Fontaine, 1990; Doran, 1992; Dagosto, 1994; Gebo and Chapman, 1995; Remis, 1995), partly because of the fact that much of the data on positional behavior are not particularly amenable to statistical testing (e.g., Mendel, 1976; Cant, 1988; Hunt, 1992).

1.1. The Problems

In the spirit of the Conference *Primate Locomotion 1995*, which marked the 30-year anniversary of the original Primate Locomotion symposium, this paper discusses some methodological problems in studies of positional behavior that warrant further investigation (see also Dagosto and Gebo, this volume). These problems (described below), which became apparent during the course of my comparative field study of the positional behavior of three platyrrhine species, *Pithecia pithecia*, *Chiropotes satanas*, and *Cacajao calvus*, of the tribe Pitheciini, all deal with the classification of positional behaviors into discrete categories. *Pithecia* and *Chiropotes* were studied more intensively over a longer time period than was *Cacajao*; therefore, only examples from the former two species are included here.

First, establishing a positional classification and subsequently using the categories to collect data is difficult due to two factors: some observed behaviors do not lend themselves well to a one- or two-word descriptive category, and there are often subtle gradations of one behavioral category into another that may result simply from a minor change in support characteristics. For example, the behavior described as quadrupedal walk grades into pronograde clamber (Cant, 1988) as the substrate changes from a single support to multiple ones. Likewise, with increasing inclination of a support, quadrupedal walk grades into the behavior called climb (defined below), and a sitting posture grades into a vertical cling.

Second, it is important to categorize and define behaviors in an anatomically meaningful way (e.g., Jolly, 1965), so as to allow, for example, inferences to be drawn about the positional behavior of fossil primates. Categories should reflect such information as limb segment positioning and forces thought to be transmitted through limbs (Schön and Schön, 1987), and number of cheiridia or body parts in contact with supports (e.g., Hunt, 1992).

Third, consistent terminology should be used for the categories and should be carefully defined (Hunt et al., 1996). For example, "climb" is a persistent problem category, defined differently in various studies. Rose (1979) and Cartmill (1985) define climb as progression up or down a single support with vertical or steeply sloping surfaces (the definition followed in this paper), while Fleagle and Mittermeier (1980) and Gebo (1992) define it more broadly and include irregular gaits on multiple supports as well as bridging behaviors.

Lastly, the problem that constitutes the primary focus of this paper is with the variation exhibited within and between species in the use of traditionally defined behavioral categories. One behavioral category may not sufficiently describe the variation that individual conspecifics exhibit in their use of that behavior. That is, a single behavioral category can often encompass several finer categories that differ from one another depending upon the influences of body orientation and/or support characteristics. Conversely, variation in body orientation, support use, and morphology can result in differences in the manner in which species exhibit the "same" behavior.

Human observers draw arbitrary boundaries between classes of behaviors. How do we know that the boundaries that we create are biologically relevant, and how do we distinguish them? It is suggested here that we recognize this problem in the classification of positional behavior, particularly with regard to the biological function of various positional behaviors and their influence upon anatomical features.

1.2. Qualitative Aspects of Positional Categories: Are Finer Classifications Necessary?

Categories such as sit, quadrupedal walk, and leap provide us with the primary means of communicating about basic positional behaviors. Each of these categories, however, can be further subdivided into more fine-grained categories. Previous workers (e.g., Garber, 1980) have noted the importance of these finer categories; however, in the analysis of positional behavior, these behaviors are eventually pooled due to the sheer unmanageability of many small categories.

I propose that these finer categories of behavior be investigated for two main reasons. First, we need to determine whether or not minor differences in body position or support use influence the forces to which primate bodies are subjected. Biomechanical aspects of positional behavior, such as the influence of gravity, types of stresses, force application, force absorption, and friction are highly dependent upon body orientation, characteristics of supports used, and obviously, morphology. Second, the primary contexts in which these behaviors are used often differ from one another, which may indicate different functions for them. A difference in function would presumably have a profound influence on the evolution of the animals, and therefore warrants further investigation.

In this paper, I will use my field data on *Pithecia* and *Chiropotes* to demonstrate that limitations exist in the traditional manner of classifying positional behavior into coarse-grained categories. By breaking down coarse-grained behavioral categories into finer-grained ones, I have identified associations of body orientation and support characteristics with positional behaviors that would not have been possible without these fine-grained categories. However, further investigation of the influence of, for example, different *types* of sitting behavior on postcranial anatomy is in order. Since a major goal of studies of positional behavior is to create a data base from which to interpret fossil remains, any information that will allow us to make more detailed inferences about the behavior of extinct species provides an important contribution.

2. STUDY SITES AND METHODS

Between 1989 and 1991 I collected data on the positional behavior of *Pithecia pithecia* and *Chiropotes satanas* in Venezuela, where they are naturally allopatric. The study animals were well habituated and observed on separate islands in Guri Lake (for description of study sites see Pernía, 1985; Parolin, 1993; Walker, 1993, 1996).

The data on positional behavior were collected using focal animal instantaneous sampling at two minute intervals (after Garber, 1980; for a detailed description of methods see Walker, 1993, 1996). Videotaping was utilized to assist in the initial categorization of behaviors, and some drawings based on captured video frames are used to illustrate behavioral categories in this paper.

In this study, the problem of independence of consecutive samples was first dealt with in the field during data collection by noting the samples that involved a continuation of behavior from the previous data point (i.e., if the focal animals had not moved at least one body length between data points). The (presumably) non-independent samples were then eliminated before performing statistical tests. This is similar to the *post facto* "pooling" of instantaneous samples into artificial bouts when consecutive samples were the same (Hunt, 1992). The latter method, however, may eliminate samples that are, in fact, independent. The frequency data were examined for significant differences, using a Chi-square test. Differences were deemed significant if the probability value of the Chi-square statistic was 0.01 or less.

Below, I present a few commonly used positional behaviors of *Pithecia* and *Chiropotes*, and demonstrate how differences exist within coarse-grained (traditionally used) behavioral categories due to the influence of body orientation, support characteristics and support use, and morphological features. The results are organized into two main sections. In the first, I present several examples that deal primarily with differences within a species in the use of a single behavior, demonstrating the influence of various maintenance activities and support characteristics on body orientation. For some of these behaviors (i.e., sit) quantitative data are presented; for others (i.e., quadrupedal stand and quadrupedal walk) anecdotal observations are discussed. Data for this section are drawn from a three-month subset of observations on *Pithecia* and *Chiropotes*. In the second section, one behavioral example is presented that deals with differences between the two species in their use of a positional behavior traditionally represented by a single behavioral category. This data set is based upon fifteen months of observations on *Pithecia* and ten months on *Chiropotes*.

3. WITHIN-SPECIES DIFFERENCES IN POSITIONAL CATEGORIES

During postures such as sit and quadrupedal stand, the orientation of the craniocaudal body axis can be either perpendicular or parallel to the support (Figures 1 and 2). These two body orientations influence how the center of gravity is maintained over the base of support, which in turn affects the use of the limbs in weight-bearing and grasping.

During perpendicular sitting (Figure 1, left), the body axis is more or less perpendicular to the substrate and the back is sometimes flexed. Perpendicular sit poses few problems for the individual in terms of balance. Body weight is balanced on the soles of the feet, usually with the added support of the haunches. The forelimbs are often not involved in weight-bearing, leaving the hands free for other activities. During parallel sitting (Figure 1, right), the body axis is aligned with the support; therefore, the base of support is narrower and the body is subject to rolling to the side. The feet grasp the branch, with the soles on one side and digits on the other, in front of the haunches and the haunches rest on the support. In parallel sit, the forelimbs often play a weight-bearing role, with the hands used to grasp the support or a nearby branch. While grasping, the pollex may oppose the

Figure 1. Sit in *Pithecia*. (*Left)* perpendicular sit, (*right*) parallel sit.

other digits, or grips may be taken between digits II and III, as noted previously by other researchers (e.g., Erikson, 1957; Hill, 1960; Kay, 1990).

3.1. Influence of Different Maintenance Activities

Perpendicular and parallel sits differ in their occurrence in *Pithecia* during the various activities of feeding, traveling, and resting (Table 1). Overall, the perpendicular sit is

Figure 2. Stand in *Chiropotes. Left*, perpendicular stand; *right*, parallel stand.

Table 1. Variation in the percentage of perpendicular vs. parallel
sit during various maintenance activities in *Pithecia*[1]

Maintenance activity	N	Perpendicular sit	Parallel sit
Feeding	297	56.9	28.9
Traveling	33	5.9	3.9
Resting	306	37.2	67.2

[1]The differences in sit types are significant (χ^2=53.814, df=2, P<0.001).

the more common of the two sitting postures throughout all activity contexts. Perpendicular sit is a more frequent feeding than resting posture while the opposite is true for parallel sit, which is used more often during resting than during feeding (Table 1).

Quadrupedal standing occurs on horizontal or low-angled supports. As with sitting, the two main types of body orientation with reference to the long axis of the support are perpendicular and parallel, each posing distinct problems of balance. During perpendicular standing (Figure 2, left) the craniocaudal axis of the body is aligned perpendicular to the long axis of the support, resulting in an unstable posture. The back is somewhat flexed, and the feet often grasp the support with the hallux opposed to the remaining digits, while the hands grasp with all digits in line. In this position a small portion of the body is directly over the base of support and an individual is subject to pitching forward or backward. The parallel stand (Figure 2, right) is a relatively more stable posture (at least on solid supports), due to its larger base of support and the ease with which the body is maintained directly over it. The long axis of the body is parallel to the support and the limbs are primarily extended and adducted. Hand grips again may be taken between digits II and III. The primary balance problem to be dealt with is rolling to either side.

Chiropotes often uses both types of stand, while *Pithecia* was rarely observed in perpendicular stand. Observations indicate that perpendicular stand often serves as a momentary pause within a travel bout, and sometimes as a leap takeoff or landing position, while parallel stand may be a longer-term posture during any of the maintenance activities of feed, travel or rest.

3.2. Influence of Support Characteristics

3.2.1. Support Size. Pithecia uses different sized supports during perpendicular vs. parallel sit (Table 2). For example, on the smallest supports (under 2 cm in diameter) parallel sit was not observed to occur while perpendicular sit was observed in 7.5% of sitting samples (Table 2). On smaller branches, the haunches are less likely to support body

Table 2. Variation in the percentage of perpendicular vs. parallel
sit on supports of various diameters in *Pithecia*[1]

Support diameter	N	Perpendicular sit	Parallel sit
<2 cm	30	7.5	0
2-5 cm	310	49.8	48.3
6-10 cm	174	28.1	26.7
>10 cm	37	4.8	7.8
Mix [2]	79	9.8	17.2

[1]The differences in sit types are significant (χ^2=26.356, df=4, P<0.001).
[2]Mix, multiple supports of various diameters.

weight; the weight may be distributed entirely through the soles of the feet. During perpendicular sit, the hallux may be in line with other pedal digits (if the support is relatively large) or opposed to the others (if the support is relatively small). Parallel sit occurred more often than perpendicular sit on supports greater than 10 cm in diameter (7.8% vs. 4.8%) and on supports of mixed diameter (17% vs. 9.8%). These differences contribute to the overall statistical significance of the data sets (Table 2).

Support size influences quadrupedal stand and quadrupedal walk in similar ways due to problems of balance as support size decreases. On smaller supports (particularly flexible ones), the perpendicular stand is more often observed than is the parallel stand, due to the increased risk of rolling to the side. During both standing and walking, limb flexion and abduction act to lower the center of gravity for increased stability on smaller supports. The grasp taken and the degree of wrist pronation depend upon support size as well, with pronation increasing as the central weight-bearing axis moves laterally in the hand.

3.2.2. Support Inclination. Angled supports and multiple supports of mixed inclination are used approximately equally during both perpendicular and parallel sit. On horizontal supports parallel sit occurs more often than does perpendicular sit (39.0% vs. 30.9%), whereas when on deformable supports, the opposite is true, with perpendicular sit being more common (12.2% vs. 3.5%; Table 3). On angled supports, problems of reduced friction can be alleviated behaviorally; for example, by the use of more than one support for grasping.

Support inclination influences quadrupedal stand and quadrupedal walk in terms of introducing additional problems of balance and loss of friction. With increasing support inclination, perpendicular stand is observed less frequently than is parallel stand, and during the latter the animal typically faces in the upward direction of the support. The animal's center of gravity shifts with increasing inclination of a support, becoming displaced either forward or backward depending upon the direction of the angle. In order to maintain the center of gravity over the base of support, the limbs tend to be more flexed and abducted. Friction becomes more important, since there is a potential for slippage; steep inclinations increase the shear force exerted between the feet and the branch (Cartmill, 1985). During quadrupedal walk, the limbs take on different roles than needed on a horizontal branch, where they act as simple struts. The roles of the fore- and hindlimbs diverge: forelimbs are under tension while hindlimbs are used for propulsion. Therefore, in species with more frequent use of highly angled supports, longer hindlimbs may provide advantages in dealing with the greater demands of propulsion.

Table 3. Variation in the percentage of perpendicular vs. parallel sit on supports of various inclinations in *Pithecia*[1]

Support inclination[2]	N	Perpendicular sit	Parallel sit
Horizontal	214	31	39
Angled	280	44	46
Deformable	57	12	3
Mix	81	13	12

[1]The differences in sit types are significant (χ^2=26.356, df=4, P<0.001).
[2]Horizontal (0-20°), angled (21-70°), deformable (under animal's body weight), mix (supports of mixed inclinations).

4. BETWEEN-SPECIES DIFFERENCES IN POSITIONAL CATEGORIES: THE SPECIAL CASE OF LEAPING

Leaping is the behavior that epitomizes qualitative differences within categories of positional behavior, made evident, for example, by the different forms of leaping in the prosimian vertical clingers and leapers (e.g., Oxnard, 1984) as well as various species of callitrichids (Garber, 1991). In the genus *Saguinus*, differences in limb ratios correlate with differences in the type and frequency of leaping behavior (Garber, 1991). Obvious differences in the form of leaping and associated postcranial morphology are also apparent in *Pithecia* and *Chiropotes*; each species has very different ways of dealing with problems of take-off and landing forces. Below, the influence of differences in body orientation, support characteristics, and morphology are shown to affect the biomechanical aspects of leaping behavior.

Leaping plays an important role in the positional repertoire of both *Pithecia* and *Chiropotes*, with total frequencies of 40% and 25% of travel samples, respectively. *Pithecia* shares numerous behavioral and morphological specializations with the prosimian vertical clingers and leapers (Fleagle and Meldrum, 1988; Walker, 1993, in revision), while *Chiropotes* is relatively unspecialized in its leaping behavior and associated morphology.

4.1. Influence of Body Orientation

Pithecia tends to maintain an orthograde body orientation throughout all leap phases, even when taking off from a horizontal support (Figure 3). In virtually all leaps,

Preparatory   Take-off   Midflight   Landing

Figure 3. Leap phases in *Pithecia* (*top*) and *Chiropotes* (*bottom*).

Table 4. Variation in the percentage of support diameters used for leap take-off and landing in *Pithecia* and *Chiropotes*

Support diameter	Take-off[1]		Landing[2]	
	Pithecia N=887	*Chiropotes* N=407	*Pithecia* N=821	*Chiropotes* N=366
<2 cm	4.1	20.1	10.6	42.4
2-5 cm	41.1	33.8	41.3	17.5
6-10 cm	38	25.1	32.3	16.7
11-15 cm	10.2	6.9	9.3	7.1
>15 cm	1.9	1.5	1.5	1.1
Mix	4.7	12.6	5	15.2

[1]The differences in support diameters are significant (χ^2=126.24, df=5, P<0.001).
[2]The differences in support diameters are significant (χ^2=222.97, df=5, P<0.001).

the lower body swings forward for the hindlimbs to contact the landing support first. In longer leaps between vertical supports, the back and hindlimbs flex to pull the body into a tuck in midleap, as in *Tarsius* (Peters and Preuschoft, 1984). In contrast, *Chiropotes'* body orientation remains pronograde throughout take-off, midleap and landing (Figure 3). A common feature is that both species crouch before take-off; this extreme limb flexion increases the time over which propulsive extension occurs, and it potentiates the muscles to be used in propulsion (e.g., Cavanagh, 1977). These differences between the two species in body orientation result from differences in characteristics of the supports used for take-off and landing, and from morphology (see below).

4.2. Influence of Support Characteristics

4.2.1. Support Size. Characteristics of the supports used by the species differ for both takeoff and landing (Table 4). *Pithecia* most often uses supports between 2 and 10 cm in diameter for takeoff, while *Chiropotes* also often uses those less than 2 cm (Table 4). For landing, differences are also apparent. Most landings by *Pithecia* are onto landing supports between 2 and 10 cm in diameter, while for *Chiropotes* the most frequent landing supports are multiple terminal branches of less than 2 cm in diameter.

4.2.2. Support Inclination. Overall differences between the species in the inclination of takeoff and landing supports were significant (Table 5). *Pithecia* prefers vertical or near vertical supports for both takeoff and landing (Table 5), partly because of their propensity for taking off from a clinging posture. *Chiropotes*, on the other hand, most often uses horizontal or angled supports for take-off. *Chiropotes* typically travels rapidly by quadrupedal locomotion to the terminal branches, then uses this forward momentum in leap take-off. For landing, deformable supports are most often used by *Chiropotes*.

4.3. Influence of Morphology

Several of the anatomical features characteristic of specialized leapers can be distinguished in *Pithecia*, particularly at the proximal and distal femur, the vertebral column, and the tarsal and metatarsal bones (Fleagle and Meldrum, 1988). *Pithecia* frequently takes off from a stationary posture rather than during locomotion. The velocity of *Pithecia*'s leaps are greater than *Chiropotes'* (Walker, in revision); *Pithecia*'s relatively

Table 5. Variation in the percentage of support inclinations used for leap take-off and landing in *Pithecia* and *Chiropotes*

Support inclination[1]	Take-off[2]		Landing[3]	
	Pithecia N=892	*Chiropotes* N=408	*Pithecia* N=828	*Chiropotes* N=367
Horizontal	13	24.5	8.2	14.2
Angled	26.2	29.1	28.8	19.8
Vertical	45.9	1	41.4	2.5
Deformable	10.7	41.2	19.9	62.4
Mix	4.2	4.2	1.7	1.1

[1]Horizontal (0-20°), angled (21-70°), vertical (71-90°), deformable (under animal's body weight), mix (supports of mixed inclinations).
[2]The differences in support inclinations are significant (χ^2=329.03, df=4, P<0.001).
[3]The differences in support inclinations are significant (χ^2=287.49, df=4, P<0.001).

longer hindlimbs apply propulsive force to the takeoff substrate for a longer time than is seen in *Chiropotes*. At landing, *Pithecia*'s long hindlimbs contact the landing substrate first, and flex to absorb the compressive forces. The relatively unspecialized (in terms of leaping) *Chiropotes* deals with these forces by landing with all four limbs onto deformable supports.

For both *Pithecia* and *Chiropotes*, we can best describe the aforementioned behavior as leaping, but as demonstrated, the obvious qualitative and quantitative differences cannot be ignored. In fact, the leaping specializations of *Pithecia* affect its entire behavioral repertoire (Walker, 1993, in revision).

5. CONCLUSIONS

The traditional way of examining positional behavior in terms of the frequencies of standard behavioral categories continues to be an important tool for understanding the relationship between form and function. Thus, this paper in no way seeks to undermine or reduce the importance of previous or continuing studies of functional morphology, which have identified important correlates between form and function in numerous species (e.g., Fleagle, 1977; Rodman, 1979; Ward and Sussman, 1979). However, there are limitations to the standard methods used in classifying positional behavior and it is here suggested that we should attempt to deal with these limitations.

The use of fine-grained categories as demonstrated here has brought to light, for example, two clear types of sitting behavior that are associated with different substrates and with different activity contexts (e.g., resting vs. feeding). Additional studies need to be conducted for these different behaviors in order to investigate anatomical correlates and their biomechanical advantages. By identifying these correlates, we can strengthen inferences about the positional behavior of extinct species from their fossilized remains. The fact that the contexts in which these behaviors are used often differ from one another may reflect a fundamental difference in biological role (Bock and von Wahlert, 1965). We need to examine the environmental and behavioral basis for these differences, which influence the evolution of the postcranial skeleton. Another limitation in the traditional positional classifications, as demonstrated for leaping, is that simply using the term "leap" for the behavior exhibited by *Pithecia* and by *Chiropotes* does not provide a detailed description of how the behavior is conducted by the two species in terms of biomechanics, substrate use and body orientation.

In future studies of positional behavior, greater detail in classification of behavior should be a primary goal, and can be assisted by the use of technological advances in video and interfacing computer programs. Following are a few suggestions in the construction and use of finer-grained positional behavior classifications for field studies.

In the initial categorization of positional behavior in the field, video recording should be used if possible in order to describe in greater detail the orientation of limbs to trunk, body to substrate, etc., which will result in a more anatomically meaningful classification for biomechanical analyses. Developing an ethogram in this way should take relatively little additional time before data collection begins. In my own study, the finer behavioral categories, once identified, were relatively easy to observe and record in the field; thus, it should not be necessary to collect all data with video recorders to analyze later.

Video recording is not only useful in the initial categorization of behaviors, but to later convey this information in presentations and publications. With somewhat greater emphasis on graphically presenting behaviors, such as with stills from video frames, more efficient comparisons among species can be made, and biomechanical analyses conducted. Short sequences of behavior could even easily be presented on the internet (Dagosto, pers. comm.), with references in the original publication.

With continued collection of positional behavior data using carefully devised and defined categories, and with the use of new techniques, we will continue to obtain a better understanding of how primates most efficiently feed, travel and rest in their natural habitats, and the circumstances that led to the evolution of their positional adaptations.

ACKNOWLEDGMENTS

I would like to thank the organizers of the symposium on Primate Locomotion 1995 for inviting me to participate: Drs. Elizabeth Strasser, John G. Fleagle, Henry McHenry and Alfred Rosenberger. I give special thanks to Elizabeth Strasser, Marian Dagosto and two anonymous reviewers for their very insightful comments on an earlier version of this manuscript. This research was supported by grants from the National Science Foundation (BNS 89–13349), Wenner Gren Foundation for Anthropological Research, and the Graduate Center of the City University of New York. This paper is dedicated to my husband, whose memory continues to be an inspiration to me.

REFERENCES

Ashton EH, and Oxnard CE (1963) The musculature of the primate shoulder. Trans. Zool. Soc. London *29*:553–650.

Bock WJ, and von Wahlert G (1965) Adaptation and the form-function complex. Evolution *19*:269–299.

Boinski S (1989) The positional behavior and substrate use of squirrel monkeys: ecological implications. J. Hum. Evol. *18*:659–678.

Cant JGH, and Temerin LA (1984) A conceptual approach to foraging adaptations in primates. In PS Rodman and JGH Cant (eds.): Adaptations for Foraging in Nonhuman Primates. New York: Columbia University Press, pp. 249–279.

Cant JGH (1988) Positional behavior of long-tailed macaques (*Macaca fascicularis*) in northern Sumatra. Am. J. Phys. Anthropol. *76*:29–37.

Cartmill M (1985) Climbing. In M Hildebrand, DM Bramble, KF Liem, and DB Wake (eds.): Functional Vertebrate Morphology. Cambridge: Belknap Press, pp. 73–88.

Cartmill M, and Milton K (1977) The lorisiform wrist joint and the evolution of "brachiating" adaptations in the
Hominoidea. Am. J. Phys. Anthropol. 47:249–272.

Cavanagh GA (1977) Storage and utilization of elastic energy in skeletal muscle. Exercise Sports Sci. Rev.
5:89–129.

Crompton RH (1984) Foraging, habitat structure and locomotion in two species of Galago. In PS Rodman and
JGH Cant (eds.): Adaptations for Foraging in Nonhuman Primates. New York: Columbia University Press,
pp. 73–111.

Dagosto M (1994) Testing positional behavior of Malagasy lemurs. A randomization approach. Am. J. Phys. Anthropol. 94:189–202.

Doran D (1992) Comparison of instantaneous and locomotor bout sampling methods: a case study of adult male
chimpanzee locomotor behavior and substrate use. Am. J. Phys. Anthropol. 89:85–100.

Erickson GE (1957) The hands of New World primates, with comparative functional observations on the hands of
other primates. Am. J. Phys. Anthropol. 15:446.

Fleagle JG (1977) Locomotor behavior and skeletal anatomy of sympatric Malaysian leaf-monkeys (Presbytis obscura and Presbytis melalophos). Am. J. Phys. Anthropol. 46:297–308.

Fleagle JG, and Meldrum DJ (1988) Locomotor behavior and skeletal morphology of two sympatric pitheciine
monkeys, Chiropotes satanas and Pithecia pithecia. Am. J. Primatol. 16:227–249.

Fleagle JG, and Mittermeier RA (1980) Locomotor behavior, body size, and comparative ecology of seven Surinam monkeys. Am. J. Phys. Anthropol. 52:301–314.

Fontaine R (1990) Positional behavior in Saimiri boliviensis and Ateles geoffroyi. Am. J. Phys. Anthropol.
82:485–508.

Garber PA (1980) Locomotor Behavior and Feeding Ecology of the Panamanian Tamarin (Saguinus oedipus geoffroyi, Callitrichidae, Primates). Ph.D. Dissertation, Washington University.

Garber PA (1991) A comparative study of positional behavior in three species of tamarin monkeys. Primates
32:219–230.

Gebo DL (1992) Locomotor and postural behavior in Alouatta palliata and Cebus capucinus. Am. J. Primatol.
26:277–290.

Gebo DL, and Chapman CA (1995) Positional behavior in five sympatric Old World monkeys. Am. J. Phys. Anthropol. 97:49–76.

Hill WCO (1960) Primates: Comparative Anatomy and Taxonomy, Vol. IV. Edinburgh: Edinburgh University
Press.

Hunt KD (1992) Positional behavior of Pan troglodytes in the Mahale Mountains and Gombe Stream National
Parks, Tanzania. Am. J. Phys. Anthropol. 87:83–106.

Hunt KD, Cant JGH, Gebo DL, Rose MD, Walker SE, and Youlatos DL (1996) Standardized descriptions of primate locomotor and postural modes. Primates 37:363–387.

Jolly CJ (1965) Origins and Specialisations of the Long-faced Cercopithecoids. Ph.D. Dissertation, University of
London.

Kay RF (1990) The phyletic relationships of extant and fossil Pitheciinae (Platyrrhini, Anthropoidea). In JG
Fleagle and AL Rosenberger (eds.): The Platyrrhine Fossil Record. New York: Academic Press, pp.
175–208.

Kinzey WG (1967) Preface to symposium on primate locomotion. Am. J. Phys. Anthropol. 26:115–118.

Mendel F (1976) Postural and locomotor behavior of Alouatta palliata on various substrates. Folia Primatol.
26:36–53.

Morbeck ME (1976) Leaping, bounding and bipedalism in Colobus guereza: a spectrum of positional behavior.
Yrbk. Phys. Anthropol. 20:408–420.

Napier JR (1962) Monkeys and their habitats. New Scientist 295:84–88.

Napier JR (1963) Brachiation and brachiators. Symp. Zool. Soc. Lond. 10:183–195.

Oxnard, C (1984) The Order of Man. New Haven: Yale University Press.

Parolin P (1993) Forest inventory in an island of Lake Guri, Venezuela. In W Barthlott, CM Naumann, K Schmidt-
Loske, and KL Schuchmann (eds.): Animal-Plant Interactions in Tropical Environments, Proc. German
Soc. Tropical Ecology, Bonn, pp. 139–147.

Pernia JE (1985) Mapa de Fisiografia y Vegetacion del Area de Inundacion de la Tercera Etapa del Embalse el
Guri - Estado Bolivar. Internal manuscript: CVG - Electrificacion del Caroni (EDELCA), Venezuela.

Peters A, and Preuschoft H (1984) External biomechanics of leaping in Tarsius and its morphological and kinematic consequences. In Niemitz C (ed.): Biology of Tarsiers. New York: Gustav Fisher Verlag, pp.
227–255.

Prost J (1965) A definitional system for the classification of primate locomotion. Am. Anthropol. 67:1198–1214.

Remis M (1995) Effects of body size and social context on the arboreal activities of lowland gorillas in the Central African Republic. Am. J. Phys. Anthropol. *97*:413–434.

Ripley S (1967) The leaping of langurs: a problem in the study of locomotor adaptation. Am. J. Phys. Anthropol. *26*:149–170.

Ripley S (1977) Gray zones and gray langurs: is the "semi"- concept seminal? Yrbk. Phys. Anthropol. *20*:376–394.

Rodman PS (1979) Skeletal differentiation of *Macaca fascicularis* and *Macaca nemestrina* in relation to arboreal and terrestrial quadrupedalism. Am. J. Phys. Anthropol. *51*:51–60.

Rose MD (1974) Postural adaptations in New and Old World monkeys. In FA Jenkins (ed.): Primate Locomotion. San Diego: Academic Press, pp. 75–93.

Rose MD (1979) Positional behavior of natural populations: some quantitative results of a field study of *Colobus guereza* and *Cercopithecus aethiops*. In ME Morbeck, D Preuschoft and N Gomberg (eds.): Environment, Behavior and Morphology: Dynamic Interactions in Primates. New York: Gustav Fischer Verlag, pp. 75–94.

Rosenberger AL, and Stafford BJ (1994) Locomotion in captive *Leontopithecus* and *Callimico*: a multimedia study. Am. J. Phys. Anthropol. *94*:379–394.

Schön Ybarra MA, and Schön III MA (1987) Positional behavior and limb bone adaptations in red howling monkeys (*Alouatta seniculus*). Folia Primatol. *49*:70–89.

Walker SE (1993) Positional Adaptations and Ecology of the Pitheciini. Ph.D. Dissertation, City University of New York.

Walker SE (1996) Evolution of positional behavior in the saki/uakaris (*Pithecia, Chiropotes, Cacajao*). In M Norconk, AL Rosenberger and P Garber (eds.): Advances in Primatology: Adaptive Radiations of Neotropical Primates. New York: Plenum, pp. 335–367.

Walker SE (in revision) Leaping behavior of *Pithecia pithecia* and *Chiropotes satanas*, and the importance of leaping in primate locomotion. Submitted to Am. J. Phys. Anthropol., 41 ms. pages.

Ward SC, and Sussman RW (1979) Correlates between locomotor anatomy and behavior in two sympatric species of *Lemur*. Am. J. Phys. Anthropol. *50*:575–590.

3

PATTERNS OF SUSPENSORY FEEDING IN *ALOUATTA PALLIATA, ATELES GEOFFROYI,* AND *CEBUS CAPUCINUS*

David J. Bergeson

Campus Box 1114
Department of Anthropology
Washington University
St. Louis, Missouri 63130

1. INTRODUCTION

The positional behavior of a primate is not an end in itself, but is instead a means to many ends (Andrews and Groves, 1976; Charles-Dominique, 1990; Cant, 1992). One of the most important of these is food acquisition. Compared to other orders of mammals, primates consume a wide variety of food resources, most of which can be subsumed under the categories of fruit, leaves, seeds, exudates, and insect prey (Harding, 1981; Sussman, 1987). These food resources often present alternative demands to a primate consumer and probably greatly influence patterns of substrate preference and canopy use. Although data suggest that broad correlations do not exist between diet and positional behavior, it is likely that specific positional behaviors are particularly important for specific resources in the diets of many primate species (Fleagle and Mittermeier, 1980; Fleagle, 1984). The identification of such relationships is critical if we are to interpret the positional behaviors of extant primates, or the reconstructed positional behaviors of extinct primates. Only when we know the ecological context of locomotor modes and feeding postures can we begin to address questions concerning their adaptive significance (Fleagle, 1979).

It is likely that suspension has a relatively high adaptive significance in many primate species. In several primates, such as the hylobatids and atelines, suspension is a relatively frequent positional behavior, particularly during feeding. The evolutionary and biomechanical significance of frequent positional behaviors is potentially high, as there is a relatively large opportunity for energy conservation if a frequent positional behavior is made more efficient, and there is a relatively critical need for shaping the skeleton and muscle to prevent fatigue or injury (Hunt, 1992). Frequency is obviously not the only criterion for estimating the influence of specific positional behaviors, as those locomotor

Primate Locomotion, edited by Strasser *et al.*
Plenum Press, New York, 1998

modes or postures that an animal most commonly performs do not necessarily represent the behaviors for which the animal has been selectively adapted. Instead, physical stress is also an important criterion: the anatomical adaptations required for the efficient performance of a particular positional mode are determined by the muscular effort required for it and the associated stress in the musculoskeletal system (Hunt, 1992). The degree to which suspension imparts a large amount of physical stress to the musculoskeletal system and requires many anatomical specializations for efficient performance is well established (e.g., Stern and Oxnard, 1973; Andrews and Groves, 1976; Cartmill and Milton, 1977; Fleagle, 1983; Swartz et al., 1989; Takahashi, 1990; Rose, 1993; Swartz, 1993; Lemelin, 1995).

An additional criterion to use when estimating the evolutionary significance of a positional behavior is its behavioral or ecological context. If the positional behavior is used in very restricted behavioral or ecological contexts, it is possible that the positional behavior has an adaptive relationship with these contexts (Clutton-Brock and Harvey, 1979; Fleagle, 1979; Hunt, 1992). This is particularly true if the contexts have relatively high survival values (Prost, 1965). Suspension has been suggested to be particularly common in specific behavioral and ecological contexts and, by virtue of these associations, has been suggested to have an adaptive relationship with these contexts. Specifically, it has been suggested that suspension is adaptively linked with frugivory (Napier and Napier, 1967; Andrews and Groves, 1976; Mittermeier and Fleagle, 1976; Altmann, 1989), feeding among small branches (Napier, 1967; Fleagle and Mittermeier, 1980; Cartmill, 1985; Cant, 1992), and feeding in the periphery of a tree crown (Avis, 1962; Napier, 1967; Carpenter and Durham, 1969; Grand, 1972; Mendel, 1976). In this paper, I examine the effect of diet, branch size, and crown location on the positional behavior during feeding and foraging of *Alouatta palliata*, *Ateles geoffroyi*, and *Cebus capucinus*. Specifically, I test whether suspensory feeding and foraging positions are associated with fruit resources, small branches, and the crown periphery.

2. METHODS

Data on *Alouatta palliata*, *Ateles geoffroyi*, and *Cebus capucinus* were collected from July, 1993 to July, 1994 at Santa Rosa National Park and the La Selva Biological Station in Costa Rica. Data were collected at two minute intervals using an instantaneous, focal animal sampling method. Among the data collected were activity (feeding and foraging), positional behavior, diet, support size and number of supports used, and location in tree crown (Bergeson, 1996).

Feeding was defined as holding or processing a food item, and foraging was defined as actively searching for food. With the exception of the howling monkeys at Santa Rosa, individual monkeys were not marked and thus were identifiable only to broad age and sex classes.

A system of 25 locomotor modes and postures was used to classify the positional behavior of focal animals (Bergeson, 1996). The goal of this paper is not to thoroughly document the positional behavior of *Alouatta*, *Ateles*, and *Cebus*, but rather to examine the effects of support characteristics and diet on major categories of locomotion and posture. Thus, the rates of only the four most common positional categories are examined in this paper: inverted bipedal, suspension, quadrupedal stand, and sit. In addition, an inclusive "other" category was defined for the remaining 21 categories. Locomotion, which was relatively uncommon during feeding, is included in the "other" category. The posture of the focal animal was classified as inverted bipedal when the animal supported itself facing down a branch by wrapping its tail around the branch and extending its legs. The forearms were not used to support the body and were usually free to manipulate food items, if necessary. This posture was typically used on, but was not restricted to, relatively large vertical or oblique branches, and is illustrated in Figure 2E of Mittermeier and Fleagle (1976). When a focal animal used the inverted bipedal posture on horizontal branches, the legs of the animal were extended out from the side of the branch, and the body of the animal was perpendicular to the main axis of the branch. The inverted bipedal posture was distinguished from suspension by virtue of hindlimb and trunk extension. Suspension was defined as a below branch position, with any combination of the hindlimbs, tail and forelimbs used in grasping. The limbs that were acting in tension to support the animal were recorded (e.g., 2 forelimbs and tail). In suspension, unlike the inverted bipedal posture, the hindlimbs were under tension or hung free (they were not under compression), and the trunk was not extended out from the support. The posture of the focal animal was classified as quadrupedal stand when the pronograde animal was supported above-branch by compression of three or four limbs. This posture typically occurred on a relatively horizontal branch or branches. Posture was classified as sit when the weight of the animal was supported only by the ischia. Typically the trunk was orthograde, the legs were flexed, and arms supported little or no weight.

The kind of resource items that were being eaten by the focal animal at the sample point were recorded. These resource items included fruit, leaves, flowers, insects, vine fruit, vine leaves, vine stems, vine flowers, pith, vertebrates, shoots or twigs, and other (including unknown).

The diameter of the support(s) used by the focal animal was estimated in centimeters. When a focal animal used more than one support, usually the supports were approximately the same size. On the few occasions when the supports were not the same size, the size of the support that supported the most weight was recorded. The diameter estimations were later corrected using equations derived from the regression of the actual diameters of a sample of 102 branches (as measured repeatedly by a tape) on the estimated diameters of these branches. The following branch size categories were used in data analyses: small (1–2 cm in diameter), medium (3–6 cm), large (7–15 cm) and very large (over 15 cm). The number of supports used by the focal animal was recorded according to the following categorization: Single – the focal animal was supported by one branch or support; Collateral – the focal animal was supported mainly by one support, with some minor help from one or two collateral supports; and Multiple – the focal animal was supported equally by at least two supports of identical characteristics.

Data were collected on the crown use of focal animals by dividing trees into four concentric crown areas. Each crown area corresponded to one fourth of the distance from the trunk to the periphery of the tree crown. Crown area one included the trunk and the innermost one-fourth of the branch lengths. Crown areas two and three were more peripheral than crown area one. Crown area four was the terminal branch region, and corresponded to the distal one-fourth of the tree crown. Crown location was estimated without reference to support size; that is, although branches in the crown periphery were generally small, it was possible for an animal to use large branches and still be in crown area four.

Results are presented in this paper as the proportion of all two minute samples in which each positional behavior pattern occurred. For the purposes of data analyses, all instantaneous samples from each species were pooled to form three data sets corresponding to *Alouatta*, *Ateles*, and *Cebus* positional behavior samples. Between-site variation in positional behavior was low compared to interspecific variation, and the data patterns discussed in this paper occurred at both Santa Rosa and La Selva. Over one hundred hours of data were collected on each species, including 797 *Alouatta* feeding and foraging samples, 764 *Ateles* feeding and foraging samples, and 1192 *Cebus* feeding and foraging samples.

The statistical significance of data patterns was evaluated using randomization procedures. Unlike conventional parametric methods, randomization tests do not make any assumptions about the distribution of the population from which the data were collected. Instead, they create distributions of the test statistic from permutations of the observed data (e.g., Manley, 1991; Potvin and Roff, 1993; Dagosto, 1994). Significance levels are determined by the percentage of test statistic values equal to or more extreme than the value from the original data set. I used Manley's (1991) two sample randomization test where the test statistic used in the randomization loops, i.e., the mean differences between groups, is equivalent to the Students t statistic. When examining interspecific differences in categorical data (such as locomotion or posture), the use rate of each category was the test statistic in the randomization loops. In this project, 1000 permutations were used to estimate the probability value (P). I use these tests to determine the probability that the observed differences in rates occur if no relationship exists between the variables being tested. If the observed difference occurs less than one time out of 100 (i.e., $P<0.01$), then I consider the difference to be statistically significant.

In this paper, I chose not to evaluate the context specificity of positional behaviors using the percent samples of a positional behavior that occur in a particular area. Instead, I use the percentage of feeding and foraging context samples that is achieved using a particular positional behavior as a measure of context specificity. For example, when addressing the question of whether suspension is associated with the crown periphery, I do not calculate the percent of suspension samples that occur in the crown periphery, and compare this with the percent of suspension samples that occur in the crown interior. This comparison is affected by the distribution of feeding in the tree crown: most suspension occurs in the tree periphery because most feeding occurs in the tree periphery. Instead, I examine the association of suspension and crown use by calculating the rate of suspension in different areas of the crown. If the rate of suspension in the crown periphery is significantly higher than the rate of suspension in other parts of the tree crown, then I would conclude that this difference in rates is not random. I would further conclude that suspension is likely a positional strategy that is particularly suited to solve the problems associated with feeding and foraging in the crown periphery.

3. RESULTS

3.1. Diet

Previous studies have indicated that *Alouatta*, *Ateles*, and *Cebus* differ significantly in their patterns of resource use (e.g., Chapman, 1987). Similar patterns were exhibited by the primates in this study (Table 1). *Alouatta* relied primarily on leaves, although it also frequently consumed fruit and flowers. The primary resource in the diet of *Ateles* was fruit, which was supplemented by flowers, vine resources, and leaves. Like *Ateles*, *Cebus* relied on fruit resources throughout the year. Unlike *Ateles*, however, *Cebus* fed heavily on invertebrate prey, often descending to the ground in search of insects in the leaf litter. *Cebus* also was observed feeding on vertebrate prey such as mice, birds, and young coati.

It is clear that feeding and foraging were postural activities in all three species (Table 2). In *Alouatta*, feeding and foraging were achieved primarily above-branch, in a sit-

Table 1. Diet of *Alouatta palliata*, *Ateles geoffroyi*, and *Cebus capucinus* at Santa Rosa National Park and La Selva Biological Station. Data are percentages of each species diet

Diet	*Alouatta*	*Ateles*	*Cebus*
Fruit	25.7	59.9	43.4
Leaves	46.9	7.7	0.0
Flowers	22.2	12.4	0.0
Vine fruit	1.9	10.4	0.9
Vine leaves and stems	2.9	8.5	0.0
Invertebrate fauna	0.0	0.0	49.2
Other	0.4	1.1	6.5

ting or quadrupedal standing position. Below branch suspension occurred in almost 13% of feeding and foraging samples. In *Ateles*, suspension and sitting were the most important postures used when feeding and foraging, although the inverted bipedal posture and quadrupedal stand were also frequently used. Like *Alouatta*, *Cebus* usually fed or foraged by sitting or standing quadrupedally; however, *Cebus* differed from *Alouatta* in that it used suspensory postures relatively rarely.

The form of suspensory feeding and foraging postures differ greatly between howling monkeys and spider monkeys: the former almost always use tail only suspension (40.9% of suspensory feeding samples) or tail plus two legs suspension (34.4%), while spider monkeys are capable of suspending from any combination of arms, legs, and tail, particularly tail only suspension (40.7% of suspensory feeding samples) and one arm plus tail (35.9%).

None of the species used suspension more often when feeding and foraging on fruit than when feeding and foraging on other resources (Table 3). Nonetheless, in all three species different positional behaviors were used to acquire different dietary items. In *Alouatta*, the rate of suspension was highest when feeding on leaves and was lowest when feeding on fruit (Table 4). In addition, the rate of quadrupedal standing was significantly higher when it fed on fruit than on other items. *Ateles* also modified its positional behavior with respect to resource items, but again, the rate of suspension was not higher when feeding on fruit (Table 4). Instead, the most significant trends were associated with feeding on flowers: *Ateles* had a lower rate of suspension and a higher rate of sitting. In *Cebus*, suspension was relatively rare when feeding and the rate was not affected by resource type (Table 4). The rate of the inverted bipedal posture was significantly higher when feeding

Table 2. The positional behaviors used while feeding and foraging. Data are percent total feeding samples in each species

Positional behavior	*Alouatta* N=735	*Ateles* N=700	*Cebus* N=1127
Inverted bipedal posture	8.6	13.1	7.0
Quadrupedal stand	13.5	12.6	25.1
Sit	56.8	25.7	39.2
Suspension	12.7	33.3	2.7
Other	8.4	15.3	26.0

Table 3. Rates of suspensory feeding and foraging on
fruit versus all other resources

Species	Fruit	Other resources	P
Alouatta	8.5	14.3	0.043
Ateles	33.0	32.5	0.936
Cebus	2.8	2.7	1.000

on fruit than on other resources and the rate of sitting was relatively low when feeding on invertebrate fauna.

3.2. Branch Size

In all three species, smaller supports were the focus for feeding and foraging (Table 5). In *Alouatta*, over sixty percent of feeding and foraging samples occurred on small supports, and almost thirty percent occurred on medium supports. Feeding and foraging rarely occurred on large or very large supports. In *Ateles*, over half of feeding and foraging samples occurred on small supports, and many occurred on medium supports. *Cebus* displayed a pattern of support use similar to that of *Alouatta*.

Table 4. The effect of diet on the positional behavior. Data are percent total positional behavior
samples while feeding or foraging on each resource

Species	Positional behavior	Fruit	Leaves	Flowers	Vine fruit	Vines leaves & stems	Invert. fauna	Other
Alouatta[1]		N=200	N=385	N=159				
	Inverted bipedal	10.5	7.4	7.5				
	Quadrupedal stand	23.5	9.4	9.4				
	Sit	49.5	58.4	63.5				
	Suspension	8.5	16.5	9.4				
	Other	8.0	8.3	10.2				
Ateles[2]		N=397	N=53	N=85	N=69	N=63		
	Inverted bipedal	10.6	13.2	16.9	17.4	22.2		
	Quadrupedal stand	14.1	3.8	13.5	15.9	9.5		
	Sit	24.2	32.1	40.4	14.5	19.0		
	Suspension	33.0	41.5	19.1	33.3	44.4		
	Other	18.2	9.4	10.1	18.9	4.9		
Cebus[3]		N=457					N=519	N=60
	Inverted bipedal	10.5					5.4	3.3
	Quadrupedal stand	21.9					28.7	19.7
	Sit	42.0					33.1	65.6
	Suspension	2.8					2.5	1.6
	Other	22.8					30.3	9.8

[1]Significant Differences: Inverted Bipedal - none; Quadrupedal Stand - Fruit vs. Leaves (P<0.001), Fruit vs. Flowers (P<0.001); Sit - none; Suspension - Fruit vs. Leaves (P=0.003).

[2]Significant Differences: Inverted Bipedal - none; Quadrupedal Stand - none; Sit - Flowers vs. Vine Fruit (P<0.001), Flowers vs. Vine Leaves and Stems (P<0.001), Flowers vs. Fruit (P=0.002); Suspension - Fruit vs. Flowers (P=0.005), Leaves vs. Flowers (P=0.005), Flowers vs. Vine Leaves and Stems (P=0.002).

[3]Significant Differences: Inverted Bipedal - Fruit vs. Other (P=0.001), Fruit vs. Invert. Fauna (P=0.007); Quadrupedal Stand - none; Sit - Invert. Fauna vs. other (P<0.001); Suspension - none.

Table 5. Support sizes used during feeding and foraging. Data are percentages of total feeding samples for each species

Branch size	*Alouatta* N=703	*Ateles* N=602	*Cebus* N=999
Small	63.3	58.2	64.2
Medium	27.3	33.8	25.9
Large	8.4	6.9	7.8
Very Large	1.0	1.1	2.1

Only in *Alouatta* was suspension significantly more common on small branches than on larger branches (Table 6). Branch size affected the positional behavior of *Alouatta* differently than it affected the positional behavior of *Ateles* or *Cebus*. Unlike Mendel (1976), I did not find that howling monkeys used suspension almost exclusively when feeding on small supports. Instead, sitting remained the dominant howling monkey feeding posture, even on very small branches (Table 7). Nonetheless, the rate of suspension was higher on small branches than on larger branch sizes. The distributions of branch sizes used by *Alouatta* during above-branch feeding (quadrupedal stand and sit) and suspension suggest that this is because not only was suspension preferred on small branches, but also suspension was avoided on large branches (Figure 1, top). *Alouatta* rarely used suspension when on branches over ten centimeters in diameter. However, the distributions of branch sizes used during above and below branch feeding differed greatly in the small branch range, and confirm the association of suspension with small branches. In *Alouatta*, the mean branch size used during suspension was significantly smaller than the mean branch size used during above-branch feeding. The mean branch sizes are relatively unaffected by extreme values, as the mean branch size used during suspension was significantly smaller than the mean branch used during above-branch feeding, even after excluding two extreme values over 15 cm. In *Alouatta*, sitting was used at different rates on different branch sizes. *Alouatta* sat more often on medium and large branches than on small and very large branches, where suspension and the inverted bipedal posture were used, respectively (Table 7).

In *Ateles*, the rate of suspension was much lower on large and very large branches than it was on small or medium branches (Table 7). This pattern could be caused by either a preference for suspension on small branches or an avoidance of suspension on larger branches. The data in Figure 1 (bottom) suggest that the latter explanation is more likely correct than the former. The fact that the distributions are very similar in the small branch range suggests that in *Ateles*, suspension does not necessarily allow greater use of small

Table 6. Rates of suspensory feeding on small branches versus other branches. Percent feeding and foraging samples in branch sizes in which suspension was used

Species	Small branches	All other branches	P
Alouatta	17.9	5.2	<0.001
Ateles	31.4	31.2	1.000
Cebus	3.0	1.7	0.303

Table 7. The effect of branch size on the positional behavior while feeding and foraging. Data are percent total positional behavior samples on each branch size category

Species	Positional behavior	Small	Medium	Large	Very large
Alouatta[1]	Inverted bipedal	7.2	8.1	10.3	71.4
	Quadrupedal stand	15.2	9.2	8.6	28.6
	Sit	50.8	68.7	70.7	0.0
	Suspension	18.0	4.9	6.9	0.0
	Other	8.8	9.1	3.5	0.0
Ateles[2]	Inverted bipedal	11.2	13.4	22.0	42.9
	Quadrupedal stand	15.3	8.9	4.9	0.0
	Sit	24.2	31.2	39.0	57.1
	Suspension	31.4	35.2	17.1	0.0
	Other	17.9	11.3	17.0	0.0
Cebus[3]	Inverted bipedal	5.9	9.3	15.8	14.3
	Quadrupedal stand	27.4	17.5	19.7	47.6
	Sit	37.6	48.6	43.4	28.6
	Suspension	3.1	2.0	1.3	0.0
	Other	26.0	22.7	19.8	9.5

[1]Significant Differences: Inverted Bipedal - Small vs. Very Large (P<0.001), Medium vs. Very Large (P<0.001), Large vs. Very Large (P<0.001); Quadrupedal Stand - none; Sit - Small vs. Medium (P<0.001), Small vs. Large (P=0.006), Small vs. Very Large (P=0.001), Medium vs. Very Large (P<0.001), Large vs. Very Large (P<0.001); Suspension - Small vs. Medium (P<0.001).
[2]Significant Differences: Inverted Bipedal - Small vs. Very Large (P=0.001), Medium vs. Very Large (P=0.002); Quadrupedal Stand - none; Sit - Small vs. Very Large (P=0.004); Suspension - none.
[3]Significant Differences: Inverted Bipedal - Small vs. Large (P=0.006); Quadrupedal Stand - Small vs. Medium (P=0.003), Medium vs. Very Large (P=0.003); Sit - Small vs. Medium (P=0.004); Suspension - none.

branches, and instead, more precisely, above-branch postures allow greater use of very large branches. In *Ateles*, the mean branch size used during suspension was smaller than the mean branch size used during above-branch feeding (quadrupedal stand and sit). This measure of context specificity is potentially misleading, however, as the mean branch size used during above-branch feeding is greatly affected by four samples of branch sizes over 15 cm. If these four samples are excluded, the difference in means is no longer significant. One likely reason why suspension was not associated with small branches in *Ateles* is their use of multiple branches. When data only from single branch samples are analyzed, the rate of suspension was much higher on small branches (Table 8). *Ateles* also displayed other trends in positional behavior with respect to branch size. Although the data in Table 7 are suggestive that the rate of the inverted bipedal posture increased with branch size category, only the contrasts between small and very large branches and between medium and very large branches were significant. Sitting was more common on very large than on small branches.

Suspensory feeding and foraging was relatively rare in *Cebus*, and was not affected by branch size (Table 7). Quadrupedal standing was common on both small and very large branches, and the rate of sitting was not as high on small branches as on medium branches. As in *Alouatta* and *Ateles*, the inverted bipedal posture was strongly associated with larger branches.

3.3. Crown Location

In all three species, the crown periphery was the most common site of feeding and foraging (Table 9). The atelines were particularly restricted to the crown periphery when

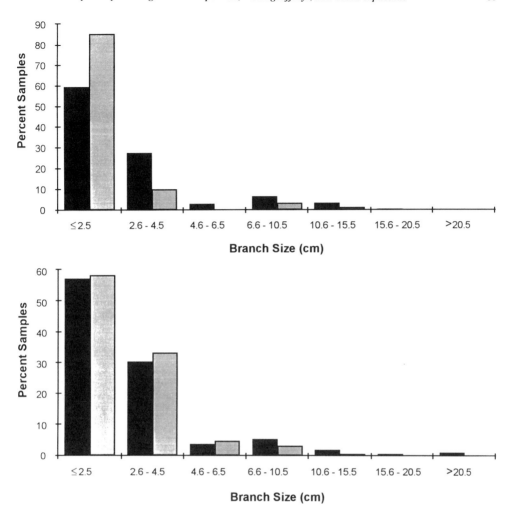

Figure 1. Branch sizes of above-branch postures and suspension. Black bars, above-branch postures. Hatched bars, suspension. (*Top*) Howling monkeys: mean above-branch branch size = 3.1 cm, mean suspension branch size = 2.2 cm, P<0.001. (*Bottom*) Spider monkeys: mean above-branch branch size = 3.4 cm, mean suspension branch size = 2.8 cm, P=0.028.

Table 8. The relationship between suspension and branch size on single branches. Data are percent suspension samples in each branch size category

Branch size	*Alouatta*	*Ateles*	*Cebus*
1-2 cm	28.6	75.4	5.5
3-6 cm	6.4	42.0	1.3
7-15 cm	8.7	22.2	1.6
> 15 cm	0.0	0.0	0.0

Table 9. Crown use during feeding and foraging. Data
are percent total feeding samples in each crown area

Crown Area	*Alouatta* N=769	*Ateles* N=716	*Cebus* N=1019
Area 1	2.5	5.0	17.6
Area 2	3.1	5.7	13.2
Area 3	13.7	14.9	18.4
Area 4	80.7	74.4	50.8

feeding and foraging, as *Alouatta* and *Ateles* fed and foraged in the periphery in over 80%
and 70% of samples, respectively. *Cebus* fed and foraged over a much greater portion of
the tree crown than did *Ateles* or *Alouatta*, although the periphery was nonetheless the
most common site of feeding and foraging.

Although for each species crown area did affect the positional behaviors used to acquire resources, in none of the species did the rate of suspensory feeding positions increase in the crown periphery (Table 10). A more detailed analysis of the effect of crown
location on the positional behavior of feeding and foraging suggests different patterns. In
Alouatta, suspensory feeding did not increase in the periphery, and the inverted bipedal
posture was closely associated with the trunk region (Table 11). *Ateles* displayed similar
patterns, as the rate of quadrupedal standing but not suspension was elevated in the periphery (Table 11). Sitting was most common in crown area 2, and the inverted bipedal
posture was closely associated with crown area 1. The positional behavior of *Cebus* during
feeding and foraging was not as affected by crown position as was the positional behavior
of the atelines (Table 11). Nonetheless, one significant trend that was found in the atelines
was also found in *Cebus*: during feeding and foraging the rate of the inverted bipedal posture was significantly higher in the trunk region.

4. DISCUSSION

4.1. Positional Behavior and Diet

The hypothesis that suspension is adaptively linked with frugivory has traditionally
been supported by data from the hylobatids: data collected on the positional behavior of
gibbons and siamangs suggest that suspensory feeding positions are more common when
they feed on fruit than when they feed on leaves (Chivers, 1974; Fleagle, 1976). In other
apes that frequently use suspensory postures, however, suspension is not particularly asso-

Table 10. Rates of suspensory feeding in crown area 4
versus all other crown areas. Data are percent total
feeding and foraging samples in each area
in which suspension used

Species	Crown area 4	All other areas	P
Alouatta	13.8	9.2	0.158
Ateles	30.7	37.5	0.107
Cebus	3.5	2.2	0.334

Table 11. The effect of crown location on the positional behavior while feeding and foraging. Data are the percent total positional behavior samples in each crown area

Species	Positional behavior	Crown area 1	Crown area 2	Crown area 3	Crown area 4
Alouatta[1]	Inverted bipedal	63.2	13.0	6.9	6.9
	Quadrupedal stand	10.5	0.0	10.9	14.6
	Sit	15.8	47.8	69.3	55.9
	Suspension	5.3	17.4	7.9	13.8
	Other	5.2	21.8	5.0	8.8
Ateles[2]	Inverted bipedal	41.7	23.7	13.7	10.1
	Quadrupedal stand	11.1	2.6	3.9	15.3
	Sit	8.3	42.1	27.5	26.0
	Suspension	36.1	28.9	41.2	30.7
	Other	2.8	2.7	13.7	17.9
Cebus[3]	Inverted bipedal	19.4	3.8	4.3	5.8
	Quadrupedal stand	20.0	25.2	20.5	27.4
	Sit	41.1	42.0	49.7	35.5
	Suspension	2.3	3.1	1.6	3.5
	Other	17.2	25.9	23.9	27.8

[1] Significant Differences: Inverted Bipedal - 1 vs. 2 (P<0.001), 1 vs. 3 (P<0.001), 1 vs. 4 (P<0.001); Quadrupedal Stand - none; Sit - 1 vs. 3 (P<0.001), 1 vs. 4 (P<0.001), 3 vs. 4 (P=0.007); Suspension - none.
[2] Significant differences: Inverted Bipedal -1 vs. 3 (P<0.001), 1 vs. 4 (P<0.001); Quadrupedal Stand - 3 vs. 4 (P=0.001); Sit - 1 vs. 2 (P<0.001); Suspension- none.
[3] Significant Differences: Inverted Bipedal - 1 vs. 2 (P<0.001), 1 vs. 3 (P<0.001), 1 vs. 4 (P<0.001); Quadrupedal Stand - none; Sit - 3 vs. 4 (P=0.002); Suspension - none.

ciated with frugivory. Cant (1987), for example, found that female *Pongo pygmaeus* used suspensory feeding postures in 59% of leaf feeding samples and 43% of fruit feeding samples. Hunt (1992) found that the rates of suspensory feeding in *Pan troglodytes* was about ten percent for both fruit and leaf feeding samples. Less is documented about the relationship between suspensory feeding and resource use in New World monkeys, although Bicca-Marques and Calegaro-Marques (1993) found a small difference in the rate of suspension by *Alouatta caraya* when it fed on fruit and leaves (18% vs. 14%).

In this study, the positional behavior of food acquisition was affected by resource use in each species. The effect of resource use was not as strong as was the effect of crown location or support size. This reinforces suggestions that the positional behavior of arboreal primates is not directly determined by resources, but instead it is affected indirectly by the location of these resources in the tree crown and the characteristics of the specific supports that contain these resources (e.g., Fleagle, 1984; Walker, 1993, 1996). Although there were trends in each species, in none of the species was suspension associated with fruit. Both *Alouatta* and *Ateles* preferred to sit while feeding on flowers, which were obtained relatively rarely in suspensory positions. Bicca-Marques and Calegaro-Marques (1993) also found that *Alouatta caraya* preferentially used sitting postures when feeding on flowers. One possible explanation for this relationship between sitting and feeding on flowers is the increased manipulation required when harvesting flowers. Flowers occur in the same location as the fruits that later replace them. But when feeding on flowers, the primates in this study inspected and manipulated the flowers much more than they inspected and manipulated fruit when feeding on it. The tightest association between positional behavior and resource in *Cebus* was between the inverted bipedal posture and fruit.

Mittermeier and Fleagle (1976) emphasized that suspended postures are important feeding adaptations for spider monkeys. By suspending itself by its tail alone, or by its tail and any combination of its hindlimbs, a spider monkey can increase its feeding sphere 150% of that available when sitting or standing quadrupedally. Suspension is much more frequent in the atelines than in other platyrrhines, and it is clear that the feeding sphere of atelines is increased by the possession of a prehensile tail. This increase, however, is associated with all parts of the tree crown, and is not restricted to the periphery of the tree. Similarly, this increase is not solely associated with frugivory, but is associated with all resources. Prehensile tail use, like suspensory feeding, is common in all parts of the tree crown, and occurs frequently regardless of resource type (Bergeson, 1996). The prehensile tail is not only used for suspensory feeding in the crown periphery, but instead serves several biological roles in each species. These roles include increasing the feeding sphere via suspensory postures, increasing the ability to balance on small branches, freeing the hands during feeding activities, increasing the ability to feed as well as forage in the crown periphery, increasing the ability to feed and move on vertical supports, and increasing the ability to cross gaps between trees (Bergeson, 1996).

4.2. Positional Behavior and Branch Size

The mechanics of arboreal movement dictate that small branches present unique problems of access, and it is reasonable to expect arboreal primates to modify their positional behavior in order to compensate for these problems. The stability of an animal located above a branch is largely determined by the ratio of the size of the animal to the size of the branch (Napier, 1967; Cartmill, 1985). On small branches, small displacements by an arboreal primate can lead to frequent pitching and toppling, because small movements cause its center of mass to move lateral to the branch surface. Because these problems of stability on small branches can be avoided if an arboreal animal hangs beneath a branch, many researchers have predicted that suspensory positions should be more common on small branches (e.g., Napier, 1967; Fleagle and Mittermeier, 1980; Cant, 1992).

Suspension is not the only way for an arboreal primate to solve the problems of balance that are associated with moving and feeding on small branches. Another way is to use more than one branch to support above-branch movement (e.g., Cartmill and Milton, 1977; Fleagle and Mittermeier, 1980; Cartmill, 1985; Fleagle, 1985). By distributing its body weight over many branches, an arboreal primate has greater control over pitching and rolling movements. In addition, the weight of a large arboreal primate is less likely to break several small branches than one small branch.

Given that suspension and multiple branch use are alternate solutions to the same problems, which solution is adopted more frequently by *Alouatta*, *Ateles*, and *Cebus*? Data in this study suggest that multiple branches are used more often than suspension to solve problems associated with small branch use. In all three species, feeding on small branches was most often achieved using quadrupedal, above-branch positions on multiple branches (Table 12). Suspension on a single branch was relatively rare. In fact, *Ateles* used single branches in less than 25% of feeding samples on small branches. The single branch model used to predict the influence of support size on positional behavior (e.g., Napier, 1967) is heuristically useful, but often disregards more common solutions.

Although spider monkeys use suspension more often than do howling monkeys, howling monkeys are more restricted to small branches when using suspension than are spider monkeys (Figure 1). One possible explanation for this restriction is the limitation imposed by hindlimb suspension on howling monkeys, and the relative versatility of forelimb suspension in spider monkeys. While spider monkeys frequently hang from their

Table 12. Positional solutions on small branches during feeding. Data are percent total positional behavior samples on small branches

Positional solutions	*Alouatta*	*Ateles*	*Cebus*
Above-branch (Single Br.)	18.6	5.7	30.3
Above-branch (Multiple/Collateral Br.)	61.5	51.9	66.5
Suspension (Single Br.)	7.6	18.6	1.7
Suspension (Multiple/Collateral Br.)	12.3	23.8	1.5

arms, howling monkeys almost always hang from their legs and/or tail. It is possible that the long arms and mobile shoulder of *Ateles* not only allow spider monkeys to use suspensory positions more often than howling monkeys, but also enable spider monkeys to use a wider variety of supports during suspension. On medium branches, the rate of suspension in spider monkeys is over six times higher than the rate of suspension in howling monkeys (Table 7). Perhaps this is because the maximum diameter of a branch that a howling monkey can efficiently grasp with its foot is smaller than the maximum diameter of a branch that a spider monkey can efficiently hang from with its forelimbs.

4.3. Positional Behavior and Crown Location

It is clear that the characteristics of tree branches vary with respect to the branch's position in the tree crown. Although trees exhibit a wide variety of crown shapes and branching patterns, the structures of most trees are variations of the same pattern: large, vertical, inflexible boles give rise to outwardly-radiating smaller, obliquely-inclined flexible branches. Branches at the crown periphery are relatively very small, flexible, and are oriented at many angles (e.g., Horn, 1971; Garber, 1980; Bertram, 1989). As the characteristics of tree branches differ with respect to crown position, it is reasonable to expect that the positional behavior of arboreal animals is affected by crown position (Ripley, 1967). In an influential paper, Grand (1972) focused on the mechanical problems associated with moving and feeding in terminal branches. Grand pointed out that by suspending under a branch, a gibbon increases its feeding sphere 100% over that achieved in an above-branch position. In addition, the deformation of terminal branches brings fruit closer to a below-branch gibbon, and moves fruit farther away from an above-branch macaque. Other authors have suggested that suspensory positions are particular useful and common in the crown periphery (e.g., Avis, 1962; Napier, 1967; Carpenter and Durham, 1969; Grand, 1972; Mendel, 1976). Fleagle (1980) has pointed out that sitting while feeding is a common approach to terminal branch feeding, and should probably be viewed as a functionally different, but not necessarily a more efficient, method of foraging. Nonetheless, suspensory feeding has been associated with the crown periphery to the extent that it has been assumed or implied that large primates cannot obtain food in the crown periphery unless they hang while feeding (e.g., Hollihn, 1984).

Although location in the tree crown significantly affected the positional behavior of each species in this study, the rate of suspensory feeding was not higher in the tree periphery than in other areas of the tree. Instead, the tightest association was between the inverted bipedal posture and crown area one (the trunk region). The inverted bipedal posture, characterized by an upside down body posture, hindlimb extension, and high tension prehensile tail use, was frequently used by each species when feeding in the trunk region. The inverted bipedal posture is possible only in prehensile-tailed animals, and is likely an important component of the manipulative, destructive, visually-oriented foraging strategy of capuchin monkeys. It is significant that while in an inverted bipedal posture the hands remain free and are thus able to manipulate food items. This freedom of the hands may play a significant role in the foraging of *Cebus*, allowing capuchins to more easily access palm fruit near the trunk, and allowing them to more efficiently rip apart and search branches for invertebrate prey.

5. SUMMARY

Suspension has been suggested to be particularly common in certain behavioral and ecological contexts, and by virtue of these associations, has been suggested to have an adaptive relationship with these contexts. Specifically, it has been suggested that suspension is adaptively linked with frugivory, with feeding among small branches, and with feeding in the periphery of a tree crown. In general, the data presented in this paper do not support these suggestions. In all three species, suspension was common regardless of resource, and was not used more often when feeding and foraging on fruit. Branch size did affect suspension in *Alouatta*, in that suspension was more common on small branches than on large branches, but this pattern was not exhibited in *Ateles* or *Cebus*. Finally, while in each species crown position did affect the positional behaviors used to acquire resources, in none of the species did the rate of suspensory feeding positions increase in the crown periphery.

Although data on the frequencies of particular positional behaviors are important, these alone do not tell us what specific problems are solved by each positional behavior (e.g., Cant, 1992), or the ecological consequences of specific positional behaviors (e.g., Charles-Dominique, 1990; Garber, 1992; Hunt, 1992). Instead, data on the context of positional behaviors are critical if we are even to begin interpreting the positional behavior of extant primates (e.g., Garber, 1992; Hunt, 1992), or reconstructing the positional behavior of extinct primates (e.g., Fleagle, 1976, 1983; Cartmill and Milton, 1977; Rose, 1993). This paper is one example of how quantitative data on the specific contexts of positional behaviors can be used to identify new ecological relationships (e.g., the inverted bipedal posture and trunks), confirm or reject proposed ones (e.g., suspension is associated with small branches in howling monkeys, but not spider or capuchin monkeys), or identify functional relationships between anatomy, behavior, and environment (e.g., forelimb suspension in spider monkeys allows them to use a greater variety of branches than does hindlimb suspension in howling monkeys).

ACKNOWLEDGMENTS

Funding for this project was provided by: the L.S.B. Leakey Foundation, the Organization for Tropical Studies, Sigma Xi, and National Science Foundation Grant SBR-9307631. I am grateful to the National Park Service of Costa Rica for allowing me to work in Santa Rosa National Park from 1993 to 1994, and am grateful to the La Selva Biological Station, Roger Blanco, Rodrigo Morera Avila, Orlando Vargas, and Sue Bergeson for their help with this project. I thank B.W. Bergeson, J. Fleagle, E. Strasser, and four anonymous reviewers for their constructive comments on an earlier version of this paper.

REFERENCES

Altmann S (1989) The monkey and the fig: A socratic dialogue on evolutionary trends. Am. Sci. *77*:256–263.

Andrews P, and Groves C (1976) Gibbons and brachiation. Gibbon and Siamang *4*:167–218.

Avis V (1962) Brachiation: The crucial issue for man's ancestry. Southw. J. Anthropol. *18*:119–148.

Bergeson DJ (1996) The positional behavior and prehensile tail use of *Alouatta palliata, Ateles geoffroyi,* and *Cebus capucinus.* Ph.D. Dissertation, Washington University.

Bertram JEA (1989) Size-dependent differential scaling in branches: The mechanical design of trees revisited. Trees *4*:241–253.

Bicca-Marques J, and Calegaro-Marques C (1993) Feeding postures in the black howler monkey, *Alouatta caraya.* Folia Primatol. *60*:169–172.

Cant JGH (1987) Positional behavior of female Bornean orangutans. Am. J. Primatol. *12*:71–90.

Cant JGH (1992) Positional behavior and body size of arboreal primates: A theoretical framework for field studies and an illustration of its application. Am. J. Phys. Anthropol. *88*:273–283.

Carpenter C, and Durham N (1969) A preliminary description of suspensory behavior in nonhuman primates. Proceedings of the Second International Congress Primatology, Atlanta, GA Volume 2. New York: Karger. pp. 147–154.

Cartmill M (1985) Climbing. In M Hildebrand, D Bramble, K Liem, and D Wake (eds.): Functional Vertebrate Morphology. Cambridge: Belknap Press. pp. 73–88.

Cartmill M, and Milton K (1977) The lorisiform wrist joint and the evolution of "brachiating" adaptations in the Hominoidea. Am. J. Phys. Anthropol. *47*:249–272.

Chapman CA (1987) Flexibility in diets of three species of Costa Rican primates. Folia Primatol. *49*:90–105.

Charles-Dominique P (1990) Ecological adaptations related to locomotion in primates: An introduction. In F Jouffroy, M Stack and C Niemitz (eds.): Gravity, Posture, and Locomotion in Primates. Firenze: Editrice II Sedicesimo, pp. 19–31.

Chivers DJ (1974) The siamang in Malaya: A field study of a primate in tropical rain forest. Contrib. Primatol. *4*:1–331.

Clutton-Brock TH, and Harvey PH (1979) Comparison and adaptation. Proc. R. Soc. Lond. B *205*:547–565.

Dagosto M (1994) Testing positional behavior of Malagasy lemurs: A randomization approach. Am. J. Phys. Anthropol. *94*:189–202.

Fleagle JG (1976) Locomotion and posture of the Malayan siamang and implications for hominoid evolution. Folia Primatol. *26*:245–269.

Fleagle JG (1979) Primate positional behavior and anatomy: Naturalistic and experimental approaches. In ME Morbeck, H Preuschoft, and N Gomberg (eds.): Environment, Behavior, and Morphology: Dynamic Interactions in Primates. New York: Gustav Fischer, pp. 313–325.

Fleagle JG (1980) Locomotion and posture. In DJ Chivers (ed.): Malayan Forest Primates. New York: Plenum Press, pp. 191–207.

Fleagle JG (1983) Locomotor adaptations of Oligocene and Miocene hominoids and their phyletic implications. In R Ciochon and R Corruccini (eds.): New Interpretations of Ape and Human Ancestry. New York: Plenum Press, pp. 301–324.

Fleagle JG (1984) Primate locomotion and diet. In D Chivers, B Wood, and A Bilsborough (eds.): Food Acquisition and Processing in Primates. New York: Plenum Press, pp. 105–117.

Fleagle JG (1985) Size and adaptation in primates. In W Jungers (ed.): Size and Scaling in Primate Biology. New York: Plenum Press, pp. 1–19.

Fleagle JG, and Mittermeier RA (1980) Locomotor behavior, body size, and comparative ecology of seven Surinam monkeys. Am. J. Phys. Anthropol. *52*:301–314.

Garber PA (1980) Locomotor behavior and feeding ecology of the Panamanian tamarin. Ph.D. Dissertation, Washington University.

Garber PA (1992) Vertical clinging, small body size, and the evolution of feeding adaptations in the Callitrichinae. Am. J. Phys. Anthropol. *88*:469–482.

Grand TI (1972) A mechanical interpretation of terminal branch feeding. J. Mammal. *53*:198–201.

Harding R (1981) An order of omnivores: Nonhuman primate diets in the wild. In R Harding and G Teleki (eds.): Omnivorous Primates: Gathering and Hunting in Human Evolution. New York: Colombia University Press, pp. 191–209.

Hollihn U (1984) Bimanual suspensory behavior: Morphology, selective advantages and phylogeny. In H Preuschoft, D Chivers, W Brockelman, and N Creel (eds.): The Lesser Apes. Edinburgh: Edinburgh University Press, pp. 85–95.

Horn HS (1971) The Adaptive Geometry of Trees. Princeton: Princeton University Press.

Hunt KD (1992) Positional behavior of *Pan troglodytes* in the Mahale mountains and Gombe Stream National Parks, Tanzania. Am. J. Phys. Anthropol. *87*:83–105.

Lemelin P (1995) Comparative and functional myology of the prehensile tail in New World monkeys. J. Morph. *224*:1–18.

Manly BFJ (1991) Randomization and Monte Carlo Methods in Biology. New York: Chapman and Hall.

Mendel FC (1976) Postural and locomotor behavior of *Alouatta palliata* on various surfaces. Folia Primatol. *26*:36–53.

Mittermeier RA, and Fleagle JG (1976) The locomotor and postural repertoires of *Ateles geoffroyi* and *Colobus guereza*, and a reevaluation of the locomotor category semibrachiation. Am. J. Phys. Anthropol. *45*:235–256.

Napier JR (1967) Evolutionary aspects of primate locomotion. Am. J. Phys. Anthropol. *27*:333–342.

Napier JR, and Napier P (1967) A Handbook of Living Primates. New York: Academic Press.

Potvin C and Roff D (1993) Distribution-free and robust statistical methods: Viable alternatives to parametric statistics? Ecology *74*:1617–1628.

Prost JH (1965) A definitional system for the classification of primate locomotion. Am. Anthropol. *67*:1198–1214.

Ripley S (1967) The leaping of langurs: A problem in the study of locomotor adaptation. Am. J. Phys. Anthropol. *26*:149–170.

Rose MD (1993) Locomotor anatomy of Miocene hominoids. In D Gebo (ed.): Postcranial Adaptation in Nonhuman Primates. DeKalb: Northern Illinois University Press, pp. 252–272.

Stern JT, and Oxnard C (1973) Primate Locomotion: Some Links with Evolution and Morphology. Basel: Karger.

Sussman RW (1987) Species-specific dietary patterns in primates and human dietary adaptations. In W Kinzey (ed.): The Evolution of Human Behavior: Primate Models. Albany: State University of New York Press, pp. 151–179.

Swartz SM (1993) Biomechanics of primate limbs. In D Gebo (ed.): Postcranial Adaptation in Nonhuman Primates. DeKalb: Northern Illinois University Press, pp. 5–42.

Swartz SM, Bertram JEA, and Biewener AA (1989) Telemetered *in vivo* strain analysis of locomotor mechanics of brachiating gibbons. Nature *342*:270–272.

Takahashi LK (1990) Morphological basis of arm-swinging: Multivariate analyses of the forelimbs of *Hylobates* and *Ateles*. Folia Primatol. *54*:70–85.

Walker SE (1993) Positional adaptations and ecology of the Pitheciini. Ph.D. Dissertation, City University of New York.

Walker SE (1996) The evolution of positional behavior in the saki/uakaris (*Pithecia*, *Chiropotes* and *Cacajao*). In M Norconk, A Rosenberger, and P Garber (eds.): Advances in Primatology: Adaptive Radiations of Neotropical Primates. New York: Plenum Press, pp. 335–367.

4

WITHIN- AND BETWEEN-SITE VARIABILITY IN MOUSTACHED TAMARIN (*SAGUINUS MYSTAX*) POSITIONAL BEHAVIOR DURING FOOD PROCUREMENT

Paul A. Garber

Department of Anthropology
109 Davenport Hall
University of Illinois
Urbana, Illinois 61801

1. INTRODUCTION

Despite evidence of marked seasonal changes in diet, day range, and patterns of habitat utilization in several primate species, little is currently known regarding intraspecific variability in locomotor behavior. In general, morphologists have tended to emphasize the constraints that anatomy places on positional behavior and have classified primate locomotion into a small set of broad categories such as leaping, suspension, and arboreal or terrestrial quadrupedalism. These categories have descriptive and functional value, but they typically have failed to include both the specific environmental context in which the behavior takes place (e.g., size, orientation, and rigidness of the support; location within the tree crown, etc.), or the relationship of positional behavior to the activity pattern of the animal (e.g., differences in patterns of posture and locomotion when exploiting particular food types). If one assumes that animals exploit substrates that allow them to move efficiently through the arboreal canopy, then there should be consistent associations between postural and locomotor behavior, support type, and activity pattern (Prost, 1965). How limited or how variable these associations are within a species is determined by *physiological factors* influencing body size and the energetic and mechanical costs of movement (Tuttle and Watts, 1985; Cant, 1987; Steudel, 1990, 1996; Doran, 1993); *environmental factors* affecting the structure of the arboreal canopy, and the spatial relationship of support types to each other and to the food resources exploited (Crompton, 1984; Garber, 1984, 1993a; Grand 1984; Pound, 1991; Cant, 1992; Bergeson, 1996); and *social factors* affecting individual spacing and differential access to resources of particular age, sex, or status classes (Hunt, 1992; Remis, 1995) (Figure 1).

Primate Locomotion, edited by Strasser *et al.*
Plenum Press, New York, 1998

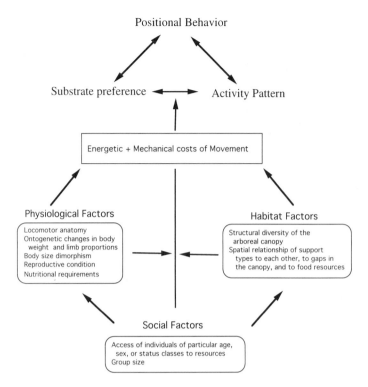

Figure 1. Schematic representation of factors that influence within species variability in positional behavior.

Recent field studies indicate that individuals of the same species may exhibit marked variation in their positional repertoire (Gebo and Chapman, 1995a,b; Dagosto, 1995). For example, Gebo and Chapman (1995a) report that in *Colobus badius* the frequencies of positional behaviors such as quadrupedalism, leaping and climbing vary considerably between years, during different seasons of the same year, and particularly in forests of different structure. In addition, Cant (1987) offers evidence of significant intersexual differences in positional behavior in adult orangutans. Male orangutans are at least twice as heavy as females orangutans. Based on observations of one fully adult male and at least 3 adult females exploiting 2 large *Ficus* trees, he concluded that males used larger supports and more above-branch postures whereas females were more frequently found on smaller branches and employed suspensory postures (Cant, 1987). In Cant's study, sex differences in body size and preferential access to parts of the tree crown were associated with differences in food harvesting techniques. Similarly, Remis (1995) reports significant differences in selection of support size and the use of suspensory postures between female and male lowland gorillas. In other species, however, patterns of positional behavior may vary minimally during different seasons of the year (Richard, 1978; Dagosto, 1995), at different field sites (see Table 6 in Garber and Pruetz, 1995), or between individuals of larger and smaller body sizes (Gebo and Chapman, 1995b). If specific knowledge of the ecological and anatomical factors that influence positional behavior in living primates is a pre-requisite for reconstructing positional behavior and habitat utilization in extinct primates, then field studies must identify those aspects of the positional repertoire that are highly sensitive to differences in forest structure and resource distribution from those that are more conservative and occur in relatively equal frequencies across a species range. In this regard our goal is to link morphology to ecology, or what has been called 'ecomorphology', through patterns of behavior (Ricklefs and Miles, 1994). The morphological design of a species results in a range of locomotor patterns. Regardless of whether this range is narrow or wide, however, it is bounded and we must be able to identify these boundaries in order to evaluate the relationship between musculoskeletal anatomy, positional behavior, and the manner in which a species exploits its environment.

In this study I examine intraspecific variability in positional behavior during feeding and foraging in two groups of moustached tamarins (*Saguinus mystax*) inhabiting forests that differ in vertical stratification, crown shape, and tree density. *Saguinus mystax* is one of the largest callitrichine species. Body weight of wild-trapped adult males averages 562 gm and body weights of adult females averages 583 gm (Garber and Pruetz, 1995). Differences in body weight, body length (crown to rump), armspan, and legspan between males and females are less than 5% (Table 1). The absence of dimorphism in this species eliminates the potential confounding effects of body weight and size in understanding behavioral variability.

Previous studies of locomotion in moustached tamarins indicate that travel is dominated by slow and rapid forms of quadrupedal progression, and by leaping (Garber, 1991). In this species, positional behavior during travel appears to be highly conservative. Garber and Preutz (1995) found only minor differences in patterns of substrate preference and locomotion among 2 groups of moustached tamarins living in forests of different structure. Less, however, is known regarding variation in the tamarin positional repertoire during feeding and foraging. In several species of Old World primates (Gebo and Chapman, 1995a,b) and Malagasy prosimians (Dagosto, 1995), there is evidence of greater variability in locomotor and postural behavior during feeding than during traveling. Given the conservative nature of travel in *Saguinus mystax*, I examine evidence of intraspecific variability in positional behavior and substrate preference during food procurement.

2. METHODS

The positional behavior and feeding ecology of moustached tamarins were studied at two field sites in the Amazon Basin of northeastern Peru. The Padre Isla field site is a protected biological reserve located approximately 10 km north of the city of Iquitos (3°44'S, 73°14'W). The island is comprised of a series of forested areas separated by narrow lagoons. The study group contained 5 adults (2 males and 3 females) and its home range was dominated by areas of low secondary forest, gallery forest, and cultivated fields (Garber et al., 1993). *Saguinus mystax* is the only species of nonhuman primate on Padre Isla. Data at this field site were collected from June through December 1990. This period encompassed the major dry season (July through September) and the early wet season (October through December).

The second field site, Río Blanco, is a relatively undisturbed primary and secondary rainforest, located 110 km southeast of Iquitos (4°05'S, 72°10'W). At this field site, moustached tamarins form mixed species troops with saddle-back tamarins (*Saguinus fuscicollis*). Data presented on the diet and positional behavior of Río Blanco moustached tamarins (group size ranged from 10–15 individuals) were collected from members of a mixed species troop from October through December 1984 (data were also collected on this group from March through September, but are presented elsewhere [Garber 1993b]). The home range of the study troop was dominated by high forest (51%), low seasonally

inundated forest (25%) and palm swamps (12%) (Norconk, 1986). In addition to moustached and saddle-back tamarins, at least 8 other primate species are found at this field site
(*Cebuella pygmaea, Callicebus moloch, Aotus* sp., *Cacajao calvus, Cebus apella, Cebus
albifrons, Pithecia aequatorialis*, and *Saimiri sciureus*). Information on the ecology of this
area has been published elsewhere (Norconk, 1986; Garber, 1993b). Identical methods of
data collection and identical definitions of food items consumed, forest architecture, positional behavior, activity patterns (feeding, foraging), and branch size and orientation were
used at each site.

Data on the tamarin diet were collected using a 2-minute focal animal time sampling
technique. Food items were classified as ripe or fleshy fruits, legumes (a dry fruit containing several seeds enclosed in a pod), floral nectar (the tamarins were not observed to consume flower petals or ovaries), exudates (plant gums and saps that exude from the trunks
of trees), and insect prey. Despite their small body size, tamarins are easily observed when
feeding on plant resources. They are sloppy feeders and commonly drop husks and flower
parts onto the ground, facilitating identification of the plant part consumed. Insect feeding
presents considerably greater problems of identification. Moustached tamarins often visually scan vegetation searching for exposed prey, or glean insects attached to leaf surfaces.
Once sighted, arthropods are captured rapidly using a stalk-and-pounce technique (Garber,
1993a; Peres, 1992). In the present study differences in insect foraging techniques and associated positional behaviors were identified, but no attempt was made to distinguish between the different types of insects exploited.

Information on forest architecture was collected in order to assess how site-specific
differences in the tree density, canopy continuity, and branching patterns affect tamarin
feeding and foraging activities. Forest architecture was studied by randomly selecting 5 ×
5 m vegetation quadrates in each group's home range. The number, height, and circumference of all trees greater than 2 m in stature and ≥2 cm in diameter at breast height (dbh),
as well as the number of lianas in each quadrat were recorded. Approximately 2% (61
quadrats) of the vegetation in the 7 ha home range of the tamarin study group on Padre
Isla and 1% (200 quadrats) of the vegetation in the 40 ha home range of the tamarin group
at the Río Blanco was sampled (Garber and Pruetz, 1995).

Quantitative information on positional behavior in adult moustached tamarins also
was collected using a 2-minute instantaneous focal animal point sampling technique.
Climbing was defined as "progression along continuous supports using various combinations of three or more limbs" (Fleagle and Mittermeier 1980:249). *Grasping* is a postural
activity associated with prehension using any combination of two or more limbs. *Sitting*
was scored when an animal was stationary while supported on its haunches. Limb placement was variable. *Hindlimb grasp* (or grasping with hindlimbs only) is a form of below
branch feeding in which the forager is oriented vertically downward, supported solely by
the prehensile grasp of its hindfeet which are loaded in tension. *Clinging* is a postural activity in which the clawlike nails of at least three limbs are embedded into the bark. Clinging typically occurs on moderate to large vertical or sharply inclined supports. *Bipedal
crouch* represents a postural behavior in which both hindlimbs are in compression and
supporting the animal's full body weight. The knees are flexed, both arms are extended.
This posture was observed when an animal was attempting to pull a large food item, still
attached to the stem, to its mouth to feed. Locomotor behaviors were defined as follows:
Quadrupedalism was defined as pronograde progression in which all four limbs were in
compression. Typically the tamarins adopted a diagonal sequence-diagonal couplet gait
during both quadrupedal walking and quadrupedal running. *Leaping* is associated with
hindlimb propulsion that enables the forager to navigate gaps in the canopy. Leaping can

occur from a stationary position, slow quadrupedal progression, or from rapid quadrupedal bounding. During foraging, leaps typically involved movement between small, discontinuous supports. Less frequently did moustached tamarins leap from trunk-to-trunk in the understorey. *Scansorial* locomotion occurs on moderate to large vertical or sharply inclined supports. Movement is facilitated by embedding their claw-like nails into the bark.

Foraging was defined as localized movements within the crown of a tree associated with food search and pursuit. Feeding was defined as handling or ingesting prey items. For example, if a tamarin was moving from one tree branch to another and scanning a nearby substrate or moving to another part of the same tree crown that bore ripe fruit, it was scored as foraging. In contrast, if an individual was turning over leaves, holding a fruit, extracting gum from a tree wound, or masticating a food item, it was scored as feeding.

Each activity record contained information on: (1) characteristics of the substrate (a) height in the canopy, estimated in 5 meter intervals above the ground; (b) branch angle, classified as horizontal (0–15°), oblique (16–74°), and vertical (75–90°); and (c) branch diameter, classified as small (≤5 cm), medium (6–10 cm), and large (>10 cm); (2) positional behavior; and (3) activity pattern. A total of 3,283 samples of positional behavior, substrate utilization, and height in the canopy during feeding and foraging were recorded for moustached tamarins on Padre Isla, and 2,531 samples of positional behavior during feeding and foraging were recorded for moustached tamarins at the Río Blanco site.

At the Padre Isla site, we were unable to follow the study group over the course of an entire day. The data are presented in terms of a profile of positional behavior, substrate utilization, and height in the canopy across twelve 2-week periods. At the Río Blanco site, the study group was followed from dawn until dusk. The unit of study was a complete tamarin day, and for each day a profile of positional behavior, substrate utilization, and height in the canopy was compiled. Data are presented for 15 complete days over 3 consecutive 1-month periods. During each 1-month period at the Río Blanco, and each 2-week period on Padre Isla, between 4.5 and 20 hrs of quantitative data on feeding and foraging were recorded (total of 109 hours of quantitative data on Padre Isla moustached tamarins and 84 hours of quantitative data on Río Blanco moustached tamarins). Observation hours were calculated by assuming that each feeding and foraging bout occurred throughout the entire 2-minute sampling period and then multiplying the number of feeding and foraging samples by 2 minutes.

Between-site and within-site comparisons of positional behavior and substrate preference were accomplished by compiling matrices of biweekly frequencies (Padre Isla) and daily frequencies (Río Blanco) for each activity, and then converting these scores into a ranked format. Pearson's product-moment correlation coefficients were calculated on ranked data for all combinations of samples (i.e., biweekly sample 1 with biweekly sample 2, biweekly sample 1 with biweekly sample 3, etc. or day 1 with day 2, day 1 with day 3, day 2 with day 3, etc.). These values were normalized and converted to Z-scores according to a hypergeometric distribution calculated using Fisher's Exact Test. The distribution of Z-scores between samples within a site and the distribution of Z-scores between-sites was compared (Students t-test, two-tailed probability) to test for differences in patterns of positional behavior across sample periods at the same site as well as between sites.

A Chi-square goodness of fitness test was used to compare the forest height profile of each forest site with the patterns of vertical ranging exhibited by the tamarins. A Students t-test (two-tailed probability) was used to examine inter-site differences in forest architecture.

Body weights and body measurements presented in Table 1 were collected by the author from several groups of wild-caught and released tamarins during the course of this

Table 1. Body weights and body measurements of wild-trapped adult female and male moustached tamarins (*Saguinus mystax*)[1]

Sex	N	Body weight (g)	Body length (cm)	Armspan (cm)	Legspan (cm)
Female[2]	48	583 ± 63	24.4 ± 0.8	31.3 ± 1.1	37.9 ± 1.4
Male	46	562 ± 59	24.6 ± 0.8	31.2 ± 0.8	37.8 ± 1.4

[1]Data are from Garber (1991) and Garber and Pruetz (1995).
[2]Measurements from nonpregnant, pregnant, and lactating females.

and other studies (Garber and Pruetz, 1995). These data were analyzed using a Students t-Test (two-tailed probability).

3. RESULTS

3.1. Diet

At each field site, moustached tamarins exploited a similar set of resources (Table 2). Based on time spent feeding and foraging, the diet of the Padre Isla study group included 56.1% insect prey, 25.4% legumes, and 16.1% fleshy fruits. Floral nectar and plant exudates accounted for the remaining 2.3% of the diet. Virtually all observations of legume feeding on Padre Isla involved trees of the genus *Inga* (Leguminosae). At least 3 species of *Inga* (*Inga edulis*, *Inga alba*, and *Inga* sp.) were common in the study troop's home range.

At the Río Blanco site, insects (44.4%), fleshy fruits (39.9%), and legumes (7.3%) accounted for 92% of feeding time (Table 2). In this forest, fleshy fruits were available throughout most of the year (Garber, 1993b) and were the principle source of readily available energy. The exploitation of legumes was more seasonal than on Padre Isla, but did account for 24.5% of total plant feeding time during the late wet season months of May and June (these results are reported in Garber, 1993b). At the Río Blanco field site, nectar accounted for 6% of feeding time. Overall, the tamarin diet at the Río Blanco was more diverse and species-rich than the tamarin diet on Padre Isla.

Table 2. Dietary patterns of moustached tamarins at two study sites in northeastern Peru[1]

Food type	Padre Isla	Río Blanco
Insects	56.1	44.4
Fleshy Fruits	16.1	39.9
Legumes	25.4	7.3
Nectar	1.1	6.0
Exudates	1.2	2.2

[1]Data from Padre Isla are based on 3283 instantaneous focal animal point samples from June through November, 1990. Data from Rio Blanco are based on 2531 instantaneous focal animal point samples from October through December, 1984.

3.2. Forest Structure

Data collected on forest structure provide evidence of significant inter-site differences. The number of trees per hectare (1170/ha at Padre Isla compared to 4240/ha at Río Blanco; t=14.5, df=259, P<0.0001) and the number of lianas per hectare (640/ha at Padre Isla compared to 1280/ha at Río Blanco; t=3.2, df=259, P<0.001) were significantly fewer at Padre Isla than on Río Blanco. On Padre Isla, trees were smaller and characterized by crowns in which growth in height exceeded growth in width. In these trees, main highway branches tended to be vertical or oblique rather than horizontal in orientation. Over 63% of the trees sampled were less than 5 meters in height, and only 6.6% were over 15 meters in height. At the Río Blanco site, taller trees with large horizontal boughs and wide spreading crowns were more common. In this forest, 20% of sample trees exceeded 15 meters in height. A vertical height profile of both forests indicate that Río Blanco tamarins exploited a forest of significantly greater stature (χ^2=148.6, df=4, P<0.001; Figure 2) and a more continuous canopy than did Padre Isla tamarins.

3.3. Vertical Use of Canopy Levels

Figure 2 provides a comparison of the height profile of each forest with tamarin vertical ranging activities during feeding and foraging. Differences between these distributions are significant (χ^2=8403, df=4, P<0.0001), and indicate that Padre Isla moustached tamarins exhibited a clear preference for exploiting resources in the mid to upper levels of the forest canopy. For example, despite the fact that trees between 11–15 meter in height accounted for only 4.6% of the trees sampled at Padre Isla, 33.5% of feeding and foraging occurred in this level of the canopy.

An identical pattern of tamarin vertical height preference was found at the Río Blanco study site. In this forest, tall trees were also a focus of tamarin feeding activities

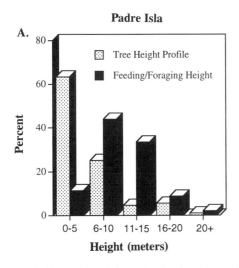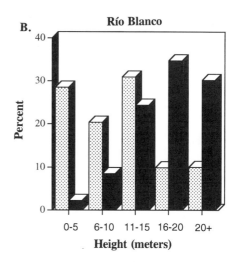

Figure 2. Comparison of the vertical height profile of the study group's home range and the feeding and foraging height preferences of that group. (A) Data for Padre Isla are based on height estimates of 194 trees in the home range of the study group. (B) Data for the Río Blanco represent height estimates of 1198 trees in the home range of this study group.

Table 3. Percentages and ranked order of positional behaviors during feeding and foraging[1]

Positional behavior	Padre Isla		Río Blanco	
	%	Rank	%	Rank
Climb/Grasp	33.5	1	43.5	1
Sit	32.5	2	26.5	2
Hindlimb grasp	15.1	3	8.2	4
Quadrupedalism	11.5	4	12.9	3
Cling	3.0	5	2.1	6
Leap	2.1	6	4.1	5
Bipedal crouch	1.8	7	1.6	7
Scansorial	0.3	8	0.6	8

[1]Data from Padre Isla are based on 3183 records of positional behavior. Data from the Río Blanco are based on 2531 records of positional behavior.

(Figure 2). Trees exceeding 15 meters in height accounted for 20% of the forest profile but fully 65% of feeding and foraging time (χ^2=3953; df=4, P<0.001).

3.4. Positional Behavior

During food procurement, climbing/grasping and sitting dominated the moustached tamarin positional repertoire. Table 3 presents a profile and description of locomotor and postural behaviors adopted by *S. mystax* during feeding and foraging. On Padre Isla, climbing and grasping using a combination of at least 3 limbs accounted for 33.5% of locomotor and postural activities (Table 3). Sitting was the next most common posture (32.5%), followed by grasping with hindlimbs only (15.1%), quadrupedal progression (11.5%), and a range of relatively infrequent behaviors such as clinging, jumping, crouching bipedally, and scansorial locomotion. These latter 4 positional behaviors accounted for only 7.2% of feeding and foraging activities.

Table 4. Variation in the percentage of feeding and foraging time spent in positional behaviors by *Saguinus mystax*[1]

Positional category	Padre Isla	Río Blanco
Climb/Grasp	34.0 ± 7.0	43.5 ± 7.8
	21.6 - 47.0	34.3 - 61.4
Sit	32.5 ± 8.6	26.5 ± 5.9
	17.7 - 47.7	14.4 - 33.3
Hindlimb grasp	15.1 ± 7.7	8.2 ± 5.8
	8.0 - 33.5	0 - 24.3
Quadrupedalism	11.5 ± 3.2	12.9 ± 5.7
	7.2 - 16.2	2.7 - 24.9
Cling	3.0 ± 1.6	2.1 ± 1.8
	0.5 - 5.4	0 - 6.7
Leap	2.1 ±1.5	4.1±2.3
	0.7 - 5.8	0.8 - 7.5
Bipedal crouch	1.8 ± 1.0	1.6 ± 1.9
	0.8 - 3.9	0 - 7.4
Scansorial	0.3 ± 0.3	0.6 ± 1.0
	0 - 0.9	0 - 3.5

[1]Data are mean ± 1 S.D. (top row), range (bottom row).

At the Río Blanco field site, moustached tamarins exhibited a similar positional repertoire to that used by the Padre Isla group (Table 3). In this taller and less disturbed rainforest, climbing and grasping continued to be the most common positional behaviors, accounting for 43.5% of food procurement activities. Sitting (26.5%), quadrupedal progression (12.9%), and grasping solely with hindlimbs (8.2%) represented the next 3 most common positional activities.

Overall there was marked similarity in the relative percentage and pattern of positional activities between sites. Between-site differences in moustached tamarin positional behavior were largely the result of the greater use of combinations of fore- and hindlimb grasping/climbing at the Río Blanco, and a greater use of grasping with hindlimbs only on Padre Isla. If climbing/grasping and grasping with hindlimbs only are combined into a single prehensile positional category, then during feeding and foraging these behaviors accounted for 48.6% of the positional repertoire on Padre Isla and 51.7% at the Río Blanco. Even when grasping with hindlimbs only is scored as a distinct postural category, a ranking of 8 postural and locomotor behaviors (Table 3) indicates limited between-site differences in the moustached tamarin positional repertoire.

3.5. Temporal Variation in Patterns of Positional Behavior

Given limited evidence for major between-site differences in the relative percentages of each positional behavior, is there evidence of significant levels of daily or biweekly variation among group members inhabiting the same site? Table 4 provides information on within-site variation in postural and locomotor behavior. The data from Padre Isla indicate that across any 2 sample periods the frequencies of a particular postural or locomotor behavior could vary by a factor of 2–4. For example, during sample period 2, sitting accounted for 17.7% of the positional repertoire, whereas during sample period 12, sitting occurred 47.7% of the time. Similarly, grasping with hindlimbs was observed 8.0% of the time during a given sample period and 33.5% in another sample period. A similar range of variation in posture and locomotion between sample periods was found at Río Blanco (Table 4). These data can be interpreted as evidence of substantial within-site variation in positional behavior.

In contrast, analyzing the tamarin positional repertoire in terms of a ranking of postural and locomotor activities for each sample period offers support for limited within-site variability. An analysis using ranked data focuses on patterns of behavior and relative frequencies rather than absolute frequencies. The rankings for each of the twelve 2-week periods for the Padre Isla tamarins are presented in Table 5. These data indicate that certain positional behaviors, such as grasping/climbing and sitting dominated feeding and foraging throughout the study, whereas other modes of posture and locomotion were always observed to occur in moderate frequency or infrequently. This pattern was highly consistent between two-week periods. For example, a comparison of Pearson's product-moment correlation coefficients calculated on ranked data for all combinations of biweekly samples fails to indicate any significant differences in positional rankings between sample periods 1–6 and sample periods 7–12 (t=1.2, df=14, P=0.23).

A similar analysis of correlation coefficients based on patterns of positional behavior during feeding and foraging at the Río Blanco field site also revealed evidence of limited within-site variability. As indicated in Table 5, climb/grasp was the most common positional behavior during each sample day. Sitting ranked second during 12 of 15 full day follows and third on the remaining 3 sample days. Quadrupedal progression or grasping with hindlimbs only were ranked third on the remaining 12 days. On no sample day did leap-

Table 5. Ranking of the positional repertoire in *Saguinus mystax* during feeding and foraging[1]

Positional behavior	Padre Isla				Río Blanco			
	Rank 1-2	Rank 3-4	Rank 5-6	Rank 7-8	Rank 1-2	Rank 3-4	Rank 5-6	Rank 7-8
Climb/Grasp	**11**	1	0	0	**15**	0	0	0
Sit	**9**	3	0	0	**12**	3	0	0
Hindlimb grasp	3	**9**	0	0	1	**10**	2	2
Quadrupedalism	0	**12**	0	0	2	**12**	1	1
Cling	0	0	**10**	2	0	0	**11**	4
Leap	0	0	**8**	4	0	1	**12**	2
Bipedal crouch	0	0	4	**8**	0	1	1	**13**
Scansorial	0	0	0	**12**	0	0	2	**13**

[1]Data from Padre Isla represent 12 2-week periods. Data from Rio Blanco represent 15 full day follows. The bold values serve to emphasize the consistent rank of that behavior in the tamarin positional repertoire.

ing, clinging, scansorial locomotion, or bipedal standing rank among the top 4 positional behaviors. A comparison of patterns of positional behavior across sample periods (Days 1–7 vs. Days 8–15) indicated no significance rank differences in 8 locomotor and postural categories used by moustached tamarins at the Río Blanco (t=1.1, df=20, P=0.27).

3.6. Association between Insect Foraging and Positional Behavior

Figure 3 presents a between-site comparison of insect capture techniques used by moustached tamarins. These techniques are (1) gleaning of prey attached to small branches and leaf surfaces; (2) visual scanning for exposed and mobile prey which, once located, are quickly seized; and (3) exploitation of insects refuging or concealed on trunks and other large vertical supports. Each capture technique is associated with an alternative set of positional behaviors. Gleaning involves active searches and manipulation of leaves using climbing, grasping, and leaping locomotion. During visual scanning, tamarins typically adopt a sitting posture or stand quadrupedally and move their heads from side to side and/or up and down searching for prey. Once detected, prey are pursued using rapid quadrupedal movement and seized or pinned by the hands. The capture of trunk refuging insects involves inspection of crevices in the bark during vertical clinging and traveling on

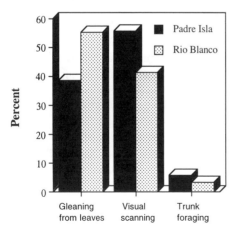

Figure 3. Between-site comparison of insect foraging techniques in *Saguinus mystax*. The three capture techniques: gleaning, scanning and striking, and trunk foraging are described in the text.

trunks using scansorial locomotion. As indicated in Figure 3, 38.6% of insect foraging by Padre Isla tamarins was associated with gleaning nonmobile or attached prey from leaf substrates. In this forest, tamarins relied heavily on visual scanning and pouncing to acquire insect prey (55.7%). In contrast, gleaning nonmobile or attached prey was the primary insect foraging technique observed in Río Blanco moustached tamarins and accounted for over 55.3% prey captures (Figure 3). Foraging for prey on large vertical trunks occurred infrequently at each site. It remains uncertain whether site-specific differences in insect foraging techniques reflect differences in prey type and prey availability, or differences in capture rates associated with vegetation density and prey visibility. These differences in foraging techniques may, however, help to account for much of the inter-site variability in tamarin positional behavior.

3.7. Substrate Preference

Resources exploited by moustached tamarins were located principally on small obliquely-oriented supports in the mid- and upper-levels of the tree crown. At the Padre Isla field site, 55.4% of foraging substrates were small in diameter and 78% were angled supports (Figure 4). Medium-sized supports accounted for 38.7% of the positional repertoire, and large supports were used infrequently (5.7%).

A similar pattern of support size was recorded for the Río Blanco study group. At this site, 68.8% of feeding and foraging time occurred on small branches (Figure 4). Medium-sized supports accounted for 23.3% of food gathering activities, and large diameter supports 8.8%.

There was, however, evidence of considerable between-site variation in the frequency with which branches of different angles or orientation were utilized. Although obliquely angled branches were exploited most frequently at each field site, observations of the Río Blanco tamarins indicated that horizontal and vertical substrates were used in considerably higher frequency, accounting for 33.7% and 25.2% of feeding/foraging activities respectively (Figure 4). On Padre Isla, vertically oriented branches were rarely vis-

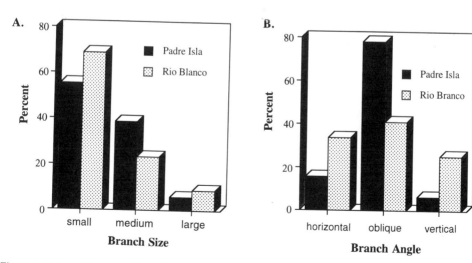

Figure 4. Between-site comparison of substrate utilization in moustached tamarins during feeding and foraging. Comparative data are presented on support size (a) and support orientation (b). See text for definitions of branch size and branch orientation categories.

ited and accounted for only 6.3% of substrates exploited. Similarly, use of horizontal supports on Padre Isla was approximately one-half that recorded at the Río Blanco (15.6% vs. 33.7; Figure 4).

3.8. Temporal Variability in Patterns of Substrate Utilization

A ranking of the data on branch size and branch angle offers additional evidence for consistency in patterns of substrate use in moustached tamarins (Table 6). On Padre Isla, small supports dominated feeding and foraging in 9 of 12 sample periods. Medium supports were the most common substrate exploited in 3 samples and the second most common in 9 samples. Large branches ranked last in support use in each of the 12 2-week periods. Use of branch angles by Padre Isla moustached tamarins followed a similar pattern. In 12 of 12 2-week sample periods obliquely-oriented supports were the most commonly used platform for feeding and foraging. In 10 of 12 sample periods, horizontal supports were the second most common feeding and foraging substrate. Vertical supports were rarely used.

Daily variation in use of supports of small, medium, and large size by Río Blanco moustached tamarins was also highly conservative. Despite the fact that on any given day, small branches accounted for between 55–71% of the substrates exploited and large branches accounted for 2–19%, a ranking of these data indicate a consistent pattern. Small branches were selected in highest frequency on each of the 15 sample days. Medium branches ranked second on 14 of 15 days and tied for second on the remaining day. Large branches were exploited least frequently on all days (except for one day in which these supports were tied for last).

In contrast to the pattern observed on Padre Isla, Río Blanco moustached tamarins were significantly more variable in their use of branches of different orientation (t=3.5, df=65, P<0.001). Daily values for the use of horizontal supports at the Río Blanco ranged from 31–45%. Use of oblique supports ranged from 32–54%, and for vertical supports these values ranged from 6–34%. A comparison of the daily rankings indicate that horizontal supports ranked highest or tied for highest on 46% (7/15) of the sample days, oblique supports ranked highest or tied for highest on 66% (10/15) of the sample days, and vertical supports tied for the highest rank on 6% (1/15) of the sample days. Daily variation

Table 6. Ranking of the positional repertoire in *Saguinus mystax* during feeding and foraging[1]

Substrate	Padre Isla			Río Blanco		
	Rank 1	Rank 2	Rank 3	Rank 1	Rank 2	Rank 3
Small	**9**	3	0	**15**	0	0
Medium	3	**9**	0	0	**15**	0
Large	0	0	**12**	0	1	**14**
Oblique	**12**	0	0	**10**	4	1
Horizontal	0	**10**	2	7	**8**	0
Vertical	0	2	**12**	1	0	**14**

[1]Data from Padre Isla represent 12 2-week periods. Data from Río Blanco represent 15 full day follows. Column totals for the Río Blanco may not always equal 15. This is the due to several cases of tied rankings. When two categories were tied for the same rank, they were both assigned to that rank. The emboldened values serves to emphasize the consistent rank of those behaviors in the tamarin positional repertoire.

in support orientation was associated with differences in the time spent exploiting particular food types. At this site there was a significant negative correlation in the frequency of time spent on oblique branches each day and percent time feeding and foraging on ripe fruit on that day (R=-0.56, P=0.029). Given the relatively large size of many fruits eaten by moustached tamarins (Garber, 1986), these primates frequently may select above-branch feeding postures on low angled platforms in an attempt to increase stability and free the hands to hold and manipulate food items.

4. DISCUSSION

Tamarins are small bodied New World monkeys that exploit a diet composed principally of insects, ripe fruits, legumes, floral nectar, and plant gums. These resources are located in different parts of the canopy, and access to each requires alternative patterns of postural and locomotor behavior. For example, several species of tamarins exploit resources (exudates, insects, and small vertebrate prey) that are located on tree trunks in the forest understorey (Garber, 1992, 1993a). Trunk foraging appears to be an important element of callitrichine feeding ecology and is facilitated by the evolution of elongated, laterally compressed, clawlike nails on all manual and pedal digits excluding the hallux. Clawlike nails enable these small bodied monkeys to cling to and travel on large vertical and sharply angled supports that are too large to be spanned by their tiny hands and feet. In contrast, when exploiting fruits, legume pods, and more mobile insect prey, tamarins concentrate their foraging activities on a network of small-to-medium sized branches located in the periphery of the tree crown (Garber, 1984, 1993a).

In this study, data are presented on within- and between-site variability in moustached tamarin positional behavior during food procurement. Tamarins must make behavioral adjustments in response to daily, weekly, and monthly changes in the availability and distribution of food resources, as do all foragers. These adjustments involve decisions regarding *what* to eat and *where* to look for food, and include the selection of particular microhabitats to exploit within their range, the selection of particular trees or food patches within these microhabitats, and the selection of particular foraging substrates or zones within the crown of a feeding tree. Given differences in the architecture of tree crowns (Horn, 1974; Halle et al., 1978) and associations between certain food types and the arboreal substrates on which they are found (Garber, 1984; Crompton, 1984; Boinski, 1989), how variable or flexible is positional behavior during feeding? In taxa such as *Propithecus verreauxi* (Richard 1978), *Propithecus diadema* (Dagosto, 1995), and *Eulemur fulvus* (postural behavior; Dagosto, 1995) there is little evidence of seasonal variation in positional behavior associated with diet. In other species such as *Galago senegalensis* (Crompton, 1984), *Otolemur crassicaudatus* (Crompton, 1984), *Colobus badius* (Gebo and Chapman, 1995a), *Eulemur rubriventer* (Dagosto, 1995), and *Varecia variegata* (Dagosto, 1995) such differences are more evident. In these more 'variable' species, the degree to which flexibility in positional behavior is determined principally by differences in forest structure, differences in foraging strategies, and/or differences in locomotor anatomy remains unclear.

Despite exploiting forests that differed significantly in tree density, canopy structure, canopy height, liana density, plant species composition, and presence of potential primate competitors (Garber and Pruetz, 1995), patterns of positional behavior in the 2 moustached tamarin study groups were highly conservative. A rank order correlation of 8 postural and locomotor behaviors at each field site revealed marked similarity in the relative breakdown of positional activities. In addition, the rankings of positional behavior within each site varied minimally between individual days and over the course of several weeks. Overall, variability in the moustached tamarin positional repertoire did not differ significantly between field sites. This occurred despite the fact that the Río Blanco moustached tamarins formed cohesive and stable mixed species troops with saddle-back tamarins (*Saguinus fuscicollis*). On Padre Isla, saddle-back tamarins were absent. Several studies of mixed species troops of moustached tamarins and saddle-back tamarins (Norconk, 1986; Garber, 1991; Peres, 1991) indicate clear differences in patterns of positional behavior and vertical use of the canopy. Data presented in the present study support the contention that the presence or absence of saddle-back tamarins at these study sites had little effect on positional behavior in moustached tamarins.

On Padre Isla, climbing/grasping ranked first or tied for first on 8 of 12 biweekly samples and ranked second in 3 of the remaining samples. Sitting ranked second in 6 of 12 sample periods and was first or tied for first in 4 biweekly samples. Small supports were used in greatest frequency in 9 of the 12 biweekly samples and ranked second in each of the 3 remaining observation periods. A pattern of limited behavioral variability was also documented at the Río Blanco. At this site, climbing/grasping was the most frequent positional behavior during each of 15 full day follows. Sitting was the next most frequent posture, ranking second on 12 of 15 days. Small supports were the most commonly used feeding and foraging platform on each sample day and medium supports were the second most common on 14 of 15 days. Overall, patterns of positional behavior and substrate preference that were commonly used by moustached tamarins on Padre Isla were also commonly used by moustached tamarins at the Río Blanco. Rare behaviors on Padre Isla were rare at the Río Blanco.

There was evidence of more pronounced between-site variation in support orientation and foraging and feeding height. The home range of the Padre Isla study group was characterized by a young and highly disturbed forest in which tree crowns were shorter, and exhibited a branching pattern dominated by vertically and obliquely-oriented boughs. During feeding and foraging almost 80% of the substrates used by moustached tamarins on Padre Isla were oblique in orientation. In contrast, the Río Blanco field site was characterized by older, taller, and more mature forest trees, many of which exhibited widespreading crowns and a large number of major horizontal branches. Río Blanco moustached tamarins more commonly fed and foraged at heights above 15 meters and were less consistent in their use of obliquely-oriented branches. Although oblique branches were the most frequently utilized support type (41%) by tamarins at this field site, horizontal branches were visited at a frequency equal to or greater than oblique supports on 46% of sample days. Differences in canopy architecture, tree heights, and the availability of support types probably account for these results.

Marked consistency in patterns of posture and locomotion between moustached tamarins living in distinct forest types suggests that, in this species, locomotor anatomy acts as a primary constraint on patterns of behavior and habitat utilization. The major between-site differences in positional behavior were associated with a greater frequency of sitting (32.5 vs. 26.5) and grasping with hindlimbs only postures (15.1 vs. 8.2) on Padre Isla and more climbing/grasping activities using a combination of fore- and hindlimb prehension (43.5 vs. 33.5) at the Río Blanco. These differences appear to reflect site-specific variations in foraging strategies and patterns of dietary emphasis rather than high levels of behavioral variability.

Hindlimb grasping is a form of below branch feeding in which the forager is oriented vertically downward, supported solely by the prehensile grasp of its hindfeet. In adopting this posture, the hands are free to manipulate and seize plant food and animal prey, but play no direct role in weight bearing. As indicated in Table 7, at both field sites

Table 7. Association between diet and hindlimb grasp in *Saguinus mystax*[1]

Food type	Padre Isla	Río Blanco
Legumes	36.3	21.0
Fleshy Fruits	15.7	7.7
Invertebrate	7.2	6.2
Exudates	–	1.7

[1]These data are interpreted as follows: During 36.3% of all observations of legume feeding at Padre Isla, the tamarins were found to adopt a grasping with hindlimbs only posture. Other positional behaviors, such as sitting (23.1%) and grasping with a combination of fore- and hindlimbs (29%) were also common when feeding on legumes at Padre Isla.

grasping with hindlimbs only occurred most frequently when feeding and foraging on legume pods (36% of legume feeding at Padre Isla and 21% of legume feeding at the Río Blanco). These pods are presented at the distal ends of terminal branches and often hang vertically down. Many of the pods have a fibrous or tough outer husk. The tamarins use their incisors and canines to pierce and strip the husk and then pry the pod open with both hands in order to feed. The high concentration of *Inga* trees and the greater reliance on legumes as a dietary staple by tamarins on Padre Isla help to explain between-site differences in the frequency of grasping with hindlimbs only postures. For example, if Río Blanco tamarins maintained their same pattern of positional behavior but increased their level of legume feeding to a value comparable to that of Padre Isla tamarins, then group differences in the frequency of hindlimb grasping postures would be reduced to only 3%.

Moustached tamarins spent 44–56% of feeding and foraging time exploiting insect prey. In terms of insect foraging techniques, they are described "as leaf-gleaners, capturing and flushing mobile prey directly from the midstorey foliage" (Peres, 1991:127). Orthopteran insects such as katydids, grasshoppers, and stick insects are the most common prey items consumed (Peres, 1991; Nickle and Heymann, 1996). In the present study, moustached tamarins were observed to rely on three primary methods of prey capture. The relative frequencies of each capture method varied between field sites. Moustached tamarins on Padre Isla relied more heavily on visual scanning followed by quadrupedal attack (pounce) to obtain prey (55.7%). Visual scanning typically occurred from a sitting position. At the Río Blanco, gleaning sedentary and attached prey from leafy substrates was the most common insect capture technique (55.3%). This involved leaf inspection and manipulation and was associated with climbing, grasping, and leaping behavior. The degree to which differences in prey capture techniques reflect site-specific ecological differences in the availability and abundance of prey types remains unclear. Nevertheless, shifts in the frequency of prey capture techniques did have a direct effect on positional behavior. During insectivory, sitting accounted for 38% of the positional repertoire of tamarins on Padre Isla, but only 18% of the positional repertoire of tamarins at the Río Blanco.

In this paper relationships between diet, habitat utilization, and positional behavior in wild moustached tamarins are described. These relationships were consistent among members of 2 different groups inhabiting 2 different rainforest sites in Amazonian Peru. Given that another tamarin species, *Saguinus fuscicollis*, is characterized by a somewhat greater range of behavioral variability (Garber and Davis, 1996), what factors are likely to play a primary role in limiting the expression of behavioral variability in *S. mystax*? Figure 1 outlines a set of physiological, social, and habitat variables that serve to constrain or broaden the way in which individuals within a species exploit their environment. In terms of the present study, differences in habitat such as forest structure and resource availability were found to have a direct, but limited effect on patterns of positional behavior. No attempt was made to measure age or sex-specific behavioral patterns. However, given minimal body size dimorphism (Table 1), infrequent within-group aggression (Heymann, 1996; Garber, 1997), and high levels of group cohesion and social cooperation, individual differences in patterns of habitat utilization are probably small. This remains to be verified empirically.

How might locomotor anatomy influence the expression of locomotor behavior? As suggested by Ricklefs and Miles (1994:16) "The structural complexity of morphological design may determine the breadth of performance by an individual." In this regard, it is possible that a species characterized by a higher degree of individual variation in positional anatomy (i.e., age or sex related differences in ontogenetic development or body size) might exhibit a potential for greater variability in positional behavior than would a species characterized by greater morphological consistency. In a recent study of positional anatomy in callitrichines, Garber and Davis (1996) identified a complex of traits of the upper limb that exhibited significantly greater levels of individual variation in saddle-back tamarins (S. fuscicollis) than in moustached tamarins (S. mystax). Several of the skeletal traits exhibiting increased variance were part of a single morphological complex and functionally interpreted as being consistent with field data indicating that saddle-back tamarins also exhibited greater variability in positional behavior than do moustached tamarins (Garber and Davis, 1996). It is possible that changes in morphological design or phenotypic variability of these skeletal elements have a disproportionate effect on the range of locomotor and postural behaviors exhibited. Analogous relationships between anatomical variability/stability and behavioral variability/stability in other primate lineages need to be explored.

In conclusion, in order to identify relationships between morphology, ecology, and patterns of behavior in living primates (ecomorphology) studies of positional behavior and positional anatomy need to include information on within-species variability. Although most traits in a population are characterized by "low levels of phenotypic variation" (Travis, 1994:99), the presence or absence of plasticity in certain traits may have an important adaptive function. In the case of several primate species (see review by Garber and Pruetz, 1995) including prosimians, monkeys, and apes there is evidence of marked between-site consistency in patterns of positional behavior. Within these taxa, regardless of whether the data are ranked or tallied by frequency, postural and locomotor behaviors that were rare at one site were rare at another site. Similarly, postural and locomotor behaviors that were common at one site were also common at another site. For other primate taxa, seasonal, site-specific, ontogenetic, and sex-based differences in positional behavior may be more extreme. The challenge to studies of locomotion and posture therefore, is not only to establish functional relationships between behavior and morphology, but in addition to identify those elements of the positional repertoire that are stable from those that are more variable and determine how each is influenced by habitat, physiology, the energetic and mechanical cost of movement, and the social environment (Figure 1).

ACKNOWLEDGMENTS

This study was conducted with the permission and assistance of the Proyecto Peruano de Primatología "Manuel Moro Sommo" and the Instituto Veterinario de Investigaciones Tropicales y de Altura (IVITA). We thank Dr. Enrique Montoya Gonzales, Executive Director of the Proyecto Peruano de Primatología, Filomeno Encarnación, and Carlos Ique for their support in this project. Assistance in the field was provided by Jill Pruetz, Walter Mermao and Eriberto Mermao. Helpful comments on an earlier draft of this manuscript were provided by Dr. David Bergeson, Dr. Elizabeth Strasser, and an anonymous reviewer. Input on statistical analyses of the data were provided by Dr. Steven Leigh. Funds to conduct this research were provided by the National Geographic Society, the National Science Foundation, and the William and Flora Hewlett Foundation. As always I wish to thank Sara and Jenni for being Sara and Jenni.

REFERENCES

Bergeson DJ (1996) The Positional Behavior and Prehensile Tail Use of *Alouatta palliata*, *Ateles geoffroyi*, and *Cebus capucinus*. Ph.D. Dissertation, Washington University.

Boinski S (1989) The positional behavior and substrate use of squirrel monkeys: ecological implications. J. Hum. Evol. 7:659–678.

Cant JGH (1987) Effects of sexual dimorphism in body size on feeding postural behavior of Sumatran orangutans (*Pongo pygmaeus*). Am. J. Phys. Anthropol. 74:143–148.

Cant JGH (1992) Positional behavior and body size of arboreal primates: a theoretical framework for field studies and an illustration of its application. Am. J. Phys. Anthropol. 88: 273–283.

Crompton RH (1984) Foraging, habitat structure, and locomotion in two species of *Galagos*. In PS Rodman and JGH Cant (eds.): Adaptations for Foraging in Nonhuman Primates. New York: Columbia University Press, pp. 73–111.

Dagosto M (1995) Seasonal variation in positional behavior of Malagasy lemurs. Int. J. Primatol. 16:807–834.

Doran DM (1993) Sex differences in adult chimpanzee positional behavior: The influence of body size in locomotion and posture. Am. J. Phys. Anthropol. 91:99–115.

Fleagle JG, and Mittermeier RA (1980) Locomotor behavior, body size, and comparative ecology of seven Surinam monkeys. Am. J. Phys. Anthropol. 52:301–314.

Garber PA (1984) Use of habitat and positional behavior in a Neotropical primate, *Saguinus oedipus*. In PS Rodman and JGH Cant (eds.): Adaptations for Foraging in Nonhuman Primates. New York: Columbia University Press, pp. 113–133.

Garber PA (1986) The ecology of seed dispersal in two species of callitrichid primates (*Saguinus mystax* and *Saguinus fuscicollis*). Am. J. Primatol. 10:155–170.

Garber PA (1991) A comparative study of positional behavior in three species of tamarin monkeys. Primates 32:219–230.

Garber PA (1992) Vertical clinging, small body size, and the evolution of feeding adaptations in the Callitrichinae. Am. J. Phys. Anthropol. 88:469–482.

Garber PA (1993a) Feeding ecology and behaviour of the genus *Saguinus*. In AB Rylands (ed.): Marmosets and Tamarins: Systematics, Ecology and Behaviour. Oxford: Oxford University Press, pp. 273–295.

Garber PA (1993b) Seasonal patterns of diet and ranging in two species of tamarin monkeys: Stability versus variability. Int. J. Primatol. 14:1–22.

Garber PA (1997) One for all and breeding for one: Cooperation and competition as a tamarin reproductive strategy. Evol. Anthropol. N:135–147.

Garber PA, Pruetz JD, and Issacson J (1993) Patterns of range use, range defense, and intergroup spacing in moustached tamarin monkeys (*Saguinus mystax*). Primates 34:1–22.

Garber PA, and Pruetz JD (1995) Positional behavior in moustached tamarin monkeys: effects of habitat on locomotor variability and locomotor stability. J. Human Evol. 28:411–426.

Garber PA, and Davis LC (1996) Intraspecific variability in anatomy and positional behavior in two tamarin species. Am. J. Phys. Anthropol. Supplement 22:110–111 (abstract).

Gebo DL, and Chapman CA (1995a) Habitat, annual, and seasonal effects on positional behavior in red colobus monkeys. Am. J. Phys. Anthropol. 96:73–82.

Gebo DL, and Chapman CA (1995b) Positional behavior in five sympatric Old World monkeys. Am. J. Phys. Anthropol. 97:49–76.

Grand TI (1984) Motion economy within the canopy: Four strategies for mobility. In PS Rodman and JGH Cant (eds.): Adaptations for Foraging in Nonhuman Primates. New York: Columbia University Press, pp. 54–72.

Halle F, Oldemann RA, and Tomlinson PB (1978) Tropical Trees and Forests: An Architectural Analysis. New York: Springer-Verlag.

Heymann EW (1996) Social behavior of wild moustached tamarins, *Saguinus mystax*, at the Estación Biológica Quebrada Blanco, Peruvian Amazonia. Am. J. Primatol. *38*:101–113.

Horn HS (1974) The Adaptive Geometry of Trees. New Jersey: Princeton University Press.

Hunt KD (1992) Social rank and body size as determinants of positional behavior in *Pan troglodytes*. Primates *33*:347–357.

Nickle DA, and Heymann, EW (1996) Predation on Orthoptera and other orders of insects by tamarin monkeys, *Saguinus mystax mystax* and *Saguinus fuscicollis nigrifrons* (Primates: Callitrichidae), in north-eastern Peru. J. Zool. Lond. *239*:799–819.

Norconk MA (1986) Interactions between species in a Neotropical forest: mixed-species troops of *Saguinus mystax* and *Saguinus fuscicollis* (Callitrichidae). Ph.D. Dissertation, University of California, Los Angeles.

Peres CA (1991) Ecology of mixed-species groups of tamarins in Amazonian terra firme forests. Ph.D. Dissertation, University of Cambridge.

Peres, CA (1992) Prey-capture benefits in a mixed-species group of Amazonian tamarins, *Saguinus fuscicollis* and *S. mystax*. Behav. Ecol. Sociobiol. *31*: 339–347.

Pound J (1991) Habitat structure and morphological patterns in arboreal vertebrates. In S Bell, E McCoy, and H Mushinsky (eds.): Habitat Structure: The Physical Arrangement of Objects in Space. New York: Chapman and Hall, pp. 109–119.

Prost J (1965) A definitional system for the classification of primate locomotion. Am. Anthropol. *67*:1198–1214.

Remis M (1995) Effects of body size and social context on the arboreal activities of lowland gorillas in the Central African Republic. Am. J. Phys. Anthropol. *97*:413–434.

Ricklefs RE, and Miles DB (1994) Ecological and evolutionary inferences from morphology: An ecological perspective. In PC Wainwright and SM Reilly (eds.): Ecological Morphology: Integrative Organismal Biology. Chicago: University of Chicago Press, pp. 13–41.

Richard AF (1978) Behavioral Variation: Case Study of a Malagasy Lemur. Lewisburg, PA: Bucknell University Press.

Steudel K (1990) The work and energetic cost of locomotion I. The effects of limb mass distribution in quadrupeds. J. exp. Biol. *154*:273–285.

Steudel K (1996) Limb morphology, bipedal gait, and the energetics of hominid locomotion. Am. J. Phys. Anthropol. *99*:345–356.

Travis J (1994) Evaluating the adaptive role of morphological plasticity. In PC Wainwright and SM Reilly (eds.): Ecological Morphology: Integrative Organismal Biology. Chicago: University of Chicago Press, pp. 99–122.

Tuttle RH, and Watts DP (1985) The positional behavior of adaptive complexes in *Pan gorilla*. In S Kondo (ed.): Primate Morphophysiology, Locomotor Analyses and Human Bipedalism. Tokyo: University of Tokyo Press, pp. 261–288.

5

LOCOMOTION, SUPPORT USE, MAINTENANCE ACTIVITIES, AND HABITAT STRUCTURE

The Case of the Tai Forest Cercopithecids

W. Scott McGraw

Department of Anatomy
New York College of Osteopathic Medicine
New York Institute of Technology
Old Westbury, New York 11568

1. INTRODUCTION

Information on support use is routinely collected in locomotor field studies to establish predictive relationships between substrates and locomotor behavior (e.g., Fleagle et al., 1981). Nevertheless, the extent that the expression of specific locomotor activities (e.g., climbing, leaping, quadrupedalism) depends on particular support types remains unclear. While it is doubtful that we will be able to satisfactorily address certain adaptive questions as they pertain to habitat (e.g., in what structural context did brachiation evolve?), it is at least possible to examine how locomotor activities vary in architecturally dissimilar contexts. Such investigations could aid in explaining how the architectural properties of arboreal habitats (i.e., support characteristics) are responsible for promoting and/or limiting specific positional behaviors.

The data necessary for this type of investigation are scarce, stemming largely from the difficulty in accurately quantifying those habitat features relevant to locomotion, viz., the relative abundance, size and inclination of different support types. Indeed, analyzing the structural environment as it pertains to even a single locomotor behavior (e.g., climbing) is a complex, multivariate problem (Cartmill, 1985). Of course, most primates have multiple behaviors in their locomotor repertoires. As Stern and Oxnard (1973:1) observe, "primates stand, sit, lie, walk, run, hop, leap, climb, hang, swing, swim, and engage in other activities too numerous to mention. They may do these things often or rarely, quickly or slowly, with agility or clumsiness, on the ground or in the trees (or, with swimming, in the water), on thick branches or thin ones, on vertical, oblique or horizontal supports, with all appendages or only some." Thus, establishing a predictive relationship

Primate Locomotion, edited by Strasser *et al.*
Plenum Press, New York, 1998

between any of the above activities and some habitat characteristic would be difficult enough if habitat were a constant variable, yet many species have wide distributions throughout structurally dissimilar habitats (e.g., Gartlan and Brain, 1968; Lawes, 1990) and even single populations commonly range throughout architecturally heterogeneous forests (e.g., McGraw, 1994; Moreno-Black and Maples, 1977).

The fact that a single species may be found in structurally different environments could have profound implications for interpreting locomotor data, particularly if the frequencies of locomotor activities are significantly affected by the habitats in which they occur. If locomotor behavior reflects proximate responses to the environment rather than evolved or morphology driven tendencies (Pounds, 1991), then establishing relationships between anatomy and behavior will be made more difficult unless behavior is observed in each of an animal's possible habitats.

Recognizing this potential source of variation, a number of workers have examined how locomotor behavior is influenced by differences in arboreal architecture (Clennon and Gebo, 1996; Dagosto, 1992; Doran and Hunt, 1994; Fleagle and Mittermeier, 1980; Garber and Pruetz, 1995; Gebo and Chapman, 1995; Kinzey, 1976; McGraw, 1996a; Orndorff, 1996). The results are not conclusive. In the study by Garber and Pruetz (1995) populations of *Saguinus mystax* exhibited virtually identical locomotor profiles in two structurally distinct Peruvian forest types. These authors found that while locomotor profiles changed little between habitats, "patterns of support preference were more variable...and appeared to reflect site-specific differences in forest architecture (1995:411)." In another study, Gebo and Chapman (1995) examined the positional behavior of *Colobus badius* inhabiting three regions of Uganda's Kibale Forest with different logging histories. These authors observed a number of significant differences in the frequencies that quadrupedalism, climbing, and leaping occurred in primary, secondary and pine forest and they related this variability to the size and abundance of gaps between adjacent trees.

Fleagle and Mittermeier (1980) examined the locomotion of seven monkeys ranging throughout edge, liana, mountain savanna and high forest within the Raleighvallen-Voltzberg Nature Reserve in Central Surinam. Although locomotor variation between forest types *per se* was not their primary focus (nor were the structural properties of the habitats quantified), these authors found that with the exception of leaping in liana forest, "correlations between other locomotor patterns and utilization of forest types are very low..."(1980:311). While these data do not necessarily imply locomotor changes between habitats, they do demonstrate the conservative nature of locomotion in a diverse group of monkeys.

Kinzey (1976) studied a group of titi monkeys (*Callicebus torquatus torquatus*) ranging throughout a number of Peruvian vegetation zones. The structural properties of these zones were not quantified; however, Kinzey (1976:468) described each zone in detail and concluded that, "patterns of locomotion and posture differed among different vegetations." In particular, elevated frequencies of vertical leaping were observed primarily in forest areas characterized by short, densely packed trees with few palms and lianas.

In a more recent study, Doran and Hunt (1994) compared the locomotion of one chimpanzee population at two sites (Gombe Stream and Mahale Mountains, Tanzania), two chimpanzee subspecies at three sites (Tai Forest, Gombe and Mahale) and two chimpanzee species at four sites (Tai, Gombe, Mahale and Zaire's Lomako Forest). Doran and Hunt demonstrated, among other things, that when the degree of arboreality is accounted for, no significant differences in locomotion were observed between the Tai and the Tanzanian chimpanzee subspecies. These findings are remarkable considering the differences between habitats: Gombe is characterized as a thicket/woodland or semi-deciduous forest, the Mahale Mountain site is a closed forest or woodland (Collins and McGrew, 1988) while the Tai site consists of lowland, evergreen forest (Boesch and Boesch, 1983). Although the structural properties of the sites were not systematically quantified, there is little doubt that differences exist. Indeed, differences in the "openness" of the East and West African forest canopies have been used to explain variation in other adaptations, most notably in the degree and manner that monkeys are hunted by Tai and Gombe chimpanzees respectively (Boesch, 1994).

This author recently presented data on the overall locomotor profiles (combined traveling and foraging) of five cercopithecid monkeys ranging in the Ivory Coast's Tai Forest (McGraw, 1996b). Despite significant variation in the relative number of boughs, branches and twigs at the same heights in two forest types, the locomotor profile of each species did not change with habitat. Further, in four of the five species, locomotor consistency was maintained due to monkeys selecting the same set of supports in each of two forest types.

The evidence for whether primates alter their locomotion in response to the structural characteristics of different forest types is, therefore, equivocal. Data from both the Neotropics and the Paleotropics suggest that New and Old World primates can exhibit conservative locomotor behavior even if their associated support use changes (Doran and Hunt, 1994; Garber and Pruetz, 1995). On the other hand, the opposite case can be made in light of the findings of Kinzey (1976) and Gebo and Chapman (1995), which imply that, in some instances, locomotor frequencies do change in response to habitat heterogeneity. Clearly, further analysis is warranted.

The first study on monkey locomotion in the Tai Forest (McGraw, 1996b) examined *overall* locomotor behavior; the locomotor profiles of five cercopithecids were considered without controlling for maintenance activities. Differences that could exist between locomotion during *traveling* compared to locomotion during *feeding* (foraging) may, therefore, have been obscured. Various authors have shown that primates not only change their locomotor behavior during traveling and foraging respectively, but that they also choose different support types during these activities (Cant, 1987; Doran, 1993; Fleagle and Mittermeier, 1980; Gebo, 1992; Hunt, 1992; Rose, 1977). If true, then these behavioral adjustments should be accounted for. This paper expands the original analysis by treating locomotion and support use during traveling separately from locomotion and support use during foraging across two forest types. Results are compared with those from the previous Tai study as well as similar analyses on other primates. Finally, the implications of these data and the problems associated with collecting information on the availability of supports are discussed.

2. METHODS

The methods in this study, summarized below, are described in detail in McGraw (1996b). Data were collected in the Ivory Coast's Tai Forest, approximately twenty-five kilometers from the Liberian border. The core study area, divided into 100 × 100 m cells, covers approximately 2 and 1/2 km² of moist, evergreen rain forest. Within this area, two forest subtypes are recognized: disturbed and undisturbed forest. Disturbance refers to any phenomenon, man-made or natural, which disrupts the vertical profile of the canopy. These phenomena, which include tree falls, may begin as "holes" in the forest extending through all canopy layers, which later become filled with thicker, colonizing, secondary growth (Brokaw, 1982). In the Tai Forest, canopy gaps such as these are important elements of the habitat profile because, as Jans et al. (1993:258) report, "forest turnover time is long (*ca* 240 years) compared to other tropical moist forests, resulting in a less dynamic, fine textured mosaic consisting of small eco-units." In some areas, disturbance is also ascribed to two major trails and an abandoned road that border the core study area. In addition, there is evidence that a few trees were felled within the study grid as recently as ten years ago.

Three cercopithecid species are discussed in this analysis: *Colobus polykomos* (western black and white colobus), *Colobus badius* (red colobus) and *Cercopithecus diana* (diana monkey). Behavioral data were taken on adult females at three minute time points and the same individual was not sampled within 15 minutes of itself to assure independence of data points (see Dagosto, 1994; Mendel, 1976; Walker, 1993). Adult females were chosen because there are more females in each species' social group; the potential sampling pool is thus greatly increased because the observer has more independent data points to chose from (see Janson, 1984, 1990). Every three minutes, the locomotor activity, maintenance activity (traveling or foraging), support type, height in forest (estimated in m) and location (grid cell) of an adult female from a given species was recorded (see McGraw, 1996b). Five locomotor activities were recognized: quadrupedal walking, quadrupedal running, leaping, climbing, and arm swinging (after Fleagle, 1977; Figures 1–3). Traveling is defined as directed, usually uninterrupted movement between major food sources and/or sleeping trees. Foraging is locomotion during feeding usually, though not always, confined to single or contiguous trees. Supports were classified as one of three types: boughs (large supports usually greater than 10 cm in diameter in which grasping with hands or feet is not possible), branches (medium sized supports between 2 and 10 cm in diameter permit-

Figure 1. *Colobus badius* (red colobus monkey) leaping between twigs.

Figure 2. *Colobus polykomos* (western black and white colobus monkey) running on a bough.

ting grasping by hands and feet) and twigs (small flexible terminal branches usually less than 2 cm in diameter) (after Fleagle, 1976 and Fleagle and Mittermeier, 1980). To determine if intraspecific locomotor differences exist between the two forest types, G-tests (R x C contingency tests) with William's correction were performed on raw data (Sokal and Rohlf, 1981).

A canopy survey was conducted to quantify the relative abundance of different sized substrates at different heights in each forest type. Each 100 × 100 m grid cell sampled within the core study area was designated as either undisturbed, if the area showed no visible signs of structural disruption at any height, or disturbed, if tree falls or human activity were sufficient to influence the architecture[*] of the forest. Using a calibrated spotting scope and rangefinder, the vertical profile of the canopy was sampled at six points within each grid cell. Sampling involved counting the number of different sized supports around a focal support (within a three m sampling field) at descending ten m intervals. In total, 36,000 square m of forest were sampled. The number of supports in each forest type was compared using a three-factor (support type, height, forest type) log linear model (Sokal and Rohlf, 1981).

To determine if monkeys chose the same general supports in each of the two forest types, the interaction of three variables (chosen support type, height, forest type) was examined simultaneously. A significant result from the interaction of these variables indicates that a monkey used different supports in each forest type. A non-significant result

---

[*] Architecture refers to the forms of stems and their derivatives (e.g., branches) as opposed to canopy physiognomy, canopy organization and canopy texture (see Parker, 1995).

Figure 3. *Cercopithecus diana* (diana monkey) climbing a trunk.

Table 1. Distribution of support types (twig, branch, bough) in undisturbed and disturbed forests[1]

Height interval	Twig Undist.	Twig Dist.	Branch Undist.	Branch Dist.	Bough Undist.	Bough Dist.	Total Undist.	Total Dist.
>40	85	48	29	7	–	–	114	55
31-40	429	314	200	156	58	13	687	483
21-30	582	748	264	427	127	101	973	1276
11-20	626	1526	295	503	108	35	1029	2064
0-10	607	1031	374	827	26	13	1007	1871
Total	2329	3667	1162	1920	319	162	3810	5749

[1]Support type data are summed frequencies at 10 m height intervals. A three factor (forest type, height interval, support type) log-linear model of support differences analysis is significant: Interaction, G(Williams) = 70.328 [Critical χ^2 (0.05) = 15.5].

from the three-way interaction does not necessarily indicate similar support use and an additional test of *conditional independence* is required, i.e., does use of support types (factor A) at different heights (factor B) differ given two forest types (factor C)?

3. RESULTS

Table 1 summarizes the results of the canopy survey. Two general points characterize both forest types: twigs are the most abundant support type at every level and boughs are (by far) the least common. There are important differences, however, between the forest types. Disturbed forest is generally denser at lower levels than undisturbed forest. Disturbed forest contains more (1) twigs, (2) branches, and (3) total stems, both overall and within each 10 m interval. This forest type has fewer boughs, particularly at lower levels. For example, between 11 and 20 m, disturbed forest has 2/3 fewer boughs as undisturbed forest despite having twice as many total supports in the same 10 m interval. These differences, as revealed by the Williams-corrected G value (70.328), are highly significant.

Table 2 details each species locomotor profile in both forest types. In general, the locomotor behavior of these monkeys conforms to the patterns seen among other primates (e.g., Cant, 1988; Fleagle and Mittermeier, 1980; Gebo and Chapman, 1995): leaping and quadrupedal running are more common during traveling, and climbing is more common during foraging. The major exception is leaping in *Colobus polykomos*; although the black and white colobus leaped less often during traveling than foraging in undisturbed forest, it leaped more often during traveling than foraging in disturbed forest.

Table 2. Locomotor behavior during traveling and foraging in undisturbed and disturbed forests[1]

Species	Locomotor behavior	Undisturbed forest		Disturbed forest	
		Traveling	Foraging	Traveling	Foraging
Colobus polykomos[2]	Arm swing	–	–	–	–
	Climb	7.9	21.9	6.9	23.6
	Leap	7.9	12.5	16.5	13
	Run	41.3	15.6	40	14.4
	Walk	42.9	50	36.6	49
Colobus badius[3]	Arm swing	3.7	6.2	2.5	4.2
	Climb	12	24.5	12.3	19.8
	Leap	19.4	14	21.6	15.6
	Run	8.7	3.5	9.7	5.9
	Walk	56.2	51.8	53.9	54.5
Cercopithecus diana[4]	Arm swing	–	–	–	0.1
	Climb	2.8	20.5	9.1	21.6
	Leap	15.7	8.6	17.4	10
	Run	27.8	8	20.5	7.8
	Walk	53.7	62.9	53	60.5

[1]For each species, the values in the traveling and foraging columns are the percentage of locomotor time spent engaged in the five locomotor behaviors.
[2]There is no significant difference in locomotion between the forest types during traveling (G[Williams]= 3.7) and during foraging (G[Williams]=0.08). For both activities the critical χ^2 value is 7.8.
[3]There is no significant difference in locomotion between the forest types during traveling (G[Williams] = 1.7) and during foraging (G[Williams]= 5.2). For both activities the critical χ^2 value is 9.5.
[4]There is no significant difference in locomotion between the forest types during traveling (G[Williams] = 5.5) and during foraging (G[Williams]= 1.2). For both activities the critical χ^2 value is 7.8.

Table 3. Support use while traveling and foraging in undisturbed and disturbed forests by *Colobus polykomos*

Height interval[1]	Support Type[2]	Undisturbed forest		Disturbed forest	
		Traveling[3]	Foraging[4]	Traveling[3]	Foraging[4]
31-40 m	Twig	15.8	27.3	17.6	30.4
	Branch	26.3	27.3	11.8	27.7
	Bough	57.9	45.4	70.6	41.9
21-30 m	Twig	20	33.3	21.2	28.7
	Branch	28	33.3	26.4	37.3
	Bough	52	33.3	52.4	34
11-20 m	Twig	21	38.5	39.4	46.4
	Branch	31.6	46.2	31.7	41.1
	Bough	47.4	15.3	28.9	12.5

[1]For each height interval, the values in the traveling and foraging columns are the percentage of locomotor time spent on each support type. Supports used at heights greater than 40 m and between 0 and 10 m were not compared across forest types due to insufficient sample sizes.
[2]The test for conditional independence during travel was insignificant ($\chi^2=5.8$) as was the test for conditional independence during foraging ($\chi^2=0.43$). The critical χ^2 value is 12.6.
[3]A three factor (forest type, height interval, support type) log-linear model of support differences analysis is insignificant: Interaction, G(Williams) = 4.3 [Critical χ^2 (0.05) = 9.5].
[4]A three factor (forest type, height interval, support type) log-linear model of support differences analysis is insignificant: Interaction, G(Williams) = 0.26 [Critical χ^2 (0.05) = 9.5].

The critical question is whether frequencies of locomotion during either traveling or foraging change across habitat types. In every instance, the test of inter-forest locomotion indicates that treating these behaviors separately reveals little difference; for no species is there a significant change in the locomotor profile during either traveling or foraging when changing forest types. This does not necessarily imply that *individual* locomotor activities are always constant across habitats. Two cases stand out, both involving movement during traveling: *C. polykomos* leaped over twice as often in disturbed forest compared to undisturbed forest and *C. diana* climbed over three times as much in disturbed forest as in undisturbed forest. Nevertheless, when locomotor behaviors are considered together (i.e., locomotor profile), monkeys did not move in significantly different ways.

Tables 3–5 show the frequencies with which each support type in each height interval is used by each species across the two forest types during traveling and during foraging. Results of both the 3-way interaction and the test of conditional independence are also indicated. Treating support use during traveling and foraging separately does not reveal a consistent trend, unlike the pattern observed when overall locomotion is examined (McGraw, 1996b). The analysis reveals that *Colobus polykomos* is the only species that uses the same supports in both forest types—regardless of the activity; both the 3-way interaction and the test for conditional independence of support use during traveling and foraging yielded non-significant results. Despite an overall lack of significance, two cases in which there was a sizable difference in the frequency with which a support was used (both involving traveling) deserve mention. Between 31 and 40 m, *C. polykomos* used 12.7% fewer boughs and 14.5% more branches in undisturbed forest than in disturbed forest. In addition, between 11 and 20 m, *C. polykomos* used 18.4% fewer twigs, but 18.5% more boughs in undisturbed compared to disturbed forest.

For *Colobus badius* (Table 4), analysis reveals that support use during traveling does not differ between forest types: the largest observed disparity was a 10% difference in the frequency that twigs were used in undisturbed forest (14.9%) compared to disturbed

Table 4. Support use while traveling and foraging in undisturbed and disturbed forests by *Colobus badius*

Height interval[1]	Support type[2]	Undisturbed forest		Disturbed forest	
		Traveling[3]	Foraging[4]	Traveling[3]	Foraging[4]
31-40 m	Twig	14.9	44.9	24.9	25.2
	Branch	16.8	31.9	17.9	30.3
	Bough	68.3	23.2	57.2	44.5
21-30 m	Twig	24.8	39.3	27.6	35.8
	Branch	26.3	33.3	24.6	35.8
	Bough	48.9	27.4	47.8	28.4
11-20 m	Twig	37.8	46.9	34.1	50.0
	Branch	31.1	32.8	34.1	32.4
	Bough	31.1	20.3	31.8	17.6

[1] For each height interval, the values in the traveling and foraging columns are the percentage of locomotor time spent on each support type. Supports used at heights greater than 40 m and between 0 and 10 m were not compared across forest types due to insufficient sample sizes.
[2] The test for conditional independence during travel was insignificant ($\chi^2=8.7$) although the test for conditional independence during foraging was significant at the 0.05 significance level ($\chi^2=16.1$). The critical χ^2 values are 12.6 for traveling and 15.6 for foraging.
[3] A three factor (forest type, height interval, support type) log-linear model of support differences analysis is insignificant: Interaction, G(Williams) = 4.3 [Critical χ^2 (0.05) = 9.5].
[4] A three factor (forest type, height interval, support type) log-linear model of support differences analysis is insignificant: Interaction, G(Williams) = 0.26 [Critical χ^2 (0.05) = 9.5].

(24.9%). On the other hand, the significant result from the test of conditional independence indicates that during foraging, this monkey chooses different supports in the two forest types. In particular, between 31 and 40 m, *C. badius* uses 20% more twigs and 21% fewer boughs in undisturbed compared to disturbed forest.

The diana monkey alters its support use the most (Table 5). Significant results from the tests of conditional independence reveal that *C. diana* chooses different supports during both traveling and foraging in both forest types. In many instances, the disparity in use of a support type at a particular height interval between forest types is substantial and highly significant (Table 5). For example, when diana monkeys travel between 21 and 30 m, they use almost 30% more boughs in undisturbed forest than in disturbed forest. This difference is magnified even further during foraging: between 21 and 30 m, 43% more boughs were used in undisturbed forest compared to disturbed forest.

4. DISCUSSION

The results of this analysis, together with those from the first study on Tai monkey locomotion (McGraw, 1996b) indicate that the locomotion of three cercopithecid monkeys, whether examined overall (combined traveling and foraging) or while controlling for maintenance behaviors (traveling versus foraging) does not change in different forest types. Although individual locomotor activities may vary, the combined locomotor profile of each monkey is statistically similar in each forest type. As such, these findings support the general conclusions of Doran and Hunt (1994), Fleagle and Mittermeier (1980) and Garber and Pruetz (1995); namely, that locomotor behavior is conservative with respect to the habitat in which it occurs.

Table 5. Support using while traveling and foraging in undisturbed and disturbed
forests by *Cercopithecus diana*

Height interval[1]	Support type[2]	Undisturbed forest		Disturbed forest	
		Traveling[3]	Foraging[4]	Traveling[3]	Foraging[4]
31-40 m	Twig	23.5	37.4	43.4	40.8
	Branch	29.5	6.3	29.2	34.8
	Bough	47	56.3	27.4	24.4
21-30 m	Twig	20.4	20.5	37.8	45.3
	Branch	26.6	13.6	38.8	31.5
	Bough	53	65.9	23.4	23.2
11-20 m	Twig	37.5	39.7	48.1	47.5
	Branch	22.5	36.5	37.3	42.8
	Bough	40	23.8	14.6	9.7

[1]For each height interval, the values in the traveling and foraging columns are the percentage of locomotor
time spent on each support type. Supports used at heights greater than 40 m and between 0 and 10 m were
not compared across forest types due to insufficient sample sizes.
[2]The test for conditional independence during travel was significant at the 0.05 significance level (χ^2=31.9)
as was the test for conditional independence during foraging (χ^2=45.4). The critical χ^2 value for both activities is e 12.6.
[3]A three factor (forest type, height interval, support type) log-linear model of support differences analysis is
insignificant: Interaction, G(Williams) = 1.9 [Critical χ^2 (0.05) = 9.5].
[4]A three factor (forest type, height interval, support type) log-linear model of support differences analysis is
insignificant: Interaction, G(Williams) = 5.3 [Critical χ^2 (0.05) = 9.5].

The data also indicate that patterns of support choice (i.e., selective use and avoidance of specific substrate types) during traveling and foraging in different forest types are
more varied and complex. When support use during traveling and foraging are examined
individually, each of the three monkey species responded in a different manner across the
two forest types. Although no monkey changed its locomotion during either maintenance
activity, one (*Cercopithecus diana*) used different supports during both activities, one
(*Colobus badius*) used different supports during foraging but not during traveling and one
(*Colobus polykomos*) used the same supports during both maintenance activities in both
forest types. This contrasts with the author's earlier study (McGraw, 1996b) in which it
was shown that locomotor consistency is maintained across structurally distinct habitats
because all monkeys select the same types of supports despite differences in their relative
availabilities. Rather, these data demonstrate that, depending on whether a monkey is finding food within a tree or traveling greater distances between feeding or sleeping sites, the
supports available can affect which supports a monkey chooses.

Although the importance of controlling for different maintenance activities is clear
(Cant, 1987; Doran, 1993; Fleagle and Mittermeier, 1980; Gebo, 1992; Gebo and Chapman, 1995; Hunt, 1992; Rose, 1977), the more interesting issue is determining why some
monkeys change their support use when others do not. For example, why might searching
for food cause a change in support use between forests?

The fact that two (*Colobus badius* and *Cercopithecus diana*) of the three species
chose different supports during foraging (Tables 4 and 5) leads one to suspect that inter-
habitat differences in the location of preferred food items was likely responsible for this
variation. While these data, per se, are not yet available, the behavior of *Colobus badius*
and *Cercopithecus diana* in the Tai Forest is now becoming more widely known (Holenweg et al., 1996; Noe and Bshary, 1997; Wachter et al., 1997). If the spatial arrangement
of food items differed between habitats, then focusing on the relationship between inter-

forest phenology and the supports used while moving within feeding patches could explain why these species (in contrast to *Colobus polykomos*) changed support types during foraging in the two forest types. In addition, attention should be given to the properties and locations of the supports containing preferred food items, not merely those that are used to get there. For example, it would be extremely useful to know how often a monkey feeds on an item attached above, below or next to the support(s) bearing the animal's weight. What criteria are used by monkeys to get into the best foraging positions? As Fleagle (1984:109) noted, "Describing and quantifying the location and abundance of primate food resources in the habitat by other then retrospective observations remains one of the major gaps in all studies of primate foraging strategies." Any compelling explanation of how monkeys manage to maintain consistent locomotor profiles while changing their support use during foraging (e.g., *Colobus badius* and *Cercopithecus diana*) should contain this information.

Why is support use during traveling less variable between forest types than support use during foraging? Two of the three species (*Colobus polykomos* and *C. badius*) chose the same substrate types for traveling in both forest types (Tables 3 and 4). This supports the idea (as others have argued before) that primates are generally conservative in their travel pathways and that routes chosen for uninterrupted movement are fairly constant and oft-frequented. Indeed, travel paths may be chosen because they provide the safest, most direct route on preferred supports (Cant, 1992). *Colobus polykomos* and *C. badius* are both large monkeys. Because large monkeys are more likely to prefer relatively larger supports (Fleagle and Mittermeier, 1980), the number of preferred supports are limited since the largest supports are always the least abundant (Table 1). These facts may, therefore, require that large monkeys select the same supports even if different forests vary in the number of each support type. This appears to be the case for *Colobus polykomos* and *C. badius*. Smaller monkeys, because of the branch to body size ratio, are able to use a wider variety of available supports and, moreover, to change their support use in response to more abundant substrates. Thus, *Cercopithecus diana* (a small monkey) is able to travel on different supports in different forest types while still using the same locomotor behaviors.

Taken together, these results certainly argue for the conservative nature of locomotion, but they also emphasize what we still do not know: Are animals that do not change their support use (e.g., *Colobus polykomos*) more "constrained" to chose a fixed support profile (e.g., Hughes et al., 1995)? If so, what are the constraining factors? What mechanisms allow a primate such as the diana monkey to change the supports it moves on, yet still maintain a consistent locomotor profile? What is the appropriate time frame (e.g., daily, weekly, seasonally) over which to sample locomotor change? These questions are the basis for a wide range of captive and field studies, all of which can contribute to the larger goal of understanding the causes of locomotor variability in order to better establish predictive relationships.

The most obvious start is to gather additional information on how other primates respond to differences in habitat. Before we can determine where the balance of evidence lies, far more field data are necessary for inter- and intra-specific comparisons. This is particularly important when one considers that the authors of no two studies discussed in this paper have quantified the habitat in the same way nor controlled for the same behavioral or habitat variables. For example, is trunk density (Garber and Pruetz, 1995), distance between canopy gaps (Cannon and Leigton, 1994a; Gebo and Chapman, 1995), or the number of different-sized supports in each canopy level (McGraw, 1996b) the most appropriate habitat variable to consider when studying the locomotion of arboreal animals? How do the hypotheses we test and the animals we study (e.g., galagos versus gorillas) dictate the architectural characteristics we focus on (see Emmons, 1995)?

Clearly, identifying the critical structural features is only one of the important tasks facing primatologists. Accurately quantifying these properties will undoubtedly be even more challenging. Support size (Fleagle and Mittermeier, 1980; Garber, 1980; Gebo, 1992), inclination (Cant, 1986; Fontaine, 1990; Gebo and Chapman, 1995; Rose, 1978), compliance (Crompton et al., 1993, Demes et al., 1995), density (Cannon and Leighton, 1994b; Garber and Pruetz, 1995; Gebo and Chapman, 1995), and spatial patterning (Kinzey, 1976; Mendel, 1976; Morbeck, 1977; Ripley, 1967) all likely play important roles in support choice and associated locomotor behaviors. This and other studies have reported information on overall support density, but quantifying the inclination and compliance of supports *in addition to those used during locomotion* (i.e., those that were avoided) requires far more rigorous methods. Such data are vital, because as Rose (1978:256–257) noted, "the degree to which particular branches were used must depend to a certain extent on the frequency with which branches of different types occur within the habitat." This requires placing data on used-supports within the broader statistical context of those supports that were ignored. For example, in an exhaustive survey of locomotion and support use in *Saimiri* and *Ateles*, Fontaine (1990:503) concluded that, "although size constraints might lead one to expect major contrasts between the two species in their use of supports, observed use rate differences for specific support types were minor relative to the strong similarities. Both *Saimiri* and *Ateles* appeared to make full use of available supports." While this may be true, to convincingly demonstrate support preference (or to assess the selective use of **any** habitat feature), it is necessary to have an independent assessment of support availability and the properties of those supports that were not selected. This kind of structural analysis will therefore require a more comprehensive research protocol.

Various methods are currently available to measure such diverse architectural variables as canopy structure (Cannon and Leighton, 1994b; Halle et al., 1978; Parker, 1993; Schaik and Mirmanto, 1985), spatial heterogeneity of tree types (Chazdon, 1996; Condit, 1996; Hart et al., 1989; Lemos and Strier, 1992; Lieberman et al., 1985; Uhl and Murphy, 1981), tree-fall gap dynamics (Brokaw, 1985; Jans et al., 1993; Martinez-Ramos et al., 1988), and vertical stratification of foliage and supports (Beadle et al., 1982; Ford and Newbold, 1971; Hubbell and Foster, 1986; Hutchinson et al., 1986; Malcolm, 1995; Parker et al., 1989, 1992). Future studies on locomotion and habitat structure will hopefully integrate these and/or other data to better understand the determinants of variability in positional behavior.

Finally, the solutions to some problems, perhaps best pursued under captive situations, will be immensely valuable. For example, what are the critical limits of support inclination and flexibility that constrain support use for animals of different sizes? When does quadrupedalism on oblique supports become climbing? How does this gradation of behaviors affect the quantification of what we see in the field? What effect does differential limb mobility have on support choice by generalized arboreal quadrupeds of varying body sizes? Are the volar pads of some cercopithecids more or less suited to deal with the frictional properties of different kinds of supports? If yes, what implications does this have for locomotion in architecturally heterogeneous habitats?

The majority of evidence discussed in this paper indicates that for some species at least, locomotion is conservative relative to the habitat in which it occurs. On the other hand, all studies demonstrate that patterns of support use can, in some cases, reflect site-specific differences. Our ability to develop predictive relationships between behavior, morphology and habitat therefore depends on the extent that we understand those factors that cause locomotion to vary as well as the processes that constrain it.

5. SUMMARY AND CONCLUSIONS

John Napier (1967:333) remarked that "locomotor adaptations have provided the principle milestones along the evolutionary pathway of primates from the Eocene to the Pleistocene." This statement echoed earlier arguments of Wood Jones (1916) and Le Gros Clark (1934), that differences in fossil and living primate post-crania could be explained largely by differential adaptations to arboreal living. To this end, the relationship between forest architecture and positional behavior has figured prominently in the evolutionary scenarios offered to explain the divergence of many primate taxa (e.g., Andrews and Aiello, 1984; Avis, 1962; Cartmill, 1985; Grand, 1972, 1984; Napier, 1967; Napier and Walker, 1967; Ripley, 1967; Rose, 1984; Stern and Oxnard, 1973; Temerin and Cant, 1983). Such scenarios are based unquestionably on a detailed understanding of the morphology and behavior of living primates as well as the habitats in which these animals live. It is therefore curious that there have been relatively few attempts to demonstrate how the locomotion of extant primates varies with or is constrained by elements of forest architecture (but see Cannon and Leighton, 1994a; Doran and Hunt, 1994; Garber and Pruetz, 1995). This information is critical because while morphological studies may show what behaviors are possible, comparative field studies document the range of probable behaviors as well as the contexts in which they occur.

This study has attempted to document the range of behaviors and contexts in which they occur by providing additional evidence on how some primates respond to habitat heterogeneity. The data reveal that although individual locomotor behaviors may vary, the locomotor profiles of three monkey species, during traveling and foraging respectively, do not change in two forest types; no species demonstrated a significant change in its locomotor profile during either maintenance activity. Patterns of support use are more variable, however. Specifically, *Colobus polykomos* did not change its support use during either maintenance activity, *Cercopithecus diana* used different supports in both forest types during both activities and *Colobus badius* used the same supports during traveling, but different supports during foraging in each forest type. Based on these results, the following conclusions can be made:

1. Locomotion is more conservative than support use; i.e., patterns of support use will change before accompanying locomotor behaviors do.
2. Controlling for maintenance activities does not change the fact that locomotor profiles are constant across different habitats: the locomotion of *Colobus polykomos*, *C. badius* and *Cercopithecus diana* does not change in two forest types whether it is examined overall (e.g., combined traveling and foraging) (McGraw, 1996b) or individually (traveling compared to foraging).

Tai Forest monkeys move in the same general manners independent of differences in their respective habitats. This fact should strengthen our ability to make predictive relationships between behavior and morphology. Nevertheless, until more detailed studies can identify the reasons why some monkeys change their accompanying support use while others do not, then establishing predictive relationships between behavior and habitat (e.g., different support types) will prove more difficult.

ACKNOWLEDGMENTS

Support for fieldwork was provided by SUNY at Stony Brook, Sigma Xi, Max Planck Institute Fur Verhaltensphysiologie, and the National Science Foundation. I thank

Ronald Noe and Bettie Sluijter for permission to work in the Tai Forest and the assistants of the Tai Forest Monkey Project for helping collect the botanical data. The comments of Elizabeth Strasser, John Fleagle, Randy Susman, Diane Doran, Brigitte Demes and three anonymous reviewers greatly improved this paper.

REFERENCES

Andrews P, and Aiello L(1984) An evolutionary model for feeding and positional behavior. In DJ Chivers, BA Wood and A Bilsborough (eds.): Food Acquisition and Processing in Primates. New York: Plenum Press, pp. 429–466.

Avis V (1962) Brachiation: The crucial issue for man's ancestry. Southwest. J. Anthropol. *18*:119–148.

Beadle CL, Talbot H, and Jarvis PG (1982) Canopy structure and leaf area index in a mature Scots pine forest. Forestry *55*:105–123.

Boesch C (1994) Hunting strategies of Gombe and Tai Chimpanzees. In RW Wrangham, WC McGrew, FBM de Waal, and P Heltne (eds.): Chimpanzee Cultures. Cambridge: Harvard University Press, pp. 77–91.

Boesch C, and Boesch H (1983) Optimization of nut cracking with natural hammers by wild chimpanzees. Behavior *3/4*:265–286.

Brokaw NVL (1982) The definition of tree fall gap and its effect on measures of dynamics of forests. Biotropica *14*:158–160.

Brokaw NVL (1985) Gap-phase regeneration in a tropical forest. Ecol. *66*:682–687.

Cannon CH, and Leighton M (1994a) Comparative locomotor ecology of gibbons and macaques: Selection of canopy elements for crossing gaps. Am. J. Phys. Anthropol. *93*:505–524.

Cannon CH, and Leighton M (1994b) A field method for making sterile morphological descriptions of tropical trees and lianas using a reliability measure. Trop. Biod. *2*:276–279.

Cant JGH (1986) Locomotion and feeding postures of spider and howling monkeys: Field study and evolutionary implications. Folia Primatol. *46*:1–14.

Cant JGH (1987) Positional behavior of female Bornean orangutans (*Pongo pygmaeus*). Am. J. Primatol. *12*:71–90.

Cant JGH (1988) Positional behavior of long-tailed macaques (*Macaca fascicularis*) in northern Sumatra. Am. J. Phys. Anthropol. *76*:29–37.

Cant JGH (1992) Postural behavior and body size of arboreal primates: A theoretical framework for field studies and an illustration of its application. Am. J. Phys. Anthropol. *88*:273–283.

Cartmill M (1985) Climbing. In M Hildebrand, D Bramble, K Liem, and D Wake (eds.): Functional Vertebrate Morphology. Cambridge: Belknap Press, pp. 73–88.

Chazdon RL (1996) Spatial heterogeneity in tropical forest structure: Canopy palms as landscape mosaics. Tree *11*:8–9.

Clennon JA, and Gebo DL (1996) Positional behavior in *Cebus capucinus*. Am. J. Phys. Anthropol. Supplement *22*:86–87 (abstract).

Collins DA, and McGrew WC (1988) Habitats of three groups of chimpanzees (*Pan troglodytes*) in western Tanzania compared. J. Hum. Evol. *17*:553–74.

Condit R (1996) Defining and mapping vegetation types in mega-diverse tropical forests. Tree *11*:4–5.

Crompton RH, Sellers WI, and Gunther MM (1993) Energetic efficiency and ecology as selective factors in the saltatory adaptation of prosimian primates. Proc. R. Soc. Lond. (Biol.) *254*:41–45.

Dagosto M (1992) Effect of habitat structure on positional behavior and substrate use in Malagasy lemurs. Am. J. Phys. Anthropol. Supplement *14*:167 (abstract).

Dagosto M (1994) Testing positional behavior of Malagasy lemurs: a randomization approach. Am. J. Phys. Anthropol. *94*:189–202.

Demes B, Jungers WL, Gross TS, and Fleagle JG (1995) Kinetics of leaping primates: Influence of substrate orientation and compliance. Am. J. Phys. Anthropol. *96*:419–429.

Doran DM (1993) Sex differences in adult chimpanzee positional behavior: The influence of body size on locomotion and posture. Am. J. Phys. Anthropol. *91*:99–116.

Doran DM, and Hunt KD (1994) Comparative locomotor behavior of chimpanzees and bonobos: Species and habitat differences. In RW Wrangham, WC McGrew, FBM de Waal, and P Heltne (eds.): Chimpanzee Cultures. Cambridge: Harvard University Press, pp. 93–108.

Emmons LE (1995) Mammals of rain forest canopies. In MD Lowman and NM Nadkarni (eds.): Forest Canopies. New York: Academic Press, pp. 199–223.

Fleagle JG (1976) Locomotion and posture of the Malayan siamang and implications for hominoid evolution. Folia Primatol. *26*:245–269.

Fleagle JG (1977) Locomotor behavior and skeletal anatomy of sympatric Malaysian leaf-monkeys (*Presbytis obscura* and *Presbytis melalophos*) Yrbk. Phys. Anthropol. *20*:440–453.

Fleagle JG (1984) Primate locomotion and diet. In DJ Chivers, BA Wood, and AL Bilsborough (eds.): Food Acquisition and Processing in Primates. New York: Plenum Press, pp. 105–117.

Fleagle JG, and Mittermeier RA (1980) Locomotor behavior, body size, and comparative ecology of seven Surinam monkeys. Am. J. Phys. Anthropol. *52*:301–314.

Fleagle JG, Mittermeier RA, and Skopec A (1981) Differential habitat use by *Cebus apella* and *Saimiri sciureus* in Central Surinam. Primates *22*:361–367.

Fontaine R (1990) Positional behavior in *Saimiri boliviensis* and *Ateles geoffroyi*. Am. J. Phys. Anthropol. *82*:485–508.

Ford ED, and Newbold PJ (1971) The leaf canopy of a coppiced deciduous woodland. I. Development and structure. J. Ecol. *59*:843–862.

Garber PA (1980) Locomotor behavior and feeding ecology of the Panamanian tamarin (*Saguinus oedipus geoffroyi*, Callitrichidae, Primates). Int. J. Primatol. *1*:185–201.

Garber PA, and Pruetz JD (1995) Positional behavior in moustached tamarin monkeys: Effects of habitat on locomotor variability and locomotor stability. J. Hum. Evol. *28*:411–426.

Gartlan JS, and Brain CK (1968) Ecology and social variability in *Cercopithecus aethiops* and *C. mitis*. In P Jay (ed.): Primates: Studies in Adaptation and Variability. New York: Holt, Rhinehart and Winston, pp. 253–292.

Gebo DL (1992) Locomotor and postural behavior in *Alouatta palliata* and *Cebus capucinus*. Am. J. Primatol. *26*:277–290.

Gebo DL, and Chapman CL (1995) Positional behavior in five sympatric Old World monkeys. Am. J. Phys. Anthropol. *97*:49–76.

Grand TI (1972) A mechanical interpretation of terminal branch feeding. J. Mammal. *53*:198–201.

Grand TI (1984) Motion economy within the canopy: Four strategies for mobility. In P Rodman and JGH Cant (eds.): Adaptations for Foraging in Nonhuman Primates. New York: Columbia University Press, pp. 54–71.

Halle F, Oldeman RAA, and Tomlinson PB (1978) Tropical Trees and Forest: An Architectural Analysis. New York: Springer-Verlag.

Hart TB, Hart JA, and Murphy PG (1989) Monodominant and species-rich forests of the humid tropics: Causes for their co-occurence. Am. Nat. *133*:613–633.

Holenweg A, Noe R, and Schabel M (1996) Waser's gas model applied to associations between red colobus and diana monkeys in the Tai National Park, Ivory Coast. Folia Primatol. *67*:125–136.

Hubbell SP, and Foster RB (1986) Canopy gaps and the dynamics of a Neotropical forest. In MJ Crawley (ed.): Plant Ecology. Oxford: Blackwell, pp. 77–95.

Hughes JJ, Ward D, and Perrin MR (1995) Effects of substrates on foraging decisions by Namib Desert gerbils. J. Mammal *76*:638–645.

Hunt KD (1992) Positional behavior of *Pan troglodytes* in the Mahale Mountains and Gombe Stream National Parks, Tanzania. Am. J. Phys. Anthropol. *87*:83–105.

Hutchinson BA, Matt DR, McMillen RT, Gross LJ, Tajchman SJ and Norman JR (1986) The architecture of deciduous forest canopy in eastern Tennessee, U.S.A. J. Ecol. *74*:635–646.

Jans L, Poorter L, van Rompaey RSAR, and Bongers F (1993) Gaps and forest zones in tropical moist forest in Ivory Coast. Biotropica *25*:258–269.

Janson CH (1984) Female choice and mating systems of the brown capuchin monkey *Cebus apella* (Primates: Cebidae). Z. Tierpsychol. *65*:177–200.

Janson CH (1990) Social correlates of individual spatial choice in foraging groups of brown capuchin monkeys, *Cebus apella*. Anim. Behav. *40*:910–921.

Kinzey WG (1976) Positional behavior and ecology of *Callicebus torquatus*. Ybk. Phys. Anthropol. *20*:468–480.

Lawes MJ (1990) The distribution of the samango monkey (*Cercopithecus mitis erythrarchus* Peters, 1852 and *Cercopithecus mitis labiatus* I. Geoffroy, 1843) and forest history in southern Africa. J. Biogeography *17*:669–680.

Le Gros Clark WE (1934) Early Forerunners of Man. London: Bailliere, Tindall and Cox.

Lemos de Sa R, and Strier KB (1992) A preliminary comparison of forest structure and use by two isolated groups of woolly spider monkeys, *Brachyteles arachnoides*. Biotropica *24*:455–459.

Lieberman M, Lieberman D, Hartshorn GS and Pevalta R (1985) Small-scale altitudinal variation in lowland tropical forest vegetation. J. Ecol. *73*:505–516.

Malcolm JR (1995) Forest structure and the abundance and diversity of neotropical small mammals. In MD Lowman and NM Nadkarni (eds.): Forest Canopies. New York: Academic Press, pp. 179–197.

Martinez-Ramos M, Alvarez-Buylla E, Sarkukhan J, and Pinero D (1988) Treefall age determination and gap dynamics in a tropical forest. J. Ecol. *76*:700–716.

McGraw WS (1994) Census, habitat preference and polyspecific associations of six monkeys in the Lomako Forest, Zaire. Am. J. Primatol. *34*:295–307.

McGraw WS (1996a) The Positional Behavior and Habitat Use of Six Monkeys in the Tai Forest, Ivory Coast. Ph.D. Dissertation. State University of New York at Stony Brook.

McGraw WS (1996b) Cercopithecid locomotion, support use, and support availability in the Tai Forest, Ivory Coast. Am. J. Phys. Anthropol. *100*:507–522.

Mendel F (1976) Postural and locomotor behavior of *Alouatta palliata* on various substrates. Folia Primatol. *26*:36–53.

Morbeck ME (1977) Positional behavior, selective use of habitat substrate and associated non-positional behavior in free-ranging *Colobus guereza* (Ruppel, 1835). Primates *18*:35–58.

Moreno-Black G, and Maples WR (1977) Differential habitat utilization of four Cercopithecidae in a Kenyan forest. Folia Primatol. *27*:85–107.

Napier JR (1967) Evolutionary aspects of primate locomotion. Am J. Phys. Anthropol. *27*:333–341.

Napier JR, and Walker A (1967) Vertical clinging and leaping - a newly recognized category of locomotor behavior of primates. Folia Primatol. *6*:204–219.

Noe R, and Bshary R (1997) The formation of red colobus-diana monkey associations under predation pressure from chimpanzees. Proc. R. Soc. London, B. *264*:253–259.

Orndorff KA (1996) Positional behavior of *Cebus capucinus* in a Costa Rican rainforest: Comparisons with published data collected in a Costa Rican dry forest. Am. J. Phys. Anthropol. supplement *22*:180–181 (abstract).

Parker GG (1993) Structure and dynamics of the outer canopy of a Panamanian dry forest. Selbyana *14*:51.

Parker GG (1995) Structure and microclimate of forest canopies. In MD Lowman and NM Nadkarni (eds.): Forest Canopies. New York: Academic Press, pp. 73–106.

Parker GG, O'Neil JP, and Higman D (1989) Vertical profile and canopy organization in a mixed deciduous forest. Vegetatio *89*:1–12.

Parker GG, Hill SM, and Kuehnel LA (1992) Access to the upper forest canopy with a large tower crane. Bioscience *42*:664–670.

Pounds JA (1991) Habitat structure and morphological patterns in arboreal vertebrates. In S Bell, E McCoy, and H Mushinsky (eds.): Habitat Structure: The Physical Arrangement of Objects in Space. New York: Chapman and Hall, pp. 109–119.

Ripley S (1967) The leaping of langurs: A problem in the study of locomotor adaptation. Am. J. Phys. Anthropol. *26*:149–170.

Rose MD (1977) Positional behavior of olive baboons (*Papio anubis*) and its relationship to maintenance and social activities. Primates *18*:59–116.

Rose MD (1978) Feeding and associated positional behavior of black and white colobus monkeys (*Colobus guereza*). In GG Montgomery (ed.): The Ecology of Arboreal Folivores. Washington DC: Smithsonian Institution, pp. 253–262.

Rose MD (1984) Food acquisition and the evolution of positional behavior: The case of bipedalism. In DJ Chivers, BA Wood, and A Bilsborough (eds.): Food Acquisition and Processing in Primates. New York: Plenum Press, pp. 509–524.

Schaik CP van, and Mirmanto E (1985) Spatial variation in the structure and litterfall of a Sumatran rain forest. Biotropica *17*:196–205.

Sokal RR, and FJ Rholf (1981) Biometry: The Principles and Practice of Statistics in Biological Research. New York: WH Freeman and Company.

Stern JT, and Oxnard CE (1973) Primate locomotion: Some links with evolution and morphology. Primatologia *4*:1–93.

Temerin LA, and Cant JGH (1983) The evolutionary divergence of Old World monkeys and apes. Am. Nat. *122*:335–351.

Uhl C, and Murphy PG (1981) Composition, structure and regeneration of a terra firma forest in the Amazon basin of Venezuela. Trop. Ecol. *22*:219–237.

Wachter B, Schabel M, and Noe R (1997) Diet overlap and poly-specific associations of red colobus and diana monkeys in the Tai National Park, Ivory Coast. Ethology *103*:514–516.

Walker SE (1993) Positional Adaptations and Ecology of the Pitheciini. Ph.D. Dissertation, City University of New York.

Wood Jones F (1916) Arboreal Man. London: Edward Arnold.

6

THE GORILLA PARADOX

The Effects of Body Size and Habitat on the Positional Behavior of Lowland and Mountain Gorillas

Melissa J. Remis

Department of Sociology and Anthropology
Purdue University
West Lafayette, Indiana 47907-1365

1. INTRODUCTION

Gorillas posses general ape adaptations for climbing yet the best studied gorillas in the Virungas depart from the arboreal, frugivorous pattern seen among the other apes. Terrestriality and folivory among gorillas have traditionally been viewed as consequences of their large body size. This study aims to separate the influences of body size from those of habitat on tree climbing by gorillas by comparing the results of 27 months of study of western lowland gorillas (*Gorilla gorilla gorilla*) at Bai Hokou, Central African Republic (Remis, 1994, 1995) to published reports on the positional behavior of mountain gorillas (*Gorilla gorilla beringei*) at Karisoke, Rwanda (Tuttle and Watts, 1985; Doran, 1996).

The first European descriptions of gorillas (*Gorilla gorilla gorilla*) by explorers and scientists in West Africa emphasized their arboreal habits (Purchas, 1625; Savage and Wyman, 1847; DuChaillu, 1861; Gregory and Raven, 1937). Nevertheless, subsequent exploration and scientific study of the terrestrial mountain gorilla (*Gorilla gorilla beringei*) in the Virungas (Akeley, 1923; Bingham, 1932; Kawai and Mizuhara, 1959; Emlen and Schaller, 1960) overshadowed the West African work and the early accounts were dismissed as fanciful exaggerations based on unreliable reports from local people.

Detailed field studies of the terrestrial and herbivorous mountain gorilla (Schaller, 1963; Fossey and Harcourt, 1977; Watts, 1985) have influenced scientific notions of the ecological significance of large body size among primates (e.g., Clutton-Brock and Harvey, 1977). Body size is one of the few attributes of animals that can be estimated from fossil remains. We want to be able to interpret its significance for the behavior of living and fossil species. Before we can understand the impact of body size on the lifestyles

Primate Locomotion, edited by Strasser *et al.*
Plenum Press, New York, 1998

of gorillas or other species, we must separate its effects from those of habitat on diet and arboreality.

Early descriptions of gorillas by anatomists (Savage and Wyman, 1847; Owen, 1859; Sanford, 1862; Morton, 1922) emphasized morphological characters associated with arboreality. Yet, the natural historian Akeley (1923) considered the arboreal aspects of gorilla morphology to be vestigial, since the adult gorillas he observed in the Virungas did not spend a significant amount of time in trees. It is not surprising that the Virunga gorillas are terrestrial, given the dwarf nature of the montane forest, the reduced canopy cover and the scarcity of tree foods (Richards, 1957; Schaller, 1963). It is surprising, however, that Akeley (1929) ignored the significance of the gorilla's specialized habitat in explaining their terrestrial adaptation. Instead, he invoked body size as the causal factor, as did others (e.g., Jones and Sabater-Pi, 1971).

Akeley (1929), Schaller (1963) and Fossey (1983) each painted a vision of mountain gorillas as terrestrial gentle giants. Meanwhile, until recently, the earlier accounts of gorillas in lowland regions using trees for feeding and nesting were discounted as anomalies, or as unscientific and invalid reports. Although they have proved difficult to study, the majority of gorillas (*Gorilla gorilla gorilla*) live in lowland tropical forests in west and central Africa. These tropical forests contain a rich complement of fauna and flora and a much higher diversity of arboreal substrates and foods, particularly fruits, than montane forest habitats (Richards, 1957; Tutin and Fernandez, 1984).

Recently, studies in tropical forests have demonstrated that lowland populations are quite distinct from mountain gorillas. Work at Lopé, Gabon, several sites in the Congo and the Central African Republic has shown that western lowland gorillas are both more arboreal and frugivorous than mountain gorillas (Fay, 1989; Fay et al., 1989; Williamson et al., 1990; Remis, 1994, 1995, in press; Nishihara, 1995; Carroll, 1996). This suggests that mountain gorilla behavior reflects an adaptation to a specialized habitat rather than a body size based species-specific pattern of obligate terrestriality.

1.1. Ape Anatomy and Behavior: The Gorilla Paradox

Gorillas are the largest living primates, with lowland males weighing, on average, 169 kg and females 71 kg (Willoughby, 1978; Jungers and Susman, 1984). Body size ranges for male mountain and lowland gorillas overlap. Female mountain gorillas may be somewhat larger than western lowland females (mountain gorilla males average 159 kg, females 98 kg, but data are few (Jungers and Susman, 1984).

Some authors have suggested that the adaptive radiation of the African apes incorporated wide variation in body size but little morphological or ecological diversity (Shea, 1983; Jungers and Susman, 1984). The African apes are frugivorous, arboreal climbers and engage in some suspensory behavior (Fleagle, 1988; Hunt, 1991), except the specialized mountain gorilla. Morphological adaptations for terrestrial knucklewalking among gorillas (Lewis, 1969; Tuttle, 1969; Tuttle and Watts, 1985), co-occur with features that aid tree climbing (Keith, 1923; Schultz, 1963; Washburn, 1973; Fleagle et al., 1981).

The discrepancy between general gorilla morphology for climbing and their terrestrial adaptation has traditionally been explained as a result of phyletic history coupled with a recent body size increase among gorillas relative to chimpanzees (Akeley, 1929; Schaller, 1963; Tuttle and Watts, 1985). Nevertheless, researchers have recently begun to acknowledge behavioral differences between subspecies that help us to understand the relationship between gorilla morphology and behavior (Fleagle, 1988; Watts, 1990; Hunt, 1991).

The terrestrial and herbivorous mountain gorilla can now be seen to depart from the arboreal, suspensory and frugivorous pattern common among other apes, including lowland gorillas. Behavioral differences between the subspecies of gorillas are correlated with anatomical differences. Mountain gorillas have relatively shorter arms and less divergent big toes than lowland gorillas (Straus, 1930; Schultz, 1934). They have more highly crested molar surfaces than the lowland forms (Uchida, 1992). Systematic data on the arboreal habits and positional behavior of lowland gorillas dispel the notion of a paradox between overall gorilla morphology and behavior and emphasize the similarities between lowland gorilla positional behavior and that of chimpanzees.

Positional behavior encompasses locomotor activities, postural modes, and manipulation of the environment (Prost, 1965) and is fundamental to lifetime reproductive success (Zihlman, 1992). Many recent positional behavior studies have built on the early work of Ripley (1967, 1979), Kinzey (1976), Morbeck (1976, 1979) and Rose (1977, 1979) that emphasized the interaction between individuals and their environment (Garber, 1984; Cant, 1992; Dagosto and Gebo, this volume). A new focus for the study of positional behavior has been assessing intraspecific variability (Garber and Preutz, 1995; Gebo and Chapman, 1995; McGraw, 1996). In addition, as Hunt (1992) and Remis (1995) have emphasized, the social realm also influences positional behavior.

Positional behavior studies of the African apes have upheld predictions concerning the relationship of body size to arboreal behavior. Large animals should face difficulties with balance, substrate size and the energetics of climbing and these should affect their frequency of climbing, size of substrates used and positional modes (Grand, 1972; Cartmill and Milton, 1977; Hunt, 1994). For example, the frequency of arboreal behaviors among pygmy chimpanzees and chimpanzees may vary with body size (Doran, 1993a,b). Arboreality among the great apes is also correlated with the amount of fruit in the diet, and lowland gorillas that eat a lot of fruit are more arboreal than mountain gorillas that do not (Williamson et al., 1990; Remis, 1994).

The influence of habitat and resulting ecological adaptation on the degree of arboreality among gorillas can be assessed by comparing the positional behavior of the western lowland and mountain subspecies. Comparisons of age and sex differences in posture and locomotion make it possible to examine the effects of body size on arboreal competence while controlling for both phylogeny and habitat (Cant, 1987a,b, 1992). Among all of the great apes, the large body size of males may constrain their positional behavior relative to females and juveniles. Specifically, males may spend more time on the ground or at lower heights of trees. When higher in trees, they may suspend more frequently, and spend less time in the small or terminal branch milieu than smaller animals (e.g., Goodall, 1977; Tuttle and Watts, 1985; Cant, 1987b, 1992; Doran and Hunt, 1994; Remis, 1995). Of course, in addition to body size, substrate use is affected by patch size, the number and types of supports available in a given tree, foraging party size, and social rank (Hunt, 1992; Remis, 1995).

2. METHODS

Data from 27 months of research on the positional behavior of western lowland gorillas at the Bai Hokou Study Site in the Central African Republic were compared to published reports on the positional behavior of the mountain gorillas at Karisoke, Rwanda collected by Watts (Tuttle and Watts, 1985) and Doran (1996). The Bai Hokou study site occupies lowland rainforest (463 m altitude) and receives, on average, 1400 mm of rain a year. Karisoke is situated in a montane rainforest (2680–3700 m altitude) and receives 1800 mm of rain per year.

During the Bai Hokou study, instantaneous samples of activities (percent time feed-ing and resting), diet, positional behavior, and substrate use were collected at one minute intervals on focal animals. Ad-libitum data were collected opportunistically on the use of trees at first encounter during contacts (see Remis, 1995 for greater detail). These data are compared to Tuttle and Watts' (1985) report of the activities and postural behavior of mountain gorillas at Karisoke. It is not possible to statistically analyze these comparisons because, although the data sets are comparable, there is no published information about the age/sex distribution of Tuttle and Watts' data. The Bai Hokou data are also compared to Doran's (1996) report of locomotion among the gorillas at Karisoke. Methods used to collect locomotor data during Doran's study of gorillas at Karisoke (Doran, 1996) were similar to those used at Bai Hokou, and the resulting data sets are analyzed by two-tailed, G tests of Independence (Sokal and Rohlf, 1981).

3. RESULTS

3.1. Gorilla Foraging Patterns

Lowland gorillas live in a spatially and temporally complex habitat. The tree canopy in lowland forests in West and Central Africa averages about 40 m or more and canopy foods (leaves, flowers and fruit) are plentiful. Not surprisingly, lowland gorillas consume many foods arboreally, especially fruit, and their daily and home ranging patterns reflect heterogeneity in the temporal and spatial distribution of foods (Table 1).

At Karisoke, trees are short (many are less than 7 m in height) and there are minimal food items available in trees to the gorillas (Fossey, 1983). Therefore, mountain gorillas have fewer incentives and opportunities to climb trees than do lowland gorillas. Most mountain gorilla foods are obtained terrestrially and are perennially available (Watts, 1984). The Karisoke gorillas do not consume the leaves of trees and eat only one kind of tree fruit (Fossey, 1983; Watts, 1984). When mountain gorillas climb trees to eat vines or epiphytes, they are often on large fallen trees and boughs only a few meters off of the ground.

3.2. Tree Climbing among Gorillas

The gorillas at Bai Hokou were never fully habituated. As a consequence, the data are biased towards arboreal sightings (only 15% of the data are of terrestrial sightings) and preclude an accurate reconstruction of percent time spent in trees. While lowland gorillas

Table 1. Gorilla foraging patterns

	Bai Hokou[1]	Karisoke[2]
Number food types	236	75
Number fruit species eaten	77	3
Percent food types obtained in trees	79	13
Day range	2.3 km	0.57 km
Annual range	18.1 km^2	9.4 km^2
Average height in trees	>20 m	<7 m

[1]Bai Hokou lowland gorilla data from Remis (1994, 1995).
[2]Karisoke mountain gorilla data from Watts (1984, 1991) and Doran (1996).

probably spend at least 20% of their time in the trees, they are probably less arboreal than chimpanzees that live in similar habitats. Adult mountain gorillas spend only 3% of their time in the trees (Tuttle and Watts, 1985).

Sex differences in arboreal substrate use among lowland gorillas are suggested by ad-libitum data on first encounters of gorillas (in trees: females 56% of samples (n=143), males 24% of samples (n=239)). Males were also recorded in the trees less frequently than females during the dry season when fruit is rare and during fruit-poor wet seasons, (instantaneous interval sampling, dry season in trees: females 91% of samples (n=89), males 64% of samples (n=188), wet season in trees: females 89% of samples (n=369), males 83% of samples (n=1053); fruit-poor wet season 1995: females 95% of samples (n=555), males 58% of samples (n=377)). In addition, females use small branches and the peripheries of trees more than do males (Remis, 1995). In all seasons, silverbacks differ from other individuals in the frequency of nesting in trees (non-silverback nests in trees: 21% of nests (n=1020), silverback nests in trees: 4% (n=211)). Similarly, the amount of time spent in the trees is correlated with body size among male and female mountain gorillas and eastern lowland gorillas (Goodall, 1977; Tuttle and Watts, 1985; Doran, 1996). At Karisoke, females spend 7% time above ground, males only 2% (Doran, 1996).

3.3. Arboreal Activity Budgets

Adult lowland gorilla arboreal activity budgets during posture are divided between feeding and resting (Bai Hokou group adults, arboreal postural activities: feed = 64%, rest = 35%, n = 777). At Bai Hokou, lone male arboreal activity budgets differ from those of group animals. They rest even more and feed less than other individuals (feed = 29%, rest = 70%, n = 475). In contrast, adult mountain gorillas spend almost all of their arboreal time feeding on bark, leaves of vines or galls, and come out of the trees to rest (Karisoke adults, arboreal postural activities: feed = 91%, rest = 9%, n = 1700 hours).

3.4. Terrestrial Positional Behavior

In general, lowland and mountain gorillas have similar terrestrial positional behavior profiles. Both lowland and mountain gorillas travel on the ground by knucklewalking and more rarely run quadrupedally or bipedally during play or display (Tuttle and Watts, 1985; Remis, 1994). While feeding on the ground, lowland gorillas at Bai Hokou stand more than do mountain gorillas. Lowland gorilla females stand even more than males and travel more frequently between food patches. Mountain gorillas appear to be able to squat and reach a large number of terrestrial foods from one location (Tuttle and Watts, 1985; Table 2).

3.5. Arboreal Feeding Postures

All gorillas spend the majority of their time in trees in postural rather than locomotor modes: sitting or squatting are common and suspensory postures are rare. Adult lowland gorillas suspend more frequently than mountain gorillas and juveniles of both subspecies suspend more than larger bodied animals (Table 3). Sex differences in arboreal posture exist among lowland gorillas. Group males sit and recline more than females, and squat less. They spend slightly less time in bipedal or suspensory postures, contrary to predictions (Cant, 1987b). Females also differ significantly from lone males, who squat less and recline more than females. When feeding arboreally, mountain gorillas squat and use

Table 2. Terrestrial feeding postures

| Posture | Bai Hokou[1] | | Karisoke[2] | |
	Females N=19	Males N=93	Females	Males
Sit	63.2	52.7	71.4	60.9
Squat	–	9.7	25.5	36.1
Biped	–	3.2	0.1	0.2
Quad. stand	36.8	31.2	2.7	2
Lie	–	3.2	0.4	1

[1]Bai Hokou lowland gorilla data from Remis (1995).
[2]Karisoke mountain gorilla data from Tuttle and Watts (1985).

tripedal postures more and sit less than lowland gorillas, probably as a result of a low density of arboreal foods.

3.6. Arboreal Locomotion

Gorillas use arboreal locomotion when climbing in or out of food trees and between feeding sites within trees. Gorillas generally travel between trees by descending and knucklewalking on the ground. For both lowland and mountain gorillas, most arboreal locomotion is quadrupedal walking and climbing, although sample sizes are small. Although there are no significant sex differences in the arboreal locomotor patterns of lowland gorillas, there are significant differences between subspecies and between male and female mountain gorillas (Table 4). Lowland gorillas vertical climb and scramble more than mountain gorillas. Lowland females engage in more bipedal locomotion than lowland males while mountain males walk bipedally in trees more than do females of their subspecies.

4. DISCUSSION

Wherever gorillas have been studied in lowland rain-forests, which contain a high number of arboreal substrates and tree foods, they use trees at a much higher frequency than do mountain gorillas. In fact, lowland gorillas have a positional repertoire as varied

Table 3. Arboreal feeding postures

| Posture | Bai Hokou[1] | | | Karisoke[2] N=1700 hrs | | |
	Juveniles N=26	Females N=269	Males N=366	Juveniles	Females	Males
Sit	34.6	62	54.6	1.2	0.8	–
Squat	3.8	27	36.1	84.1	84.6	63.1
Suspend	3.8	3	1.3	2.8	0.3	–
Biped	7.7	3.4	0.8	0.8	0.6	–
Triped	3.8	1.5	1.1	8.3	9.1	14.3
Quad. stand	42	3.8	1.9	1.9	3.7	22.6
Lie	3.8	0.4	0.3	0.9	0.8	–

[1]Bai Hokou lowland gorilla data from Remis (1995).
[2]Karisoke mountain gorilla data from Tuttle and Watts (1985).

Table 4. Arboreal locomotion

Locomotor activity	Bai Hokou[1]			Karisoke[2]	
	Females N=95	Group Males N=38	Lone Males N=41	Females N=118	Males N=35
Quadruped. walk	23.2	21.5	22	47.9	68.2
Climb	40	42.1	54	45	21.9
Scramble	24	23.7	17	–	–
Suspend	4.2	7.9	–	5.2	5.6
Biped	4.2	2.6	–	1.4	4.2
Acrobat	4.2	7.9	7	0.5	–

Tests for similarity[3]	G	P	df
Bai Hokou gorillas (BH)			
Females vs. Group Males	2.4	ns	5
Group Males vs. Lone Males	1.2	ns	4
Karisoke gorillas			
Females vs. Males	21.2	0.001	3
BH vs. Karisoke gorillas			
Females vs. Females	13.1	0.001	3
Males vs. Males	15.1	0.001	3

[1]Bai Hokou data from Remis (1994, 1995).
[2]Karisoke data from Doran (1996).
[3]G, G Test; P, probability; df, degrees of freedom; ns, not significant.

as those of the other great ape species. In contrast, arboreal activities comprise a small proportion of mountain gorilla behaviors. The impoverished positional repertoire of mountain gorillas reflects their reliance on ground foods and the scarcity of arboreal substrates and foods in their habitat. At Bai Hokou, canopy cover reaches 100% and both logged and unlogged forest areas contain, on average, 505–516 trees/ha (Carroll, 1996). In contrast, at Karisoke, even in the densest lower altitude "mixed" or *Hagenia* sp. forest areas, canopy cover is less than 50% and mean tree density is only 101.3 trees/ha. In addition, most trees produce wind-dispersed rather than fleshy fruits (Fossey, 1983; Watts, pers. comm.).

Lowland gorillas have more active terrestrial foraging patterns than do mountain gorillas, which are likely a response to habitat differences in food distribution. The lower density of herbaceous foods in lowland forest, even in areas of previous disturbance, relative to montane forest may cause lowland gorillas to move further between patches of herbs than mountain gorillas (Rogers and Williamson, 1987; Watts, 1987; Malenky et al., 1994; Remis, 1994). Further, many of the differences in arboreal postures between lowland and mountain populations are related to habitat differences in substrate availability. While lowland gorillas use the small substrates found at great heights in the canopy, the mountain subspecies make greater use of large low-lying boughs and fallen trunks.

The locomotor repertoire of lowland gorillas resembles that of chimpanzees and orangutans. Within trees, lowland gorillas scramble/clamber more than smaller bodied chimpanzees in a similar habitat. Bai Hokou gorillas scramble/clamber almost as much as orangutans, more than might be expected given the differences in locomotor anatomy between the two (Sigmon, 1974; Cant, 1987a; Table 5). As large animals, both gorillas and orangutans use scrambling/clambering behaviors to distribute their body weight over multiple substrates, especially in the periphery of trees. Among these large apes, scramble/clamber often occurs during suspension by the forelimbs because feet can reach, grasp and support weight on lower branches, resulting in quadrumanous locomotion (Cant,

Table 5. Arboreal locomotion of the great apes

	Tai[1]		Bai Hokou		Kutai[2]
Locomotor activity	*Pan* Females N=122	*Pan* Males N=103	*Gorilla* Females N=95	*Gorilla* Males N=38	*Pongo* Females N=4360
Quadruped	30.3	11.7	23.2	21.5	12
Climb	52.4	60.2	40	42.1	31.3
Scramble/Clamber	4.9	1.9	24	23.7	39.4
Suspend	7.4	5.8	4.2	7.9	11.8
Biped	0.8	5.8	2.6	4.2	–
Misc.	4.1	14.6	4.2	7.9	5.6

[1]Data from Doran and Hunt (1994).
[2]Data from Cant (1987b).

1987b). Chimpanzees employ more quadrupedalism than larger apes and large males of all species use tree-swaying and bridging to transfer between trees. Contrary to predictions (Cant, 1987b, 1992), lowland gorillas do not appear to engage in significantly less arboreal suspensory locomotion than chimpanzees (Cant, 1987b; Doran and Hunt, 1994; Remis, 1995).

The large size of gorillas may set limits on their arboreal behavior and many of the body size based predictions outlined at the outset of this paper are supported. Despite this, while lowland gorillas are more arboreal than the mountain subspecies, lowland gorilla males do not appear to weigh less than the mountain variety. Large male gorillas of all subspecies use trees, but they appear to do so less frequently than other (smaller) apes. Chimpanzees spend between 33–68% of their time in trees (Doran and Hunt, 1994), and degree of arboreality varies with habitat.

At Bai Hokou, differences in body size between male and female gorillas were associated with differences in substrate use (size, height, section and level) and modes of positional behavior. For example, females' use of smaller substrates at Bai Hokou resulted in sex differences in arboreal posture (Remis, 1995). As among smaller and less dimorphic chimpanzees, however, the most challenging results of the Bai Hokou study emerged outside the context of body size predictions and involved the effects of season, social context and patch size on substrate use (Remis, 1995; also Hunt, 1992).

At Bai Hokou, the season and the presence of other individuals, their number, and sex affect the size of the patch used and the choice of substrates once in a tree. Female gorillas use the peripheral parts of trees while feeding on fruit in the rainy season. All gorillas remain in the core of trees on large and secure substrates when feeding on leaves and bark in the dry season. In all seasons, bisexual foraging parties climb larger trees (containing a broader array of available substrates) than smaller single-sex parties or lone individuals, group males use more peripheral parts of trees than lone males, and females with males use more peripheral parts of trees than those in single sex parties (Remis, 1995). Lone males climb more, and scramble and suspend less than males in groups. Hence, the relationships between body size, diet, social context and positional behavior can be described as part of a multidimensional system by which an organism gets access to food.

This comparison of lowland and mountain gorilla positional behavior highlights the importance of the effects of habitat on intraspecific differences in positional behavior. The terrestrial repertoires of lowland and mountain gorillas are similar. Feeding is the primary impetus for climbing among all gorillas. Marked habitat differences in temporal and spatial food distribution, and the density and availability of substrates appear to be responsible for the major differences in arboreal and terrestrial foraging profiles between subspecies. Lowland gorillas spend more time in trees than mountain gorillas, obtaining a much larger proportion of their foods arboreally. This analysis suggests that the extreme habitat differences between gorilla populations may prevent maintenance of a "gorilla-typical" locomotor pattern (unlike the tamarins studied by Garber and Preutz (1995) or the monkeys studied by McGraw (1996) at the Tai Forest,). In addition, there are likely taxonomic differences in intraspecific plasticity of locomotor profiles (McGraw, 1996).

For most primates studied, habitat structure, food availability, availability of supports of different sizes, season, and social context shape positional behavior and substrate use (Dagosto, 1992; Hunt, 1992; Gebo and Chapman, 1995). We need to take these factors into account when attempting broad range intraspecific or interspecific comparisons. To fully understand the effects of body size on arboreal behavior of African apes we need to quantify habitat differences and availability of supports between populations and to conduct studies of animals in similar habitats eating similar resources. Future study of sympatric chimpanzees and gorillas in lowland forests, eating the same foods, may provide the best comparisons for understanding the ways in which the large body size of gorillas may constrain their degree of arboreality and shape interspecific differences in arboreal positional behavior.

ACKNOWLEDGMENTS

The Bai Hokou study was completed under the auspices of the Central African Ministries of Forests and Waters and Scientific Research. The support of Fulbright-IIE, National Science Foundation (with A. Richard), National Geographic Society, Wenner-Gren Foundation for Anthropological Research, World Wildlife Fund, L.S.B. Leakey Foundation, and the Mellon Fund are gratefully acknowledged. This research could not have been completed without the hard work of Louise Dion, Etienne Ndolongbe, Wonga Emile and Mokedi Priva, the logistical assistance of the Dzanga-Sangha Dense Forest Reserve Staff or the mentoring of Alison Richard. I would also like to acknowledge Warren Kinzey for helping to provide me with the opportunity to study positional behavior. David Watts kindly provided unpublished data on the trees at Karisoke. The comments of Elizabeth Strasser and three other reviewers improved this paper.

REFERENCES

Akeley CE (1923) In Brightest Africa. New York: Doubleday.
Akeley MLJ (1929) Carl Akeley's Africa. The Account of the Akeley-Eastman-Pomeroy African Hall Expedition of the American Museum of Natural History. New York: Blue Ribbon.
Bingham, HC (1932) Gorillas in a native habitat. Carnegie Inst. Wash. Publ. 426:1–66.
Cant JGH (1987a) Positional behavior of female Bornean orangutans (Pongo pygmaeus). Am. J. Primatol. 12:71–90.
Cant JGH (1987b) Effects of sexual dimorphism in body size on feeding postural behavior of Sumatran orangutans (Pongo pygmaeus). Am. J. Phys. Anthropol. 74:143–148.
Cant JGH (1992) Positional behavior and body size of arboreal primates: A theoretical framework for field studies and an illustration of its application. Am. J. Phys. Anthropol. 88:273–283.
Carroll RW (1988) Relative density, range extension, and conservation potential of the lowland gorilla (Gorilla gorilla gorilla) in the Dzanga-Sangha region of southwestern Central African Republic. Mammalia 52(3): 311–323.
Carroll RW (1996) Feeding ecology of lowland gorillas (Gorilla gorilla gorilla) in the Dzanga-Sangha Dense Forest Reserve of the Central African Republic. Ph.D. Dissertation. Yale University, New Haven, Connecticut.

Cartmill M, and Milton K (1977) The lorisiform wrist joint and the evolution of "Brachiating" adaptations in the Hominoidea. Am. J. Phys. Anthropol. *47*:249–272.

Clutton-Brock TH, and Harvey PH (1977) Species differences in feeding and ranging behavior in primates. In TH Clutton-Brock (ed.): Primate Ecology. London: Academic Press, pp. 557–579.

Dagosto M (1992) Effect of habitat structure on positional behavior and substrate use in Malagasy lemurs. Am. J. Phys. Anthropol. *14*:167.

Dagosto M, and Gebo DL (199?) Methodological issues in studying positional behavior: Meeting Ripley's challenge. In E Strasser, JG Fleagle, AL Rosenberger, and HM McHenry (eds.): Primate Locomotion: Recent Advances. New York: Plenum Press, pp. 5–29.

Doran DM (1993a) Sex differences in adult chimpanzee positional behavior: The influence of body size on locomotion and posture. Am. J. Phys. Anthropol. *91*:99–115.

Doran DM (1993b) Comparative locomotor behavior of chimpanzees and bonobos: The influence of morphology on locomotion. Am. J. Phys. Anthropol. *91*:83–98.

Doran DM (1996) The comparative positional behavior of the African apes. In WC McGrew and T Nishida (eds.): Great Ape Societies. Cambridge: Cambridge University Press, pp.213–224.

Doran DM, and Hunt KD (1994) Comparative locomotor behavior of chimpanzees and bonobos: Species and habitat differences. In RW Wrangham, WC McGrew, FB de Waal, and PG Heltne (eds.): Chimpanzee Cultures. Cambridge: Harvard University Press, pp. 93–106.

Du Chaillu P (1861) Exploration and Adventures in Equatorial Africa. London: John Murray.

Emlen JT, and Schaller GB (1960) Distribution and status of the mountain gorilla (*Gorilla gorilla beringei*) 1959. Zool. *45*:41–52.

Fay JM (1989) Partial completion of a census of the western lowland gorilla (*Gorilla g. gorilla* (Savage and Wyman)) in southwestern Central African Republic. Mammalia *53(2)*:203–215.

Fay JM, Agnagna M, Moore J, and Oko R (1989) Gorillas (*Gorilla gorilla gorilla*) in the Likouala Swamp Forests of north central Congo: Preliminary data on populations and ecology. Int. J. Primatol. *10(5)*:477–486.

Fleagle JG (1988) Primate Adaptation and Evolution. San Diego: Academic Press.

Fleagle JG, Stern JT, Jungers WL, and Susman RL (1981) Climbing: A biomechanical link with brachiation and with bipedalism. Symp. Zool. Soc. Lond. *48*:359–375.

Fossey D (1983) Gorillas in the Mist. Boston: Houghton Mifflin Co..

Garber PA (1984) Use of habitat and positional behavior in a neotropical primate, *Saguinus oedipus*. In PS Rodman and JGH Cant (eds.): Adaptations for Foraging in Nonhuman Primates: Contributions to an Organismal Biology of Prosimians, Monkeys and Apes. New York: Columbia Univ. Press, pp. 112–133.

Garber PA, and Pruetz JD (1995) Positional behavior in moustached tamarin monkeys: Effects of habitat on locomotor variability and locomotor stability. J. Hum. Evol. *28*:411–426.

Gebo DL, and Chapman CA (1995a) Positional behavior in five sympatric Old World Monkeys. Am. J. Phys. Anthropol. *97(1)*49–76.

Gebo DL, and Chapman CA (1995b) Habitat, annual, and seasonal effects on positional behavior in red colobus monkeys. Am. J. Phys. Anthropol. *96*:73–82.

Goodall AG (1977) Feeding and ranging in Kahuzi gorillas. In TH Clutton-Brock (ed.): Primate Ecology: Studies of Foraging and Ranging in Lemurs, Monkeys and Apes. London: Academic Press, pp. 450–479.

Grand TI (1972) A mechanical interpretation of terminal branch feeding. J. Mammal. *53(1)*:198–201.

Gregory WK, and Raven HC (1937) In Quest of Gorillas. New Bedford: Darwin Press.

Hunt KD (1991) Positional behavior in the Hominoidea. Int. J. Primatol. *12(2)*:95–118.

Hunt KD (1992) Social rank and body size as determinants of positional behavior in *Pan troglodytes*. Primates *33(3)*:347–357.

Hunt KD (1994) Body size effects on vertical climbing among chimpanzees. Int. J. Primatol. *15(6)*:855–865.

Jones C, and Sabater-Pi PJ (1971) Comparative ecology of *Gorilla gorilla* (Savage & Wyman) and *Pan troglodytes* (Blumenbach) in Rio Muni, West Africa. Bibl. Primatol. *13*:1–96.

Jungers WL, and Susman RL (1984) Body size and skeletal allometry in African apes. In Susman RL (ed.): The Pygmy Chimpanzee. Evolutionary Biology and Behavior. New York: Plenum Press, pp. 131–178.

Kawai M, and Mizuhara H (1959) An ecological study of the wild mountain gorilla (*Gorilla gorilla beringei*): A report of the Japan Monkey Center second gorilla expedition, 1959. Primates *2*:1–42.

Keith A (1923) Man's posture: Its evolution and disorder. Brit. Med. J. *1*.451–672.

Kinzey WG (1976) Positional behavior and ecology in *Callicebus torquatus*. Yrbk. Phys. Anthropol. *20*:468–481.

Lewis OJ (1969) The hominoid wrist joint. Am. J. Phys. Anthropol. *30*: 251–268.

McGraw WS (1996) Cercopithecid locomotion, support use, and support availability in the Tai Forest, Ivory Coast. Am. J. Phys. Anthropol. *100(4)*:507–522.

Malenky RK, Kuroda S, Vineberg EO, and Wrangham RW (1994) The significance of terrestrial herbaceous foods for bonobos, chimpanzees, and gorillas. In RW Wrangham, WC McGrew, FB de Waal, and PG Heltne (eds.): Chimpanzee Cultures. Cambridge: Harvard University Press, pp. 59–75.

Morbeck ME (1976) Leaping, bounding and bipedalism in *Colobus guereza*: A spectrum of positional behavior. Yrbk. Phys. Anthropol. *20*:408–420.

Morbeck ME (1979) Forelimb use and positional adaptation in *Colobus guereza*: Integration of behavioral, ecological and anatomical data. In ME Morbeck, H Preuschoft, and N Gomberg (eds.): Environment, Behavior and Morphology: Dynamic Interpretations. New York: Gustav-Fischer, pp. 95–118.

Morton DJ (1922) The evolution of the human foot. Am. J. Phys. Anthropol. *5*:305–336.

Nishihara T (1995) Feeding ecology of western lowland gorillas in the Nouabale-Ndoki National Park, Congo. Primates *36(2)*:151–168.

Owen R (1859) On the *Gorilla* (*Troglodytes gorilla,* Sav.). Proc. Zool. Soc. Lond. *381*:1–23.

Prost J (1965) A definitional system for the classification of primate locomotion. Am. Anthropol. *67(5)*:1198–1214.

Purchas S (1625) The Strange Adventures of Andrew Battell. In Purchas, His Pilgrims. Contayning a History of the World, in Sea Voyages and Land Travells (4th ed.). London: William Stansby for Henry Fetherstand.

Remis MJ (1994) Feeding Ecology and Positional Behavior of Western Lowland Gorillas (*Gorilla gorilla gorilla*) in the Central African Republic. Ph.D. Dissertation. Yale University.

Remis MJ (1995) The effects of body size and social context on the arboreal activities of western lowland gorillas in the Central African Republic. Am. J. Phys. Anthropol. *97*:413–433.

Remis MJ (in press) Western lowland gorillas (*Gorilla gorilla gorilla*) as seasonal frugivores: Use of variable resources. Am. J. Primatol. *43*:N-N.

Richards PW (1957) Tropical Rain Forest. Cambridge: Cambridge University Press.

Ripley S (1967) The leaping of langurs: A problem in the study of locomotor adaptation. Am. J. Phys. Anthropol. *26*:149–170.

Ripley S (1979) Environmental grain, niche diversification and positional behavior in Neogene primates. An evolutionary hypothesis. In ME Morbeck, H Preuschoft, and N Gomberg (eds.): Environment, Behavior and Morphology: Dynamic Interpretations. New York: Gustav-Fischer, pp. 37–74.

Rogers ME, and Williamson EA (1987) Density of herbaceous plants eaten by gorillas in Gabon: Some preliminary data. Biotropica *19(3)*:278–281.

Rose MD (1977) Positional behavior of olive baboons (*Papio anubis*) and its relationship to maintenance and social activities. Primates *18(1)*:59–116.

Rose MD (1979) Positional behavior of natural populations and some quantitative results of a field study of *Colobus guereza* and *Cercopithecus aethiops*. In ME Morbeck, H Preuschoft, and N Gomberg (eds.): Environment, Behavior and Morphology: Dynamic Interpretations. New York: Gustav-Fischer, pp. 76–93.

Sanford LJ (1862) The gorilla: Being a sketch of its history, anatomy, general appearance and habits. Am. J. Sci. and Arts.*xxiii*:1–17.

Savage TS, and Wyman J (1847) Notes of the external characters and habits of *Troglodytes gorilla,* a new species of orang from the Gaboon River. Boston J. N. H. *V(IV)*:417–443.

Schaller GB (1963) The Mountain Gorilla: Ecology and Behavior. Chicago: University of Chicago Press.

Schultz AH (1934) Some distinguishing characters of the mountain gorilla. J. Mammal. *15*:51–61.

Schultz AH (1963) Relations between the lengths of the main parts of the foot skeleton in primates. Folia Primatol. *1*:150–171.

Shea BT (1983) Size and diet in the evolution of African ape craniodental form. Folia Primatol. *40*:32–68.

Sigmon BA (1974) A functional analysis of pongid hip and thigh musculature. J. Hum. Evol. *3:*161–185.

Sokal RR, and Rohlf SJ (1981) Biometry. New York: W.H. Freeman and Company.

Straus M (1930) On the foot musculature of the highland gorilla (*Gorilla beringei*). Quart. Rev. Biol. *5(3)*:261–317.

Tutin CEG, and Fernandez M (1984) Nationwide census of gorilla (*Gorilla g. gorilla*) and chimpanzee (*Pan t. troglodytes*) populations in Gabon. Am. J. Primatol. *6*:313–336.

Tuttle RH (1969) Quantitative and functional studies on the hands of the Anthropoidea. J. Morph._*128(3)*:309–364.

Tuttle RH, and Watts DP (1985) The positional behavior and adaptive complexes of *Pan (Gorilla)*. In S Kondo (ed.): Primate Morphophysiology, Locomotor Analyses and Human Bipedalism. Tokyo: Univ. of Tokyo Press, pp. 261–288.

Uchida A (1992) Intra-species variation among the great apes: Implications for taxonomy of fossil hominoids. Ph.D. Dissertation, Harvard University.

Washburn S (1973) Primate field studies. In GH Bourne (ed.): Nonhuman Primates and Medical Research. New York: Academic Press, pp. 467–485.

Watts DP (1984) Composition and variation of mountain gorilla diets in the Central Virungas. Am. J. Primatol. 7:323–356.

Watts DP (1987) Effect of mountain gorilla foraging activities on the productivity of their food plant species. Afr. J. Ecol. 25:155–163.

Watts DP (1990) Ecology of gorillas and its relation to female transfer in mountain gorillas. Int. J. Primatol. 11(1):21–46.

Willoughby D (1978) All About Gorillas. New York: A.S. Barnes and Co.

Williamson EA, Tutin CEG, Rogers ME, and Fernandez M (1990) Composition of the diet of lowland gorillas at Lope Gabon. Am. J. Primatol. 21:265–277.

Zihlman AL (1992) Locomotion as a life history character: The contribution of anatomy. J. Hum. Evol. 22:315–325.

II

MORPHOLOGY AND BEHAVIOR

INTRODUCTION TO PART II

John G. Fleagle

Throughout its development the study of primate locomotion has been characterized by its interdisciplinary focus combining ecological and behavioral observations with physiological and morphometric studies to understand both how and why primates use different patterns of locomotion and posture. In addition, studies of primate locomotion are increasingly involving experimental approaches that allow researchers to examine behavior, form and function in a controlled situation and go beyond descriptive and correlative analyses. The papers in this section demonstrate both the interdisciplinary and experimental approaches to the study of primate locomotion. Moreover, by clarifying the link between bony morphology and physiological function they provide ever widening windows into the behavior of extinct primates.

Laura MacLatchy (Chapter 7) uses 3-D imaging of hindlimb bones and *in vivo* manipulations of living animals to examine hip joint function in extant and fossil prosimians. She finds that patterns of joint surface distribution on the acetabulum and head of the femur accord well with measurements of joint mobility in living animals. Thus imaging of joint surfaces in fossils can be used to reconstruct patterns of limb excursion during locomotion.

Pierre Lemelin and Brian Grafton (Chapter 8) conducted experimental studies of hand use in golden-handed tamarins and squirrel monkeys in conjunction with kinematic analysis and morphometric studies to examine the functional consequences of nails and claws on primate digits. They found relatively few differences in the abilities of clawed tamarins and nail-bearing squirrel monkeys to grasp most food items and concluded that claws do not necessarily hinder grasping function. Likewise they found little evidence for a clear link between grasping ability and digital proportions.

In Chapter 9, Jeffrey Meldrum addresses an important, but rarely examined problem in functional morphology—the intermediate steps in the evolution of complex morphological features. He uses both morphological and behavioral observations to identify a series of behaviors among extant primates that seem to foreshadow the evolution of fully prehensile tails in platyrrhines.

Susan Larson (Chapter 10) uses kinematic analysis of limb excursions and electromyography of shoulder muscles to examine unique features of primate quadrupedal locomotion. She finds that compared with other vertebrates, which have been reported to be very conservative and uniform in forelimb muscle activity, primates are unique in using more forelimb protraction and in the recruitment of shoulder muscles during quadrupedal gaits.

Daniel Schmitt (Chapter 11) uses kinematic and force analysis to compare primate quadrupeds on terrestrial and arboreal supports. He finds that the primates use more flexed limbs on arboreal supports, but surprisingly this does not generate higher joint reaction forces as predicted and hypothesizes that primates are using compliant gaits during arboreal locomotion.

7

RECONSTRUCTION OF HIP JOINT FUNCTION IN EXTANT AND FOSSIL PRIMATES

Laura MacLatchy

Department of Anthropology
Boston University
Boston, Massachusetts 02215

1. INTRODUCTION

The strepsirhines engage in a variety of locomotor behaviors, including vertical clinging and leaping, active arboreal quadrupedalism and deliberate climbing and suspension. In each of these locomotor modes, hip abduction plays a role in enabling an animal to move in the complex, three-dimensional arboreal realm (Figure 1). For instance, hip abduction characterizes clinging postures prior to leap take-off in galagos, indriids and *Lepilemur* (Anemone, 1990; Demes et al., 1996; Demes, pers. com.). Abducted hip postures also occur among above-branch quadrupeds such as cheirogaleids and many lemurids who primarily walk, run and leap, but who also climb and engage in hindlimb suspension (Oxnard et al., 1990), both of which require moderate hip abduction. In turn, lorisids, such as the potto, frequently use highly abducted hip postures and even on the ground progress slowly with limbs abducted (Walker, 1969, 1979; Oxnard et al., 1990). Since hip abduction is such an important and variable component of hindlimb locomotion in extant small-bodied primates, the ability to reconstruct this behavior in small-bodied fossil primates should prove useful in determining their overall locomotor profile.

Investigations of the mammalian hip joint have emphasized the importance of relating joint morphology to ranges of movement, both to understand function and to reconstruct locomotion in extinct forms. This line of research has been largely limited to making qualitative comparisons of joint shape, and indirectly inferring joint mobility (e.g., Elftman, 1929; Walker, 1974; McHenry, 1975; Fleagle, 1976; Jenkins and Camazine, 1977; Harrison, 1982; Dagosto, 1983; Stern and Susman, 1983; Fleagle and Meldrum, 1988; Gebo, 1989; Anemone, 1990; Ward, 1991). Several excellent studies have investigated pelvic and femoral morphology in small bodied primates (e.g., McArdle, 1981; Fleagle and Anapol, 1992; Dagosto and Schmid, 1996), but none have focused on mobility *per se*.

Primate Locomotion, edited by Strasser *et al.*
Plenum Press, New York, 1998

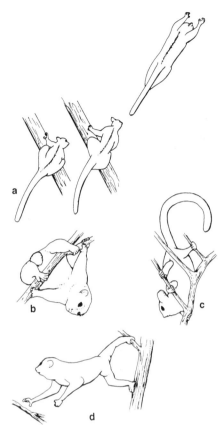

Figure 1. Examples of strepsirhine primates using hip abduction. (*a*) during preparation for leaping in *Galago* (after a figure in Charles-Dominique, 1977); (*b*) during suspension in *Perodicticus* (after a photograph in Napier and Napier, 1967); (*c*) during climbing downwards in *Microcebus* (after a drawing by Nash in Mittermeier et al., 1994); (*d*) during bridging in *Loris* (after a drawing in Schulze and Meier, 1995 and a photograph in Charles-Dominique, 1977).

In the case of anthropoids, however, two recent studies (MacLatchy, 1996; MacLatchy and Bossert, 1996) have focused on mobility (as well as loading patterns) at the hip joint and generated new analytical techniques that can be used to study hip function in strepsirhines. In these studies, three-dimensional modeling and quantification of acetabular joint surface distribution were used to infer joint loading. More orthograde species, like *Homo* and *Pongo*, were found to have expanded cranial lunate surfaces compared to more pronograde species, such as *Pan* and Old World monkeys, reflecting the cranial direction of hip joint reaction forces in orthograde species. Analogous patterns may be found among strepsirhines. Both *Galago* and *Lepilemur* frequently engage in behaviors in which the trunk is oriented vertically, such as vertical clinging and leaping, vertical climbing and bipedal hopping on the ground (Walker, 1974, 1979; Oxnard *et al.*, 1990), resulting in hip joint forces that would be pointed into the cranial part of the acetabulum (e.g., as has been demonstrated in humans (Dalstra and Huiskes, 1995)), causing a relatively expanded cranial lunate surface. Walker (1979:553) reports that mouse lemurs "...run about quadrupedally in a small branch milieu...when running the body is horizontal and aligned with the support." The generally horizontal body orientation of *Microcebus* may generate stresses on the dorsal aspect of the acetabulum, causing it to expand. The potto, with a varied locomotor repertoire that includes pronograde quadrupedalism, below-branch suspension, cantilevering, spiraling, bridging and climbing (Walker, 1969 and 1979; Charles-Dominique, 1977; Table 1), might be expected to experience more uniform

Table 1. Extant and fossil primate specimens used in analyses. AMNH, American Museum of Natural History; DUPC, Duke University Primate Center; MCZ, Museum of Comparative Zoology; PM, Peabody Museum; UMMP, University of Michigan Museum of Paleontology

Species	Skeletal specimens	N of *in vivo* specimens from DUPC	Positional behavior (reconstructed for fossil species) and/or morphological affinities
Microcebus murinus	AMNH 174385, 174415, 174424, 174430, 174431, 174499, 185627; MCZ 44843	1	arboreal quadruped, frequent leaping and hopping and occasional use of cantilever posture (Walker, 1974)
Cheirogaleus major (animation only)	AMNH 100640; MCZ 44946, 44952	none available so 2 *Cheirogaleus medius* used	arboreal quadruped; runs and scurries; leaps rarely (Walker, 1974)
Galago senegalensis	AMNH 86502, 86503, 86504, 187356, 187357, 185358; PM 8535, 8537, 10595	2	arboreal: leaps frequently; quadrupedal running and climbing (Walker, 1979; Oxnard et al., 1990)
Perodicticus potto	AMNH 52698, 52702, 52717, 86898; MCZ 25831	2	arboreal: slow quadrupedal climbing and below branch suspension; limbs habitually abducted (Walker, 1969, 1979; Charles-Dominique, 1977)
Lepilemur mustelinus	AMNH 170553, 170555, 170560, 170561, 170562; MCZ 25407, 44925, 44926, 44931	none available so 2 *Lemur fulvus collaris* used	arboreal: leaps frequently using predominantly vertical supports; walks slowly or hops bipedally on ground (Oxnard et al., 1990)
Notharctus tenebrosus	AMNH 11474, 129382		active arboreal quadruped; leaping ability; pelvic morphology similar to *Lepilemur* (Gregory, 1920; Fleagle and Anapol, 1992)
Hemiacodon gracilis	AMNH 12613		femur similar to *Galago* and *Tarsius*; pelvis similar to *Galago* and *Lemur* (Simpson, 1940; Covert, 1995); femur resembles *Microcebus* (Cartmill, 1972)
Omomys carteri	UMMP 98604		elongated talus and calcaneus suggest some leaping; hip morphology resembles *Galago* (Dagosto, 1993, 1996; Anemone, 1995)

acetabular loading than the other strepsirhines and so have a more uniformly distributed lunate surface.

The studies on anthropoids (*ibid*) also related acetabular shape to hip mobility. The hip socket was more shallow on its dorsal aspect in apes than in monkeys, permitting a wider range of abduction in the former. It is therefore expected that among strepsirhines, those species with the greatest mobility (e.g., *Perodicticus*) will have shallower hip sockets.

Mobility was further assessed by MacLatchy (1996) using three-dimensional computer animation of hip joint elements to estimate species' maximum range of abduction, an important ability among climbing primates. Apes were found to have greater ranges of abduction than monkeys, who in turn had more mobile hips than humans. The technique was used to infer mobility in the hominid *Australopithecus afarensis*, which was found to have the same level of mobility as modern humans. A similar mobility gradient should exist among strepsirhines who have varied locomotor patterns (Table 1). The slow climbing and suspensory potto, for example, is expected to have greater ranges of hip abduction than typical above-branch quadrupeds.

Once a comparative context has been established among extant strepsirhines, the hip function of fossil primates can be considered. For the fossil primates *Notharctus tenebrosus, Omomys carteri* and *Hemiacodon gracilis,* selection of a phylogenetically limited extant comparative sample is difficult because the relationships of the notharctines and omomyines to living species are not entirely clear (Beard *et al.*, 1988; Dagosto, 1988; Fleagle, 1988; Martin, 1993; Kay et al., 1997). Nevertheless, extant strepsirhines are a reasonable choice because their behavior spans several locomotor modes, because they have been described as good analogs for the fossil taxa under scrutiny, and because their hip joint elements resemble those of the fossil species (Table 1).

Each of the fossil taxa possesses characteristics of "hindlimb-dominated arboreal locomotion" as outlined by Martin (1990), including overall adaptations for arboreal leaping. *Notharctus* is a notharctine adapid from the middle Eocene of North America. It was a diurnal folivore (Fleagle, 1988) and its positional repertoire has been variously characterized as both an active arboreal quadruped and a vertical clinger and leaper (Gregory, 1920; Napier and Walker, 1967; Gebo, 1985; Szalay and Dagosto, 1980; Dagosto, 1993). *Omomys* and *Hemiacodon* are two genera of omomyine omomyids, also from the middle Eocene of North America. Both were probably nocturnal and ate insects and fruit (Covert, 1995). As with *Notharctus*, suggestions that these species typically used vertical postures (e.g., Napier and Walker, 1967) have been challenged, although leaping is thought to comprise a significant portion of their locomotor repertoire (Dagosto, 1993; Covert, 1995). Locomotor reconstructions of the three taxa have focused on whether they are best characterized as active arboreal quadrupeds or as vertical clingers and leapers, in part because of the role these categories played in early attempts to describe the locomotor mode of the ancestral primate (Napier and Walker, 1967; Cartmill, 1972). In recent years, the former characterization has become most prevalent (Gebo, 1985; Fleagle, 1988; Dagosto, 1993; Covert, 1995).

In the present study, the emphasis will be on reconstructing the abductory capabilities of the hip and how this helps to characterize the overall adaptations of the hindlimb. This will be accomplished using three-dimensional modeling techniques that enable an analysis of acetabular joint surface shape, which is used to infer excursion and loading patterns at the hip, and direct estimates of hip joint posture and maximum abduction using computer animation. In addition, hip abduction will be assessed *in vivo* in order to evaluate the accuracy of the computer animation.

2. MATERIALS AND METHODS

2.1. Specimens

The extant sample is a phylogenetically diverse one (2 cheirogaleids, 2 lorisoids and 1 lemuroid; Table 1). While this does not affect comparisons of direct measurements of mobility, inferences concerning acetabular morphology are unconstrained by phylogeny. As discussed above, this is because the relationships of the notharctines and omomyines to living species are unclear. In addition, use of Magnetic Resonance Imaging (MRI) imposed time constraints and reduced the number of museum collections that could be used, and so limited the number of taxa and specimens that could be examined.

Acetabular morphology was analyzed in four extant taxa (*Microcebus*, *Galago*, *Perodicticus* and *Lepilemur*); these four and one additional taxon (*Cheirogaleus major*) were analyzed by computer animation (Table 1), while five taxa were studied *in vivo*. Only three *Cheirogaleus major* specimens were imaged, but they are included in order to compare estimated mobility to *in vivo* data collected for *Cheirogaleus medius*. No *Lepilemur* are present in captivity, so *in vivo* estimates could not be obtained for this genus, and *Lemur* was studied instead. While the computer/*in vivo* species matches are not exact, they do allow comparisons to be made of the two methods of quantifying hip abduction.

The sacrum is necessary to orient the pelvis in three dimensional space, but no associated sacra were available for the fossil specimens. For *Notharctus*, unassociated sacra of two other *Notharctus* specimens (AMNH 11475, 11752) were used; for the omomyids, a *Galago senegalensis* sacrum was used because of size and gross morphological similarities between *Galago* and omomyid pelves.

2.2. Imaging and Image Analysis

Specimens were imaged on an 8.45 Tesla MRI system (Bruker Instruments, Billerica, MA) as described in MacLatchy and Bossert (1996). In the x and y directions, the image resolution varied between 0.08 mm (for the small *Microcebus* specimens) and 0.18 mm (for the larger *Perodicticus* specimens) and was twice as large (i.e., between 0.17 mm and 0.37 mm) in the z direction. For the femur, contiguous images were taken from the top of the greater trochanter (or the top of the femoral head, whichever was higher) to a level one-third down the length of the femoral shaft. For the pelvis, contiguous images over the entire acetabular and adjacent areas (including the ischium and pubis) were taken.

Three-dimensional images were reconstructed from digitized MRI images (MacLatchy and Bossert, 1996). Three-dimensional coordinates were interactively selected and coded according to anatomical region to facilitate modeling. The subchondral bone of the femoral head, the femoral shaft, the *fovea capitis*, the lunate surface, the ischium and the acetabular fossa were all differently coded. The long axes of the ischium and femoral shaft were determined by finding the line that bisected the center of mass of two cross-sections through the ischium and two cross-sections through the femoral shaft (MacLatchy and Bossert, 1996).

A mid-sagittal plane was created for each three-dimensional model of a pelvis. Three three-dimensional coordinates are required to define a plane. As described in MacLatchy (1996), two x, y, z coordinates were located on the pubic symphysis, which coincided with the midplane of the animal. The location of the third coordinate was determined by first measuring the perpendicular distance from the tip of the ischial spine to the mid-sagittal plane (as determined by the midplane of the sacrum) while the pelves and sacrum were articulated. In the three-dimensional reconstruction, the coordinate on the ischial spine was then projected by the measured distance (in the x-plane) to supply the third x, y, z coordinate. The plane was used to determine femoral posture during computer simulations of joint movement.

2.3. Modeling of Lunate Surface

Lunate surface distribution relates to both joint force patterns and mobility. Expansion of articular surfaces is thought to occur in areas normal to the greatest or most frequent joint forces (Latimer et al., 1987; Ward, 1991), hence the distribution of acetabular articular surface may reflect the loading environment of the joint. In the case of mobility, greater coverage of the femoral head by a deep acetabulum restricts hip movement, while less coverage by a shallow acetabulum permits a wider range of excursion (Elftman, 1929; Jenkins and Camazine, 1977; Ruff, 1988; Ward, 1991).

A detailed description of the modeling procedure is in MacLatchy and Bossert (1996). The basis for the analysis was to fit a linear least squares best-fit sphere to computer reconstructions of the acetabulum. Only the coordinates defining the lunate surface were fit to a sphere; the acetabular fossa was not included since it is not part of the weight bearing surface of the joint.

In order to quantify the distribution pattern of the subchondral bone, two axes were set up within the best-fit sphere (Figure 2). Axis 1 was defined as the line between the center of the best-fit sphere and the center of the acetabular fossa. Axis 2 was on a plane perpendicular to Axis 1 and went through the center of the best-fit sphere; the location of Axis 2 on this plane was determined by projecting a line parallel to the long axis of the ischium onto the plane. Hence, these reference axes were consistent from specimen to

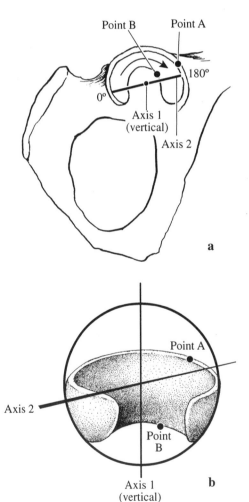

Figure 2. Modeling of lunate surface. (*a*) Lateral view of a right acetabulum, showing Axes 1 and 2 and the cranial data points. Caudal to the left, dorsal to the top. Angles from Axis 2 are measured in a clockwise direction from the caudal end (0°) of Axis 2. (*b*) Ventral view of an acetabulum, showing the best fit sphere and the same cranial data points as in *a*. The location of the acetabular rim and junction between the acetabular fossa and the lunate surface are quantified by measuring the angular displacement from Axis 1 at specific angles from Axis 2 (i.e., at caudal, dorsal and cranial locations). The data point on the rim of the acetabulum (Point A) has an angular displacement from Axis 1 of 100° and an angular displacement from Axis 2 of 165°. The data point at a junction of the articular surface and the acetabular fossa (Point B) has an angular displacement from Axis 1 of 40° and an angular displacement from Axis 2 of 165°. The total angular extent of the articular surface at this cranial location is determined by subtracting the minimum angle from Axis 1 from the maximum angle from Axis 1 (100° − 40° = 60°).

specimen. Once the axes were defined, the locations of data points on the articular surface could be localized by plotting the angle from Axis 1 (the angle subtended by the line from a point on the articular surface to the center of the sphere and Axis 1) against the angle from Axis 2 (the angle subtended by the line from a point on the articular surface to the center of the sphere and Axis 2) for data points on the articular surface. The angle from Axis 1 gives information about how far up the sides of the best fit sphere (or the walls of the acetabulum) a given point is located. The highest angles from Axis 1 occur at the acetabular rim while the lowest occur at the lunate surface/acetabular fossa junction. The angle from Axis 2 is indicative of where a given point is located relative to the circumference of the acetabulum.

The difference between the highest and lowest angles from Axis 1 at cranial, dorsal and caudal locations on Axis 2 were used to assess the relative size of the lunate surface wall in order to relate it to regional loading patterns. The caudal (or ischial) end of Axis 2 was designated as 0° and angles were measured in a clockwise direction. Dorsal measurements were taken at 75° from Axis 2, cranial measurements at 165°, and caudal measurements at 345°. The maximum angle from Axis 1 of coordinates at these same cranial, dorsal and caudal locations on Axis 2 were used to assess relative femoral coverage (or the maximum distal extent, relative to the fossa, of the lunate surface). Regional variation in coverage should have different functional implications, with cranial coverage limiting medial rotation, dorsal coverage limiting abduction and caudal coverage limiting lateral rotation (MacLatchy and Bossert, 1996).

All pelves were analyzed in this way, and the means and standard deviations of the angles from Axis 1 at the three locations were calculated for each species.

2.4. Quantification of Femoral Posture

Jenkins and Camazine (1977) quantified the location of the *fovea capitis* in carnivores and found that ambulatory carnivores have foveae that are located more dorsoposteriorly than do cursorial carnivores; they related this to a more abducted femoral posture. This study uses foveal location to infer femoral posture by quantifying the orientation of the femoral shaft relative to the midsagittal plane when the fovea is centered in the acetabular fossa. The orientation of the femoral shaft was standardized by flexing it at an angle of 90° to the long axis of the ischium (Figure 3). The 90° angle was chosen because it represents the approximate degree of flexion relative to the ischium of the femur in a standing or moderately flexed clinging posture as determined from X-rays and observations of articulated skeletons (Peabody Museum and Museum of Comparative Zoology, Harvard University).

A linear least squares best-fit sphere was fit to computer reconstructions of the subchondral bone of the femoral head (excluding the *fovea capitis*). The 3-dimensional model of the femur was placed in its corresponding hip socket so that the centers of curvature of the best-fit spheres for the head and acetabulum were coincidental. This procedure eliminated the need to assign a value for cartilage thickness, but assumes that cartilage is evenly distributed, as in studies that estimate range of movement by manually opposing the subchondral surfaces of osteological specimens and moving them throughout their range of congruence (e.g., Latimer et al., 1987; Swartz, 1989).

The midpoint of all of the *fovea capitis* coordinates was placed opposite the midpoint of all of the coordinates of the acetabular fossa. In this position, the angle that the long axis of the femoral shaft made with the midsagittal plane (in the coronal plane) was measured. This angle was deemed the "neutral posture" of the femur, or the degree of abduction of the femur in a hypothetical stance phase.

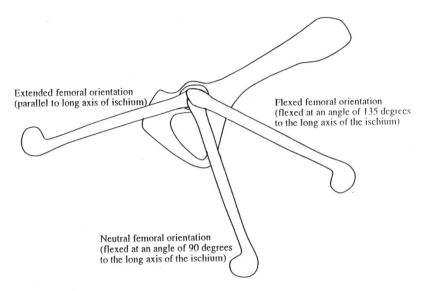

Extended femoral orientation
(parallel to long axis of ischium)

Flexed femoral orientation
(flexed at an angle of 135 degrees
to the long axis of the ischium)

Neutral femoral orientation
(flexed at an angle of 90 degrees
to the long axis of the ischium)

Figure 3. Lateral view of a right femur and innominate showing the three angles of flexion of the femoral shaft during abduction simulations.

A sphere closely approximates the shape of the femoral head and the acetabulum for all the extant species except the *Galago* femoral head, which has a posterior femoral articular surface that resembles a cylinder (Walker, 1974; McArdle, 1981; Anemone, 1990; MacLatchy, 1993). A larger portion of the femoral head extended beyond the surface of the best-fit sphere in *Galago* than in any other species, so the possibility existed that a portion of the head might contact the acetabulum during rotation of the femur about the center of the superimposed best-fit spheres. The surface of the femoral head was thus carefully monitored in simulations.

2.5. Computer Simulation of Maximum Abduction

Maximum hip joint abduction is determined by simulating joint movement with three-dimensional computer reconstructions of the femur and pelvis. The centers of curvature of the best fit spheres of the femoral head and acetabulum were made coincidental. Abduction simulations were conducted from a flexed position, in which the long axis of the femoral shaft was flexed at an angle of 135° to the long axis of the ischium, and an extended position, in which the femoral axis was parallel to the long axis of the ischium (Figure 3). The abduction simulations could have been performed at virtually any degree of flexion or extension. The 135° flexed position was chosen to represent femoral orientation during climbing during forward progression, while the extended position was chosen to represent a more acrobatic, postural position for the femur.

At least three factors may constrain abduction as performed in the computer simulations. First, the fovea and its attached ligament normally remain within the confines of the acetabular fossa in order to prevent the *ligamentum teres* from becoming wedged between the femoral head and the lunate surface (Jenkins and Camazine, 1977). Second, bony constraints such as the greater trochanter or femoral neck may contact the acetabular rim or wall and halt movement, and third, movement may be limited by the femoral subchondral bone margin. The cineradiographic study by Jenkins and Camazine demonstrated that during normal gait in carnivores, the edge of the acetabular articular surface overlaps slightly with non-articular areas of the femoral head in some species. Hence, in this analysis, acetabular surface was allowed to overlap with non-articular parts of the femur until one of the other constraints was met (see below) and movement had to cease. Overlap, when it occurred, was very small.

The simulations proceeded as follows: a femur was oriented at the appropriate degree of flexion relative to the ischium (0° or 135°) and then abducted and rotated (while maintaining the degree of flexion relative to the ischium) until the greater trochanter approached the pelvis, or the *fovea capitis* threatened to overlap with the lunate surface, or both. When abduction was limited by the greater trochanter, the femur was abducted until the greater trochanter just contacted the pelvis, and then adducted 5° in the coronal plane to provide a standardized way to account for soft tissue between the trochanter and pelvis.

2.6. *In Vivo* Quantification of Maximum Abduction

In vivo assessments of joint excursion are rare (e.g., Turnquist, 1983, 1985; Demes et al., 1996) and these data are the first on maximum abduction of the hip joint in any primate species.

Five species of strepsirhines (Table 1) were studied at the Duke University Primate Center. Two individuals of each species were anesthetized with iso-fluorane, with the exception of *Microcebus* where only one individual could be used. While anesthetized, the legs of the animals were passively manipulated and the maximum angles of abduction in highly flexed and completely extended postures were measured. These postures approximately mimicked the angles of femoral flexion (0° and 135° to the ischium) in the computer simulations. X-rays of the various postures were taken in order to determine joint configuration, but the X-rays were in the dorsal-ventral plane and so the angle of the femoral shaft relative to the ischium in the parasagittal plane could not be precisely replicated.

Turnquist (1983) found that the range of flexion-extension of the joints of caged patas monkeys was greater than that of free-ranging patas monkeys, and attributed it to lack of muscle tone and concomitant lack of resistance to passive manipulation. All of the animals used in this study were caged, but it is unlikely that this affects hip abduction, since the primary constraints to abduction are thought to be osteological.

2.7. Statistical Comparisons

Interspecific statistical comparisons among means were made using the least significant difference (LSD) test statistic for planned comparisons (Sokal and Rohlf, 1981), since species were chosen because of their locomotor diversity and it was assumed *a priori* that the functional parameters would differ. Results were considered significant if the P-values were <0.05. The LSD tests were performed using the PowerPC version of the Statview Statistical package.

3. RESULTS

3.1. Lunate Surface Morphology

The total angular extent of the lunate surface was determined in order to assess suspected regional differences in hip joint loading. The cranial lunate surface wall is more extensive than the dorsal and caudal walls in all species (Figure 4, top). Interspecific

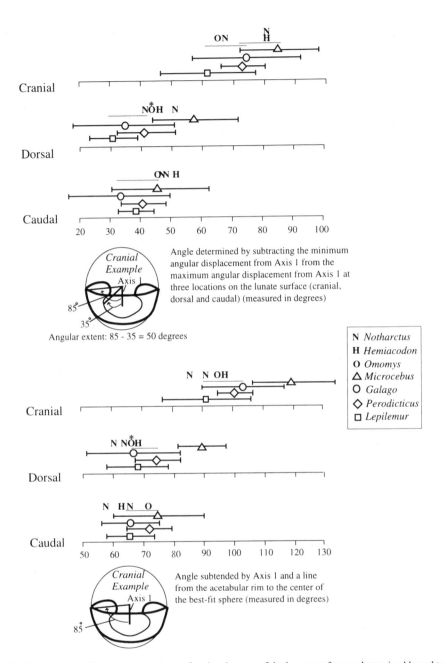

Figure 4. *(Top)* Interspecific mean comparisons of regional extent of the lunate surface as determined by subtracting the minimum angular displacement from Axis 1 (measured in degrees) from the maximum angular displacement from Axis 1 at three locations on the lunate surface (cranial, dorsal and caudal). Data points are means +/- one standard deviation. A horizontal line over extant species means indicates that there is no difference in means as determined by Fisher's LSD (P=0.05). *(Bottom)* Interspecific mean comparisons of regional femoral coverage as determined by maximum angular displacement from Axis 1 at three locations on the lunate surface (cranial, dorsal and caudal). Same conventions as in *Top*. Note that the dorsal rim of the *Omomys* acetabulum is slightly eroded and therefore the total extent of the dorsal acetabular wall and the maximum extent of the dorsal acetabular wall are probably underestimates.

differences show few distinct trends, with the notable exception of the dorsal acetabular surface, which is more extensive in *Microcebus* than in the other extant species. The fossil species group with *Galago, Perodicticus* and *Lepilemur,* with the exception of one *Notharctus* specimen with a degree of dorsal expansion that is greater than in *Galago, Perodicticus* and *Lepilemur*, though not as great as in the committed quadruped *Microcebus*.

The maximum extent of the lunate surface reflects the amount of coverage that the acetabulum provides for the femoral head. In the four extant species, the cranial aspect of the lunate surface extends significantly farther up the sides of the best-fit sphere than do the dorsal and caudal surfaces, which are similar (Figure 4, bottom). Hence this parameter and that assessing lunate surface expansion described above both indicate that across species, there is a preponderance of joint surface on the cranial aspect of the lunate surface. The quadrupedal *Microcebus* has more dorsal and cranial coverage than do *Galago, Perodicticus,* and *Lepilemur. Notharctus, Hemiacodon* and *Omomys* have a pattern of coverage like that of *Galago, Perodicticus* and *Lepilemur* with less cranial and dorsal coverage than *Microcebus*.

3.2. Ranges of Abduction

Careful monitoring of the *Galago* hip joint during computer simulations revealed that the femoral head and acetabulum never contacted each other during simulations, despite the fact that the best-fit sphere used to oppose the two joint components only loosely approximated the *Galago* femoral head. Presumably, contact was avoided because the articular cartilage is thick. A comparison of hip joint X-rays for *Galago* and *Perodicticus* indeed reveals a relatively wide interval between subchondral bone surfaces and therefore thick cartilage in the *Galago* (Figure 5). The spherical femoral modeling is thus deemed adequate for the purpose of the computer simulations.

Little variation was observed in neutral posture among the strepsirhines, with all species tending to have moderately abducted neutral postures (Figure 6, top). *Microcebus*

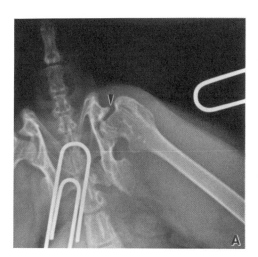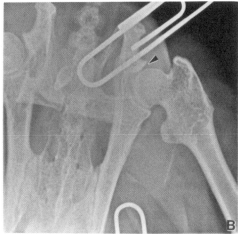

Figure 5. X-rays of hip joints of (*A*) *Galago* and (*B*) *Perodicticus*. Femora are flexed and abducted. Arrows are pointing to the caudal aspect of the acetabulum. Note the relatively "looser" fit of the femoral head in the hip socket of the galago compared to the potto.

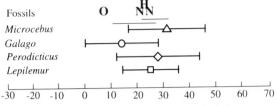

Angle of femoral shaft relative to mid-sagittal plane
in a neutral posture (degrees)

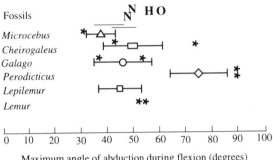

Maximum angle of abduction during flexion (degrees)

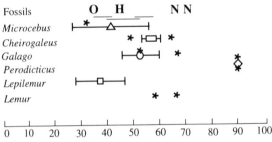

Maximum angle of abduction during extension (degrees)

Figure 6. (*Top*) Angle of femoral shaft with the midsagittal plane when the fovea is centered in the acetabular fossa and the femoral shaft is flexed at an angle of 90° to the ischial ramus. This measure is described in the text as the "neutral posture". Same conventions as in Figure 4. (*Middle*) Maximum angle of abduction during flexion. Bars are means +/- one standard deviation for computer simulations, and stars represent *in vivo* trials. (*Bottom*) Maximum angle of abduction during extension. Bars are means +/- standard deviations for computer simulations, and stars represent *in vivo* trials. No standard deviation bars are shown for *Perodicticus* because the computer program evaluates angles up to 90°, and all angles were greater than or equal to 90°.

has a more abducted neutral posture than *Galago* but otherwise no significant differences were observed (Table 2). In a neutral posture, two *Notharctus* specimens have a mean angle of abduction of 24°, similar to the other strepsirhines. *Hemiacodon* has a similar degree of abduction in a neutral posture (23°), but *Omomys* has a less abducted neutral posture (9°) than the other fossil species and all the extant species except *Galago*.

Results for mean abduction during flexion and extension demonstrate very good concordance between the computer simulations and the *in vivo* data (Figure 6). There were significant interspecific differences in the angles of maximum abduction during flexion, with the slow climber *Perodicticus* having a greater range of abduction than all other species (Figure 6, middle). Among the remaining extant species, only *Cheirogaleus* and *Microcebus* differ significantly (Table 2). The two *Lemur fulvus* specimens, like the three Eocene primates, have maximum angles of abduction that are similar to all but *Microcebus* at one extreme and *Perodicticus* at the other.

During abduction in the flexed posture, it was sometimes possible to rotate the head of the femur so that the fovea was no longer opposite the fossa, but was inferolateral to the acetabular notch; *in vivo* this would result in the *ligamentum teres* being dragged into the

Table 2. Comparisons of properties between species. All units are in degrees

Variable	Microcebus		Cheirogaleus		Galago		Perodicticus		Lepilemur		Interspecific comparisons[1]
	Mean	SD	Mean	SD	Mean	SD	Mean	SD	Mean	SD	
Cranial extent of lunate surface	84.4	13.5			73.8	18.4	72.3	7.2	60.9	16.3	M/L
Dorsal extent of lunate surface	57.3	14.5			34.2	16.6	41.0	9.5	30.9	8.1	M/G, M/P, M/L
Caudal extent of lunate surface	45.7	16.4			33.7	16.2	40.7	7.7	38.6	6.0	
Cranial femoral coverage	119.9	13.2			103.6	13.9	100.7	6.5	91.5	14.5	M/G, M/P, M/L
Dorsal femoral coverage	89.9	8.3			67.2	16.3	74.6	7.5	68.4	10.4	M/G, M/P, M/L
Caudal femoral coverage	74.7	15.0			65.9	9.7	71.9	8.9	65.5	8.4	M/G, M/P, M/L
Neutral femoral posture	32.1	12.8			14.2	12.5	27.8	15.7	25.2	10.9	M/G
Abduction during flexion	38.8	4.0	50.1	10.5	45.9	10.9	73.0	10.3	45.6	8.2	M/C, M/P, C/P, G/P, P/L
Abduction during extension	40.9	13.7	58.2	2.2	52.0	5.7	90.0	0.0	37.7	9.0	M/C, M/P, C/P, C/L, G/P, G/L, P/L

[1] Only those species with significantly different means (P<0.05) as determined by Fisher's LSD test are listed; all other pairings are not significantly different. Species are represented by the first letter of the genus name.

notch, but not over the articular surface. This orientation was regularly achieved in the potto specimens, which had wide, cranioventrally oriented acetabular notches. Other strepsirhines possess a wide notch, but their femoral necks and/or greater trochanters tended to inhibit abduction before the fovea could position itself inferolateral to the acetabular notch.

Pottos have greater hip abduction during femoral extension than do the other strepsirhines (Figure 6, bottom). There is greater variability among the other extant taxa during extension than in flexion, with *Cheirogaleus* and *Lemur* having greater mobility than mouse lemurs and *Lepilemur*, while *Galago* is intermediate. Greater variability was also observed among the fossil species: the 2 *Notharctus* specimens have mobilities that exceed all but potto. *Hemiacodon* falls within a single standard deviation of the means of *Microcebus, Galago* and *Lepilemur*, while *Omomys* is similar to *Microcebus* and *Lepilemur*.

Repeated measures of angle of abduction in the same individual yielded estimates that differed by 3.2° (or 6.5 % difference) on average (ten trials), substantially better than the 5–10° differences reported using manipulated osteological specimens (Swartz, 1989).

4. DISCUSSION

4.1. Lunate Surface Morphology

4.1.1. Extant Species. Analysis of acetabular morphology discriminates between *Microcebus* on the one hand, and *Galago, Perodicticus* and *Lepilemur* on the other. The relatively expanded cranial lunate surfaces combined with unexpanded dorsal lunate surfaces in *Lepilemur* and *Galago* seem reasonable given the frequent vertical behaviors, especially leaping, exhibited by these taxa (Table 1). Likewise, the more expanded dorsal articular surface (and inferred higher dorsal loads) in the quadrupedal *Microcebus* compared to *Galago, Perodicticus* and *Lepilemur* agrees with inferences made from locomotor data (Table 1). *Microcebus* also practices leaping, but lands on all fours after leaping, unlike *Lepilemur* and *Galago senegalensis*, which frequently land only on their hind feet (Walker, 1979; Oxnard et al., 1990).

Somewhat surprisingly, the acetabular morphology of the potto resembles that of *Lepilemur* and especially *Galago* in suggesting primarily cranially-directed loads. Although pottos use vertically (and diagonally) oriented substrates (Charles-Dominique, 1971), their varied locomotor repertoire was expected to generate relatively uniform acetabular loading. It may be that the acetabular morphology in this species is phyletic baggage from a leaping ancestor (Walker, 1969), and that joint reaction forces from slow climbing have been insufficient to cause significant acetabular remodeling. Other comparative studies of lorisid pelves have documented that morphological variation between galagos and lorises is not as great as might be expected given their divergent locomotor modes (McArdle, 1981; Fleagle and Anapol, 1992).

It is also possible that the loading environment is not adequately assessed by a consideration of "pronograde" vs. "orthograde" positional behavior, although studies on anthropoids have shown they are strongly correlated (MacLatchy and Bossert, 1996; MacLatchy, 1996). Studies of the human acetabulum have demonstrated that there are significant relationships between orthogrady and both joint force direction (cranial) and the locations of the highest density cortical and trabecular bone (cranial) (Dalstra *et al.*, 1993; Dalstra and Huiskes, 1994). It is possible, however, that the distribution of acetabular articular surface may reflect force transmission patterns as much as habitual orientation of the pelvis. Hip joint loads being transferred to the sacroiliac joint must pass through at least the lateral aspect of the cranial acetabular region, hence, regardless of trunk orientation, the cranial acetabulum may be under greater overall stress than the dorsal or caudal lunate surfaces. This would also explain why *Microcebus*, despite having more dorsal acetabular articular surface than the other species, also has more articular surface on the cranial aspect of the acetabulum than elsewhere. These issues could best be resolved with experimental data. One could insert pressure sensitive film (e.g., as in Wang *et al.*, 1995) into the acetabula of cadaver specimens and determine contact areas and pressures about the hip socket under different femoral loads. The loading environment can also be reconstructed from mechanical testing of cadaver pelves using strain gauges to measure stress distributions, and through finite element modeling (e.g., Dalstra et al., 1995).

Lunate surface morphology was used to infer mobility as well as loading. *Microcebus murinus* has more femoral coverage than do the other primates for both the cranial and dorsal surfaces. Greater cranial coverage would inhibit medial rotation while greater dorsal coverage would inhibit abduction, and, indeed, the maximum range of abduction of *Microcebus* tended to be somewhat lower than those of most of the other species. The slightly reduced abductory capabilities of *Microcebus* may be related to its small size. Jenkins has proposed that for small mammals, both terrestrial and arboreal substrates pose similar challenges (1974). These substrates necessitate a versatile locomotor pattern and significant abductory capabilities (Jenkins, 1974); however, to a mouse lemur, the small branch environment may not be as "discontinuous" as arboreal substrates are for larger primates, and may generally require less extremes of hip abduction in order to move between supports.

4.1.2. Fossil Species. The distribution of the acetabular articular surfaces of the three fossil species are like those of *Galago, Perodicticus* and *Lepilemur*. The cranial lunate surface is expansive (particularly in one *Notharctus* specimen and in *Hemiacodon*), with less articular surface on the dorsal and caudal aspects. One *Notharctus* individual, however, has an expanded dorsal lunate surface that approaches, but does not reach, the level of expansion seen in *Microcebus*. Pronounced cranial expansion could be indicative of vertically oriented behaviors, but this association is speculative since the potto, a species that frequently uses horizontal and diagonal supports as well as vertical ones (Charles-Dominique, 1971), has a similar acetabular morphology. Little dorsal femoral coverage in all three species suggests that the potential for hip abduction was substantial.

4.2. Ranges of Abduction

4.2.1. Extant Species. In a neutral posture all of the small bodied primates in this study have femoral shaft angles that are equal to or greater than the angles of abduction found in anthropoid primates (MacLatchy, 1996). The angles are similar to the degree of abduction in a stationary stance for non-cursorial small mammals such as the opossum, echidna and hamster (35–40°), the rat and ferret (25°), and the tree shrew (30°), and, with the exception of *Galago*, greater than values for more cursorial mammals such as the cat and fox (10°) (Jenkins, 1971; Jenkins and Camazine, 1977). The leapers (*Galago* and *Lepilemur*) and above-branch quadruped (*Microcebus*) do not have limbs that are more adducted in a neutral posture than does *Perodicticus*, despite the potto's frequently abducted limbs (Walker, 1969).

Simulations of maximum hip abduction agree well with *in vivo* measures. This is particularly significant because it suggests that the bony constraints used in the simulations, namely fovea and fossa locations and sizes, and greater trochanter height, are the primary determinants of maximum abduction. This is not to deny the importance of soft tissue features in joint function, but structures like the acetabular labrum are probably more important in providing joint stability, hence their absence from the models does not seem to affect the predictive power of the three-dimensional animation.

All of the strepsirhines, except the potto, had similar maximal ranges of abduction during flexion. Abduction during flexion regularly occurs during climbing, as when an animal reaches for a distant support, as well as during various feeding and resting postures. It is also associated with vertical clinging postures prior to leaping in some primates (Anemone, 1990; Demes et al., 1996).

More interspecific variability in mean maximal abduction during extension is evident. A similar study on anthropoids (MacLatchy, 1996) found that chimpanzees and Old World monkeys do not differ very much in maximal abduction during flexion, but during extension, they have different maximal angles of abduction. An abducted, extended femoral posture is probably rarer than an abducted, flexed femur in all but the potto whose legs are habitually abducted and rotated even in extension. Abduction during extension probably only occasionally occurs during active climbing, since it would require that an animal back up (e.g., descend tail first along a vertical support), or maintain a grasp on a support while the femur is extended, and is probably more likely to be limited to postural behaviors. It is hypothesized that the ability to both extend and abduct the femur is a characteristic of primates capable of varied climbing postures, the potto and other lorisids being extreme examples (Oxnard et al., 1990; Schulze and Meier, 1995; Figure 1). The extreme mobility of the potto hip is partly enabled by the wide acetabular notch, which allows considerable repositioning of the *ligamentum teres*. Although not quantified in this study, the apparent marked elevation of the femoral head above the neck may also play a role in enabling hip abduction, similar to the way in which relatively small neck size in orangutans has been proposed to facilitate abduction by providing more articular contact area on the superior aspect of the femoral head (Ruff, 1988).

4.2.2. Fossil Species. In a neutral posture, the two *Notharctus* specimens and *Hemiacodon* had a relatively abducted femur like most of the strepsirhines; *Omomys'* femoral posture was more adducted, but fell within the broad range of values exhibited by *Galago*.

For all three fossil species, maximum angles of abduction during flexion fell within the middle of the range of the extant sample and indicate significant abductory, and therefore climbing, capabilities. The two *Notharctus* specimens distinguished themselves by having a degree of abduction during extension that was equal to or greater than all but the potto, suggesting that this species may have been capable of more versatile climbing postures than the other extant strepsirhines and the omomyids. *Omomys* has a lower degree of abduction during extension than *Hemiacodon*; combined with postural data this suggests that *Omomys* may have used less abducted hindlimb postures.

4.3. Femoral Morphology

Additional information is supplied by the morphology of the femoral head of *Hemiacodon* and *Omomys*, which are similar in appearance to those of *Tarsius* and *Galago* (Simpson, 1940; Napier and Walker, 1967; Walker, 1974; MacLatchy, 1993; Anemone, 1995; Dagosto and Schmid, 1996), and which are characterized by a flattened posterior portion and a spherical anterior portion. This femoral morphology has somewhat inappropriately been described as "cylindrical" (Walker, 1974); among living primates, only

Galago and *Tarsius* have "cylindrical" femoral heads (e.g., Walker, 1974; McArdle, 1981; Anemone, 1990). The "cylindrical" characterization stems from the posterior and superior extension of the articular surface onto the neck, and the observation that the neck is relatively thick and oriented perpendicular to the shaft (Walker, 1974). This characterization, however, ignores the proximal portion of the femoral head, which consists of an anteriorly offset hemispherical cap.

The pronounced extension of the subchondral bone on to the back of the neck has been interpreted to be a postural adaptation to vertical clinging (Anemone, 1990, 1995). In the clinging posture, the posterior part of the femoral head is in contact with the caudal articular surface of the acetabulum and the femora are strongly flexed, abducted and laterally rotated. When a femur is placed in the acetabulum (either by hand or in the computer) to simulate a vertically clinging posture (the close-packed position (Figure 5)) and then extended, it is clear that the caudal aspect of the acetabulum articulates smoothly with the relatively flat posterior surface of the femoral head. Hence, during flexion and extension the posterior aspect of the femur is effectively loaded as a cylinder. Rather than a postural adaptation, I suggest that the posterior flattening of the head may provide stability during flexion/extension, as in cursorial mammals such as bovids whose femoral heads are flattened on both the anterior and posterior aspects (Kappelman, 1988). While this femoral morphology is found in tarsiers (Anemone, 1990), Miocene lorisids (Walker, 1970) as well as *Hemiacodon* and *Omomys* (Dagosto, 1993; Anemone, 1995; Dagosto and Schmid, 1996), other clinging and leaping primates, notably the indriids (Demes et al., 1996) and *Lepilemur* have more spherical femoral heads. Demes et al. (1996) have suggested that the "cylindrical" head shape may provide necessary stability during the higher substrate forces generated by smaller species during leaping. This hypothesis thus implies a significant leaping adaptation in the fossil primates that bear this morphology: *Hemiacodon, Omomys* and Miocene lorisids.[*]

There is no reason to suppose that *Galago* (or *Hemiacodon* or *Omomys*) is inordinately limited in its range of hip movement by its femoral head morphology. Most of the proximal hemispherical cap is not in contact with the acetabulum during flexion and extension. Because the anterior cap is spherical, it will permit substantial medial rotation and adduction from the laterally rotated, abducted position of the femur during flexion (MacLatchy, 1993). The composite femoral morphology suggests a joint with stability through flexion and extension but also a wide range of rotation and abduction/adduction (Gebo, 1989).

The femoral features of the omomyines are congruent with a *Lepilemur* or *Galago*-like locomotor repertoire that includes both leaping and climbing. Other workers have found that some aspects of the *Hemiacodon* skeleton, such as the ischial morphology (Fleagle and Anapol, 1992) and the tarsal bones (Simpson, 1940) lack *Galago*-like specializations for leaping. In the case of *Omomys*, Covert (1995) notes that the distal femur is anteroposteriorly deep, supporting leaping adaptations, while Dagosto (1993) interprets foot bones to be elongated and rather cheirogaleid-like, but without the extreme lengthening seen in *Tarsius* or *Galago*. Thus, there is a rather interesting parallel between *Omomys* and *Hemiacodon* and Miocene lorisids: all have the "cylindrical" femoral head, which may be associated with leaping, but lack the *Galago*-like elongation of the tarsals (Simpson, 1940; Dagosto, 1993; Walker, 1970). This implies leaping kinematics that probably

---

[*] Walker's (1970:254) initial characterization of the Miocene lorisids was that they had "...a vertical clinging and leaping locomotion like that of modern galagids." Gebo (1989:362) subsequently described the Miocene species as having "...a more generalized locomotor pattern than the vertical clinging and leaping characteristic of galagos...". Femoral features such as a straight femoral shaft and an anteroposteriorly high and mediolaterally narrow distal femur, however, argue for leaping adaptations.

differed substantially from the extant "foot-powered jumpers" (*sensu* Gebo and Dagosto, 1988) that have a "cylindrical" femoral head morphology. Hence, the shared features of the *Hemiacodon, Omomys* and *Galago* hip joints probably imply gross similarities related to their locomotor mode, but not an equivalency in the biomechanics of their leaping.

5. SUMMARY

One of the most useful consequences of a three-dimensional approach to studying joints is that the separate elements can be brought together within the computer in order to investigate how they might have functioned as a unit. Hip posture and mobility can be directly quantified, instead of relying on intermediate proxies to infer skeletal excursion. Based on preliminary comparisons of the computer simulations and *in vivo* ranges of movement in strepsirhines, bony morphology seems to be the primary constraint to joint mobility in the hip. This bodes well for behavioral reconstruction of hip use in fossil species.

Differences in abduction during extension may be an indicator of differential ability to assume versatile suspensory, bridging and/or reaching postures, while abduction during flexion may be a more general indicator of arboreality, except in species with extremely mobile hips like *Perodicticus*. The potto had a greater range of maximum excursion than the other strepsirhines despite the fact that its acetabular anatomy and its neutral posture (and therefore its foveal morphology) did not differ from them. This extreme mobility may be enabled by the wide acetabular notch and the elevation of the femoral head above the neck.

Acetabular morphology does not distinguish between loading environments or differential mobilities in *Galago, Perodicticus* and *Lepilemur. Galago, Perodicticus* and *Lepilemur* all have expanded cranial lunate surfaces relative to the rest of the acetabulum. In all but *Perodicticus*, this is supported behaviorally by the frequent use of vertical postures. In contrast, the quadrupedal *Microcebus* also has an extensive dorsal lunate surface, indicative of relatively higher or more frequent dorsal loads. The maximum extent of the lunate walls was used to gauge femoral coverage, which is related to hip mobility. Relatively lower cranial and dorsal acetabular walls in *Galago, Perodicticus* and *Lepilemur* supports greater hip mobility than in *Microcebus*.

The three fossil species exhibit a similar acetabular morphology, one that was not characterized by substantial dorsal loading and, therefore, does not suggest restriction to committed quadrupedalism. Mobility was moderate although *Notharctus* had a greater range of maximum hip abduction than did the omomyids, and so possibly had a more versatile repertoire of hindlimb postures. Femoral head morphology in the two omomyids is proposed to be related to providing stability during flexion and extension without compromising mobility, and suggests leaping and climbing adaptations.

ACKNOWLEDGMENTS

I am grateful to A.W. Crompton and Maria Rutzmoser at the Museum of Comparative Zoology, and to Guy Musser and Ross MacPhee at the American Museum of Natural History for allowing me to study specimens under their care. Gregg Gunnell (University of Michigan) and John Alexander (AMNH) kindly permitted me to study their unpublished fossil material. I thank Ken Glander for permission to conduct research at the Duke University Primate Center, Patricia Feeser (DUPC) for performing the anesthetizations and Ted Wheeler (Duke University Medical Center) for taking the X-rays. Ultra-detailed X-ray film was provided free of charge by Kodak. Thank-you to Edward Cheal (VA Hospital, West Roxbury, MA) for use of the digitizing program "DIGIT", and especially to Debbie Burstein and PV Prasad (both of the Beth Israel Hospital, Boston, MA) for assistance with the MRI. Three-dimensional computer programs were written by William Bossert. Figures 1 and 2 were prepared by Luci Betti. Valuable comments were provided by William Bossert, Brigitte Demes, Farish Jenkins, Jr., David Pilbeam, three anonymous reviewers and, especially, Elizabeth Strasser. Generous funding was provided by the National Science Foundation (SBR 9300671), the Wenner-Gren Foundation and the Mellon Foundation.

REFERENCES

Anemone RL (1990) The VCL hypothesis revisited: Patterns of femoral morphology among quadrupedal and saltatorial prosimian primates. Am. J. Phys. Anthropol. *83*:373–393.

Anemone RL (1995) The hip of *Omomys carteri,* a North American omomyine primate from the Middle Eocene. Am. J. Phys. Anthropol. Suppl. *20*:58.

Beard KC, Dagosto M, Gebo DL, and Godinot M (1988) Interrelationships among primate higher taxa. Nature *331*:712–714.

Cartmill, M (1972) Arboreal adaptations and the origin of the Order Primates. In R Tuttle (ed.): The Functional and Evolutionary Origin of Primates. Chicago: Aldine, pp. 97–122.

Charles-Dominique P (1977) Ecology and Behaviour of Nocturnal Primates. London: Duckworth.

Covert HH (1995) Locomotor adaptations of Eocene primates: Adaptive diversity among the earliest prosimians. In L Alterman, GA Doyle, MK Izard (eds.): Creatures of the Dark: The Nocturnal Prosimians. New York: Plenum Press, pp 495–509.

Covert HH and Jolley LA (1995) The forelimb of *Omomys carteri*, a North American primate from the middle Eocene. Am. J. Phys. Anthropol. Suppl. *20*:79.

Dagosto M (1983) Postcranium of *Adapis parisiensis* and *Leptadapis magnus* (Adapiformes, Primates). Adaptational and phylogenetic significance. Folia Primatol. *41*:49–101.

Dagosto M (1988) Implications of postcranial evidence for the origin of euprimates. J. Hum. Evol. *17*:35–56.

Dagosto M (1993) Postcranial anatomy and locomotor behaviour in Eocene primates. In DL Gebo (ed.): Postcranial Adaptation in Nonhuman Primates. DeKalb: Northern Illinois University Press, pp. 199–220.

Dagosto M, and P Schmid (1996) Proximal femoral morphology of omomyiform primates. J. Hum. Evol. *30*:29–56.

Dalstra M, and Huiskes R (1995) Load transfer across the pelvic bone. J. Biomech. *28*:716–724.

Dalstra M, Huiskes R, Odgaard A, and van Erning L (1993) Mechanical and textural properties of pelvic trabecular bone. J. Biomech. *26*:523–535.

Demes B, Jungers WL, Wunderlich RE, Richmond BG, Fleagle JG, and Lemelin P (1996) Body size and leaping kinematics in Malagasy vertical clingers and leapers. J. Hum. Evol. *31*:367–388.

Elftman HO (1929) Functional adaptations in the pelvis of marsupials. Bull. Am. Mus. Nat. Hist. *58*:189–232.

Fleagle JG (1976) Locomotor behaviour and skeletal anatomy of sympatric Malaysian leaf monkeys (*Presbytis obscura* and *Presbytis melalophos*). Yrbk. Phys. Anthropol. *20*:441–453.

Fleagle JG (1988) *Primate Adaptation and Evolution*. New York: Academic Press.

Fleagle JG, and Anapol FC (1992) The indriid ischium and the hominid hip. J. Hum. Evol. *22*:285–305.

Fleagle JG, and Meldrum DJ (1988) Locomotor behaviour and skeletal morphology of two sympatric pithecine monkeys, *Pithecia pithecia* and *Chiropotes satanas*. Am. J. Primatol. *16*:227–249.

Gebo DL (1989) Postcranial adaptation and evolution in Lorisidae. Primates *30*:347–367.

Gebo DL (1985) The nature of the primate grasping foot. Am. J. Phys. Anthropol. *67*:269–277.

Gebo DL, and Dagosto M (1988) Foot anatomy, climbing, and the origin of the Indriidae. J. Hum. Evol. *17*:135–154.

Gregory WK (1920) On the structure and relations of *Notharctus*, an American Eocene primate. Mem. Am. Mus. Nat. Hist. *3*:49–243.

Harrison T (1982) Small-bodied apes from the Miocene of East Africa. Ph.D. Dissertation, University College of London.

Harvey PH, Martin RD, and Clutton-Brock TH (1987) Life histories in comparative perspective. In BB Smuts, DL Cheney, RM Seyfarth, RW Wrangham, and TT Struhsaker (eds.): Primate Societies. Chicago: University of Chicago Press, pp. 181–196.

Jenkins FA, Jr. (1971) Limb posture and locomotion in the Virginia opossum (*Didelphis marsupialis*) and in other non-cursorial mammals. J. Zool., Lond. *165*:303–315.

Jenkins FA, Jr. (1974) Tree shrew locomotion and primate arborealism. In FA Jenkins, Jr. (ed.): Primate Locomotion. New York: Academic Press, pp. 85–115.

Jenkins FA, Jr., and Camazine SM (1977) Hip structure and locomotion in ambulatory and cursorial carnivores. J. Zool., Lond. *181*:351–370.

Kappelman J (1988) Morphology and locomotor adaptations of the bovid femur in relation to habitat. J. Morphol. *198*:119–130.

Kay RF, Ross C, and Williams B (1997) Anthropoid origins. Science *275*:797–804.

Latimer B, Ohman JC, and Lovejoy, CO (1987) Talocrural joint in African hominoids: Implications for *Australopithecus afarensis*. Am. J. Phys. Anthropol. *74*:155–175.

MacLatchy LM (1993) Three dimensional quantification of hip morphology in fossil primates. J. Vert. Paleontol. *13*:48a.

MacLatchy LM (1996) Another look at the australopithecine hip. J. Hum. Evol. *31*:455–476.

MacLatchy LM, and Bossert W (1996) An analysis of the articular surface distribution of the femoral head and acetabulum in anthropoids, with implications for hip function in Miocene hominoids. J. Hum. Evol. *31*:425–453.

McArdle JE (1981) Functional morphology of the hip and thigh of the loriformes. Contrib. Primatol. *17*:1–132.

McHenry HM (1975) A new pelvic fragment from Swartkrans and the relationship between the robust and gracile australopithecines. Am. J. Phys. Anthropol. *43*:245–262.

Martin RD (1990) Primate origins and evolution: A phylogenetic reconstruction. Princeton: Princeton University Press.

Martin RD (1993) Primate origins: Plugging the gaps. Nature *363*:223–234.

Mittermeier RA, Tattersall I, Konstant WR, Meyers DM, Mast RB, and Nash S (1994) Lemurs of Madagascar. Washington: Conservation International.

Napier JR ,and Napier PH (1967) A Handbook of Living Primates. New York: Academic Press.

Napier JR, and Walker AC (1967) Vertical clinging and leaping, a newly recognized category of locomotory behaviour among primates. Folia Primatol. *6*:180–203.

Oxnard CE, Crompton RH, and Lieberman SS (1990) Animal Lifestyles and Anatomies. Seattle: University of Washington Press.

Ruff CB (1988) Hindlimb articular surface allometry in Hominoidea and *Macaca*, with comparisons to diaphyseal scaling. J. Hum. Evol. *17*:687–714.

Simpson GG (1940) Studies on the earliest primates. Bull. Am. Mus. Nat. Hist. *77*:185–212.

Schulze H, and Meier B (1995) Behavior of captive *Loris tardigradus nordicus*: A qualitative description, including some information about morphological bases of behavior. In L Alterman, GA Doyle, and MK Izard (eds.): Creatures of the Dark: The Nocturnal Prosimians. New York: Plenum Press, pp 221–249.

Sokal RR, and Rohlf FJ (1981) Biometry. San Francisco: W.H. Freeman and Company.

Stern JT, and Susman RL (1983) The locomotor anatomy of *Australopithecus afarensis*. Am. J. Phys. Anthropol. *60*:279–317.

Swartz SM (1989) The functional morphology of weight bearing: Limb joint surface area allometry in anthropoid primates. J. Zool., Lond. *218*:441–460.

Szalay FS, and Dagosto M (1980) Locomotor adaptations as reflected on the humerus of Paleogene primates. Folia Primatol. *34*:1–45.

Turnquist JE (1983) Influence of age, sex and caging on the joint mobility in the patas monkey (*Erythrocebus patas*). Am. J. Phys. Anthropol. *61*:211–220.

Turnquist JE (1985) Passive joint mobility in patas monkeys (*Erythrocebus patas*): Rehabilitation of caged animals after release into a free-ranging environment. Am. J. Phys. Anthropol. *67*:1–5.

Walker A (1969) The locomotion of the lorises, with special reference to the potto. J. E. Afr. Wildlife *7*:1–5.

Walker A (1970) Post-cranial remains of the Miocene Lorisidae of East Africa. Am. J. Phys. Anthropol. *33*:249–262.

Walker A (1974) Locomotor adaptations in past and present prosimian primates. In FA Jenkins, Jr. (ed.): Primate Locomotion. New York: Academic Press, pp. 349–381.

Walker A (1979) Prosimian locomotor behaviour. In GA Doyle, and R D Martin (eds.): The Study of Prosimian Behaviour. London: Duckworth, pp. 543–565.

Wang C, Cheng C, Chen C, Lu C, Hang Y, and Liu T (1995) Contact areas and pressure distributions in the subtalar joint. J. Biomech. *28*:269–279.

Ward CV (1991) Functional anatomy of the lower back and pelvis of the Miocene hominoid *Proconsul nyanzae* from Mfangano Island, Kenya. Ph.D. Dissertation, Johns Hopkins University, Baltimore.

8

GRASPING PERFORMANCE IN *SAGUINUS MIDAS* AND THE EVOLUTION OF HAND PREHENSILITY IN PRIMATES

Pierre Lemelin[1] and Brian W. Grafton[2]

[1]Department of Anatomy
Northeastern Ohio Universities College of Medicine
Rootstown, Ohio 44272
[2]School of Biomedical Sciences
Division of Biological Anthropology
Kent State University
Kent, Ohio 44242

1. INTRODUCTION

The Order Primates is characterized by a unique suite of cranial and postcranial features that may have evolved for visual predation on insects or exploitation of fruits in a small-branch milieu [see Cartmill (1992) for recent review]. Compared to more generalized mammals like tree shrews, primates possess a relatively larger brain with a more developed visual area and more reduced olfactory bulbs, orbits that are more approximated and more convergent with one another, and grasping hands and feet that bear nails instead of claws (Cartmill, 1970, 1972, 1974a,b, 1992; Le Gros Clark, 1959, 1963; Martin, 1986; Wood Jones, 1916).

With regard to the grasping extremities of primates, the anthropological literature has usually linked enhanced grasping abilities of the hands and feet of primates with the lack of claws. For example, Le Gros Clark (1959: 174) argued that "Compared with claws these [nails] provide a much more efficient grasping mechanism for animals which find it necessary to indulge in arboreal acrobatics, for by their greater pliability they can be adapted with much more precision to surfaces of varying shape, size, and texture". In an important contribution on primate hands, Napier (1993: 40) stated the following: "Claws are not compatible with prehensility because of the mechanical obstruction of claws overgrowing the fingertips..." In the same vein, Hershkovitz (1970) suggested that claws degenerated in primates to enable the extremities to achieve full phalangeal flexure when gripping. This character association between grasping abilities of the extremities and lack of claws can also be found in many introductory textbooks of physical anthropology. Among the evolutionary trends that distinguish primates from the rest of the mammals, Nelson and colleagues (1992: 113) wrote that primates are characterized by "Flexible hands and feet with a good deal of prehensility (grasping ability)" and added that "This feature is associated directly with the lack of claws and retention of five digits."

The ability to manipulate objects is inherent to the definition of hand prehensility. According to Napier (1961: 116–117), "A convergent hand can be termed prehensile when the digits approximate in such a manner that an object may be grasped and held securely against external influences (e.g., gravity) that may be tending to displace it." Tree shrews, many rodents and some carnivores use both hands together to bring food objects to the mouth, whereas primates can accomplish the same action with only one hand (Bishop, 1964; McClearn, 1992; Napier, 1961, 1993; Polyak, 1957). Again, the argument can be made that claws prevent prehensile grips while an animal is reaching and gripping objects with only one hand.

Tamarins represent an appropriate group of primates to test this hypothesis because of their hand anatomy and ecology. Like all callitrichids, they have hands that bear sharp claws (Hershkovitz, 1977), which are likely to have evolved from a platyrrhine ancestor with nails (Ford, 1980; Rosenberger, 1977; Thorndike, 1968). Field data revealed that tamarins are eclectic feeders that mainly eat ripe fruits and nectar, but also eat arthropods,

Figure 1. *Saguinus midas* (drawing courtesy of Stephen D. Nash). The fingers and toes of the golden-handed tamarin sport sharp claws.

especially orthopteran insects such as grasshoppers and crickets (Garber, 1980, 1984, 1988, 1992, 1993; Mittermeier and van Roosmalen, 1981; Snowdon and Soini, 1988; Soini, 1987; Sussman and Kinzey, 1984; Terborgh, 1983). Tamarins display different insect-catching strategies, but all of them appear to involve a great deal of hand use (Garber, 1993; Terborgh, 1983).

The putative association between presence of claws and the inability to achieve prehensile grips implies that tamarins have limited capabilities of grasping and manipulating using only one hand. This study aimed to test that implication in the golden-handed tamarin, *Saguinus midas* (Figure 1). Manual grasping behavior was observed in tamarins for several types of food objects under experimental conditions. For each type of food object, the ability of tamarins to accomplish single-handed grips was quantified and compared to that of a squirrel monkey (*Saimiri sciureus*), a platyrrhine with nails. Then, finger proportions were compared between callitrichids, squirrel monkeys, and a variety of clawed mammals for which some behavioral data on their manual grasping abilities were available. These comparisons were made in order to identify morphological features of the hand that may be linked with potential differences in grasping behavior.

2. MATERIALS AND METHODS

Three different types of food objects (i.e., blueberries, live crickets, and live mealworms) were presented one by one on a platform placed just outside the bottom of enclosures housing three adult male and one adult female golden-handed tamarins, or one adult male squirrel monkey. During the experiments, live crickets were kept in a cooler filled with ice so that they did move but not hop out of reach of the animals when positioned on the platform. The mesh of the enclosure wire was narrow enough so that only one forelimb could be used to reach for a food object placed on the 9 cm diameter platform. Each subject was able to reach the platform easily using protraction of one forelimb.

For all three categories of food objects, a total of 606 grips for tamarins and 60 grips for the squirrel monkey were recorded on videotape using a Panasonic AG-450 Super VHS camcorder or Sony 8 mm camcorder equipped with a 10X zoom lens and mounted on a tripod. The high-speed shutter (between 1/125 and 1/1000 sec) of the camcorder was used when possible during filming (30 frames/sec). This option was valuable for obtaining still frames with less motion blur. Each videotape was viewed and analyzed with a video cassette recorder plugged into a monitor. The frame-by-frame playback option allowed the classification of grips into successful and unsuccessful events for each food object category. A successful grip involved a reach and a secure grab of the food object, completed by retrieval toward the mouth for ingestion. A grip was considered unsuccessful when an animal reached for a food object, but either failed to grab it securely while touching it, or simply dropped it while bringing the food object to the mouth.

The ability to grasp and retain a food object with only one hand should be reflected in the proportions of the manual ray elements. Shorter metacarpals relative to the digits, longer digits relative to the metacarpals, or a combination of both provide some of the mechanical requirements to achieve prehensile grips. In animals using simultaneous convergence of all the digits during grasping, such proportions allow the digital portion of the ray to encircle and to secure an object against the palm by flexion of both metacarpophalangeal (MP) and proximal interphalangeal (IP) joints. Following Napier (1993), a phalangeal index was computed to obtain an estimate of the degree of prehensility of each finger. Lengths of the metacarpals, proximal phalanges, and middle phalanges for rays II through V were measured on skeletal specimens of callitrichids (*Saguinus midas, Saguinus geoffroyi, Leontopithecus rosalia,* and *Callithrix jacchus*), squirrel monkeys (*Saimiri sciureus*), tree squirrels (*Sciurus carolinensis*), tree shrews (*Tupaia glis*), and procyonids (*Procyon lotor* and *Potos flavus*). The phalangeal index corresponds to the sum of the length of the proximal and intermediate phalanges times one hundred and divided by the length of the corresponding metacarpal. The terminal phalanges were excluded in the calculation of the phalangeal indices for several methodological reasons. The distal phalanges of many skeletal specimens were either missing, disarticulated or covered with dry skin. Moreover, since most skeletal specimens measured for this study possess pointed and elongated distal phalanges in the shape of a claw, it was difficult to assess their true length contributing to the prehensility of the hand.

3. RESULTS

3.1. Behavioral Data

3.1.1. Comparison of Grasping Behavior and Abilities between Saguinus midas *and* Saimiri sciureus. Tamarins and squirrel monkeys used similar prehensive patterns referred to as "whole-hand control" by Bishop (1964). Such prehensive patterns, which are characteristic of prosimians and most New World monkeys (Bishop, 1964; Costello and Fragaszy, 1988), involved no independent movement of the digits and objects were picked up in a manner analogous to the human "power grips" described by Napier (1956, 1993). Although the prehensive patterns of the primates under investigation were fairly uniform, the positions of the hand while holding an object or prehensile grips (Bishop, 1964) varied according to the size and shape of the object held.

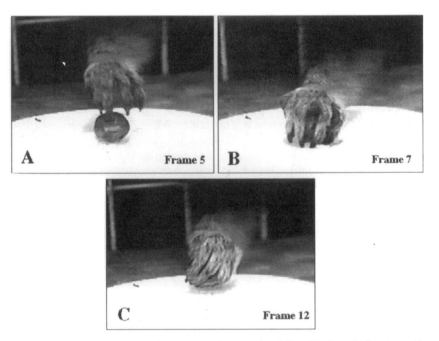

Figure 2. Series of video frames of a single grasping event involving a blueberry in *Saguinus midas.*

Blueberries were gripped in different manners by tamarins and the squirrel monkey despite using a similar "whole-hand control" prehensive pattern. In tamarins, the hand reached with the fingers flexed 90° at the proximal IP joints (Figure 2a) and dropped from above the food object for its retrieval (Figure 2 b,c). In some instances, blueberries were swept from the side instead of from above. Also, blueberries were gripped occasionally by a scissor-grip between the sides of the second and third digits. In the squirrel monkey, the hand reached for blueberries in a similar manner with the fingers flexed 90° at the proximal IP joints. Unlike tamarins, however, the thumb of the squirrel monkey played a more important role when gripping blueberries off the platform. When the fingers encircled a blueberry by flexion of both MP and proximal IP joints, the thumb was pressing laterally on the food object. As a result, the volar surface of the thumb was oriented in a different plane from that of the fingers. Although the thumb was involved extensively during grasping, there was no independent movement of the digits that characterizes precision gripping as in *Cebus apella* (Costello and Fragaszy, 1988).

Blueberries were gripped successfully in a proportion of 9 to 1 in both tamarins and the squirrel monkey (Figure 3). In tamarins, a total of 188 grasping events involving only one hand for blueberries was counted. This number was much lower for the squirrel monkey with a total of 31 events. The percentages of successful grips, however, were similar for both taxa (89.9% for tamarins and 90.3% for the squirrel monkey) (Figure 3).

Tamarins and the squirrel monkey caught live crickets using grips reminiscent of those reported for small-bodied prosimians (Bishop, 1964; Lemelin, 1996a,b; Niemitz, 1984) and *Callithrix jacchus* (Rothe, 1971). In most cases, the hand reached toward the cricket with the digits extended and adducted at the MP joints (Figure 4a). A few centimeters from the cricket, the hand opened with a very rapid movement of abduction of all the digits at their MP joints, including the thumb (Figure 4b). Thus, just before contact with the cricket, the digits were widely spread and covered the greatest area possible. At contact, all the digits converged and flexed at their MP and proximal IP joints, encircling the prey for retrieval (Figure 4c,d). Retrieval of the object toward the mouth was completed by flexion and supination of the forearm.

Like blueberries, crickets were gripped successfully in a proportion of 9 to 1 in both tamarins and the squirrel monkey (Figure 3). In tamarins, 157 single-handed grips involving crickets were counted and 89.9% of these grips were successful. The proportion of

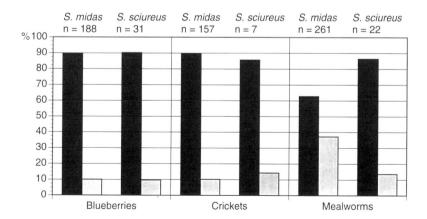

Figure 3. Proportions of successful (black) and unsuccessful (shaded) grips for three types of food objects in *Saguinus midas* and *Saimiri sciureus.*

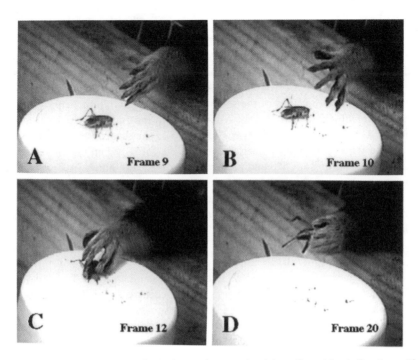

Figure 4. Series of video frames of a single grasping event involving a live cricket in *Saguinus midas.*

successful grips for the squirrel monkey was similar (85.7%). This result should be interpreted with caution, however, because of the very small sample size.

The most notable differences in grasping behavior between tamarins and the squirrel monkey were observed when mealworms were presented. In tamarins, the hand reached toward the mealworm with all the digits extended and adducted at the MP joints (Figure 5a). Prior to contact with the mealworm, the digits were more adducted at their MP joints as for crickets. Upon contact, the mealworm was brought toward the palm by several successive and rapid movements of flexion and extension of all the digits at both MP and IP joints (Figure 5b,c). Despite firm contact between the tips and claws of the central digits and the palm, mealworms had a tendency to slip away during retrieval (Figure 5d). On many occasions, successful grips involving mealworms were completed using scissor-grips between the sides of digits II and III or digits III and IV. In the squirrel monkey, the hand reached for mealworms in a manner similar to that for crickets with the digits extended and abducted at the MP joints. At contact, the fingers flexed at both proximal IP and MP joints to encircle the mealworm. As for blueberries and crickets, the thumb was positioned laterally on the food object so that the volar surface of the thumb was oriented in a different plane from that of the fingers. Unlike tamarins, the thumb played a more important role in the squirrel monkey when gripping and only one movement of flexion and extension of all the digits was necessary to retrieve successfully a mealworm from the platform.

Not surprisingly, the percentages of successful grips for mealworms differed between tamarins and the squirrel monkey (Figure 3). With a total of 261 grips, tamarins only had a 62.8% success rate for mealworms. In contrast, the squirrel monkey had an 86.4% success rate for the same type of food object for a total of 22 grips.

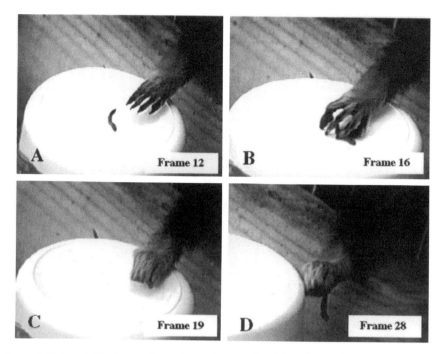

Figure 5. Series of video frames of a single grasping event involving a live mealworm in *Saguinus midas.*

3.2. Morphometric Data

3.2.1. Comparisons of Hand Proportions and Grasping Behavior between Callitrichids, Saimiri, *and Nonprimate Mammals. Saguinus midas* and other callitrichids can be distinguished from *Saimiri sciureus* not only by having claws instead of nails, but also by different hand proportions. Squirrel monkeys possess higher phalangeal indices for the central digits (III and IV) compared to those of each callitrichid taxon (pairwise comparisons using a Mann-Whitney U-test, $P < 0.05$; Figure 6; Table 1). Among callitrichids, *L. rosalia* has significantly lower phalangeal indices for all rays (Mann-Whitney U-test, $P < 0.01$; Figure 6; Table 1). This can be explained by the more important contribution of the metacarpals in total hand length in *L. rosalia* compared to *Callithrix* and *Saimiri* (Jouffroy et al., 1991). Of all anthropoids, *Leontopithecus* possesses the longest hands relative to forelimb length (Jouffroy et al., 1991) and compared to other callitrichids and *Saimiri,* *Leontopithecus* captures prey by probing through narrow holes or dense foliage (Hershkovitz, 1977; Rylands, 1989). Relatively long hands, independently of their degree of prehensility, may be more useful for probing behavior.

For most rays, primates and several nonprimate mammals form two distinctive clusters (Figure 6). Raccoons (*Procyon*) have the lowest phalangeal indices of all taxa considered in the comparative sample (Figure 6; Table 1). McClearn (1992) mentioned that *P. lotor* uses both forepaws to hold small food items despite its notorious ability to locate and handle food items with the hands. In contrast, its close relative the kinkajou (*Potos flavus*) commonly relies on single-handed grips to hold small food objects like wild figs (McClearn, 1992). Interestingly, kinkajous are well-differentiated from raccoons in terms of hand proportions (Figure 6; Table 1). For all rays, kinkajous have higher phalangeal in-

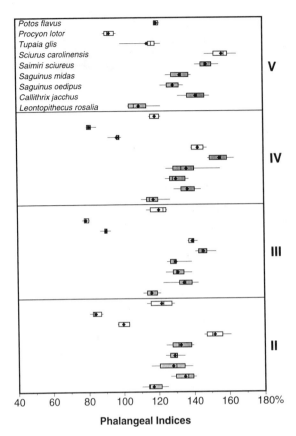

Figure 6. Box-plot charts of the phalangeal indices for manual rays II through V in nonprimate mammals, *Saimiri*, and callitrichids. The diamond represents the mean, the vertical line the median, the left and right corners of the rectangle the 25th and 75th percentile, and the left and right ends of the horizontal line the 10th and 90th percentile. The shaded rectangles represent platyrrhine taxa and open ones nonprimate mammals. See Table 1 for sample sizes.

Table 1. Descriptive statistics of the phalangeal indices for manual rays II through V of nonprimate mammals, *Saimiri*, and callitrichids[1]

Taxon	Ray II	Ray III	Ray IV	Ray V
Potos flavus	121.71	120.49	118.55	119.93
(N=3)	101.83–141.59	104.66–136.31	110–127.1	115.62–124.24
Procyon lotor[2]	84.03	79.22	81.48	92.95
(N=10, 11)	81.93–86.13	78.09–80.35	80.04–82.92	90.85–95.04
Tupaia glis	99.84	90.2	96.92	114.6
(N=10)	97.77–101.92	88.62–91.78	94.85–99	108.38–120.81
Sciurus carolinensis	151.89	139.22	142.86	156.95
(N=10)	147.6–156.19	137.63–140.8	140.23–145.49	152.27–161.63
Saimiri sciureus	132.39	145.6	155.17	147.94
(N=5)	124.46–140.32	140.07–151.12	147.57–162.77	141.65–154.24
Saguinus midas	128.91	129.94	136.34	133.11
(N=8)	125.6–132.23	125.41–134.48	126.47–146.21	128.18–138.05
Saguinus oedipus	128.03	131.12	130.44	129.19
(N=8)	120.54–135.51	126.62–135.63	126–134.88	125–133.37
Callithrix jacchus	134.72	134.56	136.85	141.97
(N=8)	129.74–139.7	128.38–140.74	132.17–141.52	136.33–147.61
Leontopithecus rosalia	116.84	116.11	117.34	109.59
(N=7)	111.55–122.12	112.56–119.67	111.66–123.01	102.6–116.58

[1]Values under the columns are (*top*) arithmetic mean and (*bottom*) the 95% confidence intervals around the mean.
[2]For ray II N=10, for rays III-V N=11.

dices, thus longer digits relative to the corresponding metacarpals, compared to raccoons (Mann-Whitney U-test, $P < 0.01$). Tree shrews (*Tupaia glis*) cluster in between raccoons and kinkajous, except for ray V (Figure 6). Like raccoons, tree shrews rarely pick up food objects with single-handed grips and both hands are used whenever possible (Bishop, 1964). Unlike other nonprimate mammals considered here, tree squirrels (*Sciurus carolinensis*) have similar or higher phalangeal indices to those of callitrichids and *Saimiri* (Figure 6; Table 1). The digits of *Sciurus* contribute close to 65% of total hand length compared to 55% or lower for *Saimiri* and *Callithrix* (Jouffroy et al., 1991). Behavioral data on grasping abilities of tree squirrels are very scanty. Typically, they use both hands to hold food items, especially when adopting a bipedal stance (Koprowski, 1994).

4. DISCUSSION

4.1. Presence of Claws and Prehensile Abilities

The results of the behavioral experiments demonstrate that despite the presence of claws on their hands, tamarins are as competent as squirrel monkeys to achieve single-handed (i.e., prehensile) grips when picking up food objects from the platform. The nailed digits of squirrel monkeys may be more efficient to grasp smaller and slippery objects such as mealworms. Nonetheless, the comparable success rates for blueberries and crickets between tamarins and squirrel monkeys do not support the general notion that claws hinder the ability of grasping food objects with only one hand.

If claws do not impede the ability to use prehensile grips, one wonders why most primates have nails. Cartmill (1974a) demonstrated that claws are superior to nails and volar skin in most arboreal situations, except for small-branch supports. Field studies by Garber (1980, 1984) and Garber and Sussman (1984) on the positional behavior of *Saguinus geoffroyi*, however, showed that fine-branch supports (less than 3 cm in diameter) are preferred when foraging for fruits and insects. Similarly, a large portion of the locomotor repertoire of *S. midas* during feeding involves quadrupedal movement on small-diameter supports (Fleagle and Mittermeier, 1980; Thorington, 1968). Among nonprimate mammals, *Caluromys,* a South American didelphid marsupial that sports sizable claws except on the hallux, also relies preferentially on small and terminal branches while moving and foraging for fruits and insects high in the canopy (Charles-Dominique et al., 1981; Rasmussen, 1990). These data suggest that claws may not hinder the ability of an animal gathering fruits and chasing insects on fine-branch supports.

Another possible explanation for nailed extremities may lie in the histological structure of the glabrous skin of primates compared to other mammals. In his essay establishing the bases of the arboreal theory of primate origins, Wood Jones (1916: 158) stressed that "Not only does the hand come to take over the crude grasping functions of the teeth and jaws, but in gradual stages it slowly but surely usurps the delicate tactile duties of the muzzle." Le Gros Clark (1959: 285) championed the arboreal theory by reiterating that "...in the evolution of the Primates, the more primitive tactile organs represented by the vibrissae have been gradually replaced by the development of the more delicately informative tactile pads on the terminal phalanges of the digits." Despite the pitfalls of the arboreal theory in explaining the adaptiveness of unique primate features (Cartmill, 1970, 1972, 1974b, 1992), Wood Jones and Le Gros Clark's observations were important: compared to most mammals, primates appear to rely extensively on the tips of their digits to acquire information about their environment.

Comparative studies from Krause (1860) and Winkelmann (1963, 1965) indicated that the glabrous skin of primates possesses a unique type of cutaneous sense organ called Meissner's corpuscle. Meissner's corpuscles are rapidly adapting mechanoreceptors [i.e., the neuronal response occurs as long as the stimulus is moving (Johansson, 1979)] located in the dermal papillae of the glabrous skin and play a vital role in tactile acuity and discrimination (Johansson, 1978, 1979; Munger and Ide, 1988; Vallbo and Johansson, 1978). Studies of humans revealed that the proportion of Meissner's corpuscles is highest at the tip of the digits compared to other areas of the hand (Johansson, 1979; Johansson and Vallbo, 1976, 1979) and that the number of Meissner's corpuscles varies with age or the frequency of manual labor (Cauna, 1959; Quilliam and Ridley, 1971). In contrast, the glabrous skin of most nonprimate mammals has a type of cutaneous sensory organ analogous to Meissner's corpuscles called "mammalian end organ" (Winkelmann, 1963, 1965). Although morphological differences have been described between mammalian end organs and Meissner's corpuscles (Winkelmann, 1965), there is little comparative neurophysiological data to support possible tactile differences between these two types of cutaneous receptors. From a comparative viewpoint, it is also interesting to note that Meissner's corpuscles have been reported for the glabrous skin of *Didelphis virginiana* (Winkelmann, 1964, 1965) and mouse toe pads (Idé, 1976).

From this evidence, it is possible that early in their evolutionary history, primates may have improved tactile faculties at the tip of their digits in response to specific manual tasks by increasing apical pad area and consequently the number of Meissner's corpuscles. It is also possible that this putative increase in area may have triggered morphological changes from a claw to a nail. No present data exist, however, to support these statements. Further study of the histological and molecular structure of the digital volar skin of primates and other mammals, as well as the relationship between the presence/absence of claws, nail size and apical pad area, could prove fruitful in clarifying the problem of the lack of claws in primates.

4.2. Hand Proportions and Prehensile Abilities

The comparison of ray proportions between *Saguinus midas* and other clawed mammals also provides insights in sorting some of the morphological characteristics that co-vary with hand prehensility. For example, *Tupaia* possesses shorter proximal and middle phalanges relative to the length of the metacarpals compared to *Saguinus* (Figure 7). If *Tupaia* uses solely one hand, a food object is secured only marginally between the palm and the digits. Again, behavioral data from Bishop (1964) show that tree shrews have difficulty in achieving prehensile grips and usually use both hands to hold food objects. In contrast, *Saguinus* has longer proximal and middle phalanges relative to the length of the metacarpals (Figure 7). When the flexed digits converge toward the palm, they can completely encircle food objects, thus providing a secure single-handed grip. The behavioral data presented above are evidence for the ability of *S. midas* to accomplish single-handed grips. Similarly, the differences in hand proportions between two closely related procyonid taxa parallel their reported differences in manual prehensile abilities. Raccoons have shorter fingers relative to the palm and usually hold small objects between both hands (McClearn, 1992). In contrast, kinkajous have longer digits relative to the palm and hold small fruits using single-handed grips (McClearn, 1992).

The case of the tree squirrels also demonstrate that factors other than relative digit length were equally important in the evolution of manual prehensility in primates. From the strict point of view of hand proportions, we would expect tree squirrels to behave like

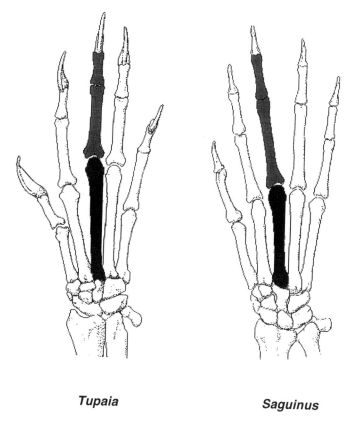

Tupaia   **Saguinus**

Figure 7. Differences in hand proportions between *Tupaia* and *Saguinus*. In tamarins, the proximal and middle phalanges (in gray) are longer relative to the metacarpals (in black) compared to tree shrews. [Modified from Hershkovitz (1977) with permission from The University of Chicago Press]

Saguinus. Comparative studies from Garber (1980) and Garber and Sussman (1984) clearly indicated that *Saguinus* and *Sciurus* are behaviorally and ecologically distinct. Unlike tamarins, tree squirrels (*S. granatensis*) avoid fine branches (less than 3 cm in diameter) during travel and foraging, and transport food objects in their mouth to larger and more stable branches for consumption (Garber, 1980; Garber and Sussman, 1984). From a kinematic point of view, it is possible that tree squirrels have more limited flexion when the fingers converge toward the palm, thus preventing prehensile grips, compared to primates with similar hand proportions.

Tree squirrels are also very different from tamarins and other primates in the anatomy of their visual system. Although tree squirrels possess an enlarged cortical representation of the binocular portion of their field of vision (Hall et al., 1971; Kaas et al., 1972), they have much less convergent and frontated orbits compared to primates, especially anthropoids (Cartmill, 1970, 1972, 1974b; Ross, 1995, 1996). Allman (1977: 27) emphasized the importance of more frontally directed orbits to improve "...the *optical quality* of the *retinal images* for the *central part* of the visual field located in front of the animal..." [italicized words are by Allman]. This is especially useful for predatory animals relying on precise eye-hand coordination to catch moving prey (Allman, 1977). A recent study by Servos et al. (1992) demonstrated that humans perform significantly better when reaching and grasping objects with one hand under binocular vs. monocular viewing conditions. On that basis, we suggest that elongation of the digits relative to the palm was probably a critical factor in the evolution of manual prehensility in early primates, *in combination* with other factors like the development of musculoskeletal features that promoted digital movement as well as neural specializations that enhanced vision quality, hand control, and tactile faculties of the digits.

ACKNOWLEDGMENTS

We would like to thank Drs. Alfred Rosenberger, John Fleagle, Henry McHenry, and Elizabeth Strasser for inviting us to contribute a chapter for this book. Dr. Patricia Wright gave us permission to film the tamarins, Dr. Patricia McDaniel and Alan Sironen from the Cleveland Metroparks Zoo granted us permission to film the squirrel monkey under their care, and Joanne Labate and the caretakers of the Cleveland Metroparks Zoo provided invaluable help during the behavioral experiments. The curators and staffs of the American Museum of Natural History (New York), Anthropologisches Institut und Museum der Universität Zürich-Irchel (Zürich), Cleveland Museum of Natural History (Cleveland), Field Museum of Natural History (Chicago), Museum of Comparative Zoology, Harvard (Cambridge), National Museum of Natural History (Washington, D.C.), and Nationaal Natuurhistorisch Museum (Leiden) gave much appreciated help and access to their skeletal collections. Stephen Nash kindly provided the artwork for Figure 1 and The University of Chicago Press gave permission to use copyrighted material for Figure 7. Finally, we are grateful to Dr. Paul Garber and especially to an anonymous reviewer who gave very thorough and useful reviews on earlier versions of this manuscript. This research was supported by grants from the National Science Foundation (SBR-9318750), Sigma-Xi, The Scientific Research Society, and the Doctoral Program in Anthropological Sciences, SUNY at Stony Brook to PL.

REFERENCES

Allman J (1977) Evolution of the visual system in the early primates. In JM Sprague, and AN Epstein (eds.): Progress in Psychobiology and Physiological Psychology. Vol. 7. New York: Academic Press. pp. 1–53.

Bishop A (1964) Use of the hand in lower primates. In J Buettner-Janusch (ed.): Evolutionary and Genetic Biology of Primates, Vol. 2. New York: Academic Press, pp. 133–225.

Cartmill M (1970) The orbits of arboreal mammals: A reassessment of the arboreal theory of primate evolution. Ph.D. Dissertation, University of Chicago.

Cartmill M (1972) Arboreal adaptations and the origin of the order Primates. In RH Tuttle (ed.): The Functional and Evolutionary Biology of Primates. Chicago: Aldine-Atheton, pp. 3–35.

Cartmill M (1974a) Pads and claws in arboreal locomotion. In FA Jenkins (ed.): Primate Locomotion. New York: Academic Press, pp. 45–83.

Cartmill M (1974b) Rethinking primate origins. Science *184:*436–443.

Cartmill M (1992) New views on primate origins. Evol. Anthropol. *1:*105–111.

Cauna N (1956) Nerve supply and nerve endings in Meissner's corpuscles. Am. J. Anat. *99:*315–350.

Charles-Dominique P, Atramentowicz M, Charles-Dominique M, Gérard H, Hladik A, Hladik CM, and Prévost MF (1981) Les mammifères frugivores arboricoles nocturnes d'une forêt guyanaise: Inter-relations plantes-animaux. Rev. Ecol. (Terre Vie) *35:*341–435.

Costello MB, and Fragaszy DM (1988) Prehension in *Cebus* and *Saimiri:* I. Grip type and hand preference. Am. J. Primatol. *15:*235–245.

Ford SM (1980) Callitrichids as phyletic dwarfs, and the place of the Callitrichidae in Platyrrhini. Primates *21:*31–43.

Fleagle JG, and Mittermeier RA (1980) Locomotor behavior, body size, and comparative ecology of seven Surinam monkeys. Am. J. Phys. Anthropol. *52:*301–314.

Garber PA (1980) Locomotor behavior and feeding ecology of the Panamanian tamarin (*Saguinus oedipus geoffroyi*, Callitrichidae, Primates). Int. J. Primatol. *1:*185–201.

Garber PA (1984) Use of habitat and positional behavior in a neotropical primate, *Saguinus oedipus*. In P Rodman and JGH Cant (eds.): Adaptations for Foraging in Nonhuman Primates. New York: Columbia University Press, pp. 112–133.

Garber PA (1988) Diet, foraging patterns, and resource defense in a mixed species troop of *Saguinus mystax* and *Saguinus fuscicollis* in Amazonian Peru. Behaviour *105:*18–34.

Garber PA (1992) Vertical clinging, small body size, and the evolution of feeding adaptations in the Callitrichinae. Am. J. Phys. Anthropol. *88:*469–482.

Garber PA (1993) Feeding ecology and behaviour of the genus *Saguinus*. In AB Rylands (ed.): Marmosets and Tamarins. Oxford: Oxford University Press, pp. 273–295.

Garber PA, and Sussman RW (1984) Ecological distinctions between sympatric species of *Saguinus* and *Sciurus*. Am. J. Phys. Anthropol. *65:*135–146.

Hall WC, Kaas JH, Killackey H, and Diamond IT (1971) Cortical visual areas in the grey squirrel (*Sciurus carolinensis*): A correlation between cortical evoked potential maps and architectonic subdivisions. J. Neurophysiol. *34:*437–452.

Hershkovitz P (1970) Notes on Tertiary platyrrhine monkeys and a description of a new genus from the late Miocene of Columbia. Folia Primatol. *12:*1–37.

Hershkovitz P (1977) Living New World Monkeys (Platyrrhini), Vol. 1. Chicago: The University of Chicago Press.

Idé C (1976) The fine structure of the digital corpuscle of the mouse toe pad, with special reference to nerve fibers. Am. J. Anat. *147:*329–356.

Johansson RS (1978) Tactile sensibility in the human hand: Receptive field characteristics of mechanoreceptive units in the glabrous skin area. J. Physiol. (London) *281:*101–123.

Johansson RS (1979) Tactile afferent units with small and well demarcated receptive fields in the glabrous skin area of the human hand. In DR Kenshalo (ed.): Sensory Functions of the Skin of Humans. New York: Plenum Press, pp. 129- 145.

Johansson RS, and Vallbo ÅB (1976) Skin mechanoreceptors in the human hand. An inference of some population properties. In Y Zotterman (ed.): Sensory Functions of the Skin in Primates. Oxford: Pergamon Press, pp. 171–184.

Johansson RS, and Vallbo ÅB (1979) Tactile sensibility in the human hand: Relative and absolute densities of four types of mechanoreceptive units in the glabrous skin area. J. Physiol. (London) *286:*283–300.

Jouffroy FK, Godinot M, and Nakano Y (1991) Biometrical characteristics of primate hands. Hum. Evol. *6:*269–306.

Kaas JH, Hall WC, and Diamond IT (1972) Visual cortex of the grey squirrel (*Sciurus carolinensis*): Architectonic subdivisions and connections from the visual thalamus. J. Comp. Neurol. *145:*273–305.

Koprowski JL (1994) *Sciurus carolinensis*. Mammal. Species *480:*1–9.

Krause W (1860) Die terminalen Köperchen der einfach sensiblen Nerven. Hanover: Hahn.

Le Gros Clark WE (1959) The Antecedents of Man. Edinburgh: Edinburgh University Press.

Le Gros Clark WE (1963) History of the Primates. An Introduction to the Study of Fossil Man. 4th ed. Chicago: The University of Chicago Press.

Lemelin P (1996a) Relationships between hand morphology and feeding strategies in small-bodied prosimians. Am. J. Phys. Anthropol. Suppl. 22:148 (abstract).

Lemelin P (1996b) The Evolution of Manual Prehensility in Primates: A Comparative Study of Prosimians and Didelphid Marsupials. Ph.D. Dissertation, State University of New York at Stony Brook.

Martin RD (1986) Primates: A definition. In B Wood, L Martin, and P Andrews (eds.): Major Topics in Primate and Human Evolution. Cambridge: Cambridge University Press, pp. 1–31.

McClearn D (1992) Locomotion, posture, and feeding of kinkajous, coatis, and raccoons. J. Mamm. *73:*245–261.

Mittermeier RA, and van Roosmalen MGM (1981) Preliminary observations on habitat utilization and diet in eight Surinam monkeys. Folia Primatol. *36:*1–39.

Munger BL, and Idé C (1988) The structure and function of cutaneous sensory receptors. Arch. Histol. Cytol. *51:*1–34.

Napier JR (1956) The prehensile movements of the human hand. J. Bone Jt. Surg. *38B:*902–913.

Napier JR (1961) Prehensility and opposability in the hands of primates. Symp. Zool. Soc. Lond. *5:*115–132.

Napier JR (1993) Hands (Revised ed.). Princeton: Princeton University Press.

Nelson H, Jurmain R, and Kilgore L (1992) Essentials of Physical Anthropology. St. Paul: West Publishing Co.

Niemitz C (1984) Synecological relationships and feeding behavior of the genus *Tarsius*. In C Niemitz (ed.): Biology of Tarsiers. Stuttgart: Gustav Fischer Verlag, pp. 59–76.

Polyak S (1957) The Vertebrate Visual System. Chicago: The University of Chicago Press.

Quilliam TA, and Ridley A (1971) The human Meissner corpuscle. J. Anat. *109:*338–339 (abstract).

Rasmussen DT (1990) Primate origins: Lessons from a neotropical marsupial. Am. J. Primatol. *22:*263–277.

Rosenberger AL (1977) *Xenothrix* and ceboid phylogeny. J. Hum. Evol. *6:*461–481.

Ross C (1995) Allometric and functional influences on primate orbit orientation and the origins of the Anthropoidea. J. Hum. Evol. *29:*201–227.

Ross C (1996) Adaptive explanation for the origins of the Anthropoidea (Primates). Am. J. Primatol. *40:*205–230.

Rothe H (1971) Some remarks on the spontaneous use of the hand in the common marmoset (*Callithrix jacchus*). Proc. 3rd Int. Congr. Primat., Zürich (1970) *3:*136–141. Basel: Karger.

Rylands AB (1989) Sympatric Brazilian callitrichids: The black tufted-ear marmoset, *Callithrix kuhli,* and the golden-headed lion tamarin, *Leontopithecus chrysomelas.* J. Hum. Evol. *18:*679–695.

Servos P, Goodale MA, and Jakobson LS (1992) The role of binocular vision in prehension: A kinematic analysis. Vision Res. *32:*1513–1521.

Snowdon CT, and Soini P (1988) The tamarins, genus *Saguinus.* In RA Mittermeier, AB Rylands, A Coimbra-Filho, and GAB Fonseca (eds.): Ecology and Behavior of Neotropical Primates, Vol. 2. Washington, D.C.: World Wildlife Fund, pp. 223- 298.

Soini P (1987) Ecology of the saddle-back tamarin *Saguinus fuscicollis illigeri* on the Rio Pacaya, northeastern Peru. Folia Primatol. *39:*11–32.

Sussman RW, and Kinzey WG (1984) The ecological role of the Callitrichidae: A review. Am. J. Phys. Anthropol. *64:*419–449.

Terborgh J (1983) Five New World Primates. Princeton: Princeton University Press.

Thorington RW, Jr. (1968) Observations of the tamarin *Saguinus midas.* Folia Primatol. *9:*95–98.

Thorndike EE (1968) A microscopic study of the marmoset claw and nail. Am. J. Phys. Anthropol. *28:*247–262.

Vallbo ÅB and Johansson RS (1978) Tactile sensory innervation of the glabrous skin of the human hand. In G Gordon (ed.): Active Touch: The Mechanism of Recognition of Objects by Manipulation. Oxford: Pergamon Press, pp. 29–54.

Winkelmann RK (1963) Nerve ending in the skin of primates. In J Buettner-Janusch (ed.): Evolutionary and Genetic Biology of Primates, Vol. 1. New York: Academic Press, pp. 229–259.

Winkelmann RK (1964) Nerve endings of the North American opossum (*Didelphis virginiana*): A comparison with nerve endings of primates. Am. J. Phys. Anthropol. *22:*253–258.

Winkelmann RK (1965) Innervation of the skin: Notes on a comparison of primate and marsupial nerve endings. In AG Lyne and BF Short (eds.): Biology of the Skin and Hair Growth. New York: American Elsevier, pp. 171–182.

Wood Jones F (1916) Arboreal Man. London: Edward Arnold.

9

TAIL-ASSISTED HIND LIMB SUSPENSION AS A TRANSITIONAL BEHAVIOR IN THE EVOLUTION OF THE PLATYRRHINE PREHENSILE TAIL

D. Jeffrey Meldrum

Departments of Biological Sciences and Anthropology
Idaho State University
Pocatello, Idaho 83209

1. INTRODUCTION

The atelines (*Ateles*, *Lagothrix*, *Brachyteles,* and *Alouatta*) are distinguished among the New World primates by the presence of a prehensile tail, equipped with a naked volar pad covered with dermatoglyphic friction skin (Geoffroy Saint-Hilaire, 1829). This adaptation plays a significant role in the definition of the feeding and locomotor niche of the atelines (Rosenberger and Strier, 1989). Atelines exhibit modifications of the sacral and caudal vertebrae (Ankel, 1972; German, 1982), caudal musculature (Lemelin, 1995) and cerebral cortical representation of the tail (Falk, 1980). The capuchin monkey (*Cebus*) also displays prehensile abilities in its relatively shorter tail, but lacks the volar pad and other distinctive caudal morphologies present in the atelines, suggesting prehensile tails evolved in parallel in *Cebus* and the atelines (Rosenberger, 1983; Lemelin, 1995).

Several explanations have been offered as proximate causes for the evolution of prehensile tails in the platyrrhine primates. These include: to accommodate increased body size (e.g., Napier, 1967; Grand, 1972), to exploit a frugivorous diet (e.g., Mittermeier and Fleagle, 1976; Cant, 1977), to reduce distance and time traveling between feeding patches (Milton, 1984; Cant, 1986), to enhance the ability to feed in terminal branches (Grand, 1972, 1984; Bergeson, 1996), and to utilize fragile forest structures (e.g., Emmons and Gentry, 1983; Christoffer, 1987). As with most evolutionary explanations, it seems most reasonable that a combination of these factors is likely responsible, since no one is completely unique to the atelines or *Cebus*, or for that matter, other prehensile-tailed mammals.

In this paper I offer a hypothesis proposing a mechanism for the evolution of prehensile tails in some New World monkeys. The mechanism is certainly not independent of

Primate Locomotion, edited by Strasser *et al.*
Plenum Press, New York, 1998

or exclusive of the explanations enumerated above, but entails a transitional behavior that may have served as a source for the key innovation of tail prehension, within the context of the influences of body size, support characteristics, foraging strategies, etc. This hypothesis springs from observations of tail-assisted hind limb suspension in atelines and other primates not generally considered as possessing a prehensile tail.

Before exploring this hypothesis directly, I will discuss the generally accepted and somewhat problematic definition of prehensile tails. Next, the distribution of prehensile abilities of the tail will be briefly reviewed. Then, the associated behavior of hind limb suspension and foot reversal will be examined and their correlated morphologies discussed. Finally, the proposed mechanism of tail-assisted hind limb suspension as a transitional behavior in the evolution of the ateline prehensile tail will be considered together with analogies drawn from non-primate taxa.

2. DEFINITION OF "PREHENSILE"

It is necessary to consider the definition and characterization of a prehensile tail before exploring this hypothesis further. Emmons and Gentry (1983) have suggested a rather circumscribed definition that is generally adopted by investigators:

> A prehensile tail is one which can support alone the weight of the suspended body; semi-prehensile tails can be wrapped around branches and support a significant part, but not all, of the body weight (p. 513).

This definition places primary emphasis on the ability of the "prehensile" tail alone to support body weight. It is implied in this definition that "semi-prehensile" tails are used in conjunction with other appendages to support the suspended body weight.

A singular emphasis on body weight support, however, may have resulted in many prehensile qualities of primate tails being overlooked or trivialized. According to Webster's New Universal Unabridged Dictionary, the adjective *prehensile* conveys the quality of seizing or grasping, "as the *prehensile* tail of some monkeys" (McKechnie, 1983). It is this connotation of grasping ability that was given preeminence in earlier discussions of New World primate prehensile tails. Consider, for example, Elliot's (1913) description of *Ateles'* prehensile tail:

> The tail is unsurpassed, if not unequaled, in its flexibility, always in motion, the tip as sensitive as that of the elephant's trunk, grasping with an unshakable firmness anything and everything it may touch, and fulfilling in the highest degree...the purposes of a fifth hand. By it, fruits or other desirable objects otherwise unattainable are seized and brought within reach of the mouth or hands, and it also can hold its possessor suspended in the air..." (pp. 21–22)

This more encompassing characterization takes into account the grasping and manipulative capabilities of primate tails that have attained the greatest development in spider monkeys and other atelines. It is the dexterity and sureness of grip endowed by the modifications to the distal end of the ateline tail combined with the increased robusticity that permits body weight to be supported unassisted. These manipulative abilities, if not the ability to support total body weight, however, are also present to a lesser degree in many other primate tails, some considered "semi-prehensile," as well as many generally considered non-prehensile. This more general connotation of prehensile will be employed throughout the remainder of this paper.

3. DISTRIBUTION OF PREHENSILE TAILS

Prehensile capabilities, in the sense of seizing or grasping, have been noted in the tails of primates typically considered as having non-prehensile tails. Early examples of such field observations include tail use in such catarrhine primates as *Cercocebus albigena* (Allen, 1925; Haddow, 1952; Tappen, 1960). Jones and Sabater Pi (1968) observed that *C. albigena* wraps the tip of the tail around branches, in a prehensile fashion, when at rest. Rose (1974) tabulated reports of the occurrence of prehensile tail use in a number of cercopithecoid as well as platyrrhine taxa generally considered non-prehensile tailed. These include species of Old World monkeys such as guenons, the already mentioned mangabeys, and macaques, and among the New World primates, *Aotus*, *Callicebus*, *Saimiri*, and the callitrichines (see Rose, 1974, Table 2, for specific references).

Observations of behavior by captive primates further reveals a diversity of prehensile capabilities. For example, Dandelot (1956) described the tail of captive *Cercopithecus lhoesti* as "semi-prehensile," noting its crooked tip and frequent twining about supports (Figure 1). Karrer (1970) also referred to the tail of captive *Macaca irus* (=*fascicularis*) as semi-prehensile and described its employment in food and object retrieval and social interaction. Erwin (1974) also noted similar but less dexterous manipulations by captive *Macaca mulatta*.

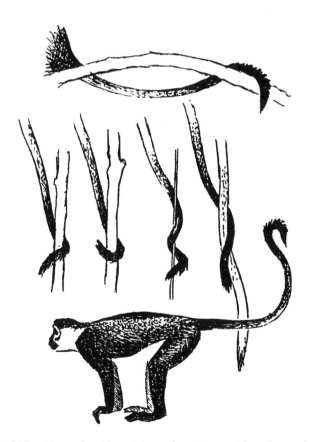

Figure 1. The tail of L'hoesti's monkey (*Cercopithecus lhoesti*) engaged in various prehensile activities (from Dandelot, 1956).

I queried primate keepers and field researchers via the Internet Bulletin Board, *Primate-Talk*, and received numerous responses relating observations of prehensile tail use by cercopithecoid monkeys. For example, KB Swartz (pers. comm.) related this observation of *M. fascicularis*:

> They would stick their tails through the grid in the floor of the cage and use their tails to retrieve dropped bits of food. I first saw it when they retrieved Froot-Loops that I was using to adapt them to my presence. They would curl the top 1/3 of the tail around the Froot-Loop and pick it up and bring it into the cage.

Also, reminiscent of Dandelot (1956) are observations by J Moore (pers. comm.):

> *C. lhoesti* at the San Diego Zoo appear to use tail semi-prehensile, as balance or brace, regularly. Tail is carried in crooked position, which remains under anesthesia (but is not skeletal, it can be bent out).

Finally, from CA Bramblett (pers. comm.):

> Infant vervet and Sykes monkeys have relatively prehensile tails. They wrap the infant's tail around mother's tail when the infant is clinging ventrally to provide additional attachment to the mother. If you hold a guenon infant in your hand, the infant's tail will wrap with a substantial grip around your arm.

This last observation is reminiscent of a published figure of an infant squirrel monkey (*Saimiri sciureus*) suspended by its tail from its keeper's finger (Rosenblum and Cooper, 1969). Furthermore, it highlights the fact that some infant monkeys are capable of greater prehensile tail use than are the adults of the same species, such as some colobines and baboons (Rose, 1974).

4. ASSOCIATED POSITIONAL BEHAVIORS AND MORPHOLOGIES

Rose (1974) made particular note of the distinction between tail use during postural behaviors and those employed during locomotor behaviors. The use of the tail in stereotyped "non-prehensile-tailed" primates was generally associated with postural activities. Prehensile tail use during locomotor activities is largely restricted to the atelines, and especially the spider monkey, in which the tail is used in concert with the arms during brachiation, and displays specializations for hyperextension associated with an orthograde suspensory posture (Lemelin, 1995). The atelines also frequently use the prehensile tail in postural activities, including hindlimb suspension (e.g., Carpenter and Durham, 1969; Mendel, 1976; Mittermeier, 1978; Fleagle and Mittermeier, 1980; Grand, 1984; Schon Ybarra, 1984; Cant, 1986; Fontaine, 1990).

During personal observations of positional behavior of the Malagasy lemurs housed at the Duke Primate Center, I became immediately impressed by the habit of many of the lemurs to posture suspended from their hind limbs. Similar observations have been made in the wild (Meldrum et al., 1997). Some taxa were particularly adept at this, including the acrobatic sifak, *Propithecus*, the black lemur, *Eulemur macaco*, and the ruffed lemur, *Varecia*. This behavior was employed on horizontal supports of various diameters, including the small-branch milieu. It was also adopted when descending large trunks head first. Further discussion of this behavior and its anatomical correlates, as an alternative strategy

Figure 2. A comparison of the hindlimb-suspension and tail-bracing postures of the bearded saki (*Chiropotes satanas*) [redrawn from van Roosmalen et al. (1981)] and the red ruffed lemur (*Varecia variegata rubra*) [redrawn from a photograph by the author].

for expanding the feeding kinesphere, are presented elsewhere (Meldrum et al., 1997). In association with hind limb suspension in *Varecia*, the tail was frequently draped over the support during hind limb suspension (Figure 2). It was not simply acting as a counterbalance, but was forcefully braced against the superior surface of the horizontal support, or braced against the side of a trunk or over adjacent supports when descending a vertical trunk head first.

Hind limb suspension is a postural behavior employed to varying degrees by a number of primate and non-primate mammals. The skeletal correlates of this behavior have been examined in a number of taxa (e.g., Jenkins and McClearn, 1984; Meldrum et al., 1997). In a study of the locomotor behavior and anatomy of two platyrrhine primates, the sympatric sakis, *Pithecia* and *Chiropotes*, a number of these skeletal correlates were observed in *Chiropotes* (Fleagle and Meldrum, 1988). Specific contrasts were made between their respective tarsal elements. Features distinguishing *Chiropotes* from *Pithecia* are associated with the enhancement of plantarflexion and supination and are comparable to those cited by Jenkins and McClearn (1984) for hindfoot reversal and/or hind limb suspension in a number of mammalian taxa. On the basis of the anatomical correlates and anecdotal field observations (van Roosmalen et al., 1981), Fleagle and Meldrum (1988) suggested that hind limb suspension may constitute an important component of *Chiropotes* positional behavior. This prediction has since been borne out through field observations by W Kinzey (pers. comm.) and Walker (1994, 1997). Just as observed in *Varecia*, a remarkably similar use of the tail during hind limb suspension was noted and figured by van Roosmalen et al. (1981) in *Chiropotes* (Figure 2). They made particular note, based on brief field observations by Mittermeier (1977), that hind limb suspensory postures by *Chiropotes* significantly increased the feeding sphere, in spite of the absence of a prehensile tail.

Ankel (1962, 1963) and German (1982) have identified a number of distinctive skeletal features of the ateline prehensile tail including large sacral foramina and neural canal, an initial caudal segment comprising additional shorter elements, a longest caudal element situated more distally in the tail, and distal vertebrae that are shortened and flattened. Meldrum and Lemelin (1991) have observed that the shortening of the distal segment elements may be present to a lesser extent in some taxa that use their tails in a semi-prehensile fashion during postural behaviors. This is suggested in a contrast between the caudal vertebrae of *Pithecia* and *Chiropotes*. Figure 3 illustrates a plot of the ratio of proximal width/length of individual caudal vertebrae for a small sample of extant and fossil platyrrhines. It demostrates the relative shortening of the distal elements in the tail of *Chiropotes* as compared to *Pithecia*. Also noteworthy is an inflexion in the curve for *Chiropotes* (n=2), caused by a relatively shorter single caudal segment roughly two-thirds

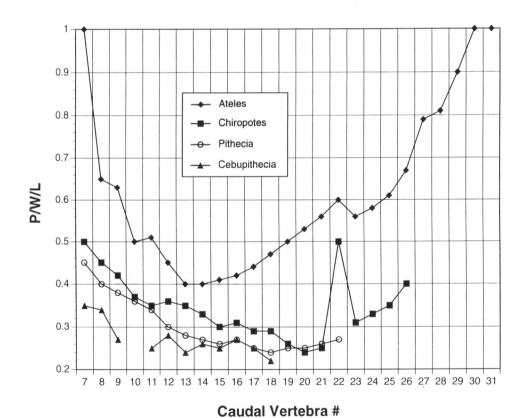

Caudal Vertebra #

Figure 3. Graph illustrating the relationship between the proximal width and length of individual caudal elements for selected platyrrhine primates (n=2 for each taxon).

down the length of the tail. It suggests a correspondence to the flexion point of the tail as it is braced over the support during hind limb suspension. A similar but less marked inflexion is present in the much more robust tail of *Ateles* (n=2) at about the same point. Further investigation is needed to establish the functional correlation of this flexion point and tail draping behavior in these and other taxa.

The holotype of the fossil pitheciine *Cebupithecia* preserves much of the tail and portions of the tarsals. This provides the opportunity to further examine the relationship between skeletal correlates of hind limb suspension and tail use in an extinct platyrrhine with a relatively long tail. The tarsals of *Cebupithecia* lack the traits correlated with hind limb suspension. *Cebupithecia* had a less robust tail, and although the distal elements are incomplete, the overall similarity to those of *Pithecia* is evident (Meldrum and Lemelin, 1991).

I have also noted similar tail use in some cercopithecoids under semi-natural conditions (Meldrum, 1989). Examination of cinefilms made by Rollinson (1975) revealed examples of tail twining and bracing during hind limb suspension or other below-branch reaching activities in guenon species, such as *Cercopithecus cephus*. For example, when attempting to reach fruits hanging from a small diameter horizontal support, a mustached monkey suspended from its hind limbs while bracing with its tail (Figure 4a). As the hind limbs became more extended and the foot more supinated, the prehensile twining of the tail about the support became more pronounced (Figure 4b).

In a second example, a mustached monkey with body oriented perpendicular to a horizontal support of moderate diameter, leaned out over the side of the support to investigate something. Its tail was draped over an adjacent support in a manner very similar to that observed in *Varecia* and *Chiropotes* (Figure 5a). As the monkey leaned farther out to the side of the primary support, displacing its center of gravity, the tail twined more prehensily around the secondary support (Figure 5b) until finally, the very tip of the tail was curled tightly about the support and the tensed tail was noticeably arched indicating forceful flexion (Figure 5c).

5. HYPOTHETICAL TRANSITIONAL BEHAVIOR

Clearly, a common utilization of the primate tail is correlated with positional behaviors that place the center of gravity other than directly above the arboreal support. These tail uses range from a non-contacting passive counterbalance, to a relatively active flexing brace, to very active modified prehensile grip. Such utilization of the tail is present in nearly all long-tailed primates to some degree, and limited prehensile use of the tail is more widespread than is often recognized. From these observations I propose the following hypothesis: that hind limb suspension, assisted by tail-bracing and/or twining, served as a transitional positional behavior in the evolution of the highly derived platyrrhine prehensile tail (Figure 6).

Indirect corroboration of this hypothesis is to be found in the proposals by Emmons and Gentry (1983) and Christoffer (1987) that the fragile vegetation and limited lianas of the Neotropics, by comparison to the Paleotropics, played a role in selection for and elaboration of prehensile tails. It is noteworthy that the evolution of prehensile tails has occurred independently in a number of neotropical mammalian families in addition to the atelines, including numerous didelphids, procyonids, mermecophagids, erethizontids, as well as several reptiles and amphibians. It is unlikely, however, that the tail serves identicle behavioral roles in all of these taxa.

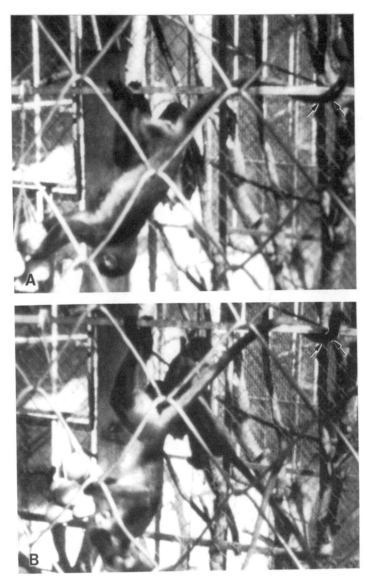

Figure 4. A mustached guenon (*Cercopithecus cephus*) reaching for fruit suspended below a horizontal support. See text for further discussion.

Of further significance is that parallels can be drawn to the distribution of prehensile-tailed mammals, including phalangerids, petaurids, burramyids and tarsipedids (Bergeson, 1996), and correlated forest types in Australia. Of particular significance to the hypothesis of tail-assisted hind limb suspension is the presence of a robust divergent hallux among the marsupials of both Australia and South America, associated with hind limb grasping/suspension and hindfoot reversal. The convergent functional morphology of hindfoot reversal in prehensile-tailed mammals such as the neotropical procyonid, *Potos*, and various didelphids, has been described by Jenkins and McClearn (1984). The combination of grasping halluces and prehensile tails in non-primate taxa would appear to lend support to the hypothesis advocated here.

Figure 5. A mustached guenon (*Cercopithecus cephus*) leaning to the side and below a large horizontal support. See text for futher discussion.

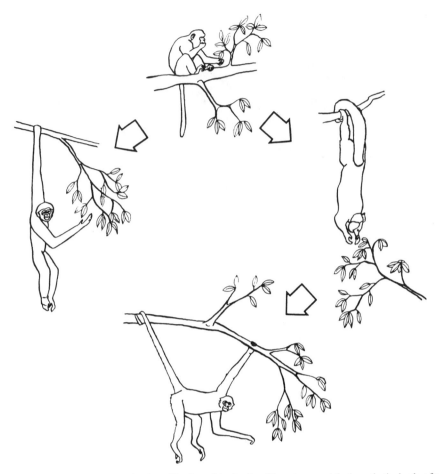

Figure 6. Representation of strategies for utilization of the feeding kinesphere, and the hypothetical role of tail-assisted hindlimb suspension as a transitional behavior in the evolution of the platyrrhine prehensile tail.

6. CONCLUSIONS

In conclusion, it is proposed that: (1) restricted body-weight-support definitions of prehensile tails may overlook many prehensile capabilities and functions of the generalized primate tail; (2) tail bracing and twining, in conjunction with hind limb suspension selected for greater caudal robusticity and prehensile capabilities in some primate lineages, especially ancestral atelines and *Cebus*; and (3) prehensile tails were initially a postural adaptation that, with the advent of the specialized volar pad of the ateline tail, permitted the tail to acheive a sufficient grip to support body weight unassisted. Only later did it evolve as an elaborated locomotor adaptation in the spider monkeys.

ACKNOWLEDGMENTS

I would like to thank the organizers of this Conference and acknowledge the numerous discussions with many individuals, which spawned these ideas and encouraged their formalization, and the anonymous reviewers who helped to focus their presentation.

REFERENCES

Allen JA (1925) Primates collected by the American Museum Congo expedition. Bull. Am. Mus. Nat. Hist. *47*: 283–499.

Ankel F (1962) Vergleichende Untersuchungen ueber die skelettmorphologie des Greifschwanzes suedamerikanischer affen (Platyrrhina). Z. Morph. Oekol. Tiere *52*:131–170.

Ankel F (1963) Zur Morphologie des Greifschwanzes bei suedamerikanishen Affen. Z. Morph. Anthropol. *53*:12–18.

Ankel F (1972) Vertebral morphology of fossil and extant primates. In RH Tuttle (ed.): Functional and Evolutionary Biology of Primates. Chicago: Aldine Press, pp. 223–240.

Bergeson DJ (1996) Positional behavior and prehensile tail use of *Alouatta palliata*, *Ateles geoffroyi* and *Cebus capucinus*. Ph.D. Dissertation, Washington University.

Cant JGH (1977) Ecology, locomotion and social organization of spider monkeys (*Ateles geoffroyi*). Ph.D. Dissertation, University of California, Davis.

Cant JGH (1986) Locomotion and feeding postures of spider and howling monkeys: Field study and evolutionary interpretation. Folia Primatol. *46*:1–14.

Carpenter CR, and Durham NM (1969) A preliminary description of suspensory behavior in primates. In HO Hofer (ed.): Recent Advances in Primatology. Proc. 2nd Intl. Congr. Primatol. *2*:147–154.

Christoffer C (1987) Body size differences between New World and Old World, arboreal tropical vertebrates: Cause and consequences. J. Biogeography. *14*:165–172.

Dandelot P (1956) Note sur le comportment de deux Cercopitheques de l'Hoest en captivité. Mammalia *20*:330–331.

Elliot DG (1913) A Review of the Primates. Vol II. New York: American Museum of Natural History.

Emmons LH, and Gentry AH (1983) Tropical forest structure and the distribution of gliding and prehensile-tailed vertebrates. Am. Nat. *121*:513–524.

Falk D (1980) Comparative study of the endocranial casts of New and Old World monkeys. In RL Ciochon and AB Chiarelli (eds.): Evolutionary Biology of the New World Monkeys and Continental Drift. New York: Plenum Press, pp. 275–292.

Fleagle JG, and Meldrum DJ (1988) Locomotor behavior and skeletal morphology of two sympatric pitheciine monkeys, *Pithecia pithecia* and *Chiropotes satanus*. Am. J. Primatol. *16*:227–249.

Fleagle JG, and Mittermier RA (1980) Locomotor behavior, body size, and comparative ecology of seven Surinam monkeys. Am. J. Phys. Anthropol. *52*:301–314.

Fontaine R (1990) Positional behavior in *Saimiri boliviensis* and *Ateles geoffroyi*. Am. J. Phys. Anthropol. *82*:485–508.

Geoffroy Saint-Hilaire I (1829) Remarques sur les caracteres generaux singes americains, et descrition d'un genre nouveau, sous le nom d'Eriode. Mem. Mus. Hist. Nat. (Paris) *17*:121–165.

German RZ (1982) The functional morphology of caudal vertebrae in New World monkeys. Am. J. Phys. Anthropol. *58*:453–459.

Grand TI (1972) A mechanical interpretation of terminal branch feeding. J. Mammal. *53*:198–201.

Grand TI (1978) Adaptations of tissue and limb segments to facilitate moving and feeding in arboreal folivores. In GG Montgomery (ed.): The Ecology of Arboreal Folivores. Washington, DC: Smithsonian Institution Press, pp. 231–242.

Grand TI (1984) Motion economy within the canopy: Four strategies for mobility. In PS Rodman and JGH Cant (eds.): Adaptations for Foraging in Nonhuman Primates. New York: Columbia University Press, pp. 54–72.

Haddow AJ (1952) Field and Laboratory studies on an African monkey, *Cercopithecus ascanias schmidti* Matschie. Proc. Zool. Soc. Lond. *122*:297–394.

Jenkins FA Jr, and McLearn D (1984) Mechanisms of hind foot reversal in climbing mammals. J. Morph. *182*:197–219.

Jones C, and Sabater Pi J (1968) Comparative ecology of *Cercocebus albigena* (Gray) and *Cercocebus torquatus* (Kerr) in Rio Muni, West Africa. Folia Primatol. *9*:99–113.

Lemelin P (1995) Comparative and functional myology of the prehensile tail in New World monkeys. J. Morph. *224*:351–368.

McKechnie JL, ed. (1983) Webster's New Universal Unabridged Dictionary, 2nd Edition. New York: Simon and Schuster.

Meldrum DJ (1989) Terrestrial adaptations in the feet of African cercopithecines. Ph.D. Dissertation, State University of New York at Stony Brook.

Meldrum DJ, and Lemelin P (1991) Axial skeleton of *Cebupithecia sarmientoi* (Pitheciinae, Platyrrhini) from the middle Miocene of La Venta, Colombia. Am. J. Primatol. *25*:69–89.

Meldrum DJ, Dagosto M, and White J (1997) Hindlimb suspension and foot reversal: An adaptation for increased feeding kinesphere in primates and other arboreal mammals. Am. J. Phys. Anthropol. *102*:85–102.

Mendel F (1976) Postural and locomotor behavior of *Alouatta palliata* on various substrates. Folia Primatol. *36*:1–19.

Mittermeier RA (1977) Distribution, synecology and conservation of Surinam monkeys. Ph.D. Dissertation, Harvard University.

Mittermeier RA (1978) Locomotion and posture in *Ateles geoffroyi* and *Ateles paniscus*. Folia Primatol. *30*:161–193.

Napier JR, and Napier PH (1967) A Handbook of Living Primates. New York: Academic Press.

Norconk MA, and Kinzey WG (1994) Challenge of neotropical frugivory: Travel patterns of spider monkeys and bearded sakis. Am. J. Primtol. *34*:171–183.

Rollinson J (1975) Interspecific comparisons of locomotor behavior and prehension in eight species of African forest monkey: A functional and evolutionary study. Ph.D. Dissertation, University of London.

Rose MD (1974) Postural adaptations in New and Old World monkeys. In FA Jenkins Jr (ed.): Primate Locomotion. New York: Academic Press, pp. 201–222.

Rosenberger AL (1983) Tale of tails: Parallelism and prehensility. Am. J. Phys. Anthropol. *60*:103–107.

Rosenberger AL, and Strier KB (1989) Adaptive radiation of the ateline primates. J. Hum. Evol. *18*:717–750.

Rosenblum LA, and Cooper RW, eds. (1969) The Squirrel Monkey. New York: Academic Press.

Schön Ybarra MA (1984) Locomotion and postures of red howlers in a deciduous forest-savanna interface. Am. J. Phys. Anthropol. *63*:65–76.

Tappen N (1960) Problems of distribution and adaptation of the African monkeys. Curr. Anthropol. *1*:91–120.

van Roosmalen MGM, Mittermeier RA, and Milton K (1981) The bearded sakis, genus *Chiropotes*. In AF Coimbra-Filho and RA Mittermeier (eds.): Ecology and Behavior of Neotropical Primates, Vol. 1. Rio de Janeiro: Academia Brasileira de Ciencias, pp. 419–441.

Walker SE (1994) Positional behavior and habitat use in *Chiropotes satanas* and *Pithecia pithecia*. In B Thierry, JR Anderson, JJ Roeder, and N Herrenschmidt (eds.): Current Primatology, Vol. 1: Ecology and Evolution. Proc. XIV IPS Congress. Strasbourg: Universitè Louis Pasteur, pp. 195–201.

Walker SE (1996) The evolution of positional behavior in the saki/uakaris (*Pithecia*, *Chiropotes* and *Cacajao*). In M Norconk, A Rosenberger, and P Garber (eds.): Advances in Primatology: Adaptive Radiations of Neotropical Primates. New York: Plenum, pp. 335–363.

10

UNIQUE ASPECTS OF QUADRUPEDAL LOCOMOTION IN NONHUMAN PRIMATES

Susan G. Larson

Department of Anatomical Sciences
School of Medicine
State University of New York at Stony Brook
Stony Brook, New York 11794-8081

1. INTRODUCTION

Quadrupedal walking and running are certainly not the first things that come to mind when one considers unique aspects of primate locomotion. However, there is a growing body of information about how the form of quadrupedalism displayed by primates differs from that of nonprimate mammals (see Vilensky, 1987, 1989). One of the most distinctive characteristics of primate quadrupedalism is that they typically utilize a diagonal sequence/diagonal couplets walking gait pattern (i.e., foot falls in sequence: left hind, right fore, right hind, left fore, with diagonal limbs moving as a pair), in contrast to the almost universally employed lateral sequence walking gait (left hind, left fore, right hind, right fore) of nonprimate mammals (Howell, 1944; Prost, 1965, 1969; Hildebrand, 1967; Rollinson and Martin, 1981; Vilensky, 1989; Vilensky and Larson, 1989). This difference in gait pattern is not trivial, since a diagonal sequence/diagonal couplet walking gait creates a strong potential for interference between the ipsilateral hind and forelimbs (Figure 1). The potential for hind/forelimb interference is exacerbated in primates by their long limbs (due to their relatively longer limb bones, Alexander et al., 1979), and by their propensity to use relatively longer stride lengths than nonprimate quadrupeds (Vilensky, 1980; Alexander and Maloiy, 1984; Reynolds, 1987). As a result, many primate quadrupeds must regularly "overstride" during walking, that is, touch down with their hind foot ahead of their ipsilateral hand by passing it either "inside" or "outside" of the forelimb (Hildebrand, 1967; Reynolds, 1985b; Larson and Stern, 1987; see Figure 1). Another distinctive aspect of primate gait utilization is the infrequent use of a running trot (defined as diagonal limbs moving synchronously with relative stance duration of each limb less than 50%; see Hildebrand, 1967). Primates generally progress directly from a walk to a gallop, and even

Primate Locomotion, edited by Strasser *et al.*
Plenum Press, New York, 1998

Forelimb Outside

Forelimb Inside

Figure 1. Side view of a chimpanzee knuckle-walking. Figures are drawn from videotape records. The chimpanzee in the upper row is walking with its right forelimb outside the overstriding right hind limb, while the chimpanzee in the lower row is walking with its right forelimb inside the overstriding right hind limb.

when they do run[*], they do not trot (Hildebrand, 1967; Vilensky et al., 1988). Finally, primates have been shown to rely more on their hind limbs for both support and propulsion than do nonprimate mammals (Kimura et al., 1979; Kimura, 1985; Demes et al., 1992, 1994).

To date, no integrated explanation for this unique set of characteristics of primate quadrupedalism has been proposed. Based on the kinetic data indicating that primates support more weight on their hind limb than nonprimates, several researchers have attributed the differences in gait patterns between primates and nonprimates to differences in body weight distribution (Iwamoto and Tomita, 1966; Tomita, 1967; Kimura et al., 1979, Kimura, 1985; Rollinson and Martin, 1981). Vilensky (1989) and Vilensky and Larson (1989), however, argue that the actual location of the center of mass in quadrupedal primates is not dissimilar to that of other mammals, and point out that animals with markedly posteriorly placed centers of gravity, such as rabbits (some of which employ walking gaits (Dagg, 1974)), use a lateral sequence walk.

The present study will review some additional aspects of quadrupedal locomotion in primates, namely patterns of limb motion and of muscle recruitment, and attempt to link these various pieces of information into some coherent whole.

2. OBSERVATIONS

2.1. Limb Kinematics

Mammalian quadrupeds can be loosely divided into two groups: cursors (adapted for rapid walking and running; generally over 3 kg), and non-cursors (not speed adapted,

---

[*] Running in this case refers to a symmetrical gait wherein the footfalls of the two fore- and two hind feet are equally spaced, and each foot is on the ground less than 50% of the total stride interval (Hildebrand, 1967). It thus differs from galloping which is an asymmetrical gait in which the footfalls of the fore- and hind pairs of limbs are not evenly spaced.

therefore including all other quadrupedal locomotor modes; usually 3 kg or less) (Jenkins, 1971; Alexander and Maloiy, 1984). Non-cursors tend to walk with very flexed limbs, and as Jenkins (1971) documents, their limb movements also deviate considerably from parasagittal planes. Small mammals use large maximum limb excursion angles (the angle the limb passes through between foot touchdown and lift-off) (McMahon, 1975, 1984) during walking, and for the forelimb, Jenkins and Weijs (1978) and Fischer (1994) suggest that scapular rotation and translation account for most of this motion, with relatively minor contributions by motion at the glenohumeral and elbow joints. As mammals get larger, they stand and walk with more extended limbs that move in roughly parasagittal planes. They move their limbs through smaller maximum excursion angles (McMahon, 1975), and if the differences between opossums (Jenkins and Weijs, 1978), cats (Miller and Van der Meché, 1975; English, 1978a,b) and dogs (Nomura et al., 1966; Tokuriki, 1973) can be taken to reflect a size-related trend, this implies that forelimb motion involves less scapular rotation.[†]

Quadrupedal primates cover an extremely wide size range from about 60 g to about 160 kg (Jungers, 1985), bridging the size ranges of cursors and non-cursors. Other than lorisines, however, which have been studied quite extensively (Dykyj, 1980; Jouffroy et al., 1983; Jouffroy and Petter, 1990; Demes et al., 1990; Ishida et al., 1990), very few kinematic data exist for very small or very large quadrupedal primates. More data are available on quadrupedal monkeys, in the size range of approximately 3 to 20 kg, comparable to smaller cursors such as cats and dogs. Like cats and dogs, primates tend to walk with extended limbs that move roughly in parasagittal planes, although arboreal primate quadrupeds are said to use somewhat more flexed and abducted limbs than terrestrial species (e.g., Napier, 1967; Grand 1968a,b, 1984; Fleagle, 1977; Morbeck, 1979). Also like cursors, primates display only limited scapular rotation during forelimb movement during walking (generalizing from the vervet examined cineradiographically by Whitehead and Larson, 1994). Significant kinematic differences in patterns of limb excursion, however, exist between primate and nonprimate quadrupeds.

Figure 2 displays a series of schematic forelimbs (humeral plus radial/ulnar segments) at touchdown and lift-off for a variety of mammals during a walking gait. Although it would be desirable to include scapular position in this analysis, data exist for only a few taxa. What information does exist suggests a similar initial position at touchdown in most mammals including primates, with the scapular spine being approximately 45–60 degrees to horizontal (see Figure 6). Greater differences exist in scapular position at lift-off (see Nomura et al., 1966; Tokuriki, 1973; Miller and Van der Meché, 1975; English, 1978; Jenkins and Weijs, 1979; and Whitehead and Larson, 1994). The nonprimate mammalian quadrupeds in Figure 2 display a forelimb touchdown in which the glenohumeral joint remains retracted, that is, the humerus is behind a vertical line through the shoulder. The primate quadrupeds, on the other hand, begin support phase with their humerus in a protracted position, i.e., ahead of vertical. At the end of support phase, the humerus in the primates portrayed in Figure 2 is less retracted than that of the nonprimate quadrupeds. It may be that primates end support phase with a less retracted forelimb as a means of reducing some of the potential for interference between the forelimb and the ipsilateral hind limb. If the supporting forelimb is imagined as pivoting over the wrist during a step as in

---

[†] Although it has been suggested that an aclaviculate condition in cursorial mammals is related to increased anteroposterior excursion of the shoulder (e.g., Eaton, 1944; Howell, 1944), Jenkins (1974) reports that clavicular loss mainly serves to confine shoulder motion to a parasagittal plane with no change in anteroposterior shoulder excursion.

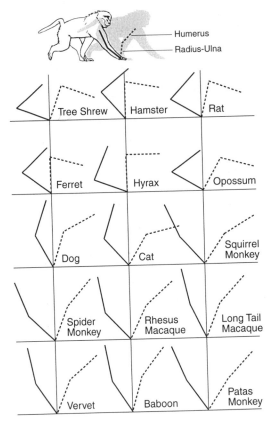

Figure 2. Left forelimb postures during a walking step in a variety of mammals (moving from left to right). All limb segments (humerus and radius-ulna) are drawn the same length in all animals. The solid lines represent the position of the limb at touchdown, and the dashed lines represent the limb at the end of support phase at lift-off. The non-cursors in the top two rows display highly flexed limb postures, whereas the larger cursors (dog, cat and primates) use more extended limb postures. In all nonprimates illustrated here, the humerus is in a retracted position at touchdown, but is protracted in all of the primates. Data for the non-cursors are from Jenkins (1971); for the dog are from Tokuriki (1973); for the cat are from Miller and Van der Meché (1975); for the squirrel monkey are from Vilensky et al. (1994), for the spider monkey are from Schmitt (1994), and for the remaining primates are from Schmitt (1995).

Figure 2, then the primate forelimb appears to pass through a more symmetrical angular excursion than in nonprimates, that is, approximately half of the excursion occurs prior to when the shoulder is over the wrist, and half occurs after. Most of the excursion of the forelimb in nonprimates occurs beyond the point when the shoulder is directly over the wrist. Unfortunately, besides the monkey species displayed in Figure 2, kinematic data on forelimb position during walking are not available for many other nonhuman primates. Extreme protraction of the forelimb at touchdown during walking is clearly characteristic of lorisines (Jouffroy et al., 1983; Jouffroy and Petter, 1990; Demes et al., 1990; Ishida et al., 1990), and personal observations indicate that a protracted humerus at touchdown is typical of all monkeys and large-bodied apes. Published figures support this conclusion (e.g., Muybridge, 1957, Plates 142 and 143; Hildebrand, 1967, Figure 5; Grand, 1968a, Figure 4; Rose, 1979, Figure 4–2; Larson and Stern, 1987, Figure 1; Larson and Stern, 1989b, Figure 2; Meldrum, 1991, Figures 4, 9, and 10). Lemelin (pers. com.) reports that a protracted forelimb is also observed in other walking quadrupedal prosimians besides lorisines.

Primate quadrupeds also begin hind limb support phase with the limb highly protracted. They end it with the hind limb equally retracted resulting in significantly greater angular excursions at the hip during walking than what is observed in nonprimates (Reynolds, 1987). The difference is most pronounced in suspensory/climbing adapted species. Reynolds (1987) concludes that large angular excursion at the hip coupled with relatively long limb segments (Alexander et al., 1979) produce the relatively long hind limb stride lengths during walking that have been reported for quadrupedal primates (Vilensky, 1980; Alexander and Maloiy, 1984; Reynolds, 1987).

It is unclear whether the more protracted forelimb position observed during a walking step in primate quadrupeds is also associated with increased forelimb angular excursion. The average angular excursion for the primates displayed in Figure 2 (65 degrees) is similar to the average for the non-cursors (66 degrees), and a little higher than the average for the cat and dog (57 degrees). McMahon (1975, 1984) has described a gradual inverse relation between body size and maximum limb excursion ($\mu\ M^{-.10}$), which would predict that the larger bodied primates should display relatively smaller limb excursion angles than the non-cursors. The relation described by McMahon, however, is for animals traveling at the speed at which the trot/gallop transition occurs, and all of the animals portrayed in Figure 2 are traveling at much slower speeds. Reports vary on how angular excursion varies with speed (see Vilensky, 1987), making it impossible at present to determine if primate forelimbs display the same increase in angular excursion as documented for primate hind limbs.

Nonetheless, the position of the primate humerus during a walking step can be related to a long forelimb stride length. Throughout support phase, the glenohumeral joint in primates is more protracted than in nonprimates, meaning that the joint is more open (obtuse angle between the scapula and humerus) thus increasing the effective length of the limb. This augments the already elongated forelimbs of primates due to relatively longer limb bones (Alexander et al., 1979), making an overall increase in limb length mainly responsible for the relatively long forelimb strides of primates.

2.2. Muscle Activity Patterns

Based on observations of electromyographic (EMG) activity of homologous limb muscles during walking in different animals, Goslow and coworkers have suggested that muscle recruitment patterns in vertebrates are quite conservative (Jenkins and Goslow, 1983; Peters and Goslow, 1983; and Goslow et al., 1989). Goslow et al. (1989) illustrated the point by contrasting the patterns of shoulder muscle recruitment in a lizard and an opossum, two rather dissimilar vertebrates (see Figure 3). Despite differences in limb orientation and shoulder structure, four major muscle groups showed similar patterns of activity: the latissimus dorsi and the pectoralis muscles acted primarily during support phase, the deltoids acted during swing phase, and the supracoracoideus - supraspinatus/infraspinatus homologue were biphasic. This result was especially surprising for the supracoracoideus - supraspinatus/infraspinatus homologue since they are configured quite differently as Figure 3 shows. Goslow and colleagues also observed certain similarities to patterns of shoulder muscle recruitment of birds (Dial et al., 1987, 1991), which led them to propose what may be called the neuromuscular conservation hypothesis: Motor patterns of homologous muscles have been maintained during the evolution of tetrapods, and a primitive organization of the neural control components has persisted in derived groups despite differences in morphology (Goslow et al., 1989).

EMG data suggest that the neuromuscular conservation hypothesis may indeed apply to the primate hind limb. Okada et al. (1978) reported that the recruitment pattern of the vastus lateralis, biceps femoris, gastrocnemius and tibialis anterior in a Japanese macaque and a spider monkey during walking were similar to data reported for cats and dogs except for small timing differences, and Kimura et al. (1979) obtained similar results on the same four muscles in a baboon and chimpanzee. Jungers et al. (1980) undertook a detailed study of quadriceps femoris activity in *Lemur fulvus*, patas and woolly monkeys,

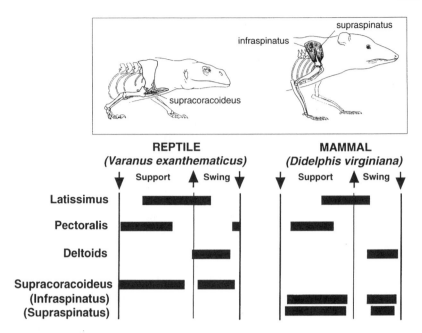

Figure 3. Electromyographic activity of homologous muscles in the Savannah Monitor lizard and opossum during walking. Bars represent the most consistent activity. In both animals, the latissimus dorsi and pectoralis muscles are active mainly in support phase, the deltoids mainly in swing phase, and the supracoracoideus - infraspinatus/supraspinatus homologues are biphasic. Redrawn from Goslow et al. (1989).

and also observed only a small delay in recruitment of the quadriceps complex relative to nonprimate patterns. Like Okada et al. (1978) and Kimura et al. (1979), Jungers et al. (1980) suggested that such timing differences could be due to the differences in footfall patterns between primates and other quadrupeds. Finally, Vangor and Wells (1983) studied 14 muscles of the hip and thigh of two spider monkeys, two woolly monkeys and two patas monkeys, and found "a surprising degree of similarity in phasic activity patterns" (p. 130) to what had been observed in nonprimates, but noted that the primate data showed a high level of variability.

When the forelimb is considered, a very different impression of the degree of conservation of primitive motor programs in primates emerges. Larson and Stern (1987) studied the pattern of shoulder muscle recruitment in chimpanzees during knuckle-walking. The authors began with the premise that chimpanzees are basically suspensory/climbing adapted primates that have had to make certain compromises in order to continue to walk quadrupedally since their forelimbs are adapted for mobility rather than stability (e.g., widened thorax with a dorsally placed scapula, large, globular humeral head, round radial head permitting a wide range of supination/pronation, distal ulna removed from articulation with the carpals, etc.). The African apes are already unique in that they walk on their knuckles, and it would seem likely that there would be other ways in which their mode of quadrupedal locomotion would be distinct.

The EMG results from that study for the muscles illustrated by Goslow as supporting the neuromuscular conservation hypothesis are presented in Figure 4 (since chimpanzees overstride, inside and outside forelimbs were analyzed separately). In the chimpanzee the latissimus dorsi and pectoral muscles acted most consistently at the end of swing,

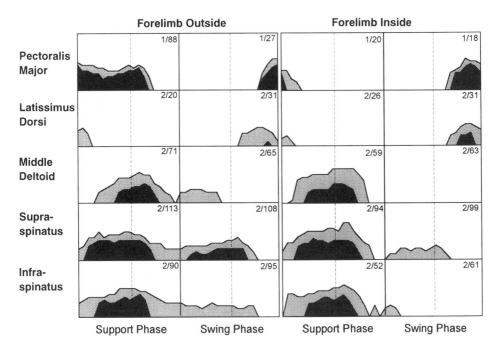

Figure 4. EMG activity matrices for chimpanzee shoulder muscles during knuckle-walking. Matrices for forelimb outside and forelimb inside steps are displayed separately. Each matrix represents a summary of the frequency and relative magnitude of activity for several steps. Blackened areas indicate highly consistent EMG activity, occurring two-thirds or more of the time. Hatched areas reflect frequent but less consistent activity, occurring one-third to two-thirds of the time. Heights of these areas reflect amplitude of activity (determined from spike height) as a percentage of the maximum amplitude observed during the recording session, which is represented by the height of the box. The numbers in each box represent the number of individuals/number of steps analyzed for each muscle. The pattern of muscle activity in the chimpanzee is different from that reported for the lizard and opossum by Goslow et al. (1989)—see Figure 3.

whereas these muscles were active through most of support phase in the lizard and opossum. Only during forelimb outside steps was the pectoralis major of chimpanzees active in support phase. Larson and Stern (1987) interpreted this activity as being due to the fact that the arm is slightly abducted during forelimb outside steps, thus requiring the action of a forelimb adductor to resist the tendency for the hand to slip out to the side. So at least during forelimb inside steps, there is a very different pattern of recruitment of these two humeral retractors in the chimpanzee compared to nonprimate vertebrates.

In Goslow's example of the lizard or opossum, the deltoids were recruited during swing phase and the supra- and infraspinatus were biphasic.[‡] Larson and Stern (1987) reported that in the chimpanzee, the middle deltoid was most consistently active during support phase of knuckle-walking, and was only occasionally active during swing phases when the forelimb was outside (Figure 3). Supraspinatus was biphasic as predicted during forelimb outside steps, but was frequently only active during support phase in forelimb inside steps. The pattern of recruitment for the infraspinatus was similar to that of the supraspinatus except that it was less frequently biphasic in forelimb outside steps showing

---

[‡] Although Goslow et al. (1989) characterized the supracoracoideus - supraspinatus/infraspinatus homologues as biphasic, they described the propulsive phase activity as more intense.

consistent activity only during support phase, and was essentially uniphasic during fore-limb inside steps.

Larson and Stern (1987) concluded from this study that the pattern of recruitment of shoulder muscles in the chimpanzee during quadrupedal locomotion, especially the absence of humeral retractor activity during support phase, was related to the chimpanzee's mobile forelimb and the need to protect it from excessive locomotor stresses. The unique pattern of muscle recruitment, therefore, reflected an aspect of their adaptation to suspensory/climbing habits, and was related to the high degree of functional differentiation between their fore- and hind limbs (Kimura et al., 1979; Kimura, 1985; Demes et al., 1992, 1994).

Stern et al. (1977) have reported that two other highly arboreal, suspensory/climbing adapted primates, spider and woolly monkeys, do not recruit the latissimus dorsi during stance phase of walking, supporting this interpretation. The pectoralis major of the woolly monkey was also inactive during support phase, although it was recruited in the spider monkey. It should be noted, however, that spider monkeys typically walk with abducted arms (Schmitt, 1994) raising the possibility that the pectoralis is again being used more as an adductor than as a humeral retractor in this species.

As a contrast to these highly arboreal primate species, Larson and Stern (1989a,b) undertook a similar set of studies of shoulder muscle recruitment in the vervet monkey, a primate that is semiterrestrial (Struhsaker, 1967; Dunbar and Dunbar, 1974; Fedigan and Fedigan, 1988), and has a dorsoventrally deep thorax, laterally placed scapula and shoulder joint morphology more like that of a cat or opossum than a chimpanzee. The EMG results, however, were more like those of the chimpanzee than like nonprimates. This was true for the humeral retractors, as well as for the deltoids and the supra- and infraspinatus (summarized in Figure 5). Unpublished EMG data recently collected in this lab on the patas monkey, arguably the most terrestrially adapted primate, demonstrated the same pattern of muscle recruitment. These results are surprising because the vervet or patas monkey forelimb is not considered to be adapted for mobility as is the chimpanzee's. Indeed, force plate studies have shown that it is only the more suspensory/climbing adapted primates that show a strong dichotomy between the forelimb and hind limb in participation in weight support and propulsion (Kimura et al., 1979; Kimura, 1985; Demes et al., 1994). This common pattern of muscle recruitment, therefore, is not simply a similar mechanical response to similar functional demands. It is the case that nearly all primates that have been studied, however, do display at least slightly higher peak vertical forces on their hind limbs than on their forelimbs, distinguishing them from all nonprimates (Demes et al., 1994). These shared characteristics of patterns of force distribution and muscle recruitment despite differences in morphology and limb use suggest that primates may display a neuromuscular conservation of their own. That is, a common motor program distinguishes the primate clade from other tetrapods. The EMG results also suggest that the most profound changes in motor control of limb motion involve the forelimb.

3. DISCUSSION

The various theories that have been offered to explain the origins of primates all share some emphasis on the use of clawless grasping extremities in an arboreal (small branch) setting to reach discontinuous supports and/or collect food. I propose that reaching out with clawless grasping extremities also forms the foundation for the various distinctive characteristics of primate quadrupedal locomotion described above.

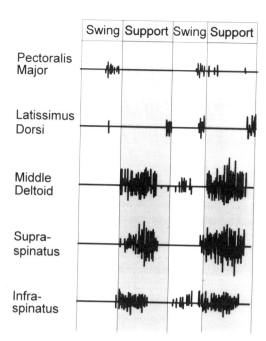

Figure 5. Representative raw EMG traces for five shoulder muscles in the vervet monkey during walking. Each trace represents two seconds of activity. Traces for the pectoralis major and the latissimus dorsi come from two different recording sessions, and the traces for the other three muscles come from a third recording session. Steps of approximately the same speed and cycle duration were selected for display, but the traces have also been graphically adjusted to simplify presentation. The pattern of muscle recruitment displayed by the vervet monkey is more similar to that of the chimpanzee than to that of nonprimate mammals.

Using clawless grasping extremities to travel or forage in an arboreal habitat, especially on small branches, requires the ability to reach out precisely to gain a secure grip on a particular branch or object. This demands versatility in the neural control of limb movement. Indeed, Georgopoulos and Grillner (1989) report that while uncomplicated locomotion on an even surface in cats requires very little supraspinal input, more complex locomotor tasks such as ladder walking are impossible after transection of the corticospinal tract. Studies of the rate of cortical neuron discharge indicate that the corticospinal input is not related to the control of either equilibrium or propulsion, but is directly involved in the correct positioning of the limbs, particularly the forelimbs. Georgopoulos and Grillner (1989) also report that the same involvement of the corticospinal system is observed in forelimb manipulatory movements. They propose that ability to use the forelimb for manipulation evolved from the ability to accurately position the limb during locomotion, and that as the supraspinal control necessary to precise locomotor limb positioning developed, so did the fine control of forelimb movements for manipulation.

Vilensky (1989) and Vilensky and Larson (1989) have proposed that these evolutionary changes in the neurological control of forelimb movements associated with the ability for precise limb positioning during locomotion and use of the forelimb for manipulation and exploration of the environment might account for the preference among primates for diagonal sequence gaits. Noting that while diagonal sequence gait is the dominate mode of quadrupedal walking among primates, spontaneous displays of lateral sequence gaits are frequently seen throughout the Order. However, although a few nonprimates habitually use diagonal sequence walking gaits, there are essentially no observations of occasional use of diagonal sequence gaits among nonprimates that habitually use lateral sequence walking gaits. Vilensky and Larson (1989) conclude that quadrupedal primates are capable of more versatility in their locomotor mode than are most nonprimate quadrupeds, which, in turn, suggests less rigidity in the control of limb movements during locomotion. Unlike cats, it has been shown that primates require supraspinal input to produce stepping motions even on flat surfaces (Eidelberg et al., 1981).

Vilensky (1989) and Vilensky and Larson (1989) propose that a greater degree of cortical control of limb movement in primates may have resulted in less dependence on spinal cord based program generators. In a study done on rabbits, Viala and Vidal (1978) demonstrated that hind limb-driving spinal circuits strongly influenced the rhythm of the forelimb circuits, and suggested that this ability of the hind limbs to control the forelimbs may be a factor in the prevalence of lateral sequence gaits in quadrupeds. Vilensky (1989) and Vilensky and Larson (1989) suggest that changes in spinal circuitry associated with increased cortical control of limb motions in primates may have simply eliminated or overridden any inherited spinal mechanisms tending to favor lateral sequence gaits. If these changes also allowed the movements of each forelimb to exert some direct influence on the ipsilateral hind limb, then the result would be a preference for diagonal sequence gaits in primates (beginning with the forelimb, diagonal sequence gaits appear as LfLhRfRh and lateral sequence gaits LfRhRfLh). Primates appear to emphasize ipsilateral fore-hind control, and nonprimates contralateral fore-hind control (or, ipsilateral hind-fore control).

Precise placement of the forelimb in a discontinuous arboreal habitat in order to reach and grasp food or supports also requires enhanced mobility at the forelimb joints. However, all else being equal, increasing the mobility of joints can only be accomplished at the expense of stability, and cannot develop if the forelimbs are subjected to large disruptive locomotor forces. Therefore, some means of limiting such forces must be developed. There are a variety of ways that joint forces engendered during locomotion can be reduced. One is to alter locomotor performance such as by moving more slowly. Another is through alteration of muscle recruitment patterns. Muscles of the forelimb that are used to help propel an animal forward among nonprimates are usually inactive during support phase in primate quadrupeds. This agrees with force plate data demonstrating that nonprimates display higher peak propulsive forces on their forelimbs, whereas primates show higher peak propulsive forces on their hind limbs (Kimura et al., 1979; Kimura, 1985; Pandy et al., 1988). Interestingly, all primates that have been studied appear to "spare" their forelimb from some of the disruptive forces associated with quadrupedal locomotion in this manner regardless of whether or not the forelimb displays particular adaptations to increased mobility. This suggests that this reduction in use of forelimb propulsive muscles is a component of the basic adaptation of reaching out with a clawless grasping hand in an arboreal setting that arose early in the evolution of primates to become characteristic of the Order.

Another way in which primate quadrupeds can limit the disruptive locomotor forces acting on their forelimbs is by supporting more of their body weight on their hind limbs, as has been demonstrated by force plate studies (Kimura et al., 1979; Kimura, 1985; Demes et al., 1992, 1994). Although this had been thought to be due to a more posterior position of the center of gravity, Reynolds (1985a,b) has argued that this alteration in weight distribution is due to an active transfer of weight to the hind limbs by the contraction of hind limb retractors. Reynolds (1985b) has shown that the hind limb retractors will be most effective at shifting weight posteriorly without generating large propulsive forces if they act when the hind limb is protracted. He thus relates the highly protracted hind limb at touchdown in primates to an increase in the effectiveness of the posterior weight shift mechanism (Reynolds, 1987). Unlike the alterations of muscle recruitment patterns, the posterior weight shift mechanism seems to be most pronounced (greatest disparity between fore- and hind limbs in weight distribution, most extreme degree of limb protraction) in those primates displaying higher degrees of forelimb mobility.

The protracted position of primate hind limbs and the accompanying larger angular excursions documented by Reynolds (1987), also contribute to increased relative hind limb stride length in primates. Demes et al. (1990) suggest that relatively long strides in lorisines may be an arboreal adaptation in that it would be a means of achieving high walking speeds without increasing stride frequency. High frequency gaits, which entail steeply increasing, relatively high peak forces, could cause branch swaying that is not only dangerous but also energy costly. Low frequency gaits would minimize such sway, which may be especially important in lorisines that rely on quiet, stealthful movements both to capture prey and avoid predators (Demes et al., 1990). Long strides with low frequency gaits would also generally increase stability by increasing hand contact time. Although long low frequency strides may have originally evolved as adaptations to cryptic habits among early primates, branch sway is unlikely to occur at the natural walking frequency of any arboreal primate (Alexander, 1991), suggesting that long low frequency strides may have been maintained in primates as a general mechanism to help maintain stability on branches.

Schmitt (1995) has proposed that long stride length in primates may also be involved in another mechanism for reducing disruptive forces on the limbs during locomotion. He suggests that certain components of a long stride, namely, longer step length, increased contact time, and protraction of the limb when coupled with joint yield increase the compliance of primate quadrupedal walking gaits. By increasing step length and contact time, peak stresses acting on the limb are reduced by increasing the time over which the reaction forces act (see McMahon, 1985, McMahon et al., 1987, and Schmitt, this volume). The highly protracted fore- and hind limbs of primates change the "angle of attack" of the limbs, thereby reducing the vertical landing speed, and the vertical stiffness of the body (McMahon et al., 1987). If limb protraction results in a greater angular excursion for a compliant limb, then the body's center of mass follows a flatter trajectory (Blickhan, 1989; Farley et al., 1993) thereby also helping to increase stability on a branch and reduce the tendency to cause branch sway. In addition, limb protraction attenuates the impact acceleration peak of the body, implying that less shock due to foot or hand strike passes upward through the body (McMahon et al., 1987).

The fact that primates do not typically display a running trot, which is a high frequency, high stiffness gait, may also be a reflection of their tendency to avoid high peak stresses on the limbs (Schmitt, 1995). Primates typically move from a walk or running walk to a gallop, which is a more compliant gait than the trot. As Preuschoft et al. (1996) have noted, asymmetric gaits such as a gallop involve longer contact times than trotting and thus limit peak substrate reaction forces. Preuschoft and Gunther (1994) suggest that another reason why primates avoid a trot is that their long relatively heavy limbs with long pendulum lengths limit the speed of the recovery stroke making a high frequency gait such as trotting difficult (see also, Preuschoft et al., 1996). In addition, and unlike most nonprimates that change gaits at fairly predictable speeds (Heglund et al., 1974), quadrupedal monkeys sometimes exhibit a walk at speeds at which they would be expected to gallop, or gallop at very low speeds when they might be expected to walk or trot. These differences are quite idiosyncratic, with two individuals of the same species displaying different patterns of gait transitions (Vilensky et al., 1990). This variability and individuality suggests that the ability to avoid a trot may be an additional by-product of a greater degree of supraspinal control of locomotion in primates.

Effective use of a clawless grasping hand in an arboreal habitat will also be enhanced by elongation of the forelimb. Cartmill (1974) has related relative forelimb elongation to the ability to hold onto a vertical support. Whereas clawed mammals can climb in trees by interlocking with the substrate, without claws, an animal must rely on friction to prevent slipping off of a branch or trunk. The ability to hold onto a large vertical support is dependent on the size of the central angle that the animal's limbs can subtend. Longer limbs can subtend larger central angles. When climbing up or down a vertical tree trunk, however, a clawless animal must retain a more secure grip with its forelimbs than with its hind limbs, since failure of the upper grip will cause it to fall backward away from the tree, whereas failure of the lower grip will result in the animal falling forward into the tree (Cartmill, 1974, 1985). In addition, clawless mammals can augment pedal friction by leaning away from the support. Thus primates not only have long limbs, but display relative forelimb elongation with increasing body size as a component of a basic adaptation to tree-dwelling habits (Cartmill, 1974; Jungers, 1985).

Other mammals such as many carnivores or ungulates also display limb elongation, but mainly through lengthening the distal elements of the limb (see Figure 6), including the fingers (Alexander et al., 1979). This is generally understood as a means of reducing the inertial properties of the limb by concentrating mass proximally thereby reducing limb momentum and the cost of reversing the limb's direction with each stride (Smith and Savage, 1956; Hildebrand, 1988; however, see Steudel, 1994). However, in so doing, these animals have sacrificed any ability to grasp with their hands or feet. The proximal and intermediate limb segments of primates have been lengthened, thus permitting the maintenance of grasping extremities.

The protracted and extended limb postures of primates further increase the effective length of their limbs (see Figure 6). However, if limb elongation were the only goal, this conceivably could have been achieved solely through lengthening the bony elements. Although extended limb postures have been recognized as a means of reducing the load arms for substrate reaction forces as body size increases to limit joint and bone stresses (Biewener, 1983, 1989, 1990) and reduce energy costs (Kram and Taylor, 1990), primate quadrupeds display the most extended limb postures at the beginning and end of stance phase when substrate reaction forces are low. They typically flex their elbow or knee through midstride, when reaction forces are high, as part of their compliant gait (Schmitt, 1995). This suggests that the unique protracted limb positions displayed at touchdown by primate quadrupeds are not related to the reduction of substrate reaction force load arms. I suggest that limb protraction, especially forelimb protraction, is more fundamentally associated with reaching out with a grasping extremity to travel and forage in the discontinuous small branch arboreal habitat. Lemelin (1996) has reported on the many parallels in hand proportions between primates and highly arboreal marsupials, such as *Caluromys*

Figure 6. Position of forelimb at touchdown in different mammals. Superimposed black lines represent scapular, humeral and radial-ulnar limb segments. The forefoot touches down ahead of the shoulder in all individuals. However, in the horse, dog, cat and capybara, this is due mainly to the elongated distal limb elements since the humerus is positioned behind a vertical line through the shoulder. In the baboon, the anterior position of the hand results from the protracted position of the humerus and extended elbow. Figures drawn from photographs in Muybridge (1957), and are not to scale.

and *Marmosa*, which also use grasping extremities to travel in a small branch habitat and to capture prey. Significantly, these arboreal marsupials also display a protracted forelimb posture during locomotion (Lemelin, pers. com.).

Finally, the protraction of the hind limb and accompanying increase in hind limb angular excursion observed in primates may have had another consequence. Schmitt and Larson (1995) suggest that the protracted position of the hind limb is also related to the occurrence of plantigrady in arboreal primates. They note that a highly protracted hind limb at touchdown results in much of the body weight being placed behind the foot. This, in turn, tends to force the heel down after touchdown. Although heel contact is not common among primates, it is seen among the species displaying the most dramatic shift in weight from the fore- to the hind limbs such as chimpanzees, orangutans, spider, woolly, and howling monkeys, and gibbons (the logical end-product of the posterior weight shift mechanism (Reynolds, 1985a,b, 1987)). Schmitt and Larson (1995) suggest that this impetus for heel contact may be a factor in the eventual evolution of the heel strike that has become characteristic of human bipedalism.

Protracted limb positions, limb elongation, and greater supraspinal control are central elements underlying the various ways in which the manner of quadrupedal locomotion of primates differs from that of nonprimates. It seems likely that these changes came about through some combination of feedback loops (see Figure 7), where reaching out for small branches using grasping hands and manually foraging for food and/or capturing prey selected for greater cortical input into the control of forelimb motion, which also enhanced the manipulatory abilities of the forelimb. Improved reaching ability requires some forelimb elongation as well as increased forelimb mobility, which can only be brought about if the necessity for limb stabilization is relaxed. This entails reducing some of the disruptive locomotor forces that affect the forelimb. In order to spare the forelimb from some of these forces, muscle recruitment patterns were changed, mechanisms for shifting weight from the forelimb to the hind limb, and for reducing the magnitude of peak forces acting

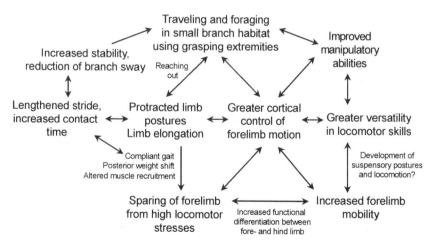

Figure 7. Interaction of factors involved in the evolution of primate quadrupedalism. Use of grasping extremities to travel and manually forage in a small branch habitat required changes in habitual limb posture and greater cortical control of limb movements. These, in turn, allowed improvement in manipulatory skills and greater versatility in locomotor abilities. Such changes, however, could only be brought about through increased joint mobility that required the development of mechanisms to limit disruptive locomotor stresses. See text for further discussion.

on the limbs by increasing limb compliance were developed. These mechanisms exploited the protracted limb postures and limb elongation that had been adopted for reaching and climbing and perhaps exaggerated their magnitude. Increasing cortical control of forelimb movements overrode spinal cord pattern generators, leading to greater versatility in gait patterns and perhaps to a preference for diagonal sequence gaits.

The reduction of peak forces, especially on the forelimb, brought about by these various mechanisms, and the morphological changes enhancing limb mobility that they permitted, have been pivotal in the development of the locomotor versatility—including various forms of climbing, scrambling, and bridging behaviors—that has come to characterize arboreal primates. These changes also constituted a major step toward the "emancipation of the forelimb" from compressive joint forces that Wood Jones (1926) and others have emphasized as the central evolutionary factor leading to the development of suspensory postures and locomotion as well as human bipedalism.

ACKNOWLEDGMENTS

I wish to thank Daniel Schmitt for many stimulating discussions about the topics contained in this paper, Jack Stern who collaborated on all of the EMG research, and Marianne Crisci for animal handling during the EMG experiments. Thanks also to Jack Stern, William Jungers and three anonymous reviewers for helpful comments on earlier versions of this paper, and to Luci Betti for preparation of some of the text figures. This material is based on work supported by the National Science Foundation under Grant SBR 9507078.

REFERENCES

Alexander R McN (1991) Elastic mechanisms in primate locomotion. Z. Morph. Anthrop. *78*:315–320.
Alexander R McN, Jayes AS, Maloiy GMO, and Wathuta EM (1979) Allometry of the limb bones of mammals from shrews (*Sorex*) to elephant (*Loxodonta*). J. Zool., Lond. *189*:305–314.
Alexander R McN, and Maloiy GMO (1984) Stride lengths and stride frequencies of primates. J. Zool., Lond. *202*:577–582.
Biewener AA (1983) Allometry of quadrupedal locomotion: The scaling of duty factor, bone curvature and limb orientation to body size. J. exp. Biol. *105*:147–171.
Biewener AA (1989) Scaling body support in mammals: Limb posture and muscle mechanics. Science *245*:45–48.
Biewener AA (1990) Biomechanics of mammalian terrestrial locomotion. Science *250*:1097–1103.
Blickhan R (1989) The spring-mass model for running and hopping. J. Biomech. *22*:1217–1227.
Cartmill M (1974) Pads and claws in arboreal locomotion. In FA Jenkins Jr. (ed.): Primate Locomotion. New York: Academic Press, pp. 45–83.
Cartmill M (1985) Climbing. In M Hildebrand, DM Bramble, KF Liem, and DB Wake (eds.): Functional Vertebrate Morphology. Cambridge: Belknap Press, pp. 73–88.
Dagg AI (1974) Running, Walking and Jumping. London: Wykeham.
Demes B, Jungers WL, and Nieschalk U (1990) Size- and speed-related aspects of quadrupedal walking in slender and slow lorises. In FK Jouffroy, MH Stack, and C Niemitz (eds.): Gravity, Posture and Locomotion in Primates. Firenze: Il Sedicesimo, pp. 175–197.
Demes B, Larson SG, Stern JT Jr., and Jungers WL (1992) The hindlimb drive of primates - theoretical reconsideration and empirical examination of a widely held concept. Am. J. Phys. Anthropol. Suppl. *14*: 69.
Demes B, Larson SG, Stern JT Jr., Jungers WL, Biknevicius AR, and Schmitt D (1994) The kinetics of "hind limb drive" reconsidered. J. Hum. Evol. *26*:353–374.
Dial KP, Goslow GE Jr., and Jenkins FA Jr. (1991) The functional anatomy of the shoulder in the European starling (*Stunus vulgaris*). J. Morph. *207*:327–344.
Dial KP, Kaplan SR, Goslow GE Jr., and Jenkins FA Jr. (1987) Structure and neural control of the pectoralis in pigeons: implications for flight mechanics. Anat. Rec. *218*:284–287.

Dunbar RIM, and Dunbar EP (1974) Ecological relations and niche separation between sympatric terrestrial primates in Ethiopia. Folia Primatol. *21*:36–60.

Dykyj D (1980) Locomotion of the slow loris in a designed substrate context. Am. J. Phys. Anthropol. *52*:577–586.

Eaton TH Jr. (1944) Modification of the shoulder girdle related to reach and stride in mammals. J. Morph. *75*:167–171.

Eidelberg E, Walden JG, and Nguyen LH (1981) Locomotor control in macaque monkeys. Brain *104*:647–663.

English AW (1978a) Functional analysis of the shoulder girdle of cats during locomotion. J. Morph. *156*:279–292.

English AW (1978b) An electromyographic analysis of forelimb muscles during overground stepping in the cat. J. exp. Biol. *76*:105–122.

Farley CT, Glasheen J, and McMahon TA (1993) Running springs: Speed and animal size. J. exp. Biol. *185*:71–86.

Fedigan L, and Fedigan LM (1988) *Cercopithecus aethiops*: A review of field studies. In A Gautier-Hion, F Bourliere, J Gautier (eds.): A Primate Radiation: Evolutionary Biology of the African Guenons. Cambridge: Cambridge Univ. Press. pp. 389–411.

Fischer MS (1994) Crouched posture and high fulcrum, a principle in the locomotion of small mammals: The example of the rock hyrax (*Procavia capensis*) (Mammalia: Hyracoidea). J. Hum. Evol. *26*:510–524.

Fleagle JG (1978) Locomotion, posture and habitat use of two sympatric leaf-monkeys in West Malaysia. In DJ Chivers and J Herbert (eds.): Recent Advances in Primatology, Vol. 1, Behavior. New York: Academic Press, pp. 331–336.

Georgopoulos AP, and Grillner S (1989) Visuomotor coordination in reaching and locomotion. Science *245*:1209–1210.

Goslow GE Jr., Dial KP, and Jenkins FA Jr. (1989) The avian shoulder: An experimental approach. Amer. Zool. *29*:287–301.

Grand T (1968a) Functional anatomy of the lower limb of the howler monkey (*Alouatta caraya*). Am. J. Phys. Anthropol. *28*:163–181.

Grand T (1968b) Functional anatomy of the upper limb. In: Biology of the Howler Monkey (*Alouatta caraya*). Bibl. Primat., No. 7, Basel: Karger, pp. 104–125.

Grand T (1984) Motion economy within the canopy: Four strategies for mobility. In P Rodman and JGH Cant (eds.): Adaptations for Foraging in Nonhuman Primates: Contributions to an Organismal Biology of Prosimians, Monkeys and Apes. New York: Columbia Univ. Press, pp. 54–72.

Heglund NC, Taylor CR, and McMahon TA (1974) Scaling stride frequency and gait to animal size: Mice to horses. Science *186*:1112–1113.

Hildebrand M (1967). Symmetrical gaits of primates. Am. J. Phys. Anthropol. *26*:119–130.

Hildebrand M (1988) Analysis of Vertebrate Structure, 3rd Ed. New York: John Wiley and Sons.

Howell AB (1944) Speed in Animals. Chicago: Univ. Chicago Press.

Ishida H, Jouffroy FK, and Nakano Y (1990) Comparative dynamics of pronograde and upside down horizontal quadrupedalism in the slow loris (*Nycticebus coucang*). In FK Jouffroy, MH Stack, and C Niemitz (eds.): Gravity, Posture and Locomotion in Primates. Firenze: Il Sedicesimo, pp. 209–220.

Iwamoto M, and Tomita M (1966) On the movement order of four limbs while walking and the body weight distribution to fore and hind limbs while standing on all fours in monkeys. J. Anthropol. Soc. Nippon *74*:228–231.

Jenkins FA Jr. (1971) Limb posture and locomotion in the Virginia opossum (*Didelphis marsupialis*) and in other non-cursorial mammals. J. Zool., Lond. *165*:303–315.

Jenkins FA Jr. (1974) The movement of the shoulder in claviculate and aclaviculate mammals. J. Morph. *144*:71–83.

Jenkins FA Jr., and Goslow GE Jr. (1983) The functional anatomy of the shoulder of the savanna monitor lizard (*Varanus exanthematicus*). J. Morph. *175*:195–216.

Jenkins FA Jr., and Weijs WA (1979) The functional anatomy of the shoulder in the Virginia opossum (*Didelphis virginiana*). J. Zool., Lond. *188*:379–410.

Jouffroy FK, and Petter A (1990) Gravity-related kinematic changes in lorisine horizontal locomotion in relation to position of the body. In FK Jouffroy, MH Stack, and C Niemitz (eds.): Gravity, Posture and Locomotion in Primates. Firenze: Il Sedicesimo, pp. 199–208.

Jouffroy FK, Renous S, and Gasc JP (1983) Etude cinéradiographique des déplacements du membre antérieur du Potto de Bosman (*Perodicticus potto*) au cours de la marche quadrupède sur une branche horizontale. Ann. Sc. Nat. Zool. 13ème sér. *5*:75–87.

Jungers WL (1985) Body size and scaling of limb proportions in primates. In WL Jungers (ed.): Size and Scaling in Primate Biology. New York: Plenum Press, pp. 345–381.

Jungers WL, Jouffroy FK, and Stern JT Jr. (1980) Gross structure and function of the quadriceps femoris in *Lemur fulvus*: An analysis based on telemetered electromyography. J. Morph. *164*:287–299.

Kimura T (1985) Bipedal and quadrupedal walking of primates: Comparative dynamics. In S Kondo (ed.): Primate Morphophysiology, Locomotor Analyses and Human Bipedalism. Tokyo: Univ. of Tokyo Press, pp. 81–104.

Kimura T, Okada M, and Ishida H (1979) Kinesiological characteristics of primate walking: Its significance in human walking. In ME Morbeck, H Preuschoft, and N Gomberg (eds.): Environment, Behavior, and Morphology: Dynamic Interactions in Primates. New York: Gustav Fischer, pp. 297–311.

Kram R, and Taylor CR (1990) Energetics of running: A new perspective. Nature 346:265–267.

Larson SG, and Stern JT Jr. (1987) EMG of chimpanzee shoulder muscles during knuckle-walking: Problems of terrestrial locomotion in a suspensory adapted primate. J. Zool. Lond. 212:629–655.

Larson SG, and Stern JT Jr. (1989a) The role of supraspinatus in the quadrupedal locomotion of vervets (Cercopithecus aethiops): Implications for interpretation of humeral morphology. Am. J. Phys. Anthropol. 79:369–377.

Larson SG, and Stern JT Jr. (1989b) The role of propulsive muscles of the shoulder during quadrupedalism in vervet monkeys (Cercopithecus aethiops): Implications for neural control of locomotion in primates. J. Motor Behavior 21:457–472.

Lemelin P (1996) The evolution of manual prehensility in primates: A comparative and functional analysis in prosimian primates and didelphid marsupials. Ph.D. Dissertation, State University of New York at Stony Brook.

McMahon TA (1975) Using body size to understand the structural design of animals: Quadrupedal locomotion. J. Appl. Physiol. 39:619–627.

McMahon TA (1984) Muscles, Reflexes and Locomotion. Princeton: Princeton Univ. Press.

McMahon TA (1985) The role of compliance in mammalian running gaits. J. exp. Biol. 115:263–282.

McMahon TA, Valiant G, and Frederick EC (1987) Groucho running. J. Appl. Physiol. 62:2326–2337.

Meldrum DJ (1991) Kinematics of the cercopithecine foot on arboreal and terrestrial substrates with implications for the interpretation of hominid terrestrial adaptations. Am. J. Phys. Anthropol. 84:273–290.

Miller S, and Van der Meché FGA (1975) Movements of the forelimbs of the cat during stepping on a treadmill. Brain Res. 91:255–269.

Morbeck ME (1979) Forelimb use and positional adaptation in Colobus guereza: Integration of behavioral, ecological and anatomical data. In M Morbeck, H Preuschoft, and N Gomberg (eds.): Environment, Behavior, and Morphology: Dynamic Interactions in Primates. New York: Gustav Fisher, pp. 95–118.

Muybridge E (1957) Animals in Motion. In LS Brown (ed.): Animals in Motion. New York: Dover. (Originally published by Chapman and Hall, London, 1899)

Napier JR (1967) Evolutionary aspects of primate locomotion. Am. J. Phys. Anthropol. 27:333–342.

Nomura S, Sawazake H, and Ibaraki T (1966) Co-operated muscular action in postural adjustment and motion in dog, from the viewpoint of electromyographic kinesiology and joint mechanics. IV. About muscular activity in walking and trot. Jap. J. Zootech. Sci. 37:221–229.

Okada M, Kimura T, Ishida H, and Kondo S (1978) Biomechanical aspects of primate quadrupedalism. In E Asmussen and K Jørgensen (eds.): Biomechanics 6A. Baltimore: Univ. Part Press, pp. 119–124.

Pandy MG, Kumar V, Berme N, Waldron KJ (1988) The dynamics of quadrupedal locomotion. J. Biomech. Engr. 110:230–237.

Peters SE, and Goslow GE Jr. (1983) From salamanders to mammals: Continuity in musculoskeletal function during locomotion. Brain Behav. Evol. 22:191–197.

Preuschoft H, and Gunther MM (1994) Biomechanics and body shape in primates compared to horses. Z. Morph. Anthropol. 80:149–165.

Preuschoft H, Witte H, Christian A, and Fischer M (1996) Size influences on primate locomotion and body shape, with special emphasis on the locomotion of 'small mammals.' Folia Primatol. 66:93–112.

Prost JH (1965) The methodology of gait analysis and gaits of monkeys. Am. J. Phys. Anthropol. 23: 215–240.

Prost JH (1969) A replication study on monkey gaits. Am. J. Phys. Anthropol. 30:203–208.

Reynolds TR (1985a) Mechanics of increased support of weight by the hindlimbs in primates. Am. J. Phys. Anthropol. 67:335–349.

Reynolds TR (1985b) Stresses on the limbs of quadrupedal primates. Am. J. Phys. Anthropol. 67:351–362.

Reynolds TR (1987) Stride length and its determinants in humans, early hominids, primates, and mammals. Am. J. Phys. Anthropol. 72:101–116.

Rollinson J. and Martin RD (1981) Comparative aspects of primate locomotion, with special reference to arboreal cercopithecines. In MH Day (ed.): Vertebrate Locomotion. Symposia of the Zoological Society of London, No. 48. London: Academic Press, pp. 377–427.

Rose MD (1979) Positional behavior of natural populations: Some quantitative results of a field study of Colobus guereza and Cercopithecus aethiops. In ME Morbeck, H Preuschoft, and N Gomberg (eds.): Environment, Behavior, and Morphology: Dynamic Interactions in Primates. New York: Gustav Fischer, pp. 74–93.

Schmitt D (1994) Forelimb mechanics as a function of substrate type during quadrupedalism in two anthropoid primates. J. Hum. Evol. *26*: 441–457.

Schmitt D (1995) A kinematic and kinetic analysis of forelimb use during arboreal and terrestrial quadrupedalism in Old World monkeys. Ph.D. Dissertation, State University of New York at Stony Brook.

Schmitt D, and Larson SG (1994) Heel contact as a function of substrate type and speed in primates. Am. J. Phys. Anthropol. *96*:39–50.

Smith JM, and Savage RJG (1956) Some locomotor adaptations in mammals. J. Linn. Soc. (Zool.) *42*:603–622.

Stern JT Jr., Wells JP, Vangor AK, and Fleagle JG (1977) Electromyography of some muscles of the upper limb in *Ateles* and *Lagothrix*. Yrbk. Phys. Anthropol. *20*:98–507.

Steudel K (1994) Locomotor energetics and hominid evolution. Evol. Anthropol. 3:42–48.

Struhsaker TT (1967) Ecology of vervet monkeys (*Cercopithecus aethiops*) in the Masai-Ambesoli Game Reserve, Kenya. Ecology *48*:891–094.

Tokuriki M (1973) Electromyographic and joint-mechanical studies in quadrupedal locomotion. I. Walk. Jap. J. Vet. Sci. *35*:433–448.

Tomita M (1967) A study on the movement patterns of four limbs in walking. 1. Observation and discussion on the two types of the movement order of four limbs seen in mammals while walking. J. Anthropol. Soc. Nippon *75*:120–146.

Vangor A, and Wells JP (1983) Muscle recruitment and the evolution of bipedality: Evidence from telemetered electromyography of spider, woolly, and patas monkeys. Ann. Sci. Nat., Zool., Paris, Ser. 13 *5*:125–136.

Viala D, and Vidal C (1978) Evidence for distinct spinal locomotion generators supplying respectively fore- and hindlimbs in the rabbit. Brain Res. *155*:182–186.

Vilensky JA (1980) Trot-gallop transition in a macaque. Am. J. Phys. Anthropol. *53*:347–348.

Vilensky JA (1987) Locomotor behavior and control in human and nonhuman primates: comparisons with cats and dogs. Neurosci. Biobehav. Rev. *11*:263–274.

Vilensky JA (1989) Primate quadrupedalism: How and why does it differ from that of typical quadrupeds. Brain Behav. Evol. *34*:357–364.

Vilensky JA, Gankiewicz E, and Townsend DW (1988) Effects of size on vervet (*Cercopithecus aethiops*) gait parameters: A cross-sectional approach. Am. J. Phys. Anthropol. *76*:463–480.

Vilensky JA, Gankiewicz E, and Townsend DW (1990) Effects of size on vervet (*Cercopithecus aethiops*) gait parameters: A longitudinal approach. Am. J. Phys. Anthropol. *81*:429–439.

Vilensky JA, and Larson SG (1989) Primate locomotion: Utilization and control of symmetrical gaits. Ann. Rev. Anthropol. *18*:17–35.

Vilensky JA, Moore AM, and Libii JN (1994) Squirrel monkey locomotion on an inclined treadmill: Implications for the evolution of gaits. J. Hum. Evol. *26*:375–386.

Whitehead PF, and Larson SG (1994) Shoulder motion during quadrupedal walking in *Cercopithecus aethiops*: Integration of cineradiographic and electromyographic data. J. Hum. Evol. *26*:525–544.

Wood Johns F (1926) The Arboreal Man. New York: Hafner.

11

FORELIMB MECHANICS DURING ARBOREAL AND TERRESTRIAL QUADRUPEDALISM IN OLD WORLD MONKEYS

Daniel Schmitt

Department of Biological Anthropology and Anatomy
Duke University School of Medicine
Box 3170 DUMC
Durham, North Carolina 27701

1. INTRODUCTION

For over a century it has been known that primates have highly mobile grasping forelimbs with the supportive functions shifted more strongly to the hindlimbs, unlike most mammals where all four limbs share a fairly equal role in weight support. Darwin (1871) was the first to recognize this distinction between forelimb and hindlimbs and to articulate its evolutionary significance. Since that point many researchers have developed theories of primate locomotor evolution that suggest that the amount of compressive weight support experienced by the forelimb of primates was gradually reduced thus facilitating the use of the forelimb in tension and then allowing its complete removal from locomotion in humans (Wood Jones, 1926; Le Gros Clark, 1959; Napier and Davis, 1959; Napier, 1967; Stern, 1976; Ripley, 1979; Reynolds, 1981, 1985a,b; Cant, 1988; Rose, 1991). Fundamental to this scenario is the belief that the change in the role of the primate forelimb is directly related to adaptations to arboreal quadrupedalism by primates.

This view of primate evolution is intuitively appealing. At present, however, it is based on untested suppositions concerning the mechanics of primate quadrupedalism. The scenario described above presupposes that arboreal locomotion requires a very different role for the forelimb than does terrestrial quadrupedalism. This is based on the fact that an arboreal environment poses mechanical challenges that are very different from a terrestrial one. Unlike the ground, tree branches are not generally level, they are finite in diameter, and the surface is discontinuous, often presenting animals with sharp directional changes and large gaps between branches (Cartmill, 1985). Additionally, food in an arboreal environment may be distributed at many levels and is often found on terminal branches. Therefore, quadrupedal locomotion in an arboreal habitat requires primates to have highly

Primate Locomotion, edited by Strasser *et al.*
Plenum Press, New York, 1998

mobile forelimbs that (1) can be held flexed in order to better maintain balance by lowering the center of gravity, (2) can be used to reach and grab in many planes, and (3) allow for rapid changes of direction and gap crossing in a discontinuous support environment (Wood Jones, 1926; Grand 1968a,b; Jenkins, 1974; Cartmill, 1985; Alexander and Ker, 1990; Kram and Taylor, 1990).

Naturalistic, anatomical, and laboratory studies support this analysis. Although detailed kinematics are unavailable from the field, qualitative locomotor data on primates in natural settings support the belief that primate arboreal quadrupeds walk with flexed forelimbs (Grand, 1968a,b; Rose, 1973; Morbeck, 1979; Cant, 1988; Dunbar, 1989). Anatomical studies suggest that arboreal quadrupedal primates have a greater range of forelimb joint motion due to decreased osteological stabilization of the forelimb joints, and reduced robusticity of the forelimb bones in comparison to those of terrestrial primates (e.g., Ashton et al., 1965, 1968; Jolly, 1967; Manaster, 1975, 1979; Feldesman, 1976; Morbeck, 1977, 1979; O'Connor and Rarey, 1979; Rodman, 1979; Bown et al., 1982; Schaffler and Burr, 1984; Schaffler et al., 1985; Harrison, 1989; Rose, 1988, 1989; Jungers and Burr, 1992, 1995). Kinetic studies of ground quadrupedalism in primarily arboreal and primarily terrestrial primate species has shown that primates have a unique distribution of vertical ground reaction forces in which the hindlimbs serve as the primary supportive organs (Kimura et al., 1979; Kimura, 1985, 1992; Reynolds, 1981, 1985a, b, Demes et al., 1992). The reduction of vertical compressive force on the forelimb has been interpreted as a means of protecting a highly mobile (and therefore inherently less stable) forelimb against disruptive locomotor stresses (Reynolds, 1981, 1985a,b).

Clearly all the presently available data supports the notion that arboreal quadrupeds have highly gracile and mobile forelimbs that are regularly held in a flexed posture. There is, however, a serious paradox in this description of primate arboreal posture. A flexed limb posture should result in higher joint and bone stresses if substrate reaction forces are equal in a crouched and extended posture as illustrated in Figure 1 (Biewener, 1982, 1983a,b, 1989, 1990). This is because the greater distance of the substrate reaction resultant to the joints yields a decreased mechanical advantage of the antigravity muscles and increases the bending moments and muscle force required to maintain this posture

a) Extended   b) Flexed

Figure 1. Schematic depiction of the effect of crouched posture on bone and joint stresses in a limb, based on models of Biewener (1989, 1990). In (a) and (b) the substrate reaction resultant force (arrow) and the limb segment lengths are identical. The only change from (a) to (b) is increased elbow flexion. This results in clearly larger moment arms and moments along the limb. This reduces the "effective mechanical advantage" of antigravity muscles, increases the force necessary to maintain posture, and increases bone and joint stresses.

(Biewener, 1982, 1983a,b, 1989, 1990). As a result of this model we might expect that arboreal primates would have more robust forelimbs with greater stabilization to prevent distraction of the shoulder or elbow. Because anatomical evidence described above (especially Manaster, 1975, 1979; Schaffler and Burr, 1984; Schaffler et al., 1985; Jungers and Burr, 1992, 1995) show that this is not the case, only two other possible solutions to this apparent conflict remain: (1) either primate quadrupeds do not use the postures described above, or (2) a change in the substrate reaction forces accompanies this change in posture thus reducing the bone and joint stresses.

2. MATERIAL AND METHODS

In an attempt to resolve this conflict I documented the kinematics and kinetics of forelimb use on terrestrial and "arboreal" supports in a group of Old World monkeys. I selected five species that included primarily arboreal, primarily terrestrial, and "semi-terrestrial" (Ripley, 1979) species (Table 1; Struhsaker, 1967; Rose, 1973, 1977; Rodman, 1979; Wheatley, 1982; Chism and Rowell, 1988; Fedigan and Fedigan, 1988; Fleagle, 1988; Dunbar, 1989; Nakigawa, 1989; Rawlins, 1993). All nine subjects were trained to walk within a plexiglas enclosure (6 m × 1 m × 1 m) on a wooden runway and on raised horizontal poles of varying diameters (Table 1). The arrangement of recording equipment is illustrated in Figure 2. During locomotion, vertical, fore-aft, and mediolateral components of the substrate reaction force data were recorded from a force plate mounted in the runway for terrestrial locomotion or attached to the poles for "arboreal" locomotion. Simultaneously, subjects were videotaped with electronically shuttered video cameras from lateral, frontal and overhead views. The methods used are described in detail elsewhere (Schmitt, 1994, 1995) and will only be summarized below.

For collecting data on terrestrial quadrupedalism the force platform was mounted in the center of a runway flush with its surface with a ½ inch gap on all sides to avoid vibrations from limbs other than the one making contact. For gathering data on "arboreal" quadrupedalism, animals were trained to run along PVC pipes coated with a nonslip surface of

Table 1. Primary habitat and number of walking steps collected for each subject
on the poles and the ground

Subject	Primary habitat	Pole (cm)					Grnd.
		1.25	2.5	6.25	8.75	11.25	
M. fascicularis (f)	arboreal[1]	–	10	11	10	10	13
M. fascicularis (m)	arboreal[1]	–	–	12	9	10	10
M. mulatta (f)	semi-terrestrial[2]	–	–	10	8	10	11
M. mulatta (f)	semi-terrestrial[2]	–	7	10	10	10	10
C. aethiops (m)	semi-terrestrial[3]	9	7	8	10	11	10
C. aethiops (m)	semi-terrestrial[3]	5	6	10	8	9	9
P. anubis (f-juvenile)	terrestrial[4]	–	10	8	12	11	10
P. anubis (m)	terrestrial[4]	–	–	6	10	5	10
E. patas (m)	terrestrial[5]	–	–	10	10	9	10

[1]Rodman, 1979; Wheatley, 1978, 1980; Cant, 1988; Fleagle, 1988.
[2]Rawlins, 1976, 1993; DeRousseau, 1988; Fleagle, 1988; Dunbar, 1989.
[3]Struhsaker, 1967; Dunbar and Dunbar 1974; Fedigan and Fedigan, 1988; Fleagle, 1988.
[4]Rose 1977; Fleagle, 1988.
[5]Kingdon, 1971; Chism and Rowell, 1988; Fleagle, 1988; Nakigawa, 1989.

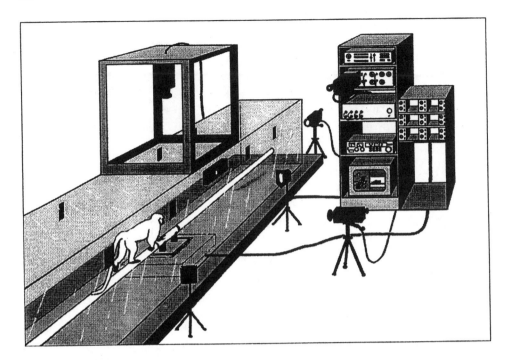

Figure 2. Diagram of the experimental enclosure, with a subject traveling on a raised horizontal pole. The force plate can be seen underneath the runway with a cord connected to the charge amplifiers. Three cameras are placed in lateral, frontal and overhead positions. Halogen lights illuminate the subject. Infrared emitters/sensors are placed along the subject's path. In the cabinet, from bottom to top, are the monitor (with split-screen image), the VCR, 100 HZ filters, oscilloscope (with camera), and special effects generator.

sand and paint. To insure collection of force data from single footfalls without influence of limbs in contact with other portions of the pole, a 30 cm-long central segment was attached to the surface of the force plate and separated from two non-instrumented segments by a 2 cm-wide gap on either side (Figure 2). Signals from the force platform were passed through a low-pass filter with a 100 Hz cut-off and displayed on an oscilloscope as deflections of three beams representing the vertical, fore-aft, and mediolateral forces (Figure 3).

Images from two of the three video cameras directed at the animal, as well as a fourth camera that recorded force traces on the oscilloscope, were combined by a special effects generator so that kinetic and kinematic data were precisely coordinated. For each animal on each substrate, lateral and frontal images of the animal crossing the plate were simultaneously displayed as a split field. Vertical, fore-aft, and transverse force curves were superimposed on the split image using previously established methods (Stern et al., 1977; Reynolds, 1981; Larson and Stern, 1989). Subject velocity was determined for each step either using visible markers on the enclosure or using a series of four infrared sensors placed at 3/4 meter intervals along the runway that triggered a series of four super-bright LED lamps visible to the lateral camera.

Using frame-by-frame playback on a VCR, steps in which peak forces were not obscured because of contact by another limb and in which the animal was traveling in a relatively straight path, with no obvious acceleration or deceleration, were selected for analysis. Angular data from the forelimb and substrate reaction force data were digitized using a microcomputer, video frame grabber board, and video analysis software.

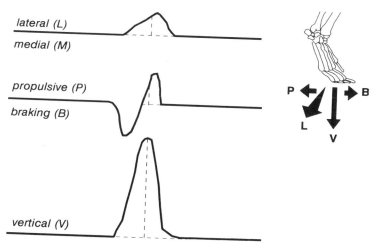

Figure 3. Schematic diagram depicting vertical, braking-propulsive, and mediolateral substrate reaction force traces. The magnitude of the three components is measured as the deflection of the oscilloscope beam and these data are used to calculate the magnitude and orientation of the SRR using the following formulae: Sagittal Resultant, $(r)=(V^2+P^2)^{0.5}$; Total SRR, $(R)=(r^2+L^2)^{0.5}$; Sagittal Angle= arc cos (V/r); Mediolateral angle = arc cos (r/R).

The substrate reaction resultant (SRR) magnitude and orientation have been found to be more useful than component forces for analyzing bone and joint morphology (see Jenkins and Weijs, 1979; Biewener, 1983b, 1989; Full and Tu, 1990; Full et al., 1991). As can be seen in Figure 3, these values were calculated by first measuring the height of the deflection from the baseline for all three substrate reaction components at the vertical peak, braking peak, propulsive peak, the braking-propulsive transition, and at midsupport (as defined by Larson and Stern [1989]). Then the magnitude (as a percentage of subject body weight) and orientation (in degrees deviation from vertical) of the SRR was calculated in a sagittal and transverse plane using formulae shown in Figure 3 and described in Schmitt (1994, 1995). Substrate reaction force data for the hindlimb were also collected in order to calculate the forelimb/hindlimb vertical force ratio on all supports.

Joint angles at the shoulder, elbow, and wrist (Figure 4) were calculated from lateral, frontal, and overhead images for the same kinetic points described above as well as for touchdown and lift-off. I identified joint centers initially on sedated subjects and relocated those points on the videotapes (Schmitt, 1995). To correct for any error caused by out-of-plane rotation of the forelimb, true intersegmental angles were calculated using an algorithm that calculates 3-D coordinate data for each joint based on the arccosine relation between the actual length of a segment and its projected length. The formulae for this procedure (Figure 4) have been described in detail in Schmitt (1994, 1995) and Chan et al. (nd).

Additionally, three other variables were measured for this analysis: (1) contact time, (2) shoulder height, and (3) vertical oscillation of the shoulder. To calculate contact time the duration of the vertical force was determined from the oscilloscope image. Shoulder height, used as a measure of "crouching", was calculated trigonometrically using the known lengths of limb segments and the angles of the joints. By creating right triangles from these data as shown in Figure 4, three vertical distances—from shoulder to elbow, el-

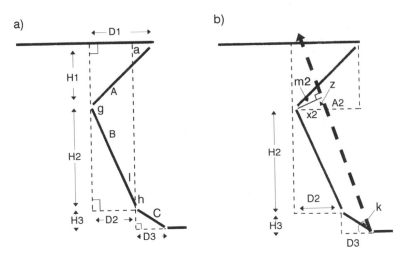

Figure 4. Diagram illustrating calculations used in the analysis. (*a*) Calculation of height (H) and contact distance (D) using midsupport as an example. A, arm length; B, forearm length; C, palm length; a, arm angle; g, elbow angle; l+h, wrist angle. All other angles and distances are calculated. For example, at midsupport, vertical distance from shoulder to elbow (H1)=A*sin(a) and total shoulder height=H1+H2+H3 where H2 and H3 are calculated in a manner similar to the calculation of H1. Similarly, horizontal distance from shoulder to elbow (D1)=A*cos(a) and total distance from point of contact to shoulder=D1+D2+D3 where D2 and D3 are calculated in a manner to the calculation of D1. (*b*) Calculation of moment arms (i.e., m2) and moments (i.e., SRR magnitude*m2) using midsupport as an example. Dashed arrow, SRR; k, sagittal angle of the SRR. Horizontal distance of the SRR from vertical (A2)=tan(k)*(H2+H3); horizontal distance of the SRR from the elbow (x2)=(D2+D3)-A2; the moment arm (m2)=x2*sin(Z). The moment at the elbow then is the total magnitude of the SRR*m2. Moment arms and moments at the shoulder are calculated in a similar manner.

bow to wrist, and wrist to ground—were calculated. These three distances were added together to determine the height of the shoulder from the substrate (expressed as a percentage of the animal's potential fully-extended height). Finally, from these data on shoulder height the vertical oscillation of the forelimb was calculated by subtracting the minimum shoulder height from the maximum shoulder height for each step.

Using methods similar to those used to calculate shoulder height, the moment arms and moments of the substrate reaction force resultant at each joint were also calculated (Figure 4). Following the method of Biewener (1983b), the SRR is placed at the head of the metacarpals. Right triangles were created in which the vertical height of the joint, the limb length, and the SRR vector could be used to calculate the distance (in centimeters) of the SRR vector from the joint and then to calculate a perpendicular distance from the SRR to the joint at any point during support phase (Figure 4). From these data it was then possible to calculate the relative moment (expressed as a percentage of body-weight multiplied by centimeters for the moment arm or %bwcm) around each joint by multiplying the magnitude of the SRR by the moment arm at each point in support phase.

The goal of this project is to compare kinetic and kinematic values during support phase of arboreal and terrestrial locomotion. Normally one would compare each value at specific points (e.g., midsupport) during support phase. Because animals differed in the timing of kinetic and kinematic points during support phase between the pole and ground, direct point-by-point comparison was difficult. To correct for this confounding factor, kinematic and kinetic values across substrates were compared by calculating the minimum, maximum, and mean values for all kinematic and kinetic variables. This was done in the following manner. Values for each step were calculated, followed by calculation of average maximum, minimum, and mean value for each substrate for each subject. The maximum, minimum, and mean values for each step were also tested for correlations with speed on each substrate using conservative nonparametric Spearman correlation methods. Data that were significantly ($P<0.05$) correlated with speed were compared across substrates within taxa, using a standard analysis of covariance (ANCOVA); the variable of interest was the dependent variable and speed was the independent (effectively held constant) variable (Sokal and Rohlf, 1981). The result is the calculation of an adjusted mean and standard error in which the effect of speed has been taken into account. Figures and tables presented in this paper show only a comparison of terrestrial travel to travel on the smallest pole on which each subject would walk.

3. RESULTS

Subjects walked at broadly overlapping speeds ranging between 0.6 - 2 m/s on the poles and ground. There were few significant differences in speed across substrates for any subject. Maximum arm protraction and retraction, maximum elbow flexion, and shoulder height all showed a significant negative correlation with speed. In contrast SRR magnitude and vertical peak force showed significant positive correlations with speed. With the exception of the patas monkey, braking forces showed significant positive correlations with speed, while propulsive forces showed a negative correlation. This latter result suggests that at higher speed subjects were decelerating. The influence of this pattern is discussed below.

Height at the shoulder indicates the position of the shoulder relative to the substrate and gives some estimation of whether a subject is "crouching" and lowering its center of gravity. As can be seen in Figure 5 all subjects showed a reduced shoulder height while traveling on an "arboreal" support. The maximum shoulder height was significantly different on the pole versus the ground for all subjects except the vervet monkeys (Figure 5). It is clear that these subjects are crouching in order to lower their center of gravity throughout at least some part of support phase on arboreal supports.

The reduced shoulder height is brought about partly by differences in the degree of protraction and retraction of the arm, which can lower the height of the shoulder at the beginning and end of support phase on arboreal supports (Table 2). Only the macaques and the young female baboon, perhaps because of her age, showed significant increases in protraction on a pole compared to the ground. This pattern reflects the arboreal habits of the macaques and probably the high mobility in the juvenile female baboon.

In contrast, maximum arm retraction shows a consistent pattern. At liftoff all subjects retracted their arm more on the pole as compared to the ground, although for the baboons and patas this substrate-related difference was nonsignificant (Table 2).

If the forelimb were a single rigid strut, the shoulder would always rise to the same maximum height during support phase. The elbow joint, however, can moderate the rise of the shoulder during support phase. The minimum elbow angle is a measure of the elbow flexion that can reduce the height of the shoulder. Figure 6 shows that all subjects increased elbow flexion on "arboreal" supports relative to the ground, although for the vervet monkeys this change was nonsignificant. This change in elbow angle from touchdown to midsupport is probably the major component in maintaining a lower vertical height throughout the middle portions of support phase.

M. fascicularis

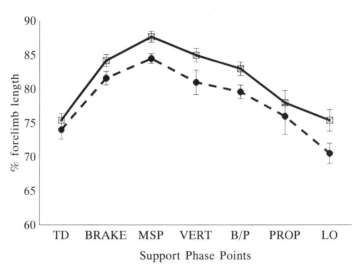

Maximum shoulder height ground vs. pole, P=0.0001

M. mulatta

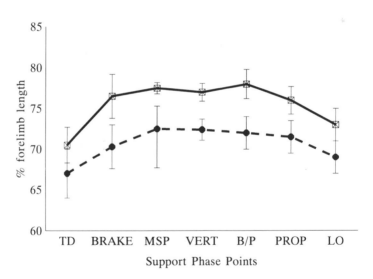

Maximum shoulder height ground vs. pole, P=0.0049

Figure 5. Displacement curves for the height of the shoulder as a percentage of forelimb length during support phase. Dashed lines with circles indicate the adjusted-y mean values on the smallest pole on which subjects would walk. In some cases (e.g., *M. fascicularis*) the smallest pole on which one subject walked was smaller than the smallest pole on which the other subject would walk. In these cases the data were pooled and, therefore, the means represent the mean of the values for the smallest pole on which each individual subject would walk. Solid lines with squares indicate adjusted-y mean values during terrestrial locomotion. Error bars indicate one standard error of the adjusted-y mean value on each substrate. Support phase is divided into touchdown (TD), braking peak (BRAKE), midsupport (MSP), vertical peak (VERT), braking/propulsive transition (B/P), propulsive peak (PROP) and liftoff (LO). Significance levels for comparisons of the pole versus the ground using ANCOVA are given at the base of each graph.

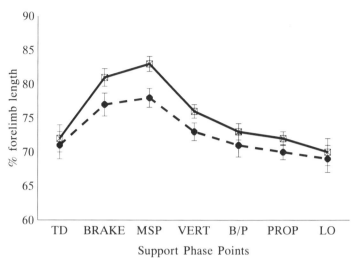

Ground and pole not significantly different.

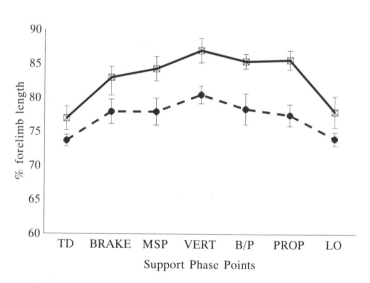

Maximum shoulder height ground vs. pole, P=0.0018

Figure 5. (*continued*)

E. patas

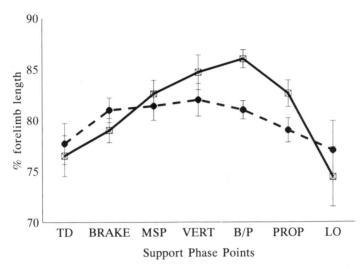

Maximum shoulder height ground vs. pole, P=0.0002

Figure 5. (*continued*)

These kinematic results clearly suggest that either joint moments and stresses are larger in arboreal quadrupedal primates or there is a change in either magnitude or orientation of the SRR. Figure 7 illustrates the pattern of SRR for subjects in this study. The maximum SRR magnitude was significantly lower on poles versus the ground for all subjects except the vervet monkeys (Figure 7). This reduction is driven primarily by a difference in vertical force on the pole versus the ground. The rhesus macaques experienced vertical peak forces that were 45% of those seen on the ground. This was the maximum reduction among all the species. This was followed by the female baboon (60%), the male baboon and *M. fascicularis* (75%), the patas (80%), and finally the vervet monkeys (93%). In contrast, most subjects showed no significant differences in fore-aft forces on the pole

Table 2. Maximum arm protraction (touchdown) and arm retraction (liftoff) angle in degrees. Pole, smallest pole on which subjects would walk. Values under Pole and Ground are the adjusted-y means and one standard error (in parentheses). P, significance values for comparisons of adjusted-y means using ANCOVA

Subject	Arm protraction			Arm retraction		
	Pole	Ground	P	Pole	Ground	P
M. fascicularis	118 (4.7)	106 (3.2)	0.04	32.5 (2.5)	40 (1.7)	0.02
M. mulatta	115 (6)	99.5 (4)	0.01	22 (4)	35 (2.5)	0.03
C. aethiops	96 (2.7)	95 (1.9)	ns	22 (1.6)	34 (1.1)	0.0001
P. anubis (f)	111 (4)	103 (3.6)	0.002	43 (1.9)	47 (2.1)	ns
P. anubis (m)	99 (2)	111.5 (3)	0.0001	38 (2.3)	44 (3)	ns
E. patas	115 (3.3)	113.5 (3)	ns	44 (1.1)	47 (2)	ns

ns, not significant

M. fascicularis

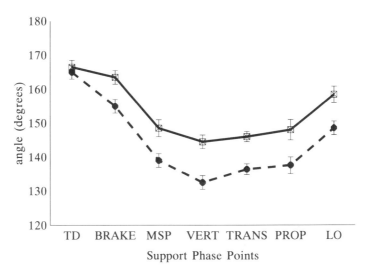

Maximum elbow flexion ground vs. pole, P=0.0016

M. mulatta

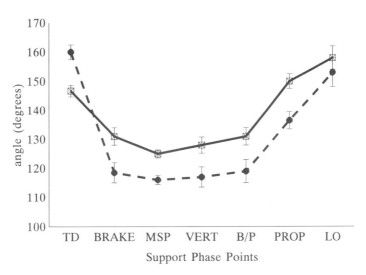

Maximum elbow flexion ground vs. pole, P=0.0004

Figure 6. Displacement curves for the anterior elbow angle during support phase. Same conventions as in Figure 5.

C. aethiops

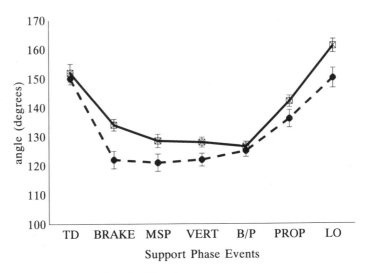

Ground and pole not significantly different

P. anubis

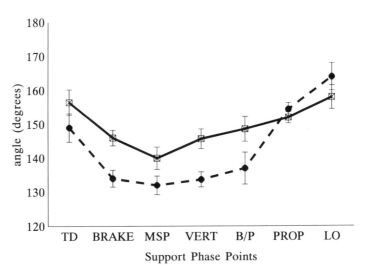

Maximum elbow flexion ground vs. pole, P=0.0297

Figure 6. (*continued*)

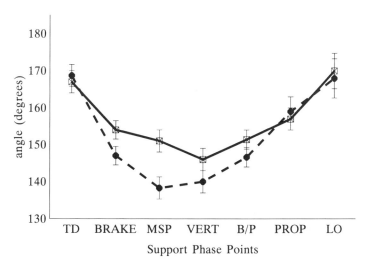

Maximum elbow flexion ground vs. pole P=0.0056

Figure 6. (*continued*)

versus the ground. This latter result may represent an interaction between a reduction in fore-aft forces due to lower accelerations of the limbs and an increase in fore-aft forces due to increased protraction/retraction of the forelimbs on a small pole.

The fact that vertical forces were different across substrates, but that fore-aft forces remained relatively unchanged has important implications. The sagittal orientation of the SRR is influenced by the relationship of the magnitude of the vertical force and the magnitude of the fore-aft forces. If the vertical force is reduced but the fore-aft force is not, the SRR vector is drawn farther away from vertical than when the vertical force is relatively high. This pattern is enhanced by those subjects where braking forces increased with speed and thus caused the SRR to be drawn more caudally at higher speeds. In all the cases, except the vervet monkeys, the ratio of the vertical/fore-aft peak forces was smaller on the pole than it was on the ground. Therefore, the SRR vector was drawn more caudally during the first half of support phase and then more cranially (Figure 8). As a result, as the limb was protracted and then flexed the SRR orientation changed at the same time and maintained nearly equivalent moment arms on the pole versus the ground.

Table 3 displays the ANCOVA results for moment arms and moments at vertical peak, the point at which SRR is generally highest in magnitude. In most subjects moment arms at the shoulder at vertical peak are not significantly different on the ground versus the pole and often they are lower on the pole than on the ground. In contrast, moment arms are generally slightly larger at the elbow at vertical peak on a pole compared to the ground. However, these differences are nonsignificant and do not lead to higher moments at vertical peak (see below). This pattern holds for most points in support phase as well. Except for the vervet monkeys, moment arms at the shoulder are consistently (though nonsignificantly) lower on the pole versus the ground except at propulsive peak, where they are always higher. Moment arms at the elbow are generally (though nonsignificantly) larger on the pole versus the ground.

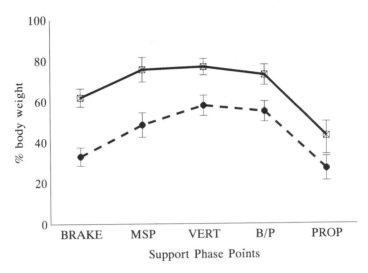

Maximum SRR ground vs. pole, P=0.0083

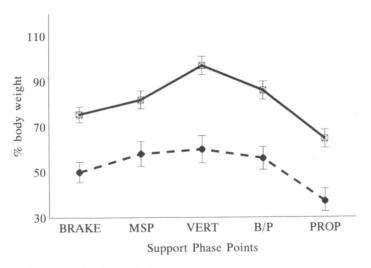

Maximum SRR ground vs. pole, P=0.0002

Figure 7. Displacement curves for the magnitude of the substrate reaction resultant as a percentage of subject body weight during support phase. Same conventions as in Figure 5.

C. aethiops

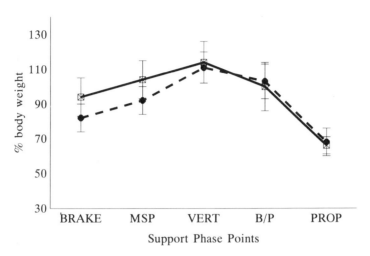

Ground and pole not significantly different.

P. anubis

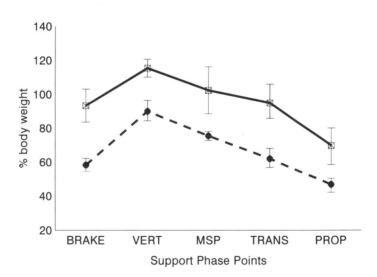

Maximum SRR ground vs. pole, P=0.0071

Figure 7. (*continued*)

E. patas

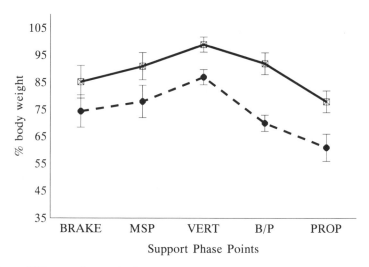

Maximum SRR ground vs. pole, P=0.0004

Figure 7. (*continued*)

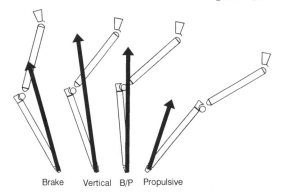

a) Representative "arboreal" step

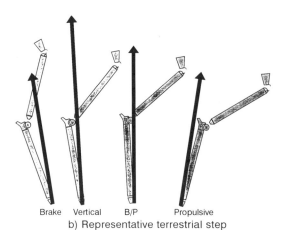

b) Representative terrestrial step

Figure 8. Schematic diagram of how limb position and SRR orientation change simultaneously on the pole ("arboreal") and the ground (terrestrial). This composite does not represent any one subject and indicates the relative orientation of the arm to the horizontal and the forearm to the arm. (*a*) The figures indicate the typical condition on a small pole. (*b*) The figures indicate the typical condition on the ground. In both cases, the points represented are the braking peak (Brake), vertical peak (Vertical), braking/propulsive transition (B/P), and the propulsive peak (Propulsive). Touchdown and liftoff were omitted from this figure because no substrate reaction forces were recorded at those points.

Table 3. Moment arms (cm) and moments (%bwcm, see text) around the shoulder and elbow at vertical peak. Pole, smallest pole on which subjects would walk. Values under Pole and Ground are the adjusted-y mean and one standard error (in parentheses)

Variable	Subject	Shoulder			Elbow		
		Pole	Ground	P	Pole	Ground	P
Moment arms at	*M. fascicularis*	6.3 (1)	6.44 (1)	ns	5.4 (0.9)	5.8 (1.2)	ns
	M. mulatta	5.2 (1.5)	6.8 (1.3)	ns	5.4 (1)	4.8 (0.9)	ns
	C. aethiops	3.5 (0.7)	3.1 (1.4)	ns	5.2 (0.5)	4.6 (1)	ns
	P. anubis (f)	3 (1.5)	5 (1)	ns	6.2 (1)	4.6 (1)	ns
	P. anubis (m)	5.6 (1.1)	5.3 (1.4)	ns	7.5 (1)	7.2 (1.6)	ns
	E. patas	4.8 (1.2)	5.8 (1.6)	ns	10.2 (1.4)	10.9 (1.3)	ns
Moments around	*M. fascicularis*	377 (127)	650 (192)	ns	289 (129)	680 (101)	0.04
	M. mulatta	437 (134)	645 (117)	ns	513 (107)	498 (93)	ns
	C. aethiops	417 (73)	558 (146)	0.02	680 (93)	395 (185)	ns
	P. anubis (f)	331 (149)	594 (132)	ns	620 (122)	480 (108)	ns
	P. anubis (m)	408 (114)	598 (92)	0.04	792 (122)	890 (134)	ns
	E. patas	285 (92)	430 (104)	ns	600 (111)	907 (124)	0.04

P, significance values for comparisons of adjusted-y means using ANCOVA.
ns, not significant

The combination of lower or equal moment arms and an SRR that is lower in magnitude yields bending moments along the forelimb that are generally lower than those on the pole compared to the ground. Table 3 illustrates this pattern in detail for the moments around the shoulder and elbow at vertical peak and Figure 9 illustrates this pattern for all points in support phase. In most cases these are not significant differences and the moments must be considered effectively equal. In cases where they were statistically significant (P<0.05) the moments were lower on the pole versus the ground. As a result of these relatively equal moments, the muscular force required to maintain a crouched posture on a pole is either less than or no greater than the force required to maintain an extended posture on the ground.

4. DISCUSSION AND CONCLUSIONS

The prediction that Old World monkeys lower their center of gravity by changing limb posture during locomotion on relatively small "arboreal" supports was supported. Therefore, if force magnitude and orientation had remained equal, the model predicted higher moment arms and moments along the forelimb. However, the moment arms at most points during support phase, with the exception of propulsive peak, are not significantly different on a small pole compared to the ground. Therefore, arboreal quadrupedalism does not necessarily engender large moments around the forelimb joints and therefore joint reaction forces are not necessarily higher on arboreal versus terrestrial supports.

It is interesting to note that vervet monkeys represent a consistent exception to the patterns described here. This may be explained by the fact the vervet monkeys were the smallest animals in my sample. It is possible that the substrates I designed were not a suitable challenge for their normal agile locomotion. It may also be the case that the very flexed elbow position that vervet monkeys maintain on both the ground and on poles better prepares these small mammals to make sudden dashes and jumps to escape predators. Such a posture would be well suited to a small quadrupedal primate that spends time both

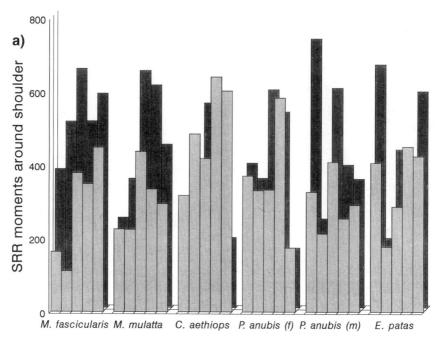

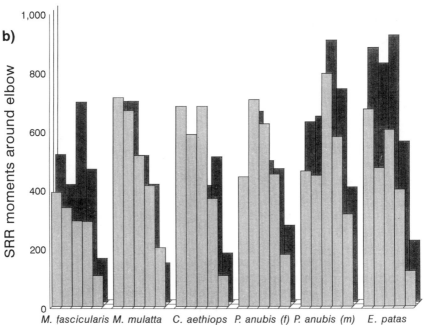

Figure 9. Adjusted-y mean values for the SRR moments around (*a*) the shoulder and (*b*) the elbow on the smallest pole (light bars) on which subjects would walk versus the ground (dark bars) at five points during support phase. For each species, from left bars indicate braking peak, vertical peak, midsupport, braking/propulsive transition, and propulsive peak. Touchdown and liftoff were omitted from this figure because no substrate reaction forces were recorded at those points. At most points SRR moments on the pole are lower than or equal to those on the ground. Table 3 contains additional data for moment arms and moments at vertical peak.

on the ground and in the trees and must leap quickly into the branches when approached by a predator (Struhsaker, 1967; Fedigan and Fedigan 1988; Fleagle, 1988).

The discovery that primate arboreal quadrupeds can maintain poorly stabilized forelimbs and a crouched posture simultaneously helps explain how the role of the forelimb could progressively change throughout primate evolution. It is also important, however, to understand the mechanism by which primates avoid the increased stresses that theoretically accompany a crouched posture. By identifying the mechanism I hope to explain how this pattern first arose among primates.

One possible way to reduce forces on the forelimb is that described by Reynolds (1981, 1985a,b). He demonstrated that arboreal primate quadrupeds walking on the ground actively shift weight posteriorly through activation of hindlimb retractors when the hindlimb is protracted. It is possible that on arboreal supports this mechanism is used to an even greater degree, thereby reducing the vertical force on the forelimb further. The data presented below examines this possibility.

Table 4 shows that the reduction of vertical peak force on the forelimbs on a small pole versus the ground is greater than the reduction in vertical peak force on the hindlimbs on a small pole versus the ground. This differential reduction supports the idea that weight is being shifted posteriorly while animals travel on the small poles. The data in Table 4, however, also demonstrate that all the animals experience an overall reduction in vertical peak force on both forelimbs and hindlimbs. The pattern of vertical peak force reduction on both forelimbs and hindlimbs is not consistent with what one might expect if these animals were actively shifting weight posteriorly in the manner described by Reynolds (1981, 1985a,b). If the hindlimb lever effects described by Reynolds (1981, 1985a,b) are the only mechanisms employed to adjust substrate reaction forces, then one would predict a reduction in peak vertical forelimb forces and an increase in peak vertical hindlimb forces on the pole versus the ground. Because this is not the case in my sample, an additional mechanism must be invoked to explain the overall reduction in vertical forces from ground to pole.

Spring-mass models of mammalian locomotion (McMahon and Greene, 1978; Taylor, 1978; McMahon, 1985; McMahon et al., 1987; McMahon and Cheng, 1990; Blickhan, 1989; Farley et al., 1993) appear to explain this pattern (Figure 10). In a basic model, the limb is seen as a massless spring with the body mass concentrated on top of the spring (Figure 10a). When this simple spring-mass system is hopping forward, the stiffness of the spring controls the contact and flight time. As the leg-spring becomes more compliant, contact time increases and height of the hop decreases. Therefore, time over which to develop

Table 4. Comparison of forelimb and hindlimb percentages for vertical peak force on the smallest pole on which subjects would walk by vertical peak force on the ground. Values under forelimb and hindlimb are the adjusted-y mean and range (in parentheses)

Subject	Forelimb	Hindlimb
M. fascicularis	74 (69–81)	86 (80–92)
M. mulatta	45 (31–56)	81 (78–88)
C. aethiops	93 (90–100)	97 (95–99)
P. anubis (f)	60 (57–70)	80 (73–87)
P. anubis (m)	74 (71–77)	82 (72–92)
E. patas	80 (77–94)	93 (88–98)

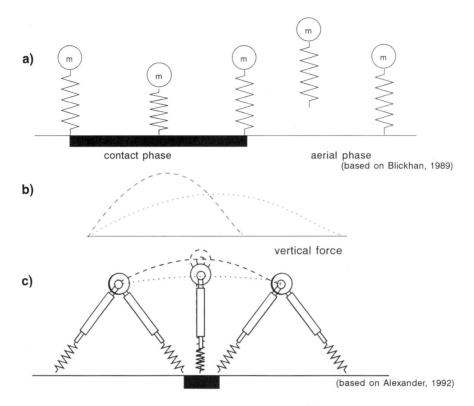

Figure 10. Simple spring-mass models of gait. (*a*) A massless leg spring, with body mass (m) mounted at one end, hopping in place. Spring compliance affects contact time. A stiffer spring than the one illustrated would remain in contact with the substrate for a shorter period of time and have a higher and longer aerial phase. The effect of this increased contact time is shown in *b*. (*b*) The vertical force of a relatively stiff spring (dashed line) has a short duration and a high peak. A more compliant spring (dotted line) yields a vertical force with a longer duration and a lower peak. (*c*) A similar spring-mass system in forward locomotion as in human walking. The dashed line indicates a relatively stiffer spring. A compliant spring (dotted line) follows a flatter path during contact and maintains longer contact time. The angle of attack also reduces the vertical landing velocity relative to the hopping spring illustrated in *a*.

force increases and the vertical peak force at any point in the contact period decreases (Figure 10b; Taylor, 1985; Blickhan,˙1989; Kram and Taylor, 1990). This model can be made more realistic by making the leg-spring walk forward in the manner of an inverted pendulum whose highest point is at midstance (Figure 10c; Blickhan, 1989; Alexander, 1992). A compliant spring reduces the rise of the mass by collapsing as the limb approaches midstance and thus provides a "flatter ride" and a longer contact time (McMahon, 1985; Blickhan, 1989; Alexander, 1992). Additionally, as the limb becomes more protracted the "angle of attack" of the leg-spring increases and the vertical landing velocity decreases thus reducing the vertical force developed by the spring (McMahon et al., 1987; Blickhan, 1989; McMahon and Cheng, 1990). Empirical data on human walking and running support this model for human locomotion. When human subjects walk or run on a compliant track or with extreme hip and knee yield during support they experience a flatter path of the center of gravity, longer contact times, and lower vertical forces (McMahon and Greene, 1979; McMahon et al., 1987; Farley, 1992; Farley et al., 1993; Schmitt et al., 1996).

Table 5. Vertical oscillation of the shoulder (maximum shoulder height - minimum shoulder height). Pole, smallest pole on which subjects would walk. Values under Pole and Ground are the adjusted-y mean and one standard error (in parentheses). P, significance values for comparisons of adjusted-y means using ANCOVA

Subject	Pole	Ground	P
M. fascicularis	5.1 (0.41)	6.9 (1.31)	0.04
M. mulatta	4.3 (0.72)	5.0 (0.98)	ns
C. aethiops	4.4 (0.89)	4.6 (0.57)	ns
P. anubis	5.3 (0.62)	6.2 (1.1)	ns
E. patas	4.4 (0.77)	7.7 (1.48)	0.02

ns, not significant

The compliant gait models described above seem to predict some of the results of my research. First, the subjects in this study experience a larger change in elbow flexion between touchdown and midsupport on a small pole than they do on the ground (Figure 6). This dramatic flexion during support phase increases the compliance of the forelimb on the pole. Second, the vertical peak force on "arboreal" supports is reduced compared to those recorded for terrestrial supports (Table 4). Finally, some of my subjects increased the angle of attack of the forelimb due to greater arm protraction at touchdown (Table 2).

To confirm that these subjects are using a "compliant" walking gait while on "arboreal" compared to terrestrial supports I examined additional indicators of compliant gait that were not included in my original analysis. Vertical oscillation of the shoulder height, a measure of the "flatness" of the rise and fall of the subject, was lower on the pole than on the ground in all subjects. This can be seen in the calculated oscillation displayed in Table 5 and in Figure 5 that depicts the path of the shoulder. This "flatter ride" is directly related to the increased flexion at the elbow. Additionally, for all subjects except the vervet monkeys hand contact is significantly longer on the small pole compared to the ground (Table 6). This latter variable is a strong indicator that the gait has become more compliant (Blickhan, 1989). As a result of increased contact time, the time available to develop force certainly increases and the peak forces throughout support phase will necessarily decrease.

The support for the hypothesis that the subjects in this study were using a compliant walking gait is based on data easily accessible from videotape without force traces. This is useful because that is the source of much additional data on primate locomotion. In addition, I calculated vertical stiffness following the methods of Farley et al. (1993). The mean vertical stiffness for the entire group was 5.78 on the ground and 3.69 on a small pole.

These data suggest that the subjects of this study reduce compressive forces on their forelimb by adopting a "compliant" walking gait. It remains to be seen whether this pattern is part of a basic primate adaptation. Although this will be the subject of future study, I believe there is evidence suggesting that compliant walking may be common in all primates when compared to the walking gaits of nonprimates (Schmitt, 1995). The crouched posture described here for Old World monkeys is distinct from the posture adopted by noncursorial small mammals described by Jenkins (1971). Differences between primate postures in general and the postures of nonprimate mammals have been noted by other authors (Reynolds, 1981, 1985a,b, 1987; Alexander and Maloiy, 1984; Vilensky, 1989; Vilensky and Larson, 1989; Fischer, 1994; Schmitt, 1995; Larson, this volume) and some of these differences suggest that primates show a unique pattern of compliant walking gait compared to nonprimate mammals.

Table 6. Contact time (in seconds) of forelimb with substrate.
Pole, smallest pole on which subjects would walk. Values
under Pole and Ground are the adjusted-y mean and
one standard error (in parentheses). P, significance values
for comparisons of adjusted-y means using ANCOVA

Subject	Pole	Ground	P
M. fascicularis	0.474 (0.023)	0.334 (0.022)	0.0001
M. mulatta	0.376 (0.03)	0.271 (0.025)	0.03
C. aethiops	0.296 (0.019)	0.262 (0.015)	ns
P. anubis	0.356 (0.055)	0.291 (0.049)	0.01
E. patas	0.586 (0.015)	0.534 (0.017)	0.01

ns, not significant

Alexander (1977) has reviewed evidence that a "stiff" walk is characteristic of humans, dogs, cats, and horses whereas a "compliant" walk is characteristic of small birds. He further speculated that noncursorial mammals may also use a compliant gait. Current evidence suggests that the limb kinematics of primates is distinct from the kinematics of noncursorial small mammals and that primates use a relatively compliant gait (Jenkins, 1971; Rollinson and Martin, 1981; Alexander and Maloiy, 1984; Reynolds, 1987; Fischer, 1994; Larson, 1997). Primates are known to have longer stride lengths and larger angular excursions than nonprimate mammals (Alexander and Maloiy, 1984; Reynolds, 1987). Animals with long stride lengths and large angular excursions have relatively low stride frequencies and longer contact times (Heglund and Taylor, 1988, Kram and Taylor, 1990). In addition, data on elbow excursion in nonprimate noncursorial mammals indicate that in many cases the elbow does not go through a period of flexion during support phase, but rather remains stable and then extends (Jenkins, 1971; Jenkins and Weijs, 1979; Fischer, 1994). Finally, primate quadrupeds protract their fore- and hindlimbs to a greater degree at touchdown than do other mammals (Reynolds, 1987; Larson, 1997). Thus, primate quadrupeds appear to have longer contact times, greater elbow yield, and an increased angle of attack of their limbs than do nonprimate mammals.

Another strong indication of a compliant gait in primates can be derived from the work of Demes and colleagues, who found that *Loris tardigradus* and *Nycticebus coucang* increased speed by protracting their hindlimbs using lateral spinal flexion, thus increasing stride length. These lorises appear to increase their gait compliance as speed increases by lowering their center of gravity and by maintaining low stride frequencies and longer stride lengths. Demes et al. (1990) postulate that these results are related to arboreal life in which an increase in stride length, angular excursion and a reduction in stride frequency reduces substrate reaction forces that tend to displace fine branches and disturb insect prey. Similar patterns of compliant locomotion were found in chameleons that travel and hunt on fine branches (Peterson, 1984).

If it is, in fact, the case that primates are unusual among mammals in adopting a compliant walking gait, the question that remains is what the selective agent of such a gait may have been for early primates. As stated in the introduction, arboreal primates must maintain a crouched posture in order to balance the relatively wide bodies on relatively narrow supports. In addition, this crouched posture allows them to change direction rapidly. In connection with this latter necessity, and the need to reach and grasp in many planes, comes the need for gracile, mobile limbs. Compliant walking gait allows both a crouched posture and mobile forelimbs. In addition, the long contact time of compliant walking may increase stability, particularly for animals grasping branches, and, therefore, further helps maintain balance. There are undoubtedly energetic costs of a compliant posture (McMahon et al., 1987). However, the advantage of increased balance and thus the ability to negotiate arboreal supports with potentially rich food sources may offset the costs (Rodman, 1979).

If this is the value of this posture, why has it evolved in primates and not other mammals? Compliant gait may reasonably be seen as part of an initial and fundamental primate adaptation to arboreal locomotion. Early primates had to develop a way of climbing vertical supports with nails. Whereas non-primate mammals can climb a vertical support by interlocking with the substrate and can "crawl" up the support with similar kinematics as in walking (Cartmill, 1974, 1985), primates need long arms to reach around the substrate (Fleagle et al., 1981; Yamazaki and Ishida, 1984; Cartmill, 1985; Hirasaki et al., 1993; Larson, this volume) and to grasp an overhead support. Primates would thus need long limbs that they were able to protract extensively (see Larson, this volume). With long arms, the need to crouch for balance was accentuated. Early primates could crouch in manner that involved protraction of the humerus and significant increase in elbow flexion during support phase, a pattern not always seen in nonprimate mammals (Larson, this volume).

Primates may have developed a "compliant walk" in order to maintain critical postural and anatomical adaptations for arboreal life on terminal branches. Compliant walking also releases the forelimb from constraints imposed by the need for a high degree of osteological stabilization. This change in both anatomy and function allowed for the evolution of the use of forelimbs in tension, a feature that is critical for the evolution of antipronograde postures and ultimately for the removal of the forelimbs from locomotion (Stern, 1976).

ACKNOWLEDGMENTS

This study was part of my thesis research conducted under the patient guidance of Dr. Susan G. Larson, to whom I am very grateful. I also wish to thank Dr. Jack T. Stern Jr. and the other members of my dissertation guidance committee, Drs. William L. Jungers, Brigitte Demes, Farish Jenkins, and Michael D. Rose for their valuable guidance and their patience. I wish to also thank Marianne Crisci for her expert animal training and assistance in data collection, Yvette Pirrone for her invaluable help in animal training, data collection and analysis, and Luci Betti-Nash for her help with illustrations. Drs. Pierre Lemelin, W. Scott McGraw, Christine Wall, and Roshna Wunderlich all have provided useful advice on this project. I am also grateful for thoughtful reviews by Drs. Jeff Meldrum, Fred Anapol, and two anonymous reviewers. This research was supported by NSF BNS 8819621, and 8904576, and SBR 9209004, and a Sigma Xi Grant-in-aid of Research. I am particularly grateful to Drs. Elizabeth Strasser, Alfred Rosenberger, Henry McHenry, and John Fleagle for organizing both the conference at which this paper was originally presented and this volume.

REFERENCES

Alexander RM (1977) Mechanics and scaling of terrestrial locomotion. In T. Pedley (ed.): Scale Effects in Animal Locomotion. London: Academic Press, pp. 93–110.
Alexander RM (1992) A model of bipedal locomotion on compliant legs. Phil. Trans. R. Soc. Lond. B *338*:189–198.

Alexander RM, and Maloiy GM (1984) Stride lengths and stride frequencies of primates. J. Zool., Lond. *202*:577–582.

Alexander RM, and Ker RF (1990) Running is priced by the step. Nature *346*:220–221.

Ashton EH, Oxnard CE, and Spence TF (1965) Scapular shape and primate classification. Proc. Zool. Soc. Lond. *145*:125–142.

Ashton EH, Healy M, Oxnard CE, and Spence T (1968) The combination of locomotor features of the primate shoulder girdle by canonical analysis. J. Zool. *147*:406–429.

Biewener AA (1982) Bone strength in small mammals and bipedal birds: Do safety factors change with body size? J. exp. Biol. *98*:289–301.

Biewener AA (1983a) Allometry of quadrupedal locomotion: The scaling of duty factor, bone curvature and limb orientation to body size. J. exp. Biol. *105*:147–171.

Biewener AA (1983b) Locomotor stresses in the limb bones of two small mammals: The ground squirrel and chipmunk. J. exp. Biol. *103*:131–154.

Biewener AA (1989) Scaling body support in mammals: Limb posture and muscle mechanics. Science *245*:45–48.

Biewener AA (1990) Biomechanics of mammalian terrestrial locomotion. Science *250*:1097–1103.

Blickhan R (1989) The spring-mass model for running and hopping. J. Biomech. *22*:1217–1227.

Bown T, Kraus M, Wing S, Fleagle J, Tiffany B, Simons E, and Vondra C (1982) The Fayum primate forest revisited. J. Hum. Evol. *11*:624–628.

Cant JGH (1988) Positional behavior of long tailed macaques (*Macaca fascicularis*) in northern Sumatra. Am. J. Phys. Anthropol. *76*:29–37.

Cartmill M (1974) Pads and claws in arboreal locomotion. In FA Jenkins (ed.): Primate Locomotion. New York: Academic Press, pp. 45–83.

Cartmill M (1985) Climbing. In DM Bramble, KF Liem and DB Wake (eds.): Functional Vertebrate Morphology. Cambridge: Belknap Press, pp. 73–88.

Chan LK, Schmitt D, and Cole TM III (nd) A method for calculation of 3-D limb position using a single videocamera. Submitted Am. J. Phys. Anthropol.

Chism J, and Rowell TE (1988) The natural history of patas monkeys. In A Gautier-Hion, F Bourlière, T Gautier, and J Kingdon (eds.): A Primate Radiation: Evolutionary Biology of the African Guenons. Cambridge: Cambridge University Press, pp. 413–437.

Darwin C (1871) The Descent of Man. New York: Appleton.

Demes B, Jungers WL, and Nieschalk U (1990) Size- and speed-related aspects of quadrupedal walking in slender and slow lorises. In F Jouffroy, M Stack, and C Niemitz (eds.): Gravity, Posture, and Locomotion in Primates. Florence: Sedicesimo, pp. 175–198.

Demes B, Larson SG, Stern JT Jr., Jungers WL, Biknevicius AR, and Schmitt D (1994) The kinetics of primate quadrupedalism: "Hindlimb drive" reconsidered. J. Hum. Evol. *26*:353–374.

Dunbar D (1989) Locomotor behavior of rhesus macaques (*Macaca mulatta*) on Cayo Santiago. J. Puerto Rican Health Sci. *8*:79–85.

Farley CT, Glasheen J, and McMahon TA (1993) Running springs: Speed and animal size. J. exp. Biol. *185*:71–86.

Fedigan L, and Fedigan LM (1988) *Cercopithecus aethiops*: A review of field studies. In A Gautier-Hion, F Bourlière, T Gautier, and J Kingdon (eds.): A Primate Radiation: Evolutionary Biology of the African Guenons. Cambridge: Cambridge University Press, pp. 413–437.

Feldesman MR (1976) The primate forelimb: A morphometric study of diversity. U. Oregon Anthropol. Papers *1*:iii–149.

Fischer MT (1994) Crouched posture and high fulcrum, a principle in the locomotion of small mammals: The example of the rock hyrax *Procavia capensis* (Mammalia: Hyracoidea). J. Hum. Evol. *26*:501–524.

Fleagle JG, Stern JT Jr., Jungers WL, Susman RL, Vangor AK, and Wells JP (1981) Climbing: A biomechanical link with brachiation and with bipedalism. Symp. Zool. Soc. Lond. *48*:359–375.

Fleagle JG (1988) Primate Adaptation and Evolution. New York: Academic Press.

Full RJ, and Tu MS (1990) Mechanics of six-legged runners. J. exp. Biol. *148*:129–146.

Full RJ, Blickhan R, and Ting LH (1991) Leg design in hexapedal runners. J. exp. Biol. *158*:369–390.

Grand TI (1968a) The functional anatomy of the lower limb of the howler monkey (*Alouatta caraya*). Am. J. Phys. Anthropol. *28*:163–181.

Grand TI (1968b) Functional anatomy of the upper limb. Biology of the howler monkey (*Alouatta caraya*). Bibl. Primatol. *7*:104–125.

Harrison T (1989) New postcranial remains of *Victoriapithecus* from the middle Miocene of Kenya. J. Hum. Evol. *18*:3–54.

Heglund NC, and Taylor CR (1988) Speed, stride frequency, and energy cost per stride: How do they change with body size and gait? J. exp. Biol. *138*:301–318.

Hirasaki E, Kumakura H, and Matano S (1993) Kinesiological characteristics of vertical climbing in *Ateles geoffroyi* and *Macaca fuscata*. Folia Primatol. *61*:148–156.

Jenkins FA Jr. (1971) Limb posture and locomotion in the Virginia opossum and other non-cursorial mammals. J. Zool., Lond. *165*:303–315.

Jenkins FA Jr. (1974) Tree shrew locomotion and the origin of primate arborealism. In FA Jenkins (ed.): Primate Locomotion. New York: Academic Press, pp. 85–116.

Jenkins FA Jr., and Weijs W (1979) The functional anatomy of the shoulder in the Virginia opossum (*Didelphis virginiana*). J. Zool., Lond. *188*:379–410.

Jolly C (1967) The evolution of the baboons. In H Vartog (ed.): The Baboon in Medical Research, Vol. II. Austin: University of Texas Press. pp. 323–338.

Jungers WL, and Burr DB (1992) Body size, long bone geometry, and locomotion in cercopithecoid monkeys. Am. J. Phys. Anthropol., Suppl. *14*:96.

Jungers WL, and Burr DB (1995) Body size, long bone geometry and locomotion in quadrupedal monkeys. Z. Morph. Anthropol. *80*:89–97.

Kimura T (1985) Bipedal and quadrupedal walking of primates: Comparative dynamics. In S Kondo, H Ishida, T Kimura, M Okada, M Yamazaki, and J Prost (eds.): Primate Morphophysiology, Locomotor Analyses and Human Bipedalism. Tokyo: University of Tokyo Press, pp. 47–58.

Kimura T (1992) Hindlimb dominance during primate high-speed locomotion. Primates *33*:465–474.

Kimura T, Okada M, and Ishida H (1979) Kinesiological characteristics of primate walking: Its significance in human walking. In ME Morbeck, H Preuschoft, and N Gomberg (eds.): Environment, Behavior, Morphology: Dynamic Interactions in Primates. New York: Gustav Fischer, pp. 297–311

Kram R, and Taylor R (1990) Energetics of running: A new perspective. Nature *346*:265–267.

Larson SG (this volume) Unique aspects of quadrupedal locomotion in nonhuman primates. In E Strasser, JG Fleagle, AL Rosenberger, and HM McHenry (eds.): Primate Locomotion: Recent Advances. New York: Plenum Press, pp. 157–173.

Larson SG, and Stern JT Jr. (1989) The role of propulsive muscles of the shoulder during quadrupedalism in vervet monkeys (*Cercopithecus aethiops*): Implications for neural control of locomotion in primates. J. Motor Behavior *21*:457–472.

Le Gros Clark WE (1959) The Antecedents of Man. Edinburgh: Edinburgh University Press.

Manaster BJ (1975) Locomotor Adaptation within the *Cercopithecus, Cercocebus*, and *Presbytis* genera: A multivariate Approach. Ph.D. Dissertation, University of Chicago.

Manaster BJ (1979) Locomotor adaptation within the *Cercopithecus* genus: A multivariate approach. Am. J. Phys. Anthropol. *50*:169–182.

McMahon TA (1985) The role of compliance in mammalian running. J. exp. Biol. *115*:263–282.

McMahon TA, and Cheng GC (1990) The mechanics of running: How does stiffness couple with speed? J. Biomech. *23*:65–78.

McMahon TA, and Greene PB (1979) The influence of track compliance on running. J. Biomech. *12*:893–904.

McMahon TA, Valiant G, and Frederick EC (1987) Groucho Running. J. Appl. Physiol. *62*:2326–2337.

Meldrum DJ (1991) Kinematics of the cercopithecine foot on arboreal and terrestrial substrates with implications for the interpretation of hominid terrestrial adaptations. Am. J. Phys. Anthropol. *84*:273–290.

Morbeck ME (1979) Forelimb use and positional adaptation in *Colobus guereza*: Integration of behavioral, ecological, and anatomical data. In M Morbeck, H Preuschoft, and N Gomberg (eds.): Environment, Behavior, and Morphology: Dynamic Interactions in Primates. New York: Gustav Fisher, pp. 95–118.

Nakagawa N (1989) Activity budget and diet of patas monkeys in Kala Maloue National Park, Cameroon: A preliminary report. Primates *30*:27–34.

Napier JR (1967) Evolutionary aspects of primate locomotion. Am. J. Phys. Anthropol. *27*:333–342.

Napier JR, and Davis PR (1959) The forelimb skeleton and associated remains of *Proconsul africanus*. Fossil Mammals of Africa #16.

O'Connor BL, and Rarey K (1979) Normal amplitudes of pronation and supination in several genera of anthropoid primates. Am. J. Phys. Anthropol. *51*:39–44.

Rawlins RG (1993) Locomotive and manipulative use of the hand in Cayo Santiago macaques (*Macaca mulatta*). In H Preuschoft and DJ Chivers (eds.): Hands of Primates. Berlin: Springer-Verlag, pp. 21–30.

Reynolds TR (1981) Mechanics of interlimb weight redistribution in primates. Ph.D. Dissertation, Rutgers University.

Reynolds TR (1985a) Mechanics of increased support of weight by the hindlimbs in primates. Am. J. Phys. Anthropol. *67*:335–349.

Reynolds TR (1985b) Stresses on the limbs of quadrupedal primates. Am. J. Phys. Anthropol. *67*:351–362.

Reynolds TR (1987) Stride length and its determinants in humans, early hominids, primates, and mammals. Am. J. Phys. Anthropol. *72*:101–115.

Ripley S (1979). Environmental grain, niche diversification, and positional behavior in Neogene primates: An evolutionary hypothesis. In ME Morbeck, H Preuschoft, and N Gomberg (eds.): Environment, Behavior, Morphology: Dynamic Interactions in Primates. New York: Gustav Fischer, pp. 37–74.

Rodman PS (1979) Skeletal differentiation of *Macaca fascicularis* and *Macaca nemestrina* in relation to arboreal and terrestrial quadrupedalism. Am. J. Phys. Anthropol. *51*:51–62.

Rollinson J, and Martin RD (1981) Comparative aspects of primate locomotion with special reference to arboreal cercopithecines. Symp. Zool. Soc. Lond. *48*:377–427.

Rose MD (1973) Quadrupedalism in primates. Primates *14*:337- 358.

Rose MD (1977) Positional behavior of olive baboons (*Papio anubis*) and its relationship to maintenance and social activities. Primates *18*:59–116.

Rose MD (1988) Another look at the anthropoid elbow. J. Hum. Evol. 17:193–224.

Rose MD (1989) New postcranial specimens of catarrhines from the middle Miocene Chinji Formation, Pakistan: Descriptions and a discussion of the proximal humeral functional morphology in anthropoids. J. Hum. Evol. *18*:131–162.

Rose MD (1991) The process of bipedalization in hominoids. In B Senut and Y Coppens (eds.): Origine(s) de la Bipédie chez les Hominidés, Cah. Paléoanthrop. Paris: Editions du CNRS, pp. 37–49.

Schaffler MB, and Burr DB (1984) Primate cortical bone microstructure: Relationship to locomotion. Am J. Phys. Anthropol. *65*:191–197.

Schaffler MB, Burr DB, Jungers WL, and Ruff CB (1985) Structural and mechanical indicators of limb specialization in primates. Folia Primatol. *45*:61–75.

Schmitt D (1994) Forelimb mechanics as a function of substrate type during quadrupedalism in two anthropoid primates. J. Hum. Evol. *26*:441–458.

Schmitt D (1995) A Kinematic and Kinetic Analysis of Forelimb Use During Arboreal and Terrestrial Quadrupedalism in Old World Monkeys. Ph.D. Dissertation, State University of New York at Stony Brook.

Schmitt D, Stern JT Jr., and Larson SG (1996) Compliant gait in humans: Implications for substrate reaction forces during australopithecine bipedalism. Am. J. Phys. Anthropol., Supplement *22*:209.

Sokal R, and Rohlf J (1981) Biometry. The Principles and Practice of Statistics in Biological Research. New York: Freeman.

Stern JT Jr. (1976) Before bipedality. Yrbk. Phys. Anthropol. *20*:59–68.

Stern JT Jr., Wells JP, Vangor AK, and Fleagle JG (1977) Electromyography of some muscles of the upper limb in *Ateles* and *Lagothrix*. Yrbk. Phys. Anthropol. *20*:498–507.

Struhsaker T (1967) Ecology of vervet monkeys (*Cercopithecus aethiops*) in the Masai-Amboseli Game Reserve, Kenya. Ecology *48*:891–904.

Taylor CR (1978) Why change gaits? Recruitment of muscles and muscle fibers as a function of speed and gait. Amer. Zool. *18*:153–161.

Taylor CR (1985) Force development during sustained locomotion: A determinant of gait, speed and metabolic power. J. exp. Biol. *115*:253–262.

Vilensky JA (1989) Primate quadrupedalism: How and why does it differ from that of typical quadrupeds? Brain Behav. Evol. *34*:357–364.

Vilensky JA, and Larson SG (1989) Primate locomotion: Utilization and control of symmetrical gaits. Ann. Rev. Anthropol. *18*:17–35.

Wheatley B (1982) Energetics and foraging in *Macaca fascicularis* and *Pongo pygmaeus* and a selective advantage for large body size in the orangutan. Primates *23*:348–363.

Wood Jones, F (1926) The Arboreal Man. New York: Hafner.

Yamazaki N, and Ishida H (1984) A biomechanical study of vertical climbing and bipedal walking in gibbons. J. Hum. Evol. *13*:563–571.

III

DATA ACQUISITION AND ANALYTIC TECHNIQUES

INTRODUCTION TO PART III

Elizabeth Strasser

Since 1965 there have been some major advances in the methods available for studying aspects of primate locomotion. The first three papers in this section illustrate some technological advances in data acquisition while the last three papers cover some analytical issues.

The paper by John Kappleman (Chapter 12) is an up-to-date review of technologies for acquiring three-dimensional data and their implications for studying primate biomechanics and locomotion. He covers software programs, various innovations in hardware, and some analytical techniques. Kappelman provides internet web site addresses where a sampling of software programs can be accessed. He discusses the strengths and pitfalls of computerized tomography, magnetic resonance imaging and laser digitizers. Finally, he describes finite element analysis and the virtual museums of tomorrow. After reading this paper, one is impatient to try everything out!

The application of one technique that Kappleman discusses, laser scanning, is elegantly demonstrated by Leslie Aiello and colleagues in Chapter 13. These authors illustrate one use of laser scanning: for interpreting fossil hominid postcranial material. They address the affiliation of OH35, a hominid tibia and fibula from Olduvai Gorge, with OH8, the foot attributed to *Homo habilis*. They examine the fossil material in the context of homologous measurements of the talocrural joint surfaces from a comparative sample of hominoids and conclude that the two Olduvai specimens not only come from different individuals, but most likely from different species. Aiello et al.'s study is a tantalizing example of what three-dimensional data acquisition and analysis holds for the future.

Another example of a technological advance is provided by Brigitte Demes (Chapter 14). Demes introduces the reader to strain gauge technology by covering its principles, theory and practice as well as the history of its application. She describes the two main areas of strain gauge use in functional morphology and illustrates them with case studies of her own. The first area is direct bone strain analysis and Demes' case-study is an *in vivo* bone strain analysis of the macaque ulna. In this study, she tests some common "assumptions inherent in the functional analysis of bone morphology" about how bones are loaded during locomotion (Demes, pg. 243). The second area is the use of strain gauges to construct custom force transducers. In Demes second case study she uses such a force transducer to measure the takeoff and landing forces generated by *Propithecus* and *Hapalemur* on a compliant pole. Strain gauge analysis and other experimental methods promise a geometric increase in our understanding of the forces and factors involved in primate locomotion.

Charles Oxnard, who was at the original symposium organized by Warren Kinzey, and his students have applied multivariate methods to literally every skeletal system in primates as well as to aspects of primate behavior and ecology. His paper in this volume (Chapter 15) is, as one reviewer commented, an appropriate successor to his 1965 contribution. Oxnard convincingly demonstrates how morphometric studies on individual anatomical units yield information about function, while morphometric studies on combinations of anatomical units shed light about evolutionary relationships. Furthermore, Oxnard presents an ingenious algebraic analogy that explains why the information content from studies of individual anatomical units differs from those on combinations of units. This paper is an elegant summary of the contributions that Oxnard and his students have made to primatology over the many years since the original conference.

Laurie Godfrey and collaborators (Chapter 16) write about the application, use, and misuse of heterochrony. They argue that the notion that ontogenetic scaling implies little to no functional-behavioral differences between ancestor and descendant is naive. It is also apparent that isometries contribute little to diagnosing heterochronic processes. Furthermore, that allometric commonality somehow necessarily implies commonality of ancestral and descendant behavioral trajectories is exposed as nonsense. To paraphrase one of the reviewers, Godfrey et al. suggest a remedy to some problems in the field that are a result of over-simplification, inconsistent jargon and sloppy thinking. This paper contributes significantly to clarifying the issues surrounding heterochrony in general and in specific application to the evolution of locomotion.

William Jungers and colleagues (Chapter 17) report on a different type of scaling. They performed scaling analyses of long bone external dimensions and quantitative estimates of long bone cross-sectional geometry based on bone mineral density scans for two sister clades: Cercopithecinae and Colobinae. They find that positive allometry in the measures of long bone strength relative to body mass is insufficient to maintain mechanical equivalence across the range of body sizes seen in cercopithecids. As a consequence, large-bodied animals have to employ behavioral and postural adjustments to make up for reduced relative bone strength. The authors also find differences in bone strength between cercopithecines and colobines, which they relate to differences in the compliance of substrates habitually used by the groups. Finally, Jungers et al. report that direct measurements of cross-sectional geometry are superior to measurements of external dimensions to estimate cross-sectional geometry for understanding long bone biomechanics. This is perhaps not an unexpected result, but given the logistical problems involved in obtaining direct measurements of cross-sectional geometry, it is comforting that Jungers et al. also validate the importance of data and inferences drawn from external diameters.

In summary, the collection of papers in this section are illustrative of advances made since 1965, as well as exciting new technologies on the horizon for all primatologists.

12

ADVANCES IN THREE-DIMENSIONAL DATA ACQUISITION AND ANALYSIS

John Kappelman

Department of Anthropology
The University of Texas at Austin
Austin, Texas 78712-1086

1. INTRODUCTION

Advances in computer technology have greatly expanded the range of topics that can be investigated within the general field of primate locomotion and the field has witnessed great strides in such areas as telemetered electromyography (Jungers and Stern, 1980). The study of the relationship between form and function in living animals remains a critical focus of functional morphology because this approach offers the only bridge to understanding the function of extinct species and the nature of evolutionary transitions. Technological advances have not, however, changed the fact that the fossil record is almost exclusively limited to skeletal remains, and studies of this material are necessarily more limited and are usually restricted to quantifying the shape and size of different skeletal elements and how these variables are related to function.

Most past studies have usually been restricted to quantifying shapes or lengths in two dimensions, or reducing complex three-dimensional geometries to two dimensions for the purpose of taking linear measurements (see Ruff, 1989), because the techniques for gathering extensive three dimensional data did not exist. Many recent advances in data acquisition build on these more traditional and time honored measurement techniques (see Martin, 1989), and the addition of the computer interface permits the rapid capture of two-dimensional linear measurements (Spencer and Spencer, 1995) and the calculation of true areal data. Some of the most exciting advances of the past few years are to be found in the acquisition of both internal and external three-dimensional shape information, with some of this work based on a necessarily limited number of landmark data points (see Richtsmeier, 1989; Corner and Richtsmeier, 1993). In addition, advances in data storage, transmission technologies, and the internet facilitate the sharing of data in ways that were never before envisioned.

Primate Locomotion, edited by Strasser *et al.*
Plenum Press, New York, 1998

2. COMPUTERS, STORAGE MEDIA, AND SOFTWARE

One of the most important technological innovations of the past twenty years is the personal computer. Computing power that was formerly in the domain of large mainframe computers is now found in the hands of nearly every researcher. Research and development during recent years has continued the trend of producing ever faster central processing units (CPUs), now with multiple CPUs, and ever larger hard drive storage devices. With regard to future forecasts, the only prediction that we can be certain of is that whatever we purchase today will be replaced by something faster with more storage capability tomorrow! Developments in read-and-write storage media have continued in lockstep as well. Optical erasable disks are presently at nearly 5 gigabytes (GB) in capacity, and the digital audio tape (DAT) technology is not far behind. The cost of write-once CD ROM technology is now within the reach of many labs, and the long life span and low per-unit cost make this medium one of the most cost effective means for backing up, sharing, and storing data. Advances in CD ROM technology with the introduction of the Digital Versitile Disk (DVD) promise to increase the storage capability of this medium by over an order of magnitude thus making it possible to archive and share nearly any size data set (Duncan and Kappelman, 1991).

The ubiquity of personal computers has in turn spawned the development of user-friendly operating systems and a nearly endless supply of software, some of which has been written specifically for the functional morphologist, but most of which is easily co-opted from other disciplines. For example, off-the-shelf software that is designed for architectural and mechanical design such as Autodesk's AutoCAD® is easily co-opted for

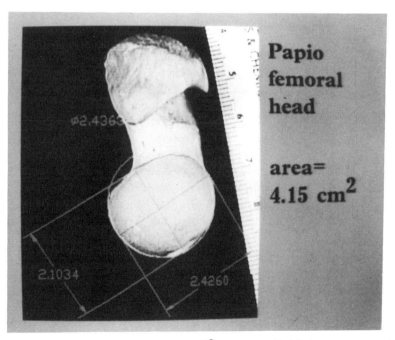

Figure 1. Video capture and input using Autodesk's AutoCAD®, a computer-aided-design program, permits a variety of traditional and true areal measurements.

such purposes as measuring phalangeal curvature (Duncan et al., 1994) or limb joint areas (Hake, 1992; Figure 1), while software packages such as the National Institute of Health's (NIH) Image®, and video image analysis (Spencer and Spencer, 1995) are more particularly designed for the needs of a biomedical researcher or functional morphologist. Other software programs such as Alias® and Wavefront® (Silicon Graphics), Soft Image® (Microsoft), or 3D Studio Max® (Autodesk) can be used to produce animations of joint and muscle movements in two and three dimensions. Sequences of actual primate locomotion captured on video can be easily downloaded to the computer using a video capture board and software such as Adobe's Premiere®, and the limb or trunk movements can be studied frame by frame with some of the packages listed above. Information about these programs can be accessed at the Internet web sites listed in Table 1.

3. TECHNIQUES OF DATA ACQUISITION

3.1. Inside and Out: Conventional and High Resolution X-Ray Computerized Tomography (CT) and MRI

Many of the new technologies for acquiring digital data have either been invented in the past few years or represent refinements of existing technologies. The primary advantage of X-rays and MRI over some of the other digitizing techniques discussed in this article is that these techniques permit the imaging of internal as well external structures. Conventional X-ray studies have witnessed a long history of use in physical anthropology (e.g., Coolidge, 1933; Weidenreich, 1940) that continues to today (e.g., Mann, 1975; Winkler et al., 1996). Even though the shadowgraph technique of passing parallel X-rays through the object has been around for many years, refinements in conventional X-ray continue and include polychromatic techniques that improve resolution by varying both the distance of the object from the X-ray source and absorption times (Wilkins et al., 1996). This new technique produces a very high level of resolution.

The development of X-ray CT began as a medical diagnostic tool in 1971 (see Taube and Adelstein 1987 for a brief history of development) and witnessed a wide number of applications in physical anthropology (e.g., Tate and Cann, 1982; Conroy and Vannier, 1984, 1987; Vannier et al., 1985; Ruff and Leo, 1986). Computerized tomography differs from conventional radiography in that the X-rays are restricted to a plane and the intensities of the beam before and after it passes through the specimen are measured. The measurement of a single slice of a specimen is completed by either rotating the object 360° on a turntable, or by rotating the X-ray source itself 360° around the object, with this latter method being critical in medical applications so as to minimize motion sickness in the patient. (This "axial" rotation of the X-ray source around the patient adds the "A" to "CT" to form the commonly used term "CAT" scan.) In order to complete a scan, the specimen is either translated vertically on the turntable or the X-ray source is moved along the long axis of the specimen. Conventional CT has a resolution on the order of 1–2 mm (or more) thickness, and this does limit the imaging of structures that are below this size, but some increase in resolution can be obtained by overlapping closely spaced images. Image reconstruction is completed by using what are usually company-specific algorithms to calculate the linear attenuation coefficient μ for each point in the specimen.

Once the image reconstruction is completed, analysis can begin on either the single slices or on the reconstructed 3-D object. The CT images can be viewed as either a digital file format on the computer screen or printed out as a series of hard copy films. Most com-

Table 1. Internet addresses for information on 3-D data acquisition hardware, analysis software, and files

Product type	Product name and description	Internet address
Analysis Software	NIH Image®, image analysis and measurement software	http://biocomp.arc.nasa.gov/3dreconstruction/software/nihimage.html
	Soft Image®, animation software	http://www.softimage.com/softimage/
	Alias®/Wavefront®, animation software	http://www.alias.com
	Autodesk's AutoCAD®, and 3D Studio® software	http://www.autodesk.com
	I-DEAS®, FEM software	http://www.sdrc.com
	ABAQUS® FEM software	http://www.abaqus.com
	Sculpt®, 3-D editing software	http://www.engr.colostate.edu/~dga/sculpt.html
	VoxBlast®, volume rendering software	http://www.vaytek.com
	General listing of 3-D reconstruction software	http://biocomp.arc.nasa.gov/3dreconstruction/software/
Hardware	High Resolution X-ray CT Lab at UT Austin	http://www.ctlab.geo.utexas.edu
	Laser Design: laser scanners	http://www.laserdesign.com/prodsurv.htm
	Digibotics, Inc.: laser scanners	http://www.digibotics.com
	Cyberware, Inc.: laser scanners	http://www.cyberware.com
	Thermoelastic stress analysis	http://www.StressPhotonics.com/
	DTM, Inc.: laser sintered 3-D printouts	http://www.dtm-corp.com
Data Sets, Files, and Animations	US National Library of Medicines: The Visible Human Project	http://www.nlm.nih.gov/research/visible/visible_human.html
	Smithsonian Laser Scanner Laboratory	http://www.digitaldarwins.sarc.msstate.edu/bvl.html
	W. W. Howell's modern human craniometric data set	ftp://utkux.utk.edu/pub/anthro/HOWTXT.ZIP
	Virtual Human Osteology Guide	http://www.dla.utexas.edu/depts/anthro/kappelman/osteolog.html
	3-D animation of human skull	http://www.dla.utexas.edu/depts/anthro/kappelman/skull.mpg
	Virtual Laboratories for Introductory Physical Anthropology	http://www.dla.utexas.edu/depts/anthro/kappelman/currdev.html
	CT generated 3-D data sets	http://biocomp.arc.nasa.gov/3dreconstruction/data/

[1]This is not an exhaustive list of all available hardware, analysis software, or Internet addresses with data sets. It is recommended that the interested user conduct a keyword search on the Internet to locate additional addresses.

panies that build medical CT scanners use proprietary file formats, and the digital images can only be viewed and analyzed on their hardware with their software. In almost all cases it is simply not possible to save and export the images in a file format that can be viewed and analyzed on a personal computer. This is a serious limitation with proprietary CT hardware and software and often serves to restrict the researcher to analyzing the data on the hospital computer, and this can be expensive and often difficult to arrange. This limitation has forced many researchers to output the image reconstructions as hard copy films and then redigitize the image (e.g., Ruff, 1989).

High resolution X-ray CT (HRXCT) (Figure 2), also known sometimes as industrial CT, offers several advantages over conventional medical CT. First, HRXCT uses a range of higher energy sources (typically 125–450 kV) than those available in medical CT, which makes the instrument capable of penetrating much denser objects including rocks and very heavily mineralized fossils. (Of course, these higher intensities also prevent this instrument from being used on living organisms.) Second, the X-ray detectors are modular, and these can be switched between linear and area detector arrays, which increases the resolution of the instrument. The combination of modular sources and detectors produces a CT system that can scan a variety of specimens across a wide range of resolutions. For example, the current resolution available on a medical CT is 1 mm (1000 μm) while that on a HRXCT ranges from 10 μm to 100 μm. This increase in resolution means that much smaller objects or structures such as trabecular bone (Figure 3) or even the developing premolar in the mandible of *Alphadon*, a tiny early Mesozoic marsupial (Cifelli et al., 1996), can be clearly imaged. One other advantage to HRXCT is that the digital files are not written in a proprietary file format, but are exportable in a variety of formats that are

Figure 2. High resolution X-ray computerized tomography (HRXCT) offers several advantages over conventional medical CT, including better resolution and non-proprietary file formats. This instrument is produced by Bio-Imaging Research, Inc. and is housed at the University of Texas at Austin (see http://www.ctlab.geo.utexas.edu).

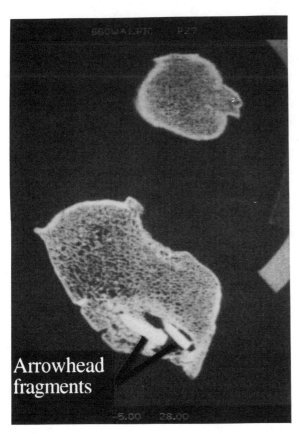

Arrowhead
fragments

Figure 3. High resolution X-ray CT image of a human distal femur showing an arrowhead that fragmented upon impact. High resolution X-ray CT provides a dramatic increase over the resolution of conventional medical CT and images tiny biological structures such as trabecular bone.

easily accessible on all personal computers. Rowe et al. (1993) have used HRXCT to produce a digital atlas of the skull of *Thrinaxodon,* which combines the original coronal sections along with digitally resliced sagittal and horizontal (transverse) sections that are presented on CD ROM in both DOS and Macintosh formats. Together these advantages combine to make HRXCT very useful scientifically.

Research with conventional medical CT can prove to be quite expensive, and access to instruments housed in hospitals or clinics usually depends upon patient throughput, which can sometimes limit its use for pure research and non-medically oriented projects. HRXCT is available to scientists on a commercial basis, but its high per-hour cost serves to greatly limit the broad application of this technology across the biological and earth sciences. In an effort to make this latter technology more widely available to the scientific community, a HRXCT laboratory has recently been established at The University of Texas at Austin. This laboratory became operational in early 1997 and is available to the research community (Rowe *et al.,* 1997). The HRXCT instrument is designed by Bio-Imaging Research, Inc. (Lincolnshire, IL) and includes 200 kV and 420 kV sources as well as linear and area detectors that offer a resolution down to 10 μm. Images can be written in a variety of file formats (e.g., TIFF, BMP, TARGA, PICT, etc.) and on a range of media (e.g., optical erasable disks, CD ROM, Zip or Jazz disks, DAT tapes) that can be accessed by all platforms of personal computers. Information on the High Resolution X-ray CT Laboratory at the University of Texas at Austin can be accessed at the Internet web site given in Table 1.

Magnetic Resonance Imaging (MRI or MR), also known as Nuclear Magnetic Resonance (NMR) imaging, is a relatively new technology that has to date seen only a few applications in studies of the functional morphology of hard tissue. Although NMR can in some cases resolve bone, its greatest utility in medical imaging has been found in the study of soft tissues. MacLatchy and Bossert (1996) and MacLatchy (1996, this volume) used a combination of conventional medical CT and NMR technologies to study the hip joint of a variety of extant and extinct primates. MacLatchy and Bossert (1996) relied on NMR for the smallest primates because it provided them with nearly an order of magnitude increase in resolution over conventional CT. (The NMR resolution of 0.15 mm is at the upper range of the resolution available for HRXCT.) Hard tissue is not easily resolved by NMR, and in their studies NMR was used to image external shape only. Given this limitation with extant material, it is unlikely that NMR will prove very effective for imaging the internal morphology of densely mineralized fossils. Other much less expensive and less invasive techniques are available for imaging external morphologies (see below).

3.2. Detailing the Surface: Laser Digitizers

Some of the most recent advances in automated 3-D imaging incorporate a laser scanner to capture the surface topography of a specimen. These scanners differ from the CT technologies discussed above because in general only the geometrical coordinates of the outside surface can be gathered. Laser surface scanners offer advantages over medical CT scanners when only surface coordinates are needed because these scanners do not in general require proprietary software, but produce data sets that can be analyzed on any personal computer platform with a variety of "off the shelf" software. Operating costs are much lower than CT and NMR because laser scanners use standard electric current, have very low maintenance costs, and do not require a licensed operator. These advantages combine to make laser scanners an affordable option for many researchers who are interested in 3-D morphology, and these instruments can now be found in many physical anthropology laboratories around the world (*e.g.,* The University of Texas at Austin, University of Liverpool, Smithsonian Institution). Laser scanners are not the only non-CT alternative to capture surface coordinates. There are many stylus operated 3-D surface data gathering systems on the market, but these are generally fairly slow, non-automated, and labor intensive, and vary dramatically in their degree of resolution. Excellent reviews of these and many other digitizing systems are given in Hartwig and Sadler (1993) and Dean (1996).

Laser scanners can be divided into two basic technologies. Both share the use of laser, but the first of these, the optical digitizer, breaks the beam of light into a plane that illuminates a single contour on the surface of the specimen. This scanner uses two CCD (charge coupled device) cameras to digitize the plane of light and capture the coordinates of this single contour. Two options are next available to complete the scan. The laser source can be translated along a plane that is parallel to the specimen with the collection of more contour data, or the specimen or the laser source can be rotated so that radial contours of data are collected. This scanning technology excels with convexly shaped specimens. An additional advantage of these systems is that a very large number of data points can be collected very rapidly (10^4 points per second) with a point spacing that varies from 10^1 -10^2 μm (see Table 1: Cyberware Inc. web site). This high speed data collection makes it possible to scan living specimens, and several scanners offer options that record the surface color of the specimen as well. Perhaps the greatest disadvantage of these systems is the distortion that results if the surface normals deviate significantly from either the plane of the laser beam or the radial direction of the scan. These distortions, generally most extreme along the edges of a specimen, often require the collection of data from separate scans of differently oriented views of the specimen. The final rendering of the digitized specimen is completed by using sophisticated "zipper" software to combine the multiple scans into a complete model. This processing of a model is computationally intensive and can require many hours.

The second laser scanning technology, known as adaptive scanning, retains the laser but keeps the beam tightly focused as a single point of light and uses a triangulation algorithm to measure the exact x and y coordinates of each point on the surface of the specimen at a fixed z coordinate or level (Figure 4). In order to eliminate the distortion resulting from mismeasured surface normals as noted above, the Digibot II® scanner (see Table 1, web site for Digibotics, Inc., Austin, TX) uses a turntable to rotate the specimen so as to bring each surface into a position that is exactly normal to the laser. Additional rotations of the specimen complete the collection of points around one horizontal contour, and the scan of the specimen is completed by translating the laser vertically though the z axis and collecting additional contours. This scanning method collects data from convex as well as concave surfaces, and creative mounting of the specimen on the turntable can produce scans of structures such as eye orbits (Table 1, Internet web site for Virtual Human Osteology Guide). Digibotics, Inc. has recently added a new lens to their system that collimates the laser beam to a much smaller diameter. This option produces a point spacing of 500 μm in the x, y, and z directions and further increases in resolution are expected. Because adaptive scanning technology collects each point separately, there is no need for

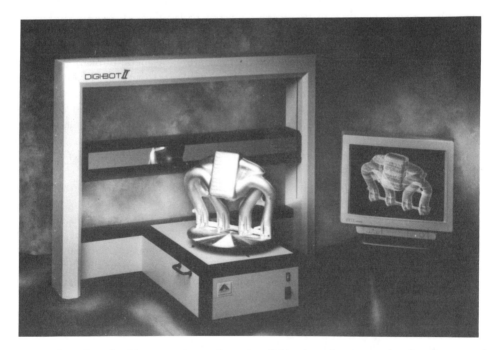

Figure 4. Three-dimensional laser scanning can be used to digitize the surface topography of nearly any specimen. The scanner shown in this figure is produced by Digibotics, Inc. (see Table 1) and permits a point spacing of 0.5 mm in the x, y, and z planes (see Figure 6). A three-dimensional animation of the human skull is available for downloading at the Virtual Human Osteology Guide web site (see Table 1).

sophisticated and computationally time-intensive "zipper" software or additional scans to completely digitize the object. The x, y, z data can be exported to a variety of software programs for wiremeshing and shading. One potential disadvantage of this system is that because each data point is collected separately, it can take several hours to scan an object at high resolution, and this type of laser scanning is not appropriate for living specimens. These scanners are, however, fully automated and do not require an operator. In the lab at The University of Texas, long scans are usually run overnight or on the weekend, and this protocol has been used to complete scans of every element of the human skeleton to produce a "virtual" human osteology guide and many other teaching applications (see Table 1).

It is commonplace for researchers who study physical objects to worry about whether or not they have collected enough data to quantify the features that they wish to describe. One important point to keep in mind when studying surfaces and shapes is that all of the scanning methods described above collect what are usually many more points than are required to mathematically describe the surface. It is not at all uncommon for laser scanners to collect on the order of 10^4 -10^6 points for a human skull. Creating a surface by triangulating or wiremeshing such a large number of points can easily produce models that are on the very edge of what today's personal computers can manipulate. Such large models often require more powerful computers such as Silicon Graphics workstations for analysis. There is no simple way to determine exactly how dense a coverage of points is required to "describe" the original surface, but there are now several software packages such as Sculpt® (see Table 1) that are designed to filter data based upon a number of user-defined criteria so as to reach an optimum between the number of data points and the surface detail of the specimen. Data point density will only increase with more powerful digitizers, and the filtering programs will in turn witness increasingly wide use.

3.3. To Have and To Hold: 3-D Printouts

Many of the digitizing technologies discussed above have come to physical anthropology directly from the fields of computer-aided engineering and manufacturing. In fact, many laser scanners are specifically designed for reverse engineering and computer-aided manufacturing (CAM). One additional area that holds great potential for physical anthropology involves the automated production of 3-D physical printouts from the computer models. These 3-D printout technologies fall under three general categories: automated milling, stereolithography (STL), and laser sintering. In each case the digitized object is written to a file format that computationally "slices" the object into thin layers. The x, y, and z coordinates of these layers are then used to produce the 3-D printout. Automated milling physically sculpts the object from wood or Styrofoam. Stereolithography builds the 3-D object layer by layer by curing epoxy, but this is often a time consuming process. Laser sintering, a process invented by DTM, Inc. (see Table 1 for web site) is the newest addition to this line up of technologies. This process uses a laser to sinter or fuse nylon or plastic powder into the first layer of the object, which, in turn, is sintered to the next layer, and so on, until the 3-D printout is completed (Figure 5). Laser sintering is a relatively rapid process and new research is focused on using more powerful lasers to sinter ceramics and metals so as to transform sintering into a rapid manufacturing process.

One direct application of the 3-D printout technologies involves digitizing fossils or extant specimens by either CT or laser scanner and using the output technologies to produce 3-D printouts of the specimens (Kappelman, 1992; Hjalgrim et al., 1995). This process could prove to be especially important in the case of very delicate fossils that could be

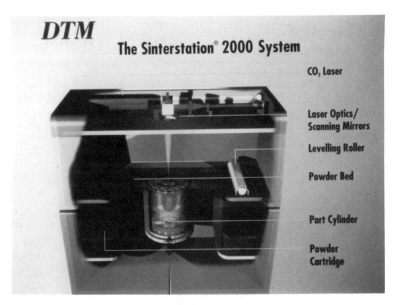

Figure 5. There are now a variety of technologies available for producing 3-D printouts of digitized files. This approach is especially useful for producing reconstructions of fragmentary fossils. The sinterstation shown in this figure is built by DTM, Inc. (see Table 1) and uses a laser to fuse or sinter plastic powder layer by layer into a 3-D model (see Figure 6). As each layer is sintered, the piston (center of the station) that holds the part translates vertically downward by a fixed distance and a new layer of powder is refreshed and leveled on the top of the piston. This new layer is next sintered to the part, and the process is repeated until the part is completed.

damaged by full preparation or even physical handling. Furthermore, given that traditional silicon or rubber molds produce only a limited number of casts, a 3-D printout of the original fossil could be used to make additional molds. It is important to note that the present high cost of these technologies makes it unlikely that traditional plaster- and plastic-based molding and casting will be completely replaced by the 3-D printout technologies at any point in the near future.

The richest potential of 3-D printout technology for physical anthropology is reached when it is combined with the versatility of computer digitizing and modeling. It is often the case that a fossil will vary in the degree of completeness of its skeletal elements, especially from either the right and left sides of its body, or element by element, and in the past a process of "informed sculpting" was used to model the missing anatomies. One useful example is seen in the case of STS 14, an early australopithecine. This specimen preserves much of its pelvic region, but the right innominate is much more complete than the left. In order to create a more accurate reconstruction of the complete pelvis, the right element was digitized with a laser scanner. After digitizing, the file was exported to Auto-CAD®, and the "mirror" function converted the right innominate into a "left" innominate. This "new" innominate was next surface meshed, saved in a STL format, and transferred to DTM's Sinterstation to produce a 3-D printout of the right innominate mirrored as a left. This 3-D nylon printout can now be combined with the true right innominate to produce a reconstruction of the pelvis (Figure 6). The range of these types of applications is endless. Kalvin et al. (1992) reconstructed a cranium from multiple fragments, and Zollikofer et al. (1995) digitized fragmentary pieces of a Neanderthal infant cranium and combined the fragments into a file for a stereolithography 3-D printout. Johanson (1995)

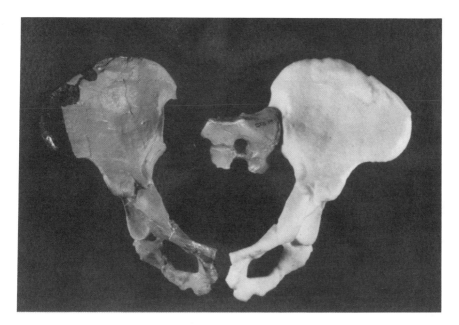

Figure 6. A reconstruction of the fragmentary pelvis of STS 14. The more complete right innominate of a cast of STS 14 was digitized with a Digibot II laser scanner (see figure 4). This file was input into AutoCAD®, mirrored to the opposite side, and printed out in 3-D on DTM's sinterstation in order to produce a complete reconstruction of the pelvis.

followed a similar approach to model a reconstruction of the cranium of *A. afarensis*. Such applications are becoming increasing common and are of great utility not only to the researcher, but also to the student because 3-D printouts allow direct tactile contact with the model and this additional sensory input is often more dramatic than simply viewing a computer model.

4. INTO THE FUTURE: DATA ANALYSIS IN THREE AND FOUR DIMENSIONS

It was noted at the beginning of this article that the study of the relationship between form and function in living animals remains a critical focus of functional morphology, not only for understanding functional relationships in extant forms but also because this is the only approach that exists for understanding extinct species (see Aiello et al., this volume). One relatively new area of analysis for investigating and understanding the relationship between form and function is the finite element method (FEM), a technique again borrowed from engineering that holds very exciting possibilities for physical anthropology. It is widely accepted that the size and shape of the various elements of the skeleton preserve a record of the genetic fingerprint as well as the past behaviors of the organism. These behaviors include the loading histories that the elements have experienced, and their shape is believed to be related to the way that the structure of the element is designed to resist or respond to these loads (Thompson, 1971; Wainwright et al., 1976; Currey, 1984; Lanyon and Rubin, 1985; Carter 1987; Carter et al., 1989). Briefly, the finite element method is used to analyze the way that an object responds to a load by first dividing its form up into a sometimes very large number of smaller but interconnected elements, each of which is assigned material properties and boundary conditions. A load is next applied to the model and the stress and strain that each element is subjected to are analyzed. These results indicate the particular regions of the element where compressive and tensile stresses are concentrated as well as the magnitude of the stress (Figure 7). Such studies are especially critical in engineering design and manufacturing because these analyses can be used to pinpoint the precise area of a load-bearing part or tool that will fail under certain loading conditions. Redesign of the part and retesting by FEM is used to produce a product that is optimally designed to resist the applied load that it will experience. A much more complete treatment of FEM can be found in Rao (1982), Huiskes and Chao (1983), and Baran (1988).

Studies of the way that transmitted loads influence the size and shape of skeletal elements share a broad number of similarities with earlier studies that used the photoelastic method of analysis (Oxnard, 1973; Ward and Molnar, 1980). Even though most photoelastic and FEM studies have until recently relied on two dimensional models only, or have in-

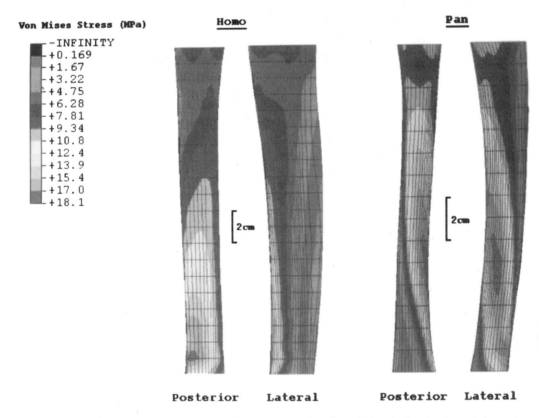

Figure 7. The finite element method is becoming a common method for analyzing the functional morphology of extant and extinct species. The internal and external surface morphologies of these two tibiae were digitized with a conventional CT, and a model consisting of many brick elements was next constructed from this file. The finite element models of these tibial shafts are from a modern human and a chimpanzee and were completed with the software program ABAQUS® (see Table 1 for web site). This particular FEM study loads both tibiae in a bipedal stance, and the contours of lighter and darker colors show areas of high and low compression, respectively.

vestigated the transmission of loads along a restricted number of planes, these studies have provided important insights into the relationship between shape and function (e.g., Oxnard, 1973, 1975, 1984; Richmond and Qin, 1996; Spears and Crompton, 1996). It is, however, the case that almost all skeletal elements have complex shapes in three dimensions and at some junction FEM studies must take this additional shape information into account. The new imaging technologies of High Resolution CT and laser scanning discussed above now permit accurate modeling of skeletal complexity on a scale from 10^1–10^6 µm that encompasses nearly the full range of vertebrate hard tissue structures. Studies that use FEM to investigate complex three-dimensional objects are becoming increasingly common (Hobatho et al., 1991; Korioth et al., 1992; Chen and Povirk, 1996; Edelman and Reeke, 1996; Ryan et al., 1996; Chen and MacLatchy, 1997; Duncan et al., 1997; Ryan, 1997; Ryan and Kappelman, 1997). Other techniques such as strain gauge studies (Lanyon and Smith, 1970; Lanyon, 1971; Lanyon et al., 1975; Finlay, 1982), thermoelastic stress studies (see Table 1), or photoelastic method (Ryan et al., 1996) provide the means for partial but nonetheless critical independent testing of FEM results.

Perhaps the most exciting and also one of the challenging aspects of applying FEM to studies of skeletal form and function is found in extending this method across the fourth dimension. Most FEM analyses are concerned with static loading models that attempt to approximate the highest loading regimes that are experienced by a structure in order to test for failure. Experiments have, however, demonstrated that most locomotor and masticatory behaviors involve complex loading cycles in which the magnitude and direction of load vary throughout the cycle (see Hannam and Wood, 1989; Korioth et al., 1992). Analogous dynamic changes are also witnessed in tracking loading regimes encountered through the ontogeny of an individual. Dynamic loading cycles are routinely modeled by FEM in engineering studies, and an example of two revolutions of a disc brake assembly can be seen at the Internet web sit "http://www.abaqus.com/applications/disk-brake/index.html." We can anticipate tests of dynamic models in studies of primate locomotion, mastication, and ontogeny in the near future (Langdon et al., 1991).

Once FEM has been used to model and study the loading parameters of a skeletal element from an extant species, the technique can be extended to fossil species. Although other methods, such as strain gauge analysis or loading rigs, offer experimental results for extant specimens that are broadly comparable to FEM, it is not possible to extend these physical methods to the study of fossils because neither the mineralized fossil itself nor cast reproductions of the fossil has the correct material properties of bone. The 3-D imaging and modeling techniques discussed above can be used to produce accurate 3-D models of the fossil, which in turn can be used in FEM testing. This combination of methods offers what is now one of the only means for testing the relationship between form and function of skeletal elements in extinct species. Furthermore, only one additional step is required to use this combination of methods to approach questions about the nature of evolutionary transitions. Once the skeletal elements from a number of extant and fossil specimens have been digitized, these files can be edited to model a hypothetical specimen that has some mix of the morphologies of the actual specimens. This process is known as "morphing" in computer animation and is commonly used to transform one cartoon character into another. For the purpose of functional morphology, "morphing" provides the means for building models with intermediate properties that can in turn be used to test hypotheses about the inferred function of intermediate morphologies. Of course, discoveries of new fossils provide the ultimate test or confirmation of the hypothetical "construction," but until that time, this combination of techniques allows the framing and testing of explicit hypotheses and offers a direct approach to questions of evolutionary transitions.

5. SHARING THE DATA: VIRTUAL MUSEUMS FOR ALL

Advances in computer technology and especially storage media and transmission links greatly facilitate the sharing of raw and processed data. Even tremendously large tables of raw data or video clips can now be easily copied to CD ROM or other media for inexpensive distribution, or transmitted via the Internet. Electronic publishing promises to even more greatly simplify the distribution of raw data, and many journals that publish hard copy formats also have Internet web sites where primary data are archived. This latter option is also available to anyone who has access to a web site. In fact, the very important modern human craniometric data set of WW Howells, was published in three volumes (Howells, 1973, 1989, 1995) has now been archived to the web (Howells, 1996: see Table 1).

It has become common practice in most disciplines for authors to report the primary data that form the basis for their observations. In fact, many journals now require that the supporting data sets be made available as a condition of publication (e.g., see *Nature* vol. 384:598). Clearly, the high costs of traditional hard copy text publishing does force some page length restrictions on published articles, but it seems reasonable to provide the actual data so that the precise details of the argument can be followed by the reader. In many cases these data are taken from specimens that are from rare or endangered species that can no longer be collected in the field, or are from very fragile fossils, or are located in museums that are difficult to access. Making measurements of these specimens widely available is a very valuable contribution in and of itself, and in many cases can significantly improve the quality and sample for any number of future projects. Once a set of standard measurements has been made, it also makes good sense to publish the actual measurements so that a global "mean" can be calculated that takes into consideration inter-observer error. A second argument can be made for making the raw data available to the scientific community because in most cases the data were collected with the aid of tax or foundation funds and in some sense then the raw data do in fact belong to the public at large. In most cases authors freely distribute the raw data once their paper has been published, but there is sometimes a tendency for authors to regard their raw data as private property. Journals in physical anthropology do not appear to have any standard policy with regard to requirements for publishing primary data, and almost any issue of the leading journals in this field will include some articles that provide a list of specimens along with actual raw measurements, while others report only transformed z-scores of species' means. The absence of any clear requirement to publish or otherwise make available the raw data appears to have, in part, contributed to a subculture of "private ownership" of the raw data.

Primary digital data sets for even a single specimen that are generated by 3-D laser scanning or High Resolution CT can easily be several orders of magnitude larger in size than nearly any data set that consists of linear, areal, or volumetric values, but these data are also easily shared via storage media or the Internet. It is perhaps the case that an even stronger argument can be made for freely sharing these sorts of data after the original investigator has published her or his observations. Unlike a set of linear measurements, whose exact landmarks may vary somewhat among researchers, 3-D laser scanning and High Resolution X-ray CT produce a nearly "virtual" copy of the original specimen, and any number of measurements can then be taken from the digital file (Kappelman, 1993). Once a high quality, high resolution 3-D scan has been completed, it is unlikely that second, third, or fourth scans using the same technology will provide much additional detail and, in the case of fragile fossils, there is little rationale for exposing the specimen to the risks associated with additional handling. In fact, predicted reductions in the future allocation of tax dollars for research may restrict spending to the point where research visits to archived collections become the exception rather than the rule. The educational and public relations benefits that would accrue from making extant and fossil collections available to the entire community as a whole, from grade schoolers to senior scientists, are difficult to quantify, but such efforts would certainly strengthen the public image of physical anthropology. At this time there is no "virtual museum" of scanned specimens, and building an Internet museum will require the coordinated efforts of many museums and curators, especially with regard to setting standards for scan resolution and file formats (see Hartwig and Sadler, 1993), and resolving the complex legal issue of who owns the "copyright" of unique specimens or the data themselves. Nonetheless, it is the case that many important and useful 3-D data sets are already available on the Internet, and any number of files can be downloaded from some of the web sites listed in Table 1. It is likely that most journals will soon take full advantage of these advances and require the publication of the raw data that can, in turn, provide the critical "cornerstones" for building the virtual museums of the future.

6. CONCLUSION

Advances in computer design, storage media, digitizing technologies, and analysis techniques have combined to open new possibilities in studies of functional morphology. Increases in the capacity and variety of digital storage media and the widespread availability of the Internet offer new possibilities for the unlimited exchange of primary data. Although the relatively new technologies of surface laser scanning and High Resolution X-ray CT produce data sets that can be used to gather simple linear measurements, the strength of these technologies lies in their ability to address questions of form in three dimensions. For example, the actual surface area of even complexly shaped three-dimensional joints or tooth cusps can be accurately measured with CAD software, thus eliminating two-dimensional approximations of these features as based on linear measurements. Analytical techniques, such as the finite element method, provide the means for assessing and testing how the size and shape of a skeletal element responds to loading, and this technique offers great promise for both understanding the functional morphology of fossil species and testing hypotheses about the nature of intermediate morphologies and evolutionary transitions. Together these advances provide new pathways for scientists to follow in fulfilling their responsibilities for making their observations widely and freely available for testing by others.

ACKNOWLEDGMENTS

I wish to thank Elizabeth Strasser, John Fleagle, Alfred Rosenberger, and Henry McHenry for inviting me to participate in the conference, "Primate Locomotion — 1995," and the Wenner-Gren Foundation for Anthropological Research as well as the Anthropology Division of NSF for funding to help defray travel and housing expenses. Special thanks are due to Henry McHenry and his family for their hospitality during the meeting. Various colleagues at UT have assisted in laboratory work and discussions, and I would like to thank Claud Bramblett, William Carlson, Alex Duncan, Reuben Reyes, Tim Rowe, Tim Ryan, Rob Scott, Ron Stearman, and Greg Weiner. I thank Leslie Aiello for discussions about the digital reconstruction of STS 14. Thanks to Paige Hake Sheehan for the use of the image in Figure 2, and Tim Ryan for the use of the image in Figure 7. My work in computer imaging has received support from NSF (DUE 9354427) and The University of Texas at Austin, and corporate support from Autodesk, Inc., Digibotics, Inc., DTM, Inc., Intel, Inc., and Tektronix, Inc. Thanks to Elizabeth Strasser and two anonymous reviewers for very helpful comments on an early draft of this paper.

REFERENCES

Aiello LC, Wood BA, Key C, and Wood C (1998) Laser scanning and palaeoanthropology: An example from Olduvai Gorge, Tanzania. In E Strasser, JG Fleagle, AL Rosenberger, and H McHenry (eds.): Primate Locomotion: Recent Advances, New York: Plenum Press, pp. 223–236.

Baran NM (1988) Finite Element Analysis on Microcomputers. New York: McGraw-Hill.

Carter DR (1987) Mechanical loading history and skeletal biology. J. Biomech. 20:1095–1109.

Carter DR, Orr TR, and Fyhrie DP (1989) Relationships between loading history and femoral cancellous bone architecture. J. Biomech. 22:231–244.

Chen X, and Povirk G (1996) Assessing errors introduced by modeling the anisotropic human mandible isotropically with finite element method. Am. J. Phys. Anthropol. Supplement 22:83.

Chen X, and MacLatchy LM (1997) The biomechanical significance of the trabecular bone in hominoid proximal femora. Am. J. Phys. Anthropol. Supplement 24:90–91.

Cifelli RL, Rowe TB, Luckett WP, Banta J, Reyes R, and Howes RI (1996) Fossil evidence for the origin of the marsupial pattern of tooth replacement. Nature 379:715–718.

Conroy GC, and Vannier MW (1984) Noninvasive three-dimensional computer imaging of matrix-filled fossil skulls by high-resolution computer tomography. Science 226:456–458.

Conroy GC, and Vannier MW (1987) Dental development of the Taung skull from computerized tomography. Nature 329:625–627.

Coolidge HJ Jr. (1933) Pan paniscus. Pygmy chimpanzee from south of the Congo River. Am. J. Phys. Anthropol. 18:1–59.

Corner BD, and Richtsmeier JT (1993) Cranial growth and growth dimorphism in Ateles geoffroyi. Am. J. Phys. Anthropol. 92:371–394.

Currey J (1984) The Mechanical Adaptations of Bones. Princeton: Princeton Univ. Press.

Dean D (1996) Three-dimensional data capture and visualization. In LF Marcus, M Corti, A Loy, GJP Naylor, and DE Slice (eds.): Advances in Morphometrics. New York: Plenum Press, pp. 53–69.

Duncan A, and Kappelman J (1991) Image manipulation, analysis, and mass storage on the microcomputer. Am. J. Phys. Anthropol. Supplement 12:69–70.

Duncan A, Kappelman J, and Shapiro LJ (1994) Metatarsophalangeal joint function and positional behavior in Australopithecus afarensis. Am. J. Phys. Anthropol. 93:67–81.

Duncan A, Podnos E, Cleghorn NE, and Kappelman J (1997) Development and analysis of 3-D finite element models of human femoral diaphyses. Am. J. Phys. Anthropol. Supplement 24:103–104.

Edelman DB, and Reeke GN Jr. (1996) Finite element analysis of hominoid proximal femora. Am. J. Phys. Anthropol. Supplement 22:101.

Finlay JB, Bourne RB, and McLean J (1982) A technique for the in vitro measurement of principal strains in the human tibia. J. Biomech. 15:723–739.

Hake P (1992) An examination of sexual dimorphism in the dentition and postcranium of Papio anubis. B.A. Senior Honors Thesis, The University of Texas at Austin.

Hannam AG, and Wood WW (1989) Relationships between the size and spatial morphology of human masseter and medial pterygoid muscles, the craniofacial skeleton, and jaw biomechanics. Am. J. Phys. Anthropol. 80:429–445.

Hartwig WC, and Sadler LL (1993) Visualization and physical anthropology. In AJ Almquist and A Manyak (eds.): Milestones in Human Evolution. Prospect Heights: Waveland Press, Inc, pp. 187–222.

Hjalgrim H, Lynnerup N, Liversage M, and Rosenklint A (1995) Stereolithography: Potential applications in anthropological studies. Am. J. Phys. Anthropol. 97:329–333.

Hobatho MC, Darmana R, Pastor P, Barrau JJ, Laroze S, and Morucci JP (1991) Development of a three-dimensional finite element model of a human tibia using experimental modal analysis. J. Biomech. 24:371–383.

Howells WW (1973) Cranial Variation in Man. A Study by Multivariate Analysis of Patterns of Differences among Recent Human Populations. Papers of the Peabody Museum of Archeology and Ethnology, 67:1–259.

Howells WW (1989) Skull Shapes and the Map. Craniometric Analyses in the Dispersion of Modern *Homo*. Papers of the Peabody Museum of Archeology and Ethnology *79*:1–189.

Howells WW (1995) Who's Who in Skulls. Ethnic Identification of Crania from Measurements. Papers of the Peabody Museum of Archeology and Ethnology *82*:1–108.

Howells WW (1996) Howell's craniometric data on the Internet. Am. J. Phys. Anthropol. *101*:441–442.

Huiskes R, and Chao EYS (1983) A survey of finite elements analysis in orthopedic biomechanics: The first decade. J. Biomech. *16*:385–409.

Johanson DC (1995) Face-to-face with Lucy's family. National Geographic *189*:96–117.

Jungers WL, and Stern JT Jr. (1980) Telemetered electromyography of forelimb muscle chains in gibbons (*Hylobates lar*). Science *208*:617–619.

Kalvin AD, Dean D, and Hublin JJ (1995) Reconstructing human fossils. IEEE Computer Graphics and Applications *15*:12–15.

Kappelman J (1992) Three-dimensional input and output of solid models for studies in functional morphology. Am. J. Phys. Anthropol. Supplement *14*:97.

Kappelman J (1993) Building an evolutionary database using digital imaging and animation software. Am. J. Phys. Anthropol. Supplement *16*:122.

Korioth TWP, Romilly DP, and Hannam AG (1992) Three-dimensional finite element stress analysis of the dentate human mandible. Am. J. Phys. Anthropol. *88*:69–96.

Langdon JH, Bruckner J, and Baker HH (1991) Paleokinematics of the australopithecine foot with three-dimensional imaging. Am. J. Phys. Anthropol. Supplement *12*:111.

Lanyon LE (1971) Strain in sheep lumbar vertebrae recorded during life. Acta. Orthop. Scand. *42*:102–112.

Lanyon LE, and Rubin CT (1984) Functional adaptation in skeletal structures. In M Hildebrand, DM Bramble, KF Liem, and DB Wake (eds.): Functional Vertebrate Morphology. Cambridge: Harvard University Press, pp. 1–25.

Lanyon LE, and Smith RN (1970) Bone strain in the tibia during normal quadrupedal locomotion. Acta. Orthop. Scand. *41*:238–248.

Lanyon LE, Hampson WGJ, Goodship AG, and Shah JS (1975) Bone deformation recorded *in vivo* from strain gauges attached to the human tibial shaft. Acta. Orthop. Scand. *46*:256–268.

MacLatchy LM (1996) Another look at the australopithecine hip. J. Hum. Evol. *31*:453–476.

MacLatchy LM (1998) Reconstruction of hip joint function in extant and fossil primates. In E Strasser, JG Fleagle, AL Rosenberger, and HM McHenry (eds.): Primate Locomotion: Recent Advances. New York: Plenum Press, pp. 111–130.

MacLatchy LM, and Bossert WH (1996) An analysis of the articular surface distribution of the femoral head and acetabulum in anthropoids, with implications for hip function in Miocene hominoids. J. Hum. Evol. *31*:425–453.

Mann AE (1975) Some Paleodemographic Aspects of the South African Australopithecines. Philadelphia: University of Pennsylvania Press.

Martin RD (1989) New quantitative developments in primatology and anthropology. Folia Primatol. *53*:1–246.

Oxnard CE (1973) Form and Pattern in Human Evolution: Some mathematical, physical, and engineering approaches. Chicago: University Chicago Press.

Oxnard CE (1975) Uniqueness and Diversity in Human Evolution: Morphometric studies of Australopithecines. Chicago: University Chicago Press.

Oxnard CE (1984) The Order of Man: A Biomathematical Anatomy of the Primates. New Haven: Yale University Press.

Rao SS (1982) The Finite Element Method in Engineering. New York: Pergamon Press.

Richmond B, and Qin Y-X (1996) Finite element methods in paleoanthropology: The case of phalangeal curvature. Am. J. Phys. Anthropol. Supplement. *22*:197.

Richtsmeier JT (1989) Application of finite-element scaling analysis in primatology. Folia Primatol. *53*:50–64.

Rowe T, Carlson WD, and Bottorff W (1993) *Thrinaxodon*: Digital Atlas of the Skull. CD ROM, 623 MB, Austin: University of Texas Press.

Rowe T, Kappelman J, Carlson WD, Ketcham R, and Denison C (1997) High resolution computed tomography: A breakthrough technology for earth scientists. Geotimes *42*:23–27.

Ruff CB (1989) New approaches to structural evolution of limb bones in primates. Folia Primatol. *53*:142–159.

Ruff CB, and Leo FP (1986) Use of computed tomography in skeletal structure research. Yrbk. Phys. Anthropol. *29*:181–196.

Ryan T (1997) A structural analysis of the cross-sectional shape of the tibia in *Homo* and *Pan* using the finite element method. M. A. Thesis, The University of Texas at Austin.

Ryan T, and Kappelman J (1997) A structural analysis of tibial shape using the finite element method. Am. J. Phys. Anthropol. Supplement *24*:202.

Ryan TM, Scott RS, Duncan A, Kappelman J, Shapiro L, Grant S, Lewis K, and Stearman R (1996) Finite element analysis using a 3-D laser scanner. Am. J. Phys. Anthropol. Supplement *22*:206.

Spears IR, and Crompton RH (1996) The mechanical significance of the occlusal geometry of great ape molars in food breakdown. J. Hum. Evol. *31*:517–535.

Spencer MA, and Spencer GS (1995) Technical note: Video-based three-dimensional morphometrics. Am. J. Phys. Anthropol. 96:443–453.

Tate JR, and Cann CE (1982) High-resolution computed tomography for the comparative study of fossil and extant bone. Am. J. Phys. Anthropol. *58*:67–73.

Taube RA, and Adelstein SJ (1987) A short history of modern medical imaging. In E Guzzardi (ed.): Physics and Engineering of Medical Imaging. Dordrecht: Martinus Nijhoff, pp. 9–40.

Thompson DW (1971) On Growth and Form. Cambridge: Cambridge University Press.

Vannier MW, Conroy GC, Marsh JL, Knapp RH (1985) Three-dimensional cranial surface reconstructions using high-resolution computed tomography. Am. J. Phys. Anthropol. *67*:299–311.

Wainwright SA, Biggs WD, Currey JD, and Gosline JM (1976) Mechanical Design in Organisms. Princeton: Princeton Univ. Press.

Ward SC, and Molnar S (1980) Experimental stress analysis of topographic diversity in early hominid gnathic morphology. Am. J. Phys. Anthropol. *53*:383–395.

Weidenreich F (1943) The skull of *Sinanthropus pekinensis*: A comparative study on a primitive hominid skull. Palaeontologia Sinica D *10*:1–485.

Wilkins SW, Gureyev TE, Gao D, Pogany A, and Stevenson AW (1996) Phase-contrast imaging using polychromatic hard X-rays. Nature *384*:335–338.

Winkler LA, Schwartz JH, and Swindler DR (1996) Development of the orangutan permanent dentition: Assessing patterns and variation in tooth development. Am. J. Phys. Anthropol. *99*:205–220.

Zollikofer CPE, Ponce de Leon MS, Martin RD, and Stucki P (1995) Neanderthal computer skulls. Nature *357*:283–285.

13

LASER SCANNING AND PALEOANTHROPOLOGY

An Example from Olduvai Gorge, Tanzania

Leslie Aiello,[1] Bernard Wood,[2*] Cathy Key,[1] and Chris Wood[2]

[1]Department of Anthropology
University College London
Gower Street
London WC1E 6BT, England
[2]Department of Human Anatomy and Cell Biology
The University of Liverpool
P.O. Box 147
Liverpool L69 3BX, England

1. INTRODUCTION

Taxonomic affiliation in Plio-Pleistocene fossil hominids is generally based on the analysis of cranial and dental material where species-specific characteristics are well recognised (Wood, 1991, 1992). With the exception of features such as the relatively long neck and small head in the australopithecine femur, taxon-specific features in the postcranial skeleton are largely unknown (Aiello and Dean, 1990). This situation is undoubtedly the result of the relative absence of associated skeletons in the Plio-Pleistocene fossil record and the general paucity of postcranial material. It follows that species-specific features in the postcranial skeleton have played only a secondary role in taxonomic studies and that it is often difficult to assign isolated postcranial fossils to specific species.

New technology is, however, making it possible to analyse isolated skeletal elements and to determine whether or not unassociated fossils could have come from a single individual, from individuals in the same species or from individuals belonging to different species. In particular, the technique of laser scanning permits detailed three-dimensional data to be gathered from complex joint surfaces. Developments in morphometrics now allow these complex three-dimensional data to be analysed in terms of both species affili-

* Current Address: Department of Anthropology, George Washington University, 2112 G Street, NW, Washington DC 20052.

Primate Locomotion, edited by Strasser *et al.*
Plenum Press, New York, 1998

ation and functional capability. For example, it is possible to compare two reciprocal complex joint surfaces, determine their degree of compatibility or congruency and on this basis, assess the probability of their association. If one of the reciprocal joint surfaces comes from a fossil of known affiliation, it is then possible to assess the probability of the other belonging to that same species.

The purpose of this paper is to illustrate the use of laser scanning in this context by addressing a particular question of hominid postcranial fossil affiliation that has been left unanswered in the literature since the 1960s. This question concerns the affiliation of OH 35, a fossil hominid tibia and fibula from Olduvai Gorge, Tanzania, which was found in level 22 of the FLK site in 1960. When these bones were first excavated, Louis and Mary Leakey believed that they belonged to the same individual as the well known OH5 '*Zinjanthropus*' skull, which had been found in 1959 in the same level of the same site (Leakey, 1971). The idea that OH35 could represent *Australopithecus boisei* was also entertained by Day (1976) and more recently by Tobias (1991). Tobias (1991) points out that OH5 would have been between 15–17 years of age at death based on modern human standards and that the tibia and fibula are from an individual of at least that age. The reason for this determination is that the distal epiphyses of the tibia and fibula are fully closed. Furthermore, recent body mass predictions for OH5 based on cranial measurements indicate a mass of between 39 and 40 kg (Aiello and Wood, 1994). The midshaft girth of the OH35 tibia (61.0 mm) suggests a roughly equivalent mass. It is virtually the same as the mean tibial midshaft girth for *Pan troglodytes* (61.7 mm, sd 5.93 mm, N=37), which has a mean body mass of 36.4 kg (female=31.1 kg, males=41.6 kg) (Harvey et al., 1986).

When OH35 was originally described by Davis (1964), he concluded that the tibia and fibula could not be assigned to any taxon with confidence. Later, the close compatibility between OH35 and the OH8 (*Homo habilis*) foot was pointed out and the possibility entertained that OH35 might belong to *Homo habilis* (Leakey et al., 1964; Leakey, 1971; Day, 1978). This interpretation was taken up and enthusiastically supported by Susman and Stern (1982) who suggested that the OH35 tibia and fibula might even come from the same *Homo habilis* individual as the OH8 foot. This is despite three points recognised by Susman and Stern (1982) that would seem to argue against this interpretation. Firstly, the site at which the OH8 foot was found (FLK NN, level 3) is some 200 m from the OH5 site (FLK). Secondly, the stratigraphic levels in which the two fossils were found are separated by 0.5 m. And thirdly, Susman and Stern (1982) believed that the OH8 foot belonged to a juvenile individual aged between about 13.7 and 13.9 years at death while the closed distal epiphyses of the tibia and fibula suggest an older individual.

The taxonomic affiliation of the OH35 tibia and fibula will be tested here by comparing the shape of the distal joint surface of the OH35 tibia to the trochlear joint surface of the OH8 talus. The main aim is to determine whether or not the shapes of these reciprocal joint surfaces are compatible enough to support the hypothesis that OH35 could have belonged to the same individual as OH8 or alternatively to the same species (*Homo habilis*).

2. MATERIALS AND METHODS

The comparative material used in this analysis consists of the reciprocal articular surfaces comprising 30 hominoid right talocrural joints, 5 males and 5 females each of adult modern humans, chimpanzees (*Pan troglodytes*) and gorillas (*Gorilla gorilla*). The modern humans are from the Spitalfields Collection housed at the Natural History Museum, London and are individuals of known sex and age. The chimpanzees and gorillas are from the Powell Cotten collection, Birchington, Kent. The fossil hominid sample is made up of OH35 and OH8 and the *Australopithecus afarensis* specimens, AL 288-1ar (a right distal tibia) and AL 288-1as (a right talus), from Hadar, Ethiopia. In both cases casts were used for the analysis. AL 288–1 was included in the analysis as an additional comparison because it is an associated joint from a single individual. All analysis was undertaken at the Department of Human Anatomy, the University of Liverpool.

The reciprocal talocrural joint surfaces were scanned using a Cyberware 3030 High Resolution Colour Laser Scanner fitted with an MM motion platform (Cyberware Laboratory Inc., Monterey, CA), running on a Silicon Graphics Indigo 2 XZ. The digitisation process was controlled by Cyberware's own "Echo" software and each complete object scan is written out to an ASCII-delimited file. A scanned object was stored as a dense cloud of three-dimensional data points to an accuracy of ± 0.1 mm. The technique is fully described in Wood et al. (submitted). The following is an abbreviated summary.

Four scans were made for each talocrural joint. In the first scan, three pieces of 1 mm lead shot were attached to the talar trochlear surface as markers to provide an orientation to the scan. This was found to be necessary because of the difficulty of locating precise reference points on the unmarked scanned image. The three pieces of lead shot were located on each talar joint surface as follows: (a) at the most posterior point on the fibular facet; (b) at the most anterior point on the fibular facet; and, (c) at the point where the plane of the anteromedial border of the tibial facet intersects the anteromedial border of the trochlear articular surface of the talus when the joint is in the close-packed, dorsiflexed, position (Figure 1a). The second, third and fourth scans were required to capture the relevant portions of the articulated talocrural joint for each pair of tibiae and tali. In the first of these scans (Scan 2), the talocrural joint was articulated in maximum dorsiflexion ('close-packed' position), and three 5 mm diameter registration spheres were attached to each of the bones of the joint in such a way that all six spheres (3 on the talus and 3 on the distal tibia) were visible in the scan. These spheres were painted matt white to ensure that they were clearly captured by the scanning process. The registration spheres were mounted on 3 cm wires so that they stand out from the bones to which they were attached (the wires were painted matt black so that the scanner did not detect them). Once the articulated joint and its 6 registration spheres were scanned, the talus and tibia were carefully separated so as not to disturb the position of either set of 3 spheres. The articular surfaces of both the talus and tibia were next scanned separately, along with their own sets of three registration spheres (Scans 3 and 4). This process thus captured both the position of the talus and tibia as an articulated unit (Scan 2) and the articular surface of each bone (Scans 3 and 4) otherwise not visible when in articulation.

It is important to emphasise that laser scanning has been used in these analyses as a data capture device. It produces a glut of data characterising the reciprocal joint surfaces and the challenge is to usefully analyse these data (Koenerink and van Doorn, 1992). We have decided here to base the analysis on landmark data rather than on surfaces. The reason for this is that proper statistical models are not yet available for surface analysis and comparison. We have defined a manageable number of landmarks that represent the tibial and talar surfaces in the form of cross-sectional curves.

In order to do this, the four scans previously described were integrated using Surfacer 4.0 software (3D Imaging International, Inc., Trumbull, CT.). The image was oriented in relation to the 3 lead shot markers that formed the Z plane. Cross-section profiles were created by projecting a series of curves (straight lines) onto the two articular surfaces of the talus and tibia point clouds in predefined planes. Five cross-sections (three anteroposteriorly and two mediolaterally) were constructed through the articulated joint relative to the triangular orientation plane of the talus (Figure 1b). Portions of the cross-section curves (Figure 2a,b) that do not articulate were removed. For the mediolateral sections,

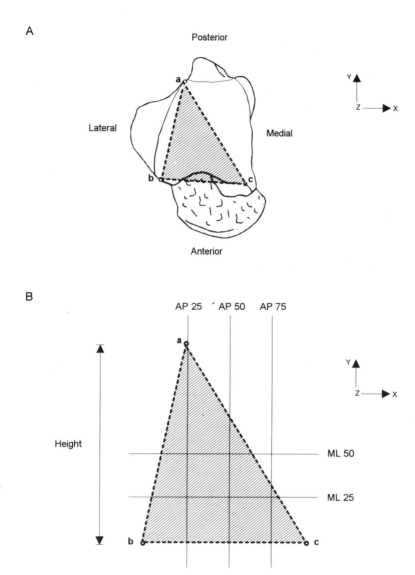

Figure 1. (*A*) Right talus in superior view. The shaded triangle, the "orientation plane", is a planar surface at Z=0 and its base (the line bc) is parallel to the X axis. Points a, b, and c represent the positions of the lead shot applied to the talus before scanning. See text for description. (*B*) Location of the 5 anteroposterior and mediolateral cross-section curves through the articulated talocrural joint. The AP50 cross-section curve is cut at the middle of the orientation plane base (i.e., at 0.5*Width). The AP25 (lateral) and AP75 (medial) curves are constructed at 25% and 75% of basal width. The ML50 cross-section curve is located at 0.5*height of the orientation plane and the ML25 curve is at 25% of height.

the articular portions of the cross-section curves were extracted by constructing a tangent across the high points of the medial and lateral crests of the talar trochlear surface, cutting the talus curve at this point and cutting the tibial curve at the locations normal to this (Figure 2a). For the anteroposterior sections, articular portions of the cross-section curves were extracted by constructing a tangent across the low points of the anterior and posterior crests of the tibial articular surface, cutting the tibial curve at this point and cutting the talus curve at the locations normal to this (Figure 2b).

The tibial and talar cross-sectional curves were analysed firstly on the basis of raw data that provide information relevant to both the size and the shape of the cross-section and secondly on the basis of size-corrected data that provide only information relevant to the shape differences in the cross-section. In order to extract the raw (size and shape) data from the cross-sections each curve was sampled at 20 even increments along its length (Figure 2c,d). This resulted in 20 co-ordinates (here termed δ co-ordinates) that measure the vertical distance of the curve from its tangent (baseline) at 20 evenly spaced locations for each tibial and talar joint surface. The congruence between the reciprocal tibial and talar cross-sections was determined by (1) squaring each of the individual co-ordinate values for the tibial and talar joint surfaces, (2) subtracting the corresponding talar co-ordinate from the tibial co-ordinate, and (3) taking the average of the summed differences over the 20 co-ordinates. It should be noted that $\delta1$ and $\delta20$ for both the tibial and the talar joint surfaces have the value of 0 and, therefore, contributed nothing to the analysis.

Size-corrected congruence data were gathered by scaling the cross-sections before the co-ordinate data were collected. This was carried out by scaling the cross-sectional curves so that in all cases point 1 was at (0,0) and point 20 was at (1,0) with the value for X proceeding from 0 to 1 in increments of 0.05 units. The congruence between the reciprocal cross-sections was then determined in the same fashion as for the non-size-corrected data.

One possible concern with this method is homogeneity of the cross-sectional curves between specimens. The curves are homogenous in the sense that they are always at a constant location in relation to the reference triangle in the Z plane. When comparing randomly associated joints, the AP75 curve projected on the tibia may not precisely match the AP75 curve as projected on the talus. As a result, the 20 points on the tibial curve will not precisely correspond to the 20 points on the talar curve. Any differences in location would be due either to size differences between the randomly associated bones (in the non-size corrected analysis) and/or to shape differences. This mismatch would be reflected in the results of the analysis and would be highly important in assessing the degree of congruency in the joint. Lack of homogeneity in this sense would be a fundamental feature of the analysis.

It is also important to point out that in this context each curve was sampled at 20 evenly-spaced points along its cord. Any mismatch between say the 5th point on a particular tibial curve and the 5th point on the corresponding talar curve that may be due to the angle of the two cords in relation to each other in the articulated joint is expected to be negligible. On the basis of radiological measurements in the living joint, Wynarsky and Greenwald (1983) and Jonsson et al. (1984) have determined that the articular cartilage shows an even thickness across the joint. Such variation in the angle of the cords in relation to each other would, therefore, be expected to be minimal and would likely be less than the resolution of the scanner (<0.5 mm).

3. RESULTS AND DISCUSSION

Congruency between the reciprocal joint surfaces was determined for four cross-sectional curves characterising the talocrural joint, two anteroposterior curves at 50% (AP50) and 75% (AP75) (most medial) and two mediolateral curves at 25% (ML25) (most anterior) and 50% (ML50). For each comparative species and each cross-sectional curve, congruency values were determined for each associated talocrural joint and for every possible intraspecific combination of tibiae and tali. This gives the range of congruency variation expected for joints from the same individual as well as that expected for a single species. In all cases except one, associated talocrural joints from single individuals were more congruent than randomly associated talocrural pairs (Figures 3, 4). The one exception is the AP75 cross-section for humans based on size-corrected data (Figure 4d). Here a single associated joint is less congruent than any of the randomly associated joints. Table 1 provides the summary congruency statistics for the comparative human, chimpanzee and gorilla samples.

In reference to the fossil talocrural joints, the results are similar in all cross-sectional comparisons based on the raw data (containing both size and shape information) (Figure 3). The OH35/8 talocrural joint consistently is less congruent than the *Australopithecus afarensis* (AL 288-1) joint. AL 288-1 shows the degree of congruency at the lower end of the range of variation expected in an associated talocrural joint (Figure 3). In relation to the human comparative sample, the OH35/8 joint also shows a degree of congruency that would be compatible with an associated joint (Figure 3, Table 2). However, in relation to the chimpanzee and gorilla comparative samples the results suggest that although the OH35/8 joint is compatible with the intraspecific variation found in these species, it is not necessarily compatible with the congruency that would be expected in an associated joint. In relation to the gorilla ML25 cross-section and the chimpanzee and gorilla AP75 cross-sections, the OH35/8 joint shows a lesser degree of congruency than any of the associated joints from single individuals (Figure 3a, gorillas; Figure 3d, chimps and gorillas; Table 2). Nevertheless, in all cases the congruency is compatible with the intraspecific variation expected from randomly associated talocrural joints.

Based on size-corrected data, the results are different (Figure 4, Table 3). The AL288-1 talocrural joint continues to be more congruent that OH35/8 and fully compatible with the congruency expected for an associated joint. However, in the case of both the AP50 and the AP75 cross-sections, the congruency shown by the OH35/8 joint is less than that of any associated joint in any of the three comparative species (Figure 4, Table 3). The same is true for the ML25 cross-section in relation to the human and gorilla comparative samples, although the congruency is compatible with an associated joint based on the chimpanzee sample. The ML50 cross-section is compatible with an associated joint based on the variation in all three of the comparative species.

The fact that the anteroposterior cross-sections give a different result than do the mediolateral cross-sections may simply reflect the differences in the magnitude of variation in congruency observed for the different cross-sections in the different comparative

———▶

Figure 2. (*A*) Mediolateral cross-section through a schematic left talocrural joint in maximum dorsiflexion. The articular portions of the cross-section curves are extracted by constructing a tangent across the high points of the medial and lateral crests of the talar trochlear surface (points 1 and 2 respectively). The talar cross-section is cut at points 1 and 2. The tibial cross-section curve (between points 3 and 4) is cut at the normals, a and b, to this tangent. (*B*) Anteroposterior cross-section through the left talocrural joint in maximum dorsiflexion. The articular portions of the cross-section curves are extracted by constructing a tangent across the low points of the anterior and posterior crests of the tibial articular surface (points 5 and 6 respectively). The tibial cross-section is cut at points 5 and 6. The talar cross-section curve (between points 7 and 8) is cut at the normals, c and d, to this tangent. (*C*) The extracted talar and tibial mediolateral cross sections. (*D*) The extracted talar and tibial anteroposterior cross-sections. The XY co-ordinates are obtained at 20 evenly-spaced increments along the length of each curve segment (only 3 indicated here for each cross-section).

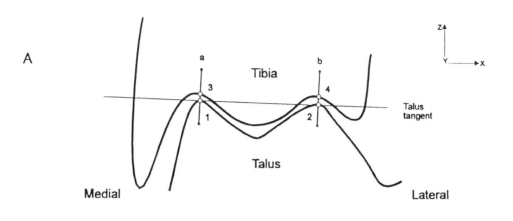

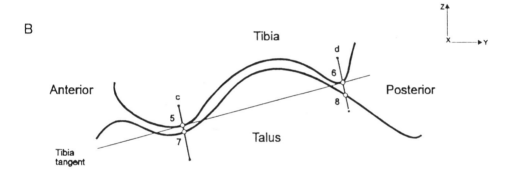

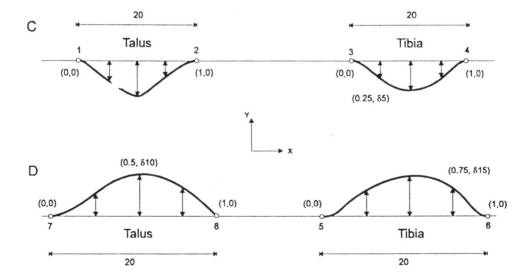

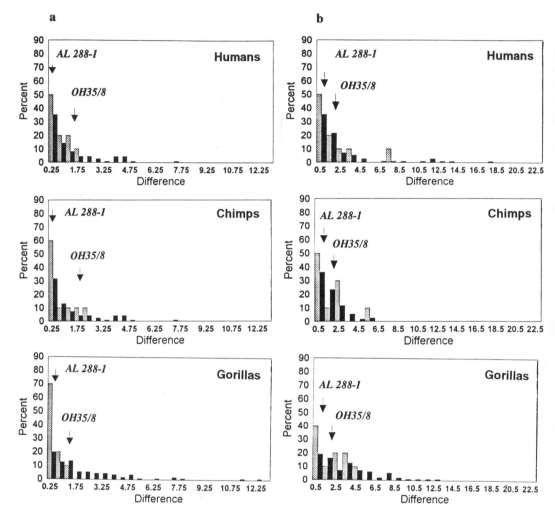

Figure 3. Congruency histograms based on raw data for humans, chimpanzees and gorillas and the fossil ta-locrural joints. (*a*) The mediolateral 25 cross-section, (*b*) the mediolateral 50 cross-section, (*c*) the anteroposterior 50 cross-section, (*d*) the anteroposterior 75 cross-section. Black bars, every possible combination of randomly associated tibiae and tali. Shaded bars, associated talocrural joints from single individuals.

species. The absolute range of variation (standard deviation) in congruency for the medio-lateral sections is much smaller than for the anteroposterior sections (Figure 5). Because of the limited range of variation in these mediolateral sections in all of the comparative samples, they may not be as good indicators as the anteroposterior sections.

The congruency values of the anteroposterior cross-sections also suggest the possi-bility that OH35 and OH8 may not belong to individuals of the same species. In relation to the human AP50 and AP75 cross-sections, the OH35/8 joint lies at the extremes of the in-traspecific variation. This is also true for the AP50 cross-section in relation to the gorilla comparative sample (Figure 4, Table 3).

The difference between the results based on the raw (size and shape) data and on the size-corrected data can most probably be explained by the fact that the OH35 distal tibia and

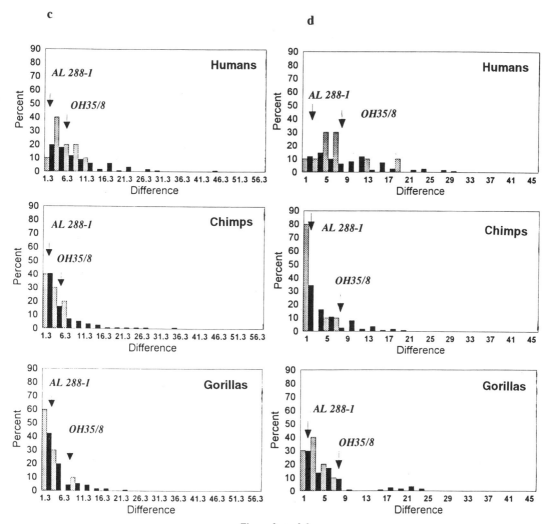

Figure 3c and d.

the OH8 talus are of a comparable size. In the raw data analysis this similarity in size obscures the differences in shape that are apparent in the size-corrected analysis. The reasonable fit between these bones that has been noted in the literature (Leakey, 1971; Susman and Stern, 1982) results from this size similarity rather than from any necessary shape similarity.

4. CONCLUSIONS

This analysis has demonstrated the use of laser scanning in the interpretation of fossil hominid postcranial material. By offering a means of analysing complex joint surfaces in three dimensions, laser scanning provides a powerful source of metrical data in addition to the traditional metrics normally taken on postcranial bones. This is particularly impor-

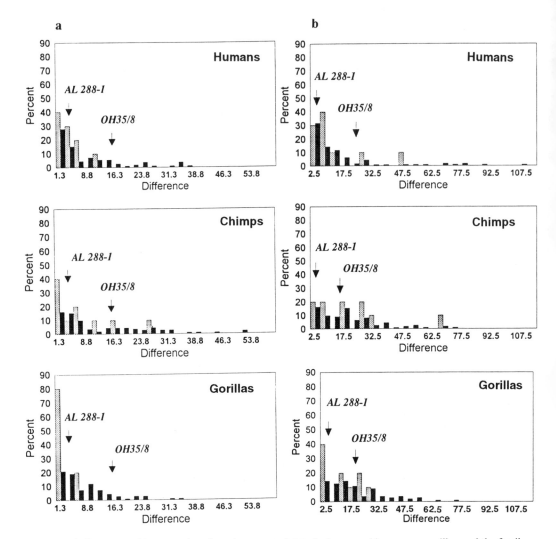

Figure 4. Congruency histograms based on size-corrected data for humans, chimpanzees, gorillas, and the fossil talocrural joints. (*a*) The mediolateral 25 cross-section, (*b*) the mediolateral 50 cross-section, (*c*) the anteroposterior 50 cross-section, (*d*) the anteroposterior 75 cross-section. Black bars, every possible combination of randomly associated tibiae and tali. Shaded bars, associated talocrural joints from single individuals.

tant for the comparative analysis of hominid postcranial fossils where traditional morphometrics involving lengths and breadths of long bone shafts and joint surfaces may not provide adequate resolution.

Although this analysis has used a relatively simple morphometric technique to compare congruency of talocrural joints, it has produced results that are entirely consistent with expectations for the known comparative samples. In particular, talocrural joints from single individuals have a closer congruency than randomly associated intraspecific pairs. Furthermore, the AL 288-1 talocrural joint in all analyses shows the close congruency expected of an associated joint from a single individual. The OH35/8 joint does not show

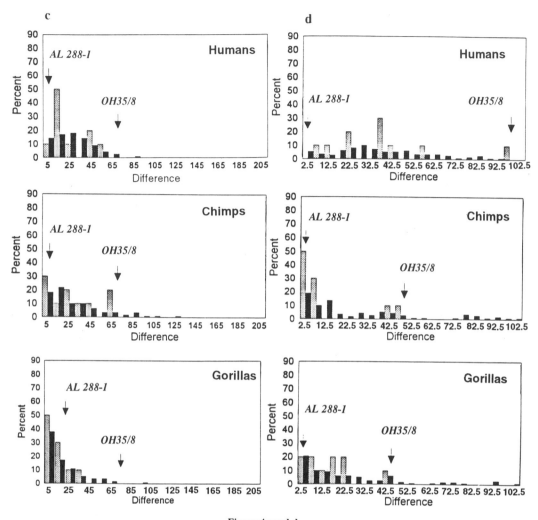

Figure 4c and d.

this same consistency. The analyses based on raw data (carrying size and shape information) suggest that OH35 and OH8 most probably represent different individuals, particularly in comparison to the congruency variation found in the gorilla and chimpanzee comparative samples, while the analyses based on size-corrected data suggest that they may even come from different species as originally thought.

While not conclusive, these results suggest that the observed similarity between OH35 and OH8 (Leakey, 1971; Susman and Stern, 1982) is primarily a size-based similarity rather than necessarily a shape-based similarity. It leaves open the possibility that the difference in joint shape between OH35 and OH8 may also indicate a difference in functional capabilities. Further studies involving more sophisticated morphometric analyses (e.g., Bookstein, 1991; Marcus et al., 1996) and larger comparative samples would be expected to provide better resolution of the taxonomic affiliation of OH35 and OH8 as well as any possible difference in functional capabilities that might be indicated by these bones.

Table 1. Congruency statistics of cross-sections for the comparative sample of humans, chimpanzees, and gorillas

	AP50		AP75		ML25		ML50	
	Mean	SD	Mean	SD	Mean	SD	Mean	SD
Shape only: Associated joint[1]								
Human	25.70	16.50	36.96	25.43	4.03	3.41	4.00	3.40
Chimpanzee	28.10	22.90	13.16	17.68	7.59	8.66	7.60	8.70
Gorilla	12.50	11.40	15.48	11.40	2.36	2.52	2.40	2.50
Shape only: Random pairs[2]								
Human	27.30	17.20	37.30	21.88	8.74	9.39	8.70	9.40
Chimpanzee	29.80	27.00	27.40	27.86	12.76	11.99	12.80	12.00
Gorilla	18.00	17.80	24.12	24.46	7.62	7.17	7.60	7.20
Size and Shape: Associated joint								
Human	6.10	3.10	7.78	5.46	0.71	0.54	0.70	0.50
Chimpanzee	3.10	2.50	1.59	2.20	0.70	0.83	0.70	0.80
Gorilla	2.20	2.20	3.12	2.11	0.46	0.45	2.10	2.30
Size and Shape: Random pairs								
Human	8.20	7.90	8.71	6.70	1.31	1.44	1.30	1.40
Chimpanzee	4.90	6.10	4.25	4.42	0.97	0.90	0.97	0.90
Gorilla	3.80	4.20	5.30	6.06	2.11	2.30	0.50	0.50

[1] Associated joints are those known to come from the same individual. N=10 for each species.
[2] Random pairs represent every possible intraspecific combination of tibiae and tali. N=90 for each species.

ACKNOWLEDGMENTS

We are grateful to Dr. Derek Howlett, Curator of the Powell Cotton Museum, Birchington, Kent for the loan of the gorilla and chimpanzee material and to Theya Molleson and the Natural History Museum, London, for the loan of the Spitalfields sample. We would also like to thank Elizabeth Strasser for the opportunity to participate in the "Primate Locomotion-1995" symposium and in that way to commemorate the many contributions of Warren G. Kinzey to the field of primate locomotion. This paper was considerably

Table 2. Congruency position of OH35/8 based on raw data[1]

Cross-Section	Human	Chimpanzee	Gorilla
AP50			
Associated joint	40.0	20.0	10.0
Random pairs	45.6	27.6	21.1
AP75			
Associated joint	20.0	0.0	0.0
Random pairs	84.0	22.2	14.4
ML25			
Associated joint	10.0	10.0	0.0
Random pairs	28.9	22.2	43.3
ML50			
Associated joint	30.0	40.0	50.0
Random pairs	28.9	26.7	56.7

[1] Values in each column are the percentage of individuals in the comparative sample with a degree of congruency that is as large as or larger than that of the OH35/8 pair. Definitions and sample sizes as in Table 1.

Table 3. Congruency position of OH35/8 based on size-corrected data[1]

Cross-Section	Human	Chimpanzee	Gorilla
AP50			
Associated joint	0.0	0.0	0.0
Random pairs	1.1	10.0	1.1
AP75			
Associated joint	0.0	0.0	0.0
Random pairs	0.0	22.2	14.4
ML25			
Associated joint	0.0	20.0	0.0
Random pairs	18.9	35.6	11.1
ML50			
Associated joint	20.0	60.0	30.0
Random pairs	27.8	56.7	44.4

[1]Definitions and sample sizes as in Table 2.

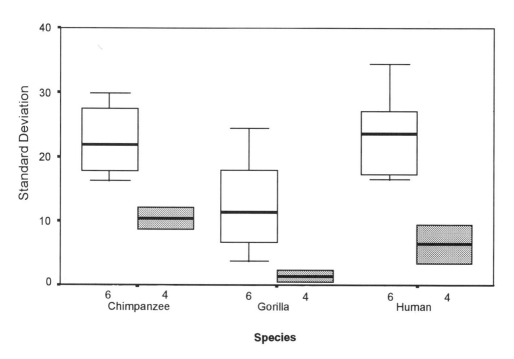

Figure 5. Boxplots illustrating the magnitude of variation (standard deviation) in congruency in both random and associated pairs of anteroposterior cross-sections (open boxes) and in both random and associated pairs of mediolateral cross-sections (shaded boxes) for chimpanzees, gorillas, and humans. Values below each boxplot is the sample size. Note that although there is variation in the standard deviation between species, for each species the magnitude of the standard deviation of the mediolateral cross-sections is always smaller than that of the anteroposterior cross-sections.

improved through suggestions made by Elizabeth Strasser, John Kappelman and two anonymous referees. The research reported in this paper was supported by The Leverhulme Trust grant F.134BB to L.A. and by NERC grant GR/H616/74 to B.W.

REFERENCES

Aiello LC, and Dean MC (1990) An Introduction to Human Evolutionary Anatomy. London: Academic Press.

Aiello LC, and Wood BA (1994) Cranial variables as predictors of hominine body mass. Am. J. Phys. Anthropol., *95:*409–426.

Bookstein FL (1991) Morphometric Tools for Landmark Data. Cambridge: Cambridge University Press.

Davis PR (1964) Hominid fossils from Bed I, Olduvai Gorge, Tanganyika. Nature *201:*967–970.

Day MH (1976) Hominid postcranial material from Bed I, Olduvai Gorge. In G Isaac and E McCown (eds.): Human Origins. California: W.A. Benjamin, pp. 363–374.

Day MH (1978) Functional interpretations of the morphology of postcranial remains of early African hominids. In CJ Jolly (ed.): Early Hominids of Africa. London: Duckworth, pp. 311–345.

Jonsson K, Fredin HO, Cedgrlund CG, and Bauer M (1984) Width of the normal ankle joint. Acta Radiologica Diagnosis *25:*147–149.

Harvey PH, Martin RD, and Clutton-Brock TH (1987) Life histories in comparative perspective. In BB Smuts, DL Cheney, RM Seyfarth, RW Wrangham, and TT Struhsaker (eds.): Primate Societies. Stuttgart: Thieme-Verlag, pp. 419–434.

Koenerink JJ, and van Doorn AJ (1992) Surface shape and curvature scales. Image and Vision Computing, *10:*557–565.

Leakey MD (1971) Olduvai Gorge: Excavations in Beds I and II, 1960–1963, Vol. 3. Cambridge: Cambridge University Press.

Leakey LSB, Tobias PV, and Napier JR (1964) A new species of the genus *Homo* from Olduvai Gorge. Nature, *202:*7–9.

Marcus LF, Corti M, Loy A, Nayler GJP, and Slice DE (1996) Advances in Morphometrics. New York: Plenum.

Susman RL, and Stern JT Jr. (1982) Functional morphology of *Homo habilis*. Nature, *217:*931–934.

Tobias PV (1991) Olduvai Gorge: The Skulls, Endocasts and Teeth of *Homo habilis*. Vol. 4. Cambridge: Cambridge University Press.

Wood BA (1991) Koobi Fora Research Project IV: Hominid Cranial Remains from Koobi Fora. Oxford: Clarendon.

Wood BA (1992) Origin and evolution of the genus *Homo*. Nature, *355:*783–790.

Wood BA, Aiello LC, Wood C, and Key C (submitted) A technique for establishing the identity of "isolated" fossil hominid limb bones. Journal of Anatomy.

Wynarsky GT, and Greenwald AS (1983) Mathematical model of the human ankle joint. Journal of Biomechanics, *16:*241–251.

14

USE OF STRAIN GAUGES IN THE STUDY OF PRIMATE LOCOMOTOR BIOMECHANICS

Brigitte Demes

Department of Anatomical Sciences
School of Medicine
State University of New York at Stony Brook
Stony Brook, New York 11794-8081

1. INTRODUCTION

The strain gauge technique is a relatively recent addition to the catalogue of experimental methods available for functional analyses, especially locomotor studies. Strain gauges track the deformation of objects they are attached to, thus allowing the reconstruction of external forces and loads that cause these deformations. They are restricted to surface use, but extrapolations allow us to reconstruct strain patterns through the object (e.g., Gross et al., 1992; see also example in Figure 8). In the field of biomechanics there are two major applications: the measurement of bone deformations and the instrumentation of force measuring devices. The data in both fields can be used in interpreting musculoskeletal morphology. Functional interpretations of bony morphology have been historically based on correlations between shape and activity. The interface is the mechanical environment into which behaviors translate and in which bone develops, maintains and/or changes its shape. The mechanical demands of particular locomotor modes are commonly derived from behavioral observations in combination with biomechanical models. Measuring the external forces acting on limbs with force transducers is a first step in testing the numerous assumptions inherent in this process. Even with this background information, however, actual loadings of a bone can only be deduced with a certain degree of plausibility. *In vivo* measurement of bone strain is currently the only method of directly determining the major loading regimes caused by the external forces acting on the bone.

In the following I will briefly describe the technique, review its applications in primate locomotor biomechanics, and give two examples of recent work of my own in this field: *in vivo* bone strains recorded on the macaque ulna (Demes et al., 1997, 1998), and a custom-designed "force pole" to measure takeoff and landing forces of leaping prosimians (Demes et al., 1995).

Primate Locomotion, edited by Strasser *et al.*
Plenum Press, New York, 1998

2. MECHANICAL BACKGROUND: STRAIN[*]

Strain is a technical term used to express deformation. It is defined as the change in length over original length or percentage change in length, and is, therefore, a dimensionless unit. Out of convenience units are nevertheless assigned, and for small deformations these are microstrain ($\mu\varepsilon$).

$$\varepsilon = \frac{\Delta L}{L} = 0.000001 = 1\mu\varepsilon$$

with ΔL = change in length and L = original length.

Strains perpendicular to any given plane are called normal strains and they are either compressive ($\varepsilon-$) or tensile ($\varepsilon+$). Strains within any given plane are called shear strains (γ).

Within any given strain field (any point on the surface where strain is measured) there is a maximum and a minimum normal strain at right angles to each other. These are called the principal strains. In most biomechancial applications, the maximum principal strain is tensile (positive sign) and the minimum principal strain is compressive (negative sign), independent from their relative magnitudes. Biaxial compressive or tensile strain is less common. The magnitude and alignment of the principal strains are indicative of the major loading regimes. In a simple, symmetrical beam, the mechanical analog most frequently used in biomechanics, axial forces cause either compressive or tensile strains in line with the long axis of the beam and with the opposite strain quality represented at right angles to the long axis. This transverse strain component is usually low and results from the change in width of the beam being tensed (decrease in width, compressive transverse strain) or compressed (increase in width, tensile transverse strain). The amount of transverse to longitudinal strain is expressed in Poisson's ratio ν, and is a function of material stiffness:

$$\nu = \frac{\varepsilon_{xx}}{\varepsilon_{yy}}$$

with ε_{yy} = longitudinal strain and ε_{xx} = transverse strain.

Poisson's ratios for bone are approximately 0.3–0.4 (Currey, 1984).

Bending results in a composite strain pattern with longitudinal tension at the convex side of bending and longitudinal compression at the concave side of bending. The transition occurs at the neutral plane of bending, which is crossed by the principal (tensile and compressive) strains at 45 degree angles (Figure 1). Shear strains (not shown in Figure 1) also result from bending because bending causes a tendency of the fibers in a beam to slide against each other. The maximum shear strains are at 45 degree angles to the normal strains, highest at the neutral plane of bending and zero in the outer fibers. Twisting results in principal strains at 45 degrees to the long axis of a symmetrical beam. These strains are highest at the surface and decrease to zero towards the centroidal neutral axis of torsion. Unlike in bending, these strains are distributed uniformly over the surface of the beam.

---

[*] Note that engineers distinguish between "true strain" and "engineering strain". Calculations of engineering strain do not take into account the change in cross-sectional dimensions of the strained object that influences its total deformation. Engineering and true strain are practically coincident when the strains are small, as is the case in most biomechanical applications.

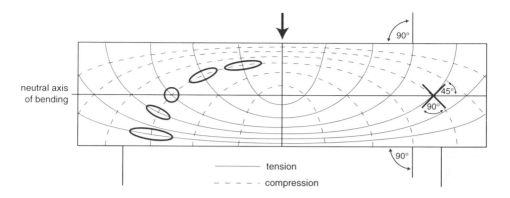

Figure 1. Principal strain trajectories on the surface of a beam loaded in three-point bending. Principal tensile strains are shown as solid lines, compressive strains as dashed lines; their relative magnitudes are visualized as ellipses or circles for several different locations on the surface (after Ramm and Wagner, 1967: 126).

Measuring the principal strains at one surface location may not suffice to resolve the loading patterns of a beam. For example, principal strains at 45 degrees to the long axis may be encountered in a twisting regime but also in bending close to the neutral axis (Figure 1). Multiple gauges are required to distinguish between twisting and bending (Gross et al., 1992). Loading regimes are also frequently superimposed, which complicates their interpretation. When measuring strain on bones the pattern is further complicated by the fact that bones are not simple beams. They are composed of more than one material (heterogeneous) and may strain differently in different directions (anisotropic); i.e., they respond with various amounts of deformation for a given external loading regime at different localities and in different directions. In addition, bones are not symmetrical.

3. STRAIN GAUGE TECHNIQUE

The most commonly used strain gauges are electric resistance gauges. They work on the basis of Kelvin's principle, which states that the electric resistance of metals is influenced by their state of strain. The gauges consist of a metal wire or foil that is, in the commonly used bonded type, fixed to a flexible base (to keep the stiffness low so that the gauge does not reinforce the object that it is bonded to). The sensitivity of the gauge is determined by the gauge factor and the length of the wire (Dove and Adams, 1964):

$$\Delta R = FR\varepsilon$$

with ΔR = change in electric resistance, F = gauge factor, R = original resistance, ε = strain.

The gauge factor is a measure of the change in resistance per unit of original resistance that will occur per unit of strain applied. The length of the wire determines the original resistance. "Good" gauges, i.e., gauges with a high sensitivity should have a high gauge factor and a high original resistance. Wires in electric resistance gauges are therefore arranged in a grid, which allows them to be long and the gauges to be small at the same time. Most commonly used strain gauges in biomechanical applications have gauge

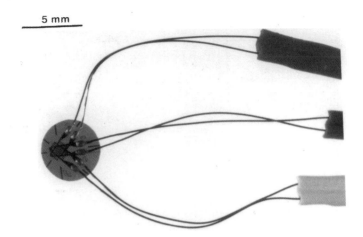

Figure 2. Rectangular rosette (Kenkyujo FRA-1-11-1L). The three elements are stacked on top of each other and attached to a larger, round epoxy backing that also holds the soldering connections to the wires from the elements. The gauge is 1 mm long, the backing 4.5 mm in diameter.

factors of around 2 and an original resistance of 120 Ω, with lengths and diameters of only a few mm (Figure 2).

The circuitry most commonly used with electric resistance gauges is a Wheatstone bridge (Figure 3). It is a circuit of four resistors, one of which is the strain gauge. A low DC voltage is applied to two contact points of the circuit, and the output voltage is measured across the other two. Any change in resistance of the gauge produces a DC offset in

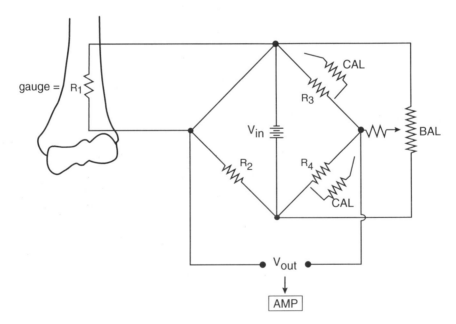

Figure 3. Wheatstone bridge circuitry with two calibration resistors (CAL) and a balancing resistor (BAL). V_{in} indicates input voltage supplied across one diagonal, V_{out} the output voltage measured across the other diagonal.

the signal voltage (because voltage is directly proportional to resistance—Ohm's law). The nominal resistance of all four resistors is originally the same (120 Ω), i.e., the output voltage is zero. However, the resistance of the gauge may change during the bonding procedure—it may be slightly strained and therefore have a higher or lower resistance than 120 Ω. The Wheatstone circuitry allows one to "balance" the circuitry *after* the gauge is in its final position.

The voltage change for a given amount of strain is determined by shunting two resistors that produce a change in output voltage corresponding to a known amount of με (e.g., ± 1000 με) with two arms of the bridge: shunt calibration. When the calibration switch is thrown they produce a change in resistance and output voltage. The measured output voltage change is used to calculate calibration factors. The output voltage is amplified and sampled either directly into a computer (which requires the transformation of the analog signal into a digital one; i.e., an A/D board) or on tape. The length of the cable from the gauge to the amplifiers should be kept short because it acts as an antenna and adds noise to the actual signal.

Single element gauges only register strain in one direction. The more frequently used rosette gauges combine three single elements arranged at known angles to each other (45 degrees: rectangular rosettes (Figure 2), 60 degrees: delta rosettes). With the information from three elements, the magnitudes and directions of the strains in a biaxial strain field can be resolved. The change in output voltage for each element is transformed into strain units using the calibration factors. Trigonometric relations as expressed in the so-called Mohr's circle of strains (Dally and Reilly, 1991; see also Biewener, 1992) are then applied to calculate strains with reference to particular axes of interest. Most frequently reported are the principal strains and their angles and the maximum shear strain. The strain in the direction of the long axis of a bone is a sometimes more intuitive vehicle to interpret deformations. The gauge orientation on the surface is prerequisite knowledge for relating strain directions to the geometry of the object the gauge is bonded to.

4. BONE STRAIN MEASUREMENTS IN PRIMATE BIOMECHANICS

Bonded metallic resistance gauges were developed in 1940 (Dove and Adams, 1964). As early as 1944, they were first applied in the measurement of bone strain. Gurdjian and Lissner (1944) surgically implanted strain gauges onto the cranial vaults of dogs and recorded deformations. They banged the heads of the anesthetized animals and registered the oscillations of the cranial vault (as well as intracranial pressure changes) in an attempt to understand head injury mechanisms. The single element gauges were about half the length of the temporal fossa of the dogs, and data reported were selected traces of strain on an oscilloscope screen. Modern rosettes are available in much smaller sizes, but reporting selected strain readings only is still common practice.

Another early *in vivo* strain study on the tibia of a dog was the first one published in an anthropological journal (Evans, 1953). The first application of strain gauge technique in *primate* biomechanics was in 1975, when Lanyon and coworkers in England implanted a rosette gauge on the anteromedial aspect of a human tibia. The subject was one of the authors. The size of the gauge was not reported but the incision to implant it was 10 cm long. Bone strains were recorded during walking and running. During the support phase of walking, the largest principal compressive strains were about 400 με, the peak principal tensile strain only slightly lower and at an angle of about 50 degrees to the long axis of the bone. The authors interpreted this pattern as the result of the combined effect of muscle pull and body weight.

Bone strain papers from Hylander's lab at Duke University began coming out in 1977. They reported strains during mastication from the mandible and facial skeleton in several species of nonhuman primates (e.g., Hylander 1977, 1979, 1984; Hylander et al., 1991; Hylander and Johnson, 1994) and represent by far the bulk of primate bone strain data. Having substantially altered and improved interpretations of primate jaw shapes, these papers demonstrate the immense importance of this approach in understanding bone adaptations.

In comparison, strain gauge analyses of the primate *locomotor* system are very limited in number. Aside from the already mentioned human tibia experiment by Lanyon and coworkers (1975), there are only a few studies on primate postcrania.

In 1977, Young et al. implanted a miniature strain transducer into the tibia of a macaque and registered strains during a variety of activities. This was more of a technical report to introduce and test the transducer, and the actual strain data (or the monkey) did not receive much attention.

Fleagle et al. (1981) measured strains on the posterior aspect of the ulna of a spider monkey with a single element gauge aligned with the long axis of the bone. The gauge registered tensile strains during the support phase of walking, which they ascribed to anterior bending of the ulna. Strains during brachiation and climbing were similar in pattern but of much lower magnitude.

In a more detailed study, Swartz et al. (1989) reported strains on the anterior and posterior surfaces of each of the long bones of gibbon forelimbs during brachiation. Hylobatid forelimb bones are very long and slender, suggesting that they experience less bending and compression than those of other animals (Kummer, 1970) and, instead, are loaded predominantly in axial tension (Swartz et al., 1989). Although this was confirmed in the experiments for the ulna, where principal strains on the anterior and posterior aspects were both tensile and in line with the long axis of the bone, the radius and humerus did not seem to experience axial loading. The radius experienced compression on the ventral midshaft and tension on the dorsal midshaft at midswing, with strain directions again in line with the long axis of the bone, i.e., a bending regime. Compressive strains were higher than tensile strains, which suggests axial compression superimposed on bending. Swartz et al. concluded that the bending is caused by the bone's curvature, which is seven times greater than that of the ulna. The gibbon radius is curved in the mediolateral plane, however, and its curvature is unlikely to be responsible for anteroposterior bending. In the anteroposterior plane the gibbon radius is straight. In addition, if bending was the predominant loading regime, the tensile and compressive peaks on opposite cortices should have occurred simultaneously—which is not the case in the representative strain traces presented in their Figure 1. Finally, the principal strains on the humerus were oriented at an angle of approximately 45 degrees to the long axis of the bone, which is typical for a torsional loading regime. These strain patterns do not support their hypothesis that compressive muscle forces and tensile external weight force will oppose each other during brachiation (in contrast to being additive in terrestrial locomotion) and thereby reduce the net strain in the skeleton. Peak strain magnitudes for the three bones are all around 1500 $\mu\varepsilon$.

Very recently, Burr et al. (1996) duplicated the Lanyon et al. (1975) experiments on the human tibia, with the subject performing more rigorous exercises. Maximum strains for zigzag runs on inclined surfaces reached nearly 2000 $\mu\varepsilon$, strains during level walking were slightly higher than those recorded by Lanyon et al.

5. CASE STUDY # 1: *IN VIVO* BONE STRAIN ANALYSIS OF THE MACAQUE ULNA

Bone strain measurements at SUNY Stony Brook's primate locomotion laboratory were started recently to test certain assumptions inherent in the functional analysis of bone morphology. When interpreting the cross-sectional geometry of long bones, for example, locomotor modes are implicitly or explicitly equated with major loading regimes: limbs that are moved predominantly in sagittal planes during locomotion will experience sagittal bending, or hind-limb bones in leaping animals will experience high loads (e.g., Demes and Jungers, 1993). For primates, specifically, a great range of loading patterns is generally assumed because they use their limbs, especially their forelimbs, in a more versatile fashion than nonprimate mammals. However plausible these assumptions may be, they can only be verified by directly measuring the deformations of bones during various activities.

The functional anatomy of the ulna and its articulations with the wrist and the humerus are difficult to interpret. The amount of weight sharing and force transmission between the two forearm bones is unknown. It is equally unclear whether and to what degree the two bones act as a unit in resisting the loads acting on the forearm; i.e., whether, in a bending regime, the bending moments would be distributed over the combined cross sections in a way that one bone experiences predominantly compression, the other one tension. The ulna and the elbow joint have, however, received considerable attention by functional morphologists. They display significant variation among primates that correlates with locomotor modes and these elements are frequently represented in the fossil record (e.g., Fleagle et al., 1975; Rose, 1988; Harrison, 1989; Richmond et al., 1998). Functional interpretations of the medial and lateral flanges that characterize the cercopithecid elbow have been offered by all of the above authors, but, as Rose points out, "a detailed functional interpretation of these features is difficult without good data on the direction in which loading forces act at the elbow at different phases of the gait cycle" (1988: 214).

Experiments were performed on an adult female rhesus macaque. The ulna offered itself because of its comparatively easy surgical accessibility and muscle-free sections. Similar experiments are currently under way for the tibia, the most accessible hind-limb bone. The experiment was performed in collaboration with the Musculo-Skeletal Research Lab (Director: Clinton Rubin) and follows their protocol (e.g., Gross et al., 1992). Prior to surgery, the animal was trained to walk while controlled by a pole attached to a neck collar. This allowed us to run the gauge wires from the animal to the animal trainer and from there to a computer. Three rosette strain gauges (Kenkyujo, 2 mm diameter) were surgically implanted on the ulna, through a single incision of 5 cm length, approximately 1/3 of the way up the shaft from the distal end in an area that required no muscle disruption (Figure 4). The bone surface was exposed by removing a small area of periosteum at each gauge site. The bone surface was then degreased and dried with isopropyl alcohol, and the gauges glued on with cyanoacrylate. The gauges were located on the (antero)medial cortex (more medial than anterior), the (antero)lateral cortex (more lateral than anterior) and the postero(lateral) cortex (more posterior than lateral). Wires from the gauges were passed through small resin flanges that were screwed onto the bone approximately 2–3 cm proximal to the gauge sites to provide strain relief[†]. The leads were then passed subcutane-

---

[†] Stress concentrations around cut holes decrease rapidly, and at a distance from the edge of the hole equal to the radius of the hole, stresses are usually equal to the applied stresses (Timoshenko, 1958).

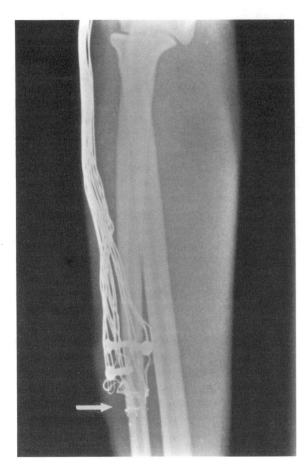

Figure 4. Anteroposterior x-ray of the macaque forearm with gauges attached. Wires are running subcutaneously to the shoulder. Bone screws hold plastic flanges (not visible) for strain relief of the wires coming from the gauges. The arrow indicates the level of the gauges.

ously to the shoulder region and surfaced through a small incision between the shoulder blades where the animal could not reach them. Here they were soldered to a connector. Connectors and wires were protected by a vest worn by the monkey.

The animal engaged in apparently normal locomotion immediately after surgery and data were collected then and again two days later. On the day following surgery the animal was sore and did not locomote normally. The animal was anaesthetized again after the second period of data collection, CT scans and X-rays were taken to verify the gauge positions (Figure 4), and the gauges were then removed. The animal recovered without complications.

Rosettes were conditioned with a 3 V excitation voltage and amplified with the Vishay Measurement Group 2110. Data were subsequently sent through a low-pass filter with a cut-off frequency of 100 Hz, A/D converted, and stored in an IBM-compatible computer. Sampling frequency was 100 Hz and sampling periods were 10 seconds long. The animal was videotaped during the experiments so that strain patterns and activity patterns could be matched (Figure 5). For synchronization, one of the strain channels was superimposed over the live animal video (see Stern et al., 1977 for more details on the method).

A wide range of locomotor activities was recorded, including walking, galloping and climbing. For sequences that were visually identified on the videotapes as representative of a particular locomotor mode, principal strains were calculated using standard formulae

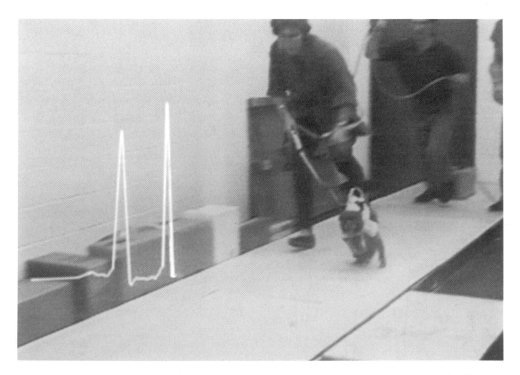

Figure 5. Experimental set-up. The wire runs from the animal's shoulder to a pole attached to its neck collar and held by the experimenter. From there it runs to the computer, being held by a second person to avoid tangling. The superimposed trace is the strain from one gauge element and was used to synchronize strain data and video images. Figure taken from video.

(Dally and Riley, 1991) in a custom-written subroutine of the Macintosh-based analysis program Igor (Wavemetrics). In addition, strain components in the direction of the long axis were calculated (longitudinal strains). Strain data for walking steps only will be discussed here.

Figure 6 presents longitudinal strains for a sequence of steps, and Figure 7 shows the tensile and compressive principal strains and the angle of the tensile (maximum) principal strain with the long axis of the bone for the three gauge sites for those steps. Peak maximum and minimum as well as longitudinal strains during the stance phase of all steps analyzed are presented in Table 1.

Peak strain magnitudes and peak strain angles during the support phase are very consistent (Table 1, Figure 7). Strain magnitudes during the swing phase are low. Strain angles throughout support phase do not change much whereas they are highly variable for the swing phase.[‡] The (antero)lateral cortex experiences longitudinal tension, the (antero)medial cortex experiences longitudinal compression (Table 1). The postero(lateral) cortex experiences comparatively low tensile strains. This distribution is compatible with mediolateral bending. The postero(lateral) gauge is probably located close to the axis of bending, which is not strictly anteroposterior but slightly oblique, from posterolateral to

---

[‡] Note, however, that angular changes are less dramatic than suggested by the traces in Figure 7, as negative (clockwise) angles correspond to positive (counterclockwise) angles minus 180 degrees.

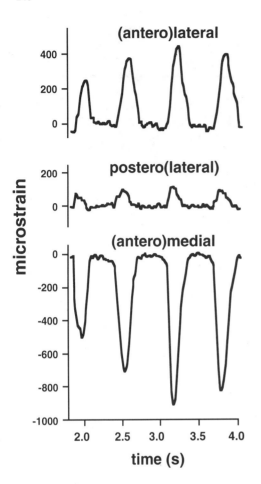

Figure 6. Longitudinal y-strains, i.e., strains in the direction of the long axis of the bone, for the three gauge sites. The (antero)lateral and postero(lateral) cortices experience tensile strains whereas the (antero)medial cortex is in compression during the support phase of walking. Strains for four consecutive steps are shown.

anteromedial. This is confirmed by the calculation of the normal strain distribution for the entire cross section using combined beam and finite element model analysis (courtesy D. Polonet; see Rybicki et al., 1977, and Gross et al., 1992 for the method). The shades of gray in Figure 8 are indicative of various amounts of tensile and compressive strains over the cross section, and the solid line corresponds to the neutral axis of bending. The deviations of the principal strains from the long axis of the bone (Table 1, Figure 7) indicate that the ulna is not loaded in pure bending during the support phase of walking, but, instead, also experiences torsion. The angles suggest superimposition of a negative (proximal clockwise) torque.

Mediolateral bending as a major component in the loading regime of the ulna is a somewhat provocative and unexpected result. The majority of *in vivo* strain studies of long bones in other species have identified anteroposterior bending as the predominant loading regime during the support phase of locomotion: sheep radius (Lanyon and Baggott, 1976), radius and tibia of horse and dog (Rubin and Lanyon, 1982), horse radius (Biewener et al., 1983), goat tibia and radius (Biewener and Taylor, 1986), and horse tibia (Biewener et al., 1988). Gross et al. (1992), on the other hand, report tension not only for the anterior, but also for a small part of the lateral cortex of the horse metacarpal, and Biewener and Bertram (1993) measured compression on the cranial (anterior) *and* medial cortices of the

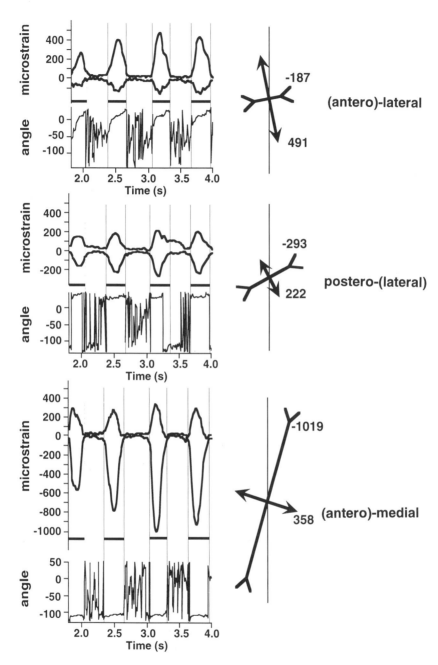

Figure 7. Principal strains and angles of the maximum principal strain with the long axis of the ulna for the three gauge sites and four consecutive steps. The arrows indicate the peak strain for one of these steps and their orientation relative to the long axis of the bone.

Table 1. Peak ulnar strains (in microstrain) for the stance phase of walking

	(Antero)medial	(Antero)lateral	Postero(lateral)
N	11	17	17
Max. principal strain	361 ± 128	495 ± 67	217 ± 44
Min. principal strain	−993 ± 150	−182 ± 28	−273 ± 31
Angle *	−107 ± 2	12 ± 2	29 ± 3
Longitudinal strain	−873 ± 138	466 ± 60	99 ± 29

*Angle of the peak maximum principal strain with the longitudinal axis of the bone; positive values indicate counterclockwise direction.

chicken tibiotarsus, a pattern similar to our findings. Finally, Figure 7a in Yoshikawa et al. (1994) indicates mediolateral bending in the tibia of dogs, however, for unnatural, bipedal locomotion. For the macaque, mediolateral bending with the medial cortex in compression is probably caused by the weight force vector that passes medially to the limb. The source of the twisting regime is less obvious. Twisting could result from forearm rotation during the support phase as well as from muscle forces. Although bending and axial loading regimes are usually emphasized in the other analyses cited above, superimposed torsion is also indicated by the deviation of the principal strain angles from the long axes of the bones. These deviations are especially large in the chick tibiotarsus (Biewener and Bertram, 1993).

Peak strain magnitudes in the range of 100 to 1000 με (Table 1)[°] are low in comparison to strains recorded during locomotion for other animals, which most commonly range in amplitude from 2000–3000 με (Rubin and Lanyon, 1982, 1984). Indeed, this latter level is considered to be beneficial for bone tissue and to be maintained through kinematic adjustments (speed, joint angles) over wide ranges of body sizes and locomotor modes ("dynamic strain similarity", Rubin and Lanyon, 1984; Rubin et al., 1994). Peak strain levels during locomotion below the 2000–3000 με bracket are reported, however, making the

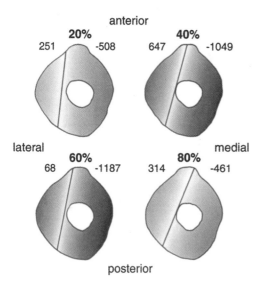

Figure 8. Distribution of normal strains in the cross section of the macaque ulna at the level of the gauge sites. The four cross sections represent the strain patterns at 20%, 40%, 60%, and 80% of support phase. The line indicates the neutral axis of bending, and the shades of gray the increasing tensile (positive) and compressive (negative) strains. The numbers are the maximum strains on the convex and concave side of bending.

° Strain magnitudes were not substantially higher during more vigorous activities low strain levels in the macaque ulna less unusual (sheep radius: Lanyon and Baggott, 1976; sheep tibia: Lanyon and Smith, 1970; human tibia: Lanyon et al., 1975; dog tibia: Bouvier and Hylander, 1984; human tibia: Burr et al., 1996). Two factors may contribute to the low strain magnitudes in the macaque ulna: (1) The radius, whose robusticity in macaques is similar to that of the ulna, may be instrumental in resisting the external forces and moments at the forearm. This is currently being explored in *in vitro* experiments in our lab where ulnar bone strain is measured with the interosseous membrane intact and, subsequently, severed. (2) The predominance of the hind limb in weight bearing and propulsion may reduce forelimb loads. Kinetic data demonstrate that forces and impulses at the macaque hind limb are higher than those at the forelimb, though this difference is not pronounced (Demes et al., 1994). Tibial bone strain data are expected to clarify this question.

Our results (Demes et al., 1997, 1998) warrant further confirmation before the strain environment of the macaque ulna during locomotion can be ultimately characterized. With regard to the functional morphology of the bone and its articulations, the strain data indicate that the radius and ulna do not act as a composite structure to resist bending but, instead, the ulna undergoes a bending regime on its own. As it is "locked" into the trochlea at the elbow joint it is likely that mediolateral bending results in mediolateral forces at this joint, which has been suggested by Schmitt's (1994) data on the orientation of the substrate reaction force. The strain data thus confirm earlier notions that the reinforcements of the cercopithecid elbow on the medial and lateral sides as described by Fleagle et al. (1975), Rose (1988), Harrison (1989) and Richmond et al. (1998) are indeed structural adaptations to counteract frontal plane forces at this joint.

6. FORCE TRANSDUCERS AND STRAIN GAUGE TECHNOLOGY

Force transducers in locomotor studies are used to quantify the forces exchanged at the contact between the moving body and the substrate. These are the forces exerted by the animal and the forces experienced by the animal, the two of them being equal in magnitude and opposite in direction. Force transducers most frequently used in human sports biomechanics and orthopedics as well as in animal locomotor research are piezoelectric force platforms (e.g., Kistler plates). They are based on the property of certain materials, like quartz, to generate an electric charge when experiencing strain. Although the responsiveness, sensitivity and precision of such platforms are high their considerable price and standard design do not recommend them for all applications. Less expensive force platforms can be built using strain gauge technology (e.g., Heglund, 1981; Biewener and Full, 1992; Nigg and Herzog, 1994).

A few studies have been performed using nonstandard force transducers to quantify substrate reaction forces. Their great advantage is the potential of mimicking natural substrates used by animals in the wild. This is especially relevant for studying arboreal locomotion. An easy way of building horizontal, branch-like force transducers is to mount bars onto force plates. This technique has been used by Ishida et al. (1990) and Nieschalk (1991) to quantify substrate reaction forces of slow and slender lorises, as well as by Schmitt (1994, this volume) to study the forces acting on the forelimbs of various monkey species.

Force transducers using strain gauge technology allow an even greater variability in design. In 1984, Yamazaki and Ishida used a pole-type force detector to analyze substrate reaction forces during vertical climbing in gibbons. The same force pole was used to compare the kinetics of vertical climbing in macaques and spider monkeys (Hirasaki et al., 1992, 1993). Bonser and Rayner (1996) constructed a force-transducing perch to record takeoff and landing forces of small birds. A force-transducing handle has been used by Chang et al. (1997) to measure superstrate forces of brachiating gibbons and chimpanzees.

7. CASE STUDY # 2: A STRAIN-GAUGED FORCE TRANSDUCER TO MEASURE LEAPING FORCES

Our force pole for measuring takeoff and landing forces of large-bodied vertical clingers and leapers (Demes et al., 1995) differs in various aspects of design and instrumentation from previous devices. Field observations in Madagascar had revealed that large-bodied indriid leapers frequently use compliant supports that sway visibly under the animals' takeoff and landing forces. Breaking a dogma in force transducer design ("good" force transducers should be rigid; i.e., have a high frequency response), we constructed a compliant force pole. It consists of an aluminum pipe instrumented with strain gauges and a PVC pipe that served as the takeoff and landing site (Figure 9). The pole is anchored in an aluminum box. Whereas the PVC pipe, mostly because of its large dimensions (diameter = 0.11 m), deforms minimally under the animal's impact, the aluminum pipe is bent considerably and initiates swaying movements of the system. These deformations are registered by the gauges. The largest deflection is proportional to the peak force during takeoff or landing.

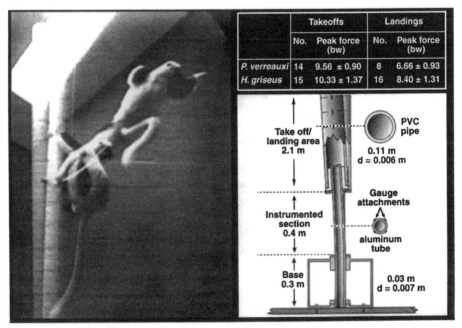

	Takeoffs		Landings	
	No.	Peak force (bw)	No.	Peak force (bw)
P. verreauxi	14	9.56 ± 0.90	8	6.66 ± 0.93
H. griseus	15	10.33 ± 1.37	16	8.40 ± 1.31

Figure 9. Sifaka leaping from force pole. The takeoff and landing (upper) part of the pole was drawn from two superimposed vidoeimages: the animal is shown prior to acceleration for takeoff and at toeoff, when the pole is maximally deflected. The dimensions of the pole and the attachment sites of the gauges as well as peak forces are indicated in the right part of the drawing.

The limitations of the described design lie in the fact that the registered deformations are the combined result of the animal's effort and the compliancy of the pole. When using rigid measuring devices, registered forces are a direct indicator of the forces generated by the animal, and force-time-curves allow one to identify, for example, periods of high accelerations and low accelerations and relate them to joint positions and muscle activities. Because of the low natural frequency of the system, particularly rapid changes in forces will be underestimated.

Resolving for the substrate force and its components was also done in a novel way. Three gauges (and a fourth as a safeguard) attached around the circumference of the aluminum pipe permit a first order approximation of the external forces. Normal strain distribution at the instrumented cross section was calculated using the same algorithm as for the above described *in vivo* bone strain experiment and the neutral axis of bending was determined. To resolve for the forces at the contact point the pole was modeled as a cantilever beam. The location of the neutral axis of bending and the height of the contact on the pole (from synchronous video) allowed to resolve for the axial forces and moments necessary to generate a measured strain distribution (Rybicki et al., 1977; Gross et al., 1992). The takeoff and landing angles of the animals at the moment of toeoff or touchdown were estimated on the video, and the resultant forces calculated from the trigonometric relationship between axial forces and angles.

Takeoff and landing forces were collected with the compliant force pole for two species of vertical clingers and leapers: *Hapalemur griseus and Propithecus verreauxi.* The pole was installed in an indoor enclosure at the Duke University Primate Center whose concrete floor allowed for solid anchorage of its base. The recording equipment was the same as used for *in vivo* bone strain data collection (see above). Only peak forces were evaluated, and they are presented here as multiples of body weight (see Demes et al., 1995 for a more detailed report).

Not unexpectedly, the peak takeoff and landing forces are higher for the smaller *Hapalemur griseus* (average weight for two animals: 1.0 kg) than *Propithecus verreauxi* (average weight for two animals: 4.0 kg; Figure 9). As smaller-bodied animals have absolutely shorter acceleration distances and relatively more muscle force available, they accelerate over short periods of time but with high pushoff forces. Larger-bodied animals, on the other hand, can make use of their absolutely longer limbs, but are restricted by a more limited force-per-unit-mass ratio; they therefore accelerate over longer periods of time and with lower pushoff forces (Demes and Günther, 1989).

Peak takeoff forces were higher than landing forces. On rigid (force platform) surfaces the landing impact is usually associated with higher peak forces, and the deceleration period is shorter than the acceleration period (e.g., Preuschoft, 1985). The elastic properties of the pole require excess force at takeoff, as part of it does not translate into leaping distance but is "used" to deform the pole. The yield of the pole at landing, on the other hand, absorbs some of the kinetic energy, increases the duration of the braking period, and, consequently, reduces the magnitude of the landing force.

Finally, the timing of the takeoff relative to the movements of the pole are of interest for the question of whether the animals take advantage of the elastic properties of the pole. The simultaneous videorecordings clearly indicate that the animals invariably take off before the elastic recoil of the pole. This is also the case for indriid leapers in their natural habitats when taking off from compliant trunks (Demes et al., 1995). It therefore appears that the vertical clingers and leapers are not able to make use of elastic substrates as power amplifiers. As they perform work on the substrates that is not returned to them, leaping from elastic supports must be less efficient than leaping from rigid supports.

8. CONCLUSIONS

Strain gauges in primate locomotor studies are used in two ways. The strain gauge technique is currently the only method that permits tracking the deformations of bone *in vivo* and relating them to animal activities. If attached to parts of the environment an animal is moving on, strain gauges can be also be used to determine the external forces acting at the contacts. The examples reviewed and provided in this chapter demonstrate the potential contributions of this technique to the field of functional morphology.

In vivo bone strain measurements provide direct evidence of bone deformation and are an invaluable tool for testing hypotheses on how bones are loaded during locomotion. The hypothesized anteroposterior bending for the macaque ulna was not confirmed by the strains measured *in vivo* on this bone, which, instead, indicated that mediolateral bending is the predominant loading regime. Major bone and joint reinforcements at the elbow are therefore to be expected in this plane.

Knowledge of the magnitude and direction of substrate reaction forces is useful in identifying the forces and, if combined with kinematic information, moments generated by and acting on an animal's musculoskeletal system. This will add significantly to our understanding of muscle distribution as well as joint and bone adaptations. In addition, it may allow evaluation of environmental variables and their influence on locomotion. In the example given here, it was demonstrated that large-bodied vertical clingers and leapers on compliant supports increase their work load and generate and endure highest forces during the takeoff, rather than the landing.

ACKNOWLEDGMENTS

The *in vivo* bone strain experiments are a combined effort of members of the Dept. of Anatomical Sciences and the Musculoskeletal Research Lab of the Dept. of Orthopaedics at SUNY, Stony Brook. Contributions of Clinton Rubin, Jack Stern, Ted Gross and Susan Larson are greatly acknowledged. Michael Hausman and David Reim assisted in the surgery, and Terry Button made X-ray and CT scans possible. David Pollonet calculated the normal strain distribution in the bone cross sections and Luci Betti prepared the artwork. Comments by Susan Larson, Brian Richmond, and three reviewers improved the manuscript. This research was supported by NSF grants DBS-920961 and SBR-9507078.

REFERENCES

Biewener AA (1992) *In vivo* measurement of bone strain and tendon force. In AA Biewener (ed.): Biomechanics - Structures and Systems. Oxford: Oxford University Press, pp. 123–147.

Biewener AA, Thomason J, Goodship A, and Lanyon LE (1983) Bone stress in the horse forelimb during locomotion at different gaits: A comparison of two experimental methods. J. Biomech. *16*:565–576.

Biewener AA, and Taylor CR (1986) Bone strain: A determinant of gait and speed? J. exp. Biol. *123*:383–400.

Biewener AA, Thomason JJ, and Lanyon LE (1988) Mechanics of locomotion and jumping in the horse (*Equus*): *In vivo* stress in the tibia and metatarsus. J. Zool., Lond. *214*:547–565.

Biewener AA, and Full RJ (1992) Force platform and kinematic analysis. In AA Biewener (ed.): Biomechanics - Structures and Systems. Oxford: Oxford University Press, pp. 45–73.

Biewener AA, and Bertram JEA (1993) Skeletal strain patterns in relation to exercise training during growth. J. exp. Biol. *185*:51–69.

Bonser RHC, and Rayner JMV (1996) Measuring leg thrust forces in the common starling. J. exp. Biol. *199*:435–439.

Bouvier M, and Hylander WL (1984) *In vivo* bone strain on the dog tibia during locomotion. Acta Anat. *118*:187–192.

Burr DB, Milgrom C, Fyhrie D, Forwood M, Nysaka M, Finestone A, Hoshaw S, Saiag E, and Simkin A (1996) *In vivo* measurement of human tibial bone strain during vigorous activity. Bone *18*:405–410.

Chang YH, Bertram JEA, and Ruina (1997) A dynamic force and moment analysis system for brachiation. J. exp. Biol. *200*:3013–3020.

Currey J (1984) The Mechanical Adaptations of Bones. Princeton: Princeton University Press.

Dally JW, and Riley WF (1991) Experimental Stress Analysis, 3rd ed. New York: McGraw-Hill.

Demes B, and Günther MM (1989) Biomechanics and allometric scaling in primate locomotion and morphology. Folia Primatol. *53*:125–141.

Demes B, and Jungers WL (1993) Long bone cross-sectional dimensions, locomotor adaptations and body size in prosimian primates. J. Hum. Evol. *25*:57–74

Demes B, Larson SG, Stern JT Jr., Jungers WL, Biknevicius AR, and Schmitt D (1994) The kinetics of primate quadrupedalism: "hindlimb drive" reconsidered. J. Hum. Evol. *26*:353–374.

Demes B, Jungers WL, Gross TS, and Fleagle JG (1995) Kinetics of leaping primates: Influence of substrate orientation and compliance. Am. J. Phys. Anthropol. *96*:419–429.

Demes B, Stern JT, Rubin CT, Larson SG, and Hausman MR (1997) Bone strain in the macaque ulna during locomotion. Am. J. Phys. Anthropol. Suppl. *24*:101.

Demes B, Stern JT Jr., Hausman MR, Larson SG, McLeod KJ, and Rubin CT (1998) Patterns of strain in the macaque ulna during functional activity. Am. J. Phys. Anthropol. *106*:87–100.

Dove RC, and Adams PH (1964) Experimental Stress Analysis and Motion Measurement. Columbus: C.E. Merrill Pub. Co.

Evans FG (1953) Methods to study the biomechanical significance of bone form. Am. J. Phys. Anthropol. *11*:413–435.

Fleagle JG, Simons EL, and Conroy GC (1975) Ape limb bone from the Oligocene of Egypt. Science *189*:135–137.

Fleagle JG, Stern JT Jr., Jungers WL, Susman RL, Vangor AK, and Wells JP (1981) Climbing: A biomechanical link with brachiation and with bipedalism. Symp. Zool. Soc. Lond. *48*:359–375.

Gross TS, McLeod KJ, and Rubin CT (1992) Characterizing bone strain distributions *in vivo* using three triple rosette strain gauges. J. Biomech. *25*:1081–1087.

Gurdjian ES, and Lissner HR (1944) Mechanism of head injury as studied by the cathode ray oscilloscope. Preliminary report. J. Neurosurgery *1*:393–399.

Harrison T (1989) New postcranial remains of *Victoriapithecus* from the middle Miocene of Kenya. J. Hum. Evol. *18*:3–54.

Heglund NC (1981) A simple design for a force-plate to measure ground reaction forces. J. exp. Biol. *93*:333–338.

Hirasaki E, Matano S, Nakano,Y, and Ishida H (1992) Vertical climbing in *Ateles geoffroyi* and *Macaca fuscata* and its comparative neurological background. In S Matano, R Tuttle, H Ishida, and M Goodman (eds.): Topics in Primatology, Vol. 3. Tokyo: University of Tokyo Press, pp. 167–176.

Hirasaki E, Kumakura H, and Matano S (1993) Kinesiological characteristics of vertical climbing in *Ateles geoffroyi* and *Macaca fuscata*. Folia Primatol. *61*:148–156.

Hylander WL (1977) *In vivo* bone strain in the mandible *of Galago crassicaudatus*. Am. J. Phys. Anthropol. *46*:309–326.

Hylander WL (1979) The functional significance of primate mandibular form. J. Morph. *160*:223–240.

Hylander WL (1984) Stress and strain in the mandibular symphysis of primates: A test of competing hypotheses. Am. J. Phys. Anthropol. *64*:1–46.

Hylander WL, Picq PG, and Johnson KR (1991) Function of the supraorbital region of primates. Archs. Oral Bio. *36*:273–281.

Hylander WL, and Johnson KR (1994) Jaw muscle function and wishboning of the mandible during mastication in macaques and baboons. Am. J. Phys. Anthropol. *94*:523–547.

Ishida H, Jouffroy FK, and Nakano Y (1990) Comparative dynamics of pronograde and upside down horizontal quadrupedalism in the slow loris (*Nycticebus coucang*). In FK Jouffroy, MH Stack, and C Niemitz (eds.): Gravity, Posture and Locomotion in Primates. Firenze: Il Sedicesimo, pp. 209–220.

Kummer B (1970) Die Beanspruchung des Armskeletts beim Hangeln. Anthrop. Anz. *32*:74–82.

Lanyon LE, and Smith RN (1970) Bone strain in the tibia during normal quadrupedal locomotion. Acta orthop. Scandinav. *41*:238–248.

Lanyon LE, and Baggott DG (1976) Mechanical function as an influence on the structure and form of bone. J. Bone Jt. Surg. *58B*:436–443.

Lanyon LE, Hampson WGJ, Goodship AE, and Shah JS (1975) Bone deformation recorded *in vivo* from strain gauges attached to the human tibial shaft. Acta orthop. Scand. *46*:256–268.

Nieschalk U (1991) Fortbewegung und Funktionsmorphologie von *Loris tardigradus* und anderen kleinen quadrupeden Halbaffen in Anpassung an unterschiedliche Habitate. Ph.D. thesis, Ruhr-Universität Bochum.

Nigg BM, and Herzog W (1994) Biomechanics of the Musculo-skeletal System. Chichester: John Wiley and Sons.

Preuschoft H (1985) On the quality and magnitude of mechanical stresses in the locomotor system during rapid movements. Z. Morph. Anthrop. *75*:245–262.

Ramm H, and Wagner W (1967) Praktische Baustatik, Teil 3, 5th ed. Stuttgart: B.G. Teubner.

Richmond BG, Fleagle JG, Kappelman J, and Swisher CC III (1998) First hominoid from the Miocene of Ethiopia and the evolution of the catarrhine elbow. Am. J. Phys. Anthropol. *105*:257–277.

Rose MD (1988) Another look at the anthropoid elbow. J. Hum. Evol. *17*:193–224

Rubin CT, and Lanyon LE (1982) Limb mechanics as a function of speed and gait. J. exp. Biol. *101*:187–211.

Rubin CT, and Lanyon LE (1984) Dynamic strain similarity in vertebrates: An alternative to allometric limb bone scaling. J. Theor. Biol. 107:321–327.

Rubin C, Gross T, Donahue H, Guilak F, and McLeod K (1994) Physical and environmental influences on bone formation. In: CT Brighton, GE Friedlaender, and J M Lane (eds.): Bone Formation and Repair. Am. Acad. Orthopaed. Surgeons, pp. 61–78.

Rybicki EF, Mills EJ, Turner AS, and Simonen FA (1977) *In vivo* and analytical studies of forces and moments in equine long bones. J. Biomech. *10*:701–795.

Schmitt D (1994) Forelimb mechanics as a function of substrate type during quadrupedalism in two anthropoid primates. J. Hum. Evol. *26*:441–457.

Stern JT Jr., Wells JP, Vangor AK, and Fleagle JG (1977) Electromyography of some muscles of the upper limb in *Ateles* and *Lagothrix*. Yrbk. Phys. Anthropol. *20*:498–507.

Swartz SM, Bertram JEA, and Biewener AA (1989) Telemetered *in vivo* strain analysis of locomotor mechanics of brachiating gibbons. Nature *342*:270–272.

Timoshenko S (1958) Strength of Materials. Part II. 3rd edition. New York: D. van Norstand Co.

Yamasaki N, and Ishida H (1984) A biomechanical study of vertical climbing and bipedal walking in gibbons. J. Hum. Evol. *13*:563–571.

Yoshikawa T, Satoshi M, Santiesteban AJ, Sun TC, Hafstad E, Chen J, and Burr DB (1994) The effects of muscle fatigue on bone strain. J. exp. Biol. *188*:217–233.

Young DR, Howard WH, and Orne D (1977) *In-vivo* bone strain telemetry in monkeys (*M. nemestrina*). J. Biomech. Eng. *99*:104–109.

15

THE INFORMATION CONTENT OF MORPHOMETRIC DATA IN PRIMATES

Function, Development, and Evolution

Charles E. Oxnard

Centre for Human Biology
Department of Anatomy and Human Biology
The University of Western Australia
WA, 6009, Australia

1. INTRODUCTION

In the 1965 Wenner Gren Symposium on Primate Locomotion organized by the late Warren Kinzey, one of the contributions was an exposition of the morphometrics of the primate shoulder (Oxnard, 1967). That study clearly demonstrated that the morphometrics of the shoulder arranged the primates on the basis of the way that the shoulder was used in the different species. Since then, morphometric studies of other individual anatomical regions have provided similar results: that species are arranged in line with functional usages of the particular anatomical regions concerned (Oxnard, 1983/1984).

However, investigations in which data sets from different anatomical regions are combined (e.g., combinations of data from several functional units of the postcranium and cranium, even from different components of the niche including behavior, ecology and diet) seem to provide for each combination, a different type of information about species groupings. That is: the groupings of the species that they produce are similar to those that result from studies of molecular evolution. They thus mirror phylogeny rather than function. As the data are the same for both individual and combined analyses, it is necessary to ask why such completely different results are obtained.

One answer may come from asking a theoretical question: what are we actually doing when we combine the information content of data taken from individual anatomical units into larger and larger data sets involving many units, and, in the end, a large portion of the whole animal? A second answer may stem from asking the practical question: how, in arranging the species in the combined studies, are the data variables clustered? We already know that the clusterings of the variables from individual units provide information of biomechanical import that helps explain the reason for the functionally adaptive separations of the species. Is there any information in the clustering of the variables from combined analyses that speaks to the reason for the evolutionary groupings of the species?

2. INFORMATION CONTENT OF LOCOMOTOR DATA: FUNCTION VERSUS PHYLOGENY

2.1. Function

Though the morphometric studies of the shoulder in the 1965 Symposium on Primate Locomotion had seemed to imply primarily locomotor usage (Oxnard, 1967) it was already starting to be recognised (e.g., Ripley, 1967; Oxnard 1967) that posture, feeding, playing, escaping and many other behaviors also utilize those same movements of the shoulder and must also be implicated in shoulder form (later reviewed more fully in Oxnard, 1983/1984).

Since then, morphometric investigations of many other individual anatomical units have been made (both by Oxnard and colleagues, and by other investigators, e.g., McHenry and Corrucini, 1975; Feldesman, 1976, 1979; Corrucini and Ciochon, 1978; Manaster, 1975, 1979; and many others since, also reviewed in Oxnard, 1983/1984). Such studies now cover every individual region of the body, those of Oxnard and colleagues specifically including the shoulder, elbow, wrist, hand, upper limb as a whole, vertebral column, hip, knee, foot, and lower limb as a whole. In every case, the results have mirrored those of the original symposium contribution on the shoulder; that is, they have provided arrangements of the species that seem to relate most obviously to the functions of the relevant anatomical part within overall animal lifestyles.

Thus, for most upper limb units, the non-human primates are arranged in a band shaped spectrum (as shown not only in Oxnard, 1967, 1968, and 1973, but also in Feldesman, 1976, and Corrucini and Ciochon, 1978 as demonstrated in Oxnard, 1983/1984). This spectrum (see figure 8.2 of Oxnard, 1983/84) extends from species that use the upper limbs in highly terrestrial milieux under compression, to those that use the upper limbs in highly arboreal niches largely under tension. Humans, not using the upper limbs at all in locomotion, lie outside the band shaped spectrum, uniquely separate from all other primates (e.g., Oxnard 1975), separate, even, from all other mammals (Oxnard, 1968).

For various lower limb regions, the non-human primates are arranged in a star-shaped spectrum (see also figure 8.2 of Oxnard, 1983/84). In the center of the star lie species from many different taxonomic groups that have in common that they are generalized arboreal quadrupeds. In the different rays of the star are highly specialized species, usually from different taxonomic groups, that are convergent in major aspects of their locomotor hindlimb function. Again, humans, with their unique lower limb dominated locomotion, bipedality, are widely separate from all the non-human primates (e.g., Oxnard, 1975, 1983/1984; Oxnard and Hoyland-Wilkes, 1993; Kidd, 1995; Kidd et al., 1996; Kidd and Oxnard, 1997).

This separateness of humans from non-human primates in the form of the various individual anatomical regions contrasts totally with the unequivocal similarity of humans and African apes in the various combined studies that employ *the same data* as the individual studies. This is a result that concurs with the various molecular findings.

The functional results from the individual anatomical regions are of especial interest for fossil assessment because they are capable of indicating, irrespective of taxonomic assignment, functional adaptation in individual fossil parts. For example, in a series of studies of different anatomical regions from 1966 to 1973 (summarized in 1975) Oxnard showed that *Australopithecus africanus* could not have been simply a biped in the human manner, the usual assessment of that species in those days. He showed it must have had a unique combination of locomotor abilities including both bipedality of a type different from that in modern humans together with arboreal propensities of one kind or another. This has been further evidenced in recent studies of *A. africanus* (e.g., Oxnard and Hoyland-Wilkes, 1993; Kidd et al., 1996, 1997). Yet it has taken many analyses by many other investigators (e.g., Prost, 1980; Tardieu, 1981; Senut, 1981; Feldesman, 1982; Jungers and Stern, 1983; Stern and Susman, 1983a, b, Susman et al., 1983; Schmidt, 1984; Zonnenveldt and Wind, 1985 and many others since that time) who all studied the more recently discovered *A. afarensis,* to gradually insert this idea into the literature. Even so, it has not been until the last few years that evidence in support of the idea of the combined terrestrial and arboreal activities of *A. africanus* (put forward as long ago as Oxnard, 1975) has been presented by some of those most strongly against the original idea. Thus, a very new study of previously undescribed australopithecine foot remains points out that *A. africanus* "while bipedal, was equipped to include arboreal, climbing activities in its locomotor repertoire…it was likely not an obligate terrestrial biped, but rather a facultative biped and climber" (Clarke and Tobias, 1995:524).

2.2. Phylogeny

In contrast to the situation in individual anatomical units (such as the shoulder, hand, hip or foot) morphometric analyses of combinations of anatomical units of the postcranium have not particularly resulted in functional understanding. Study of combinations of anatomical regions (either through [a] combinations of large numbers of specific regions: shoulder, elbow, wrist, hand, hip, knee, ankle, foot (Oxnard, 1983a), or [b] combinations of overall proportions of the entire body: upper limb, lower limb, abdomen, thorax, and head (Oxnard, 1983b), or even most recently [c] combinations of both overall bodily proportions and individual parts of the vertebral column (Milne et al., 1996 and work in progress)) seem all to arrange species in line with evolutionary relationships rather than functional convergences. Let us summarize these findings.

As might be expected of a line of work carried out over more than a thirty year period, the theoretically best combined study (i.e., an analysis of a very large number of variables taken on all units of the body in the same specimens representing the full suite of primate species) is not possible. Over the years, the specimens available for study never included quite the same samples, never quite the same specimens, never quite the same measurements, and so on. Second best sets of combined investigations have been carried out, however, and because these have been done in three quite different ways yet with the same results, considerable confidence can be placed in them.

A first set of studies employed Andrew's high dimensional displays for combining the data from the individual analyses on detailed anatomical units (such as those of the shoulder, the elbow, etc., as detailed above, Oxnard, 1983/1984). Though such analyses do take account of correlations within individual studies, they do not take account of correlations between them and therefore the "between individual studies" information is missing.

A second set of studies analyzed only a reduced number of variables representing aggregated parts of animals (e.g., measures of overall bodily form, Oxnard, 1983/1984) for all specimens and species. In this case, though the analyses take full account of correlations both within and between the individual regions, the data suffer from the deficit that, being only measures of overall bodily regions they do not contain the more detailed information inherent in the measures of the smaller anatomical units.

A third set of studies combines these two types of investigation by looking not only at detailed data from restricted specific anatomical parts (e.g., measures of the vertebrae) but also at their correlations with the cruder but none-the-less very important measures of overall bodily proportions (e.g., Milne et al., 1996, and work in progress). These last studies thus take into account both the information contained in the correlations within and between regions and the more detailed information inherent within and between measures of smaller anatomical units (the various individual vertebrae).

The net result of these various combined studies is almost always groupings of species that relate to evolutionary relationship rather than functional adaptation. Thus, each major group of the primates: lemuriformes, lorisiformes, tarsiiformes, New World monkeys, Old World monkeys, and hominoids are clearly demarcated (top frame of Figure 1 showing the result for the combination of overall body proportions). In a part of that combined study that includes only hominoids, humans are grouped with African apes, and this combined group is clearly separated from Asian apes (figure 8.19 of Oxnard, 1983/4). Such results thus concur with those of molecular evolution. Though based on the same data as the investigations of individual anatomical units, these combined results are in total contrast with the individual studies wherein what is most clearly defined are functional parallels and convergences, and where phylogenetic groups of the primates are almost totally overlapping (e.g., a study of the shoulder: bottom frame of Figure 1).

The situation for all postcranial regions is summarized in Table 1. Whenever analyses are carried out at the individual bone-joint-muscle unit or individual anatomical region level, the information about species separations that is most clearly recognizable is about FUNCTION. There must be evolutionary information within the data from these units, but it is rarely obvious. When, however, analyses are carried out on the same data, but combined to include as much of the whole postcranium as possible, thus combining many individual functional units, then the most obvious information about species separations reflects EVOLUTION. There definitely is functional information within these combined data—we know that this must be so because they contain the data of the individual analyses—but the functional information is hidden when the data are combined.

3. INFORMATION CONTENT OF NON-LOCOMOTOR DATA: FUNCTION VERSUS PHYLOGENY

3.1. Craniodental Data

A similar picture arises from our consideration of studies of the cranium (including the teeth) in a range of primate adults. Thus, from a series of investigations of measurements of the dentition and skull (Oxnard et al., 1985; Oxnard, 1987; Hayes et al., 1990, 1995, 1996; Hayes, 1994; Pan, 1998; Pan et al., in press) univariate, bivariate and multivariate examinations of individual cranial units (e.g., mandibular teeth alone, mandible alone) generally reveal groupings of species that are associated with groupings of functional variables. These variables include features such as grinding areas and cutting lengths of teeth and which are related, as might be expected, to primary dietary components or masticatory function such as degree of omnivory or herbivory. This has also been shown by a host of studies by other investigators over the years.

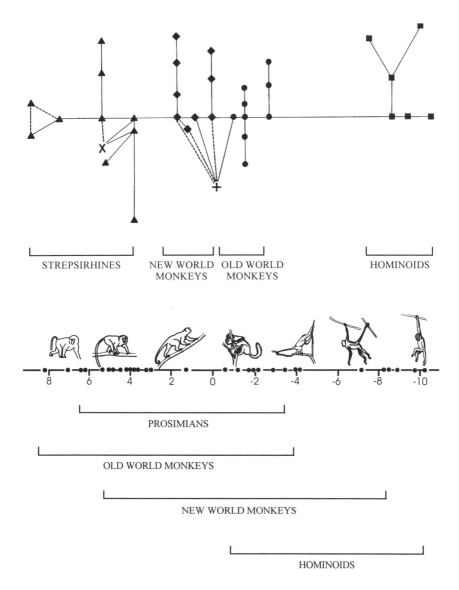

Figure 1. *Top*, an example of the separations of the species in combined studies (in this case the combination of overall body proportions as displayed in a minimum spanning tree of generalized distances - total length of tree more than 40 standard deviation units). The separations seem most to reflect phylogeny. The major groups of primates are indicated. Even the tarsier (+), long grouped as a prosimian and apparently similar in functional abilities to bush-babies (x), is well separated from what are now known as strepsirhines and is clustered closer to anthropoids (as it is by molecular data). *Bottom*, an example of the separations of species in individual studies (in this case an analysis of shoulder dimensions as displayed in the first canonical axis, units in standard deviations). The separations are related mostly to the function of the shoulder. Cartoons display arm-hanging propensities ranging from none (left) to extreme (right). The various major taxonomic groups overlap totally. Both studies employ canonical variates analyses and generalized distances.

Table 1. Summary of information content of morphometric investigations of postcranium

	Anatomical units	Functional relationship		Evolutionary relationship	
		Obvious	Not obvious	Obvious	Not obvious
Individual	Shoulder	x			x
	Arm and forearm	x			x
	Hand	x			x
	Complete upper limb	x			x
	Hip	x			x
	Knee	x			x
	Foot	x			x
	Complete lower limb	x			x
Combined	Shoulder, arm, forearm, hip, thigh, leg, foot		x	x	
	Upper limb, lower limb, thorax, abdomen, head		x	x	

The primary reason for our craniodental studies, however, was to examine data from the individual functional units in combined morphometric analyses to see whether the species were grouped in ways associated with phylogenetic relationships and sexual dimorphism. These combined results show that, indeed, the species are arranged in ways that align with their evolutionary relationships. They show furthermore, that sexual dimorphism is not a single phenomenon, essentially similar in all species except for degree and mainly related to overall size. Sexual dimorphism is, rather, different in each species, different in the various anatomical regions, differently aligned in relation to overall size, and thus far more complex than generally thought. The broad implication is that there are many different sexual dimorphisms and that these must have evolved separately since the origin of the various common ancestors of the groups concerned.

3.2. Data Pertaining to the Niche

Even studies of the primate niche concur with this idea that different kinds of information are inherent in the data, and that it takes different types of analyses to disentangle them. Thus, investigation of individual species defined through data pertaining to the niche gives information about ecological groupings of the species (the niche equivalent of function). Those same data, when applied to species combined into their higher taxonomic groups (e.g., into families) provide information pertaining to evolutionary relationships (Oxnard et al., 1990).

The aspects of the niche that were studied in these investigations include quantitative measures of locomotion, environment and diet. The multivariate statistical methods that are used are identical to those used in the parallel morphometric investigations of these same species (but of course, in these analyses, are termed nichemetrics, Crompton et al., 1987; Oxnard et al., 1990).

The results of the study of individual species show that the various niche variables (relating to locomotor activities, to the environment within which the activities are carried out, and to the dietary items that are garnered within the environment and obtained by the activities) arrange the species into groups that make locomotor, environmental and dietary sense. For example, several different groups of species are defined in relation to associations between (a) various locomotor activities (such as richochetal leaping, scurrying, and slow climbing), (b) various environmental loci (such as small branch undergrowth, main branch highways, and canopy), and (c) various dietary items (such as leaves, fruits, and animal products). Though not functional in the sense of functional morphology, these groups are extremely "functional" in the sense of ecological adaptation.

The results of the nichemetric studies are, moreover, closely concordant with the results of individual morphometric studies on the anatomies of the same series of species (compare figures 6.1 through 6.6 in Oxnard et al., 1990). This is what might have been expected; after all, it is these morphologies that exist in these niches.

The aforementioned lines of argument, however, have forced us to look further into the nichemetric studies to see if additional information might lie hidden in the associations. First, studies were undertaken to see if the clusterings of the variables in the study of individual species also made ecological or "niche" sense in a way similar to that in which clusterings of morphological variables made functional or biomechanical sense in the individual morphometric studies. This is easily seen to be so. The original publication (figure 5.12 of Oxnard et al., 1990) indicates that the variable clusters are not random or undecipherable. Thus, the locomotor variables and environmental variables are placed in each half of a circular part of the analytical space. In contrast, lying on the periphery of the circle, are each of the dietary variables. Each of these is, individually, about as far distant from each of the others as it could possibly be.

These variable relationships have now been examined more closely to see if they make ecological sense. Within the overall picture described above, smaller variable neighborhoods exist. Figures 2 and 3 show two examples. In each case a localized neighborhood of the variables makes sense in relation to the ecological adaptations outlined by the clustering of the species. Thus one neighborhood of variables (labeled at the upper part of the plot, Figure 2) includes fruit eating, living in the canopy, using many horizontal supports, and climbing. This association of variables relates to the species grouping of most bushbabies but not the two extreme leaping bushbabies that are associated more closely with the extreme leaping tarsiers. Another neighborhood of variables (Figure 3) includes leaf-eating, living on large supports, and extreme leaping (labeled on the left hand side of the plot, Figure 3). This association of variables relates to the species grouping of the various extreme leaping indriids together with *Lepilemur* and *Hapalemur*. Other variable neighborhoods make similar adaptive sense. It is evident, then, that the information content of these data not only identify niche groups of species (as in Oxnard et al., 1990) but also niche groups of variables.

This leads to the question: is there any way of examining these nichemetric data that might give information relating to evolution? Thus, though one of the niche groups contains only lorisines (all, therefore, in the same evolutionary group) several of the other niche groups contain species that are not in the same evolutionary group: for example, the grouping of the extreme leaping tarsiers along with two of the more extreme leaping bushbabies, and the grouping of indriids together with *Lepilemur* and *Hapalemur* as a second form of extreme leaping. This implies that the groups are due to adaptive convergences (niche groups) and not to evolutionary relationship.

The way in which the data cluster species into families was therefore examined. The result, given the information content about niche groups in the data, achieves some very interesting separations (Figure 4). Thus, in spite of the extreme differences in a niche sense between lorisines (mostly slow climbing insectivorous creatures) and galagines (mostly fast running and leaping frugivorous forms), this analysis achieves a closely coherent grouping of both lorisines and galagines (Lorisidae) that is widely separated from all the other species. Similarly, in spite of the marked adaptive similarities between all tarsiers and some of the bushbabies, this analysis achieves complete and utter separation of all tarsiers (Tarsiidae) from all bushbabies that are unequivocally placed together with

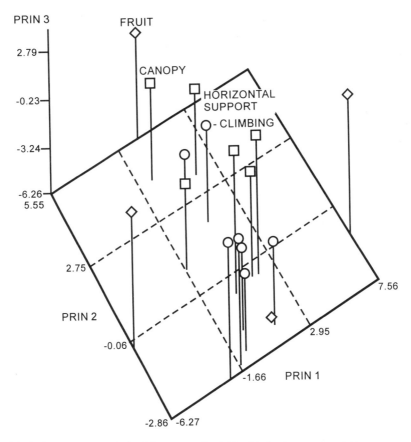

Figure 2. Principal components analysis of the groups of variables (circles, locomotor variables; squares, environmental variables; diamonds, dietary variables) responsible for the niche group separations of the species in the nichemetric studies of prosimians. A local neighborhood of specific variables is identified.

lorisines in the Lorisidae. Yet other separations of major evolutionary interest are readily evident. One of special interest is the separation of the single species *Daubentonia* as the family Daubentoniidae, from all other lemurs, especially indriids with which it has often been associated in an evolutionary sense (e.g., Tattersall and Schwarz, 1975). This melds very well with the previously determined unique morphometric separation of *Daubentonia* from all other strepsirhines (Oxnard, 1983/1984). In other words, though adaptive information is clearly present in these data, so too is information of evolutionary import. Asking the right question of the data is necessary to allow it to appear.

It might have been sufficient to have assumed that the reason for the various functional (locomotor, masticatory, ecological) arrangements of species in the individual studies is because the evolutionary information content within individual elements is swamped by the functionally adaptive parts of that element's form or make-up. Likewise, it might have been simply assessed that the reason for evolutionary information being apparently inherent in combinations of many elements might be because, whereas functionally adaptive information would not summate because function is different in each region, evolutionary information, being the same in each element, of course, would summate and swamp the different separate functional convergences (Oxnard, 1991, 1992). Only in a

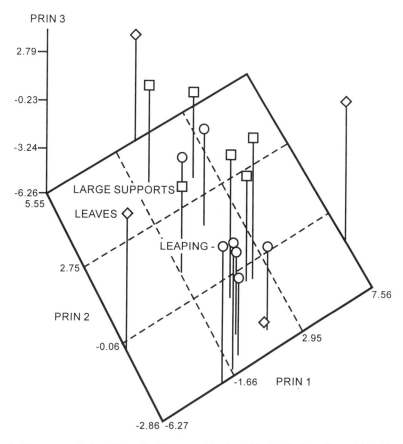

Figure 3. The same analysis as Figure 2 with a second local neighborhood of specific variables identified.

few situations where functional convergence is extreme (e.g., the similarities between bushbabies and tarsiers being dependent upon highly convergent leaping) might this not always occur (e.g., in the study of overall body proportions) and even here, the functional convergence can, in fact, be separated out (Oxnard, 1978). In an attempt to further explicate these findings, however, we can ask two further questions; first: how, in terms of theoretical thinking, might the information content of variables be arranged in combined as compared with individual studies; second, how is the information content of the variables in the combined investigations actually arranged?

4. THEORETICAL INFORMATION CONTENT OF THE DATA SETS

The question above, relating to theoretical studies of information content, can be discussed rather simply in the following manner. Let us assume that any individual morphometric analysis of an individual functional unit contains information that is partly about functional adaptation and partly about evolutionary relatedness. This means we can think of the analysis as revealing mainly functional information (f), say 5 parts, but never-

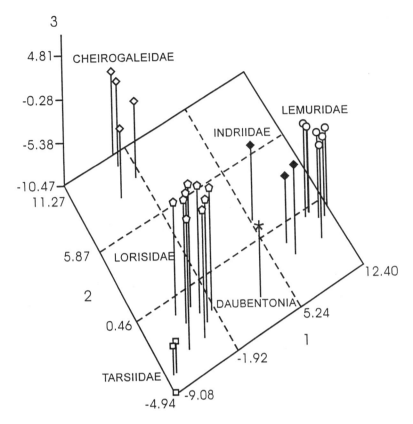

Figure 4. The arrangements of the higher taxonomic groups (families) in a canonical variates analysis (standard deviation units in each axis) of the nichemetric data.

theless also containing (if not so obviously) a lesser degree of evolutionary information (e), say 3 parts. Of course, these phenomena are really continuous; I am describing them as discrete bits to simplify matters.

The information in such an analysis might then be written as

$$= \mathbf{f1} + \mathbf{f2} + \mathbf{f3} + \mathbf{f4} + \mathbf{f5} + e1 + e2 + e3$$
$$= \text{a total of 8 units of information}$$

where the "**f**"s are overt function and the "e"s are covert evolution. This implies that "**f**" (biological function) appears as 5/8 of the total information and "e" (evolution) only 3/8 (this lesser amount may be partly why it is less easily recognised). I have used the bold format to designate the more easily recognizable portions.

Yet the above is too simplistic. It is likely that some of the evolutionary information will also be "similar to" (i.e., correlated with) some of the functional information. After all, functional adaptation is a major part of evolutionary diversity. This is the equivalent of saying that there will inevitably be at least some interactions between function and evolution. Let us assume that 2 of the "f"s and 2 of the "e"s interact. Then the information in the analysis might be rewritten as:

$$= \mathbf{f1} + \mathbf{f2} + \mathbf{f3} + \mathbf{f4}(=e) + \mathbf{f5}(=e) + e1(=\mathbf{f}) + e2(=\mathbf{f}) + e3$$
$$= \text{again, a total of 8 units of information.}$$

In this case, however, the easily identifiable biological function, **f**, appears to be a greater proportion of the total information, 7/8, exactly because it **is** readily identifiable. In contrast, the evolutionary information (e), although appearing in 5 bits out of 8 in our example, has only 1 bit (e3) in which it is clearly different from **f**. The interactive portions (f4[=e], f5[=e], e1[=f] and e2[=f]) will all be seen as **f** because of the interaction. Thus, in an individual functional unit where there are interactions between function and evolution, it is easy to see why the smaller part—evolution—may be obscured, and why the larger portion—function—may appear extremely large indeed.

If, now, we had a series of such analyses taken on different functional units (a, b, c, d, and e) the following exposition shows how, though each individual unit might greatly emphasize function, a combined analysis of all units together might sum to something different.

Thus, each individual unit analysis might give an equation like that above, so that the entire suite of analyses might look like the following:

Unit a $= \mathbf{f1a} + \mathbf{f2a} + \mathbf{f3a} + \mathbf{f4a}(=ea) + \mathbf{f5a}(=ea) + e1a(=\mathbf{fa}) + e2a(=\mathbf{fa}) + e3a$
Unit b $= \mathbf{f1b} + \mathbf{f2b} + \mathbf{f3b} + \mathbf{f4b}(=eb) + \mathbf{f5b}(=eb) + e1b(=\mathbf{fb}) + e2b(=\mathbf{fb}) + e3b$
Unit c $= \mathbf{f1c} + \mathbf{f2c} + \mathbf{f3c} + \mathbf{f4c}(=ec) + \mathbf{f5c}(=ec) + e1c(=\mathbf{fc}) + e2c(=\mathbf{fc}) + e3c$
Unit d $= \mathbf{f1d} + \mathbf{f2d} + \mathbf{f3d} + \mathbf{f4d}(=ed) + \mathbf{f5d}(=ed) + e1d(=\mathbf{fd}) + e2d(=\mathbf{fd}) + e3d$
and
Unit e $= \mathbf{f1e} + \mathbf{f2e} + \mathbf{f3e} + \mathbf{f4e}(=ee) + \mathbf{f5e}(=ee) + e1e(=\mathbf{fe}) + e2e(=\mathbf{fe}) + e3e.$

For each of these units, **f** appears to be 7/8 of the information even though f and e are actually split 5/8 and 3/8 respectively.

When, however, we add the data for each individual unit into a single combined-analysis of all units, the totals could look very different. First, the various "f"'s cannot be expected to sum beyond a single individual analysis because the functions in each unit are different. Second, in contrast, the various "e"'s (including those "e"'s that are related to "f"'s) in the individual analyses can be expected to sum in the combined analysis because the information about evolution should be the same for each unit (they are all parts of the same animal). Accordingly then, the total "**e**"'s are 25 (5 for each equation and "**e**" can now be written bold) but no single "f" is any greater than 3 (e.g., 3 "fa"'s, 3"fb"'s, 3"fc"'s and so on). Thus "**e**" (evolution) now shines out strongly at 25/40; no single "f" (function) shines out more strongly than 3/40, even though different functions total 25/40 (i.e., there is just as much functional information present in the combined analysis as in the total of individual analyses).

This line of thinking could, alone, be the reason why function is clearly evident in individual analyses, but evolution in combined analyses. There are also practical reasons, however, why the change from function to evolution might occur as individual analyses are combined. These latter reasons are not mutually exclusive with the above theoretical line of argument.

5. PRACTICAL INFORMATION CONTENT OF THE DATA SETS

Practical studies of information content depend not only on how, but on why the variables in the aggregated studies are clustered. We already know that the variables in the individual studies, in producing the functional arrangements of the species, do so through variable clusters that make functional sense and that can be shown to be related to biomechanics. What information (if any) is provided by the way the variables are clustered in the combined studies so that they produce the evolutionary relationships of the species? Are these variables, in producing the evolutionary arrangements of the species, clustered in any way that makes evolutionary sense and, if so, are the clusters related to underlying mechanisms in evolution?

This is a much more pertinent question now than it was in 1965 because much more is now known about the underlying genetic and developmental mechanisms that are responsible for adult form. This is certainly the case for the postcranium: i.e., for the limbs and trunk, experimental grafting studies in bird embryos (e.g., Wolpert et al., 1975; Wolpert and Hornbruch, 1992; Richardson et al., 1990), murine homeobox gene studies in limbs (e.g., Dolle et al., 1989) and homeobox genes for trunk axis in frogs (e.g., Ruizi and Melton, 1990). It is also true for the cranium and jaws depending upon studies of neural crest cell populations giving rise, for instance, to lower jaw morphogenetic units (e.g., Hanken and Hall 1993; Atchley, 1993; Jacobson, 1993). It could be true for studies of the niche, though as far as I am aware no underlying developmental investigations (these would be very difficult) have so far been carried out in this last arena.

A confluence of function, genetics, development and evolution is certainly the underlying biological determinant of animal structural and functional diversity. Developmental biologists, in recent years, have been able to look from the gene, through a cascade of developmental processes and a series of timing events, towards the understanding of the structure of the adult in exemplar (experimental) species (such as fruit flies, chicks and quail, rats and mice). It should equally be the case that, in the opposite vein, evolutionary biologists might be able to look from comparative studies of adult diversity, through the clusterings of variables, to the underlying genetic and developmental processes that produced them. In other words, the genetic-developmental studies and quantitative whole-organism studies should be able, each in their own right, to provide predictive hypotheses for testing by the other. In this way, *Primate Locomotion: Recent Advances*, may have a rather unexpected descendant from the *Primate Locomotion Symposium* in 1965.

5.1. Clustering of Variables in Combined Studies

5.1.1. Postcranium. The analyses of combined variables for the body overall, which clearly give evolutionary arrangements of the species (e.g., figures 8.39 and 8.40, Oxnard, 1983/84 and top frame of Figure 1), provide completely different clusters of variables to those mechanically obvious clusters of variables evident in the individual analyses of body regions (e.g., figures 7.1 and 7.2, Oxnard 1983/84 and bottom frame of Figure 1). Yet the new clusters of variables in the combined studies are not randomly arranged or uninterpretable. One cluster comprises relative lengths of all major segments of limbs and a second cluster includes relative lengths of all elements of all rays of the hands and feet. The first can also be described as a cluster of all measures of proximodistal elements of the limbs, the second as a cluster of all measures of craniocaudal elements of the only skeletal parts in this study (the cheiridia) that clearly display craniocaudal arrays. Further investigation of this combined study indicates a special variable cluster comprising all measures of the fourth manual digit (but not the fourth pedal digit), a feature related to the difference between strepsirhines and anthropoids (Table 2).

The first two clusters of variables mentioned above bear remarkable similarity to two of the developmental processes responsible for limb form. Thus, Wolpert et al. (1975) clearly demonstrated: first, a proximodistal gradient in limb segment formation and sec-

Table 2. Relationships between clusters of limb variables in diverse species of primates and developmental processes in experimental non-primates

Variable clusters	Anatomical feature	Morphological descriptor	Developmental process
1	Lengths of major segments of limbs	Measures of proximodistal elements	Proximodistal gradient
2	Lengths of all elements of hand and foot	Craniocaudal measures of cheiridia	Craniocaudal gradient
3	(not studied)	(Prediction 1)[1]	Dorsoventral gradient
4	Lengths of all parts of manual digit 4	Strepsirhine feature cf. anthropoids	(Prediction 2)[2]

[1] If morphometric studies had included dorsoventral variables, would they have formed a cluster equivalent to the dorsoventral gradient in development?
[2] If developmental studies had included strepsirhines, would there have been a special developmental process relevant to manual digit four?

ond, a craniocaudal differentiation in digit formation (Figure 5 and 6). Though the experiments were carried out on the bird wing, it is not in doubt that these principles apply to all tetrapod limbs. Further developmental studies of gradients in limb formation (e.g., Richardson et al., 1990) and studies of homeobox genes responsible for the cascade of processes involved in limb formation (Dolle et al., 1989) suggest the existence of a third, dorsoventral, spirally arranged developmental mechanism (Figure 6).

Two questions thus arise. First, if the overall studies of primate morphological diversity had included measures of a dorsoventral spiral arrangement, would they have formed a single cluster and if so, would this have been because of the underlying developmental mechanism? Unfortunately we were not percipient enough to think of that at the time. We will have to go back to obtain those data. Second, if the developmental biologists were able to carry out their studies on primates, would they find some developmental factor related specially to the fourth manual digit. This, too, has not been done and would be very difficult to do in the same manner as the prior experimental studies on birds. But, investi-

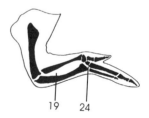

19   24

Figure 5. A summary of the grafting experiments that demonstrate the existence of a proximodistal developmental process in the bird wing (e.g., Wolpert and colleagues, 1975). *Top*, the elements of the adult bird wing labeled with two of the embryological stages responsible for them (stage 19 for the humerus, radius and ulna; stage 24 for the cheiridia). *Middle*, the effect, on final wing development, of grafting a stage 19 wing tip onto a limb bud with stage 24 removed. Stage 19 is thus represented twice in a proximodistal sequence and as a result the humerus, radius and ulna are formed twice over in a proximodistal relationship. Finger development is unaffected. *Bottom*, the effect, on final wing development, of grafting a stage 24 wing tip onto a bud with stage 19 tip removed. Stage 19 and 24 of the recipient are thus not represented and only the fingers are formed in the subsequent adult from the stage 24 of the donor.

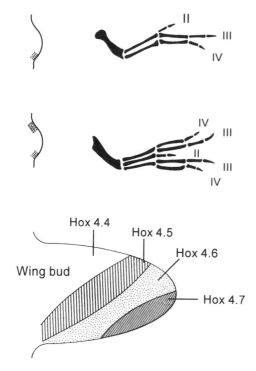

Figure 6. A summary of the grafting experiments and genetic determinations that demonstrate the existence of craniocaudal and dorsoventral developmental processes in the bird wing (Richardson et al., 1990; Dolle et al., 1989). *Top*, normal limb development in the chick embryo with caudal portion of limb bud responsible for initiating the craniocaudal arrangements of the 3 digits (II–IV) in the adult. *Middle*, grafting a portion of the caudal limb bud into a cranial position duplicates the digits in the adult in reverse order thus implying a craniocaudal process. *Bottom*, indicates the dorsoventrally rotated structure of limb materials controlled by the various homeobox genes (HOX 4.4–4.7) of the limb buds.

gations of data about the differential development of the fourth digits of hands and feet in strepsirhine and anthropoid fetuses might allow equivalent information to be obtained.

The existence of the concordances between the developmental experiments (proximodistal gradient and craniocaudal differentiation) and the whole-organism observations (proximodistal measures as separate from craniocaudal measures) is unlikely to be coincidence. It is highly likely that developmental factors determined from experiments on embryos in two or three experimental species would, if truly generalizable, have effects in the adult that would also be revealed in comparisons of the ultimate morphology of adults of large numbers of related species. It is also fascinating that each type of study (process in embryonic development—dorsoventral organization, and quantitative diversity of adult form—special arrangements of manual digit four) should be able to provide predictions for the other (Table 2).

5.1.2. Cranium Including Teeth. A similar question has been asked of the combined cranial and odontometric data in the various investigations. That is: in producing sexual and species separations that seem to be related to evolution rather than function (Oxnard, 1987) how are the variables clustered? Let us look further into that analysis.

The results seem to indicate that the variables are not clustered randomly or in noninterpretable ways. In fact, two of the variable clusters comprise the combined length dimensions of (a) both mandibular incisors and (b) all mandibular molars (Oxnard, 1987). Though these are variable clusters of lengths of teeth, they reflect the lengths of the relevant portions of the mandibular alveolar bone: (a) its incisor part, (b) its molar part. In other studies involving measures of the bones themselves (Pan, 1998) two equivalent clusters of variables have been identified: (a) one of measures at the anterior end of the jaw (the incisor part), the other of measures posteriorly (the molar part).

These two pairs of variable clusters are remarkably concordant with the findings of developmental and molecular biologists in studies of the morphogenesis of the mandible in experimental animals (e.g., Atchley, 1993). Those investigations imply that the rat jaw arises from a series of osteogenic cell populations derived from the first neural crest segment (Figure 7 and top frame of Figure 8). In addition to incisor alveolar and molar alveolar cell populations, these studies also identify populations of cells for the ramus, Meckel's cartilage, and the coronoid and angular [muscular] and condyloid [joint] elements.

Even further back in development a set of homeobox genes controls the cascade of processes from these cell populations, through the developmental units, to the eventual adult morphological components (Figure 7). Though this work has been done primarily in the rat, it is highly likely that a similar set of processes and stages is involved in all mammalian mandibles. Is it possible that the first two morphometric clusters of variables (Table 3) are discernable exactly because they are adult measures reflecting the fundamental underlying developmental alveolar components?

There are two further clusters of variables in the primate studies: (a) one of these is the lengths of all canines and (b) a second is all dimensions of the premolars (Table 3). Of course, rats do not have canines and premolars. Developmental components related to these areas could, therefore, not be recognised through experimental studies in the rat. If,

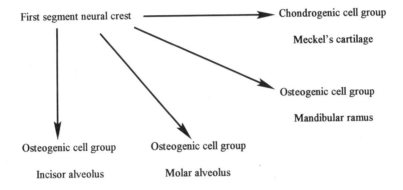

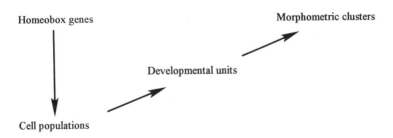

Figure 7. *Top*, the various cell groups arising from the first segment of the neural crest and giving rise eventually to individual components of the mandible. *Bottom*, the cascade of processes controlled by homeobox genes responsible for each of the anatomical units.

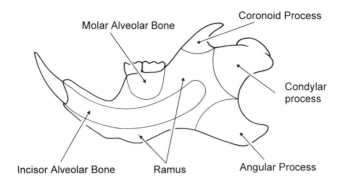

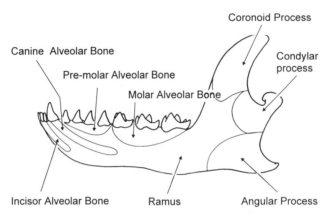

Figure 8. *Top*, the components of the mandible of the rat that are determined through various genetic mechanisms and developmental processes (after Atchley, 1993). *Bottom*, the presumed components that might be identified if similar experiments were carried out on a primate or mammal with all tooth types present.

however, the developmental studies had been carried out on a species with a full complement of types of teeth, would these additional mandibular elements (cell populations, mandibular units, adult mandibular components) have been identified? This might be represented as in the bottom frame of Figure 8, which suggests what might be the molecular

Table 3. Relationships between clusters of dental variables in morphometric studies of primate species and of developmental processes in experimental animals (rats)

Variable clusters	Anatomical feature	Morphological descriptor	Developmental process
1	Lengths of incisors	Incisor portion of jaw	Incisor alveolar bone
2	Lengths of canine	Canine portion of jaw	(Prediction 1)[1]
3	Lengths of premolars	Premolar portion of jaw	(Prediction 2)[2]
4	Lengths of molars	Molar portion of jaw	Molar alveolar bone

[1] If developmental studies could be done on mammalian species with no missing teeth, would individual cell populations, developmental units, and mandibular components still exist, or would these populations, units and components be perceived as a spectrum?

[2] If developmental studies covered an experimental primate, would separate and additional premolar cell populations, developmental units, and mandibular components exist?

and development units in, for example, a primate or other mammal with a larger complement of tooth types. This prediction from studies of whole-organism morphology is thus offered to experimental developmental biologists.

5.1.3. Niche. Although there is no developmental information yet available for the niche, it seemed worth looking at the nichemetric studies to see how the variables were clustered in that part of the study that seemed to provide information about the relatedness of species. This is the analysis (Figure 4) that separates the major groups of prosimians in a way that seems related to evolution: e.g., the clustering together of lorisines and galagines (as lorisids), the separation of tarsiers in their own group (tarsiids), and the extreme separation of *Daubentonia* from all other species (into its own family).

This particular analysis of the niche data provides the arrangements of the variables shown in Table 4. The major information that these variable clusters seem to contain relates to six evolutionary groupings and separations of taxa as described in the third column of the table.

Though these separations involve, of course, individual variables that speak to functional (ecological) adaptation, the particular ways in which the variables are combined speak to phylogenetic separations. Thus, these results confirm long established phylogenetic relationships, such as the links between bushbabies and the various lorises; they provide information relating to other relationships that have been judged equivocal in the past although generally settled now, e.g., the question of the separation of tarsiids from strepsirhines. If this is true, then other parts of the data may be evidence that can be used to test hypotheses that are not yet settled; e.g., though some workers believe that aye-ayes are merely very close relatives of indriids (e.g., Tattersall and Schwarz, 1975), these results for the Daubentoniidae are further data supporting a view that see the aye-ayes as being so different from indriids (e.g., Oxnard 1983/84) that it is most likely that they have been long separate from them.

In this case, links to developmental phenomena are not evident as they were in the morphometric studies. That may be, however, because, at least at the present time, we do not know anything about possible "developmental processes" for the behavioral-environmental-dietary complex. That these clusters of variables differ from the functional "ecological" clusters is not, however, in doubt. That they are related to phylogeny of species groups rather than ecology of individual species seems clear.

Table 4. Relationships between clusters of variables and evolutionary groups of species

Canonical axis	Variable clusters	Variables involved	Evolutionary relationships
Axis 3	loadings in 3 variables	falling, crouching, richochetal leaps, animal diet	1. separates **tarsiids** from all other groups.
Axis 2	high positive loadings in 3 variables	falling leaps, scurrying, fruit eating	2. unifies **cheirogaleids**
	high negative loadings in 3 other variables	undergrowth, vertical supports, animal diet	3. separates **cheirogaleids** from **lorisids**
Axis 1	high positive loadings in 4 variables	slow quadrupedalism, large supports, fruit, leaves	4. separates **lemurids** from all other groups; 5. clusters **galagines** and **lorisines** as **lorisids**
	high negative loadings in 4 other variables	scurrying, undergrowth, small supports, animal diet	6. separates **galagines** from **tarsiids**

Table 5. Summary of information content of all studies

Region	Information content of	Separates	Clusters
Postcranium	Individual units	Species, in relation to function (locomotion)	Variables, such as those related to biomechanics
	All regions combined	Species, in relation to evolution	Variables, such as genetic and developmental processes
Cranium and teeth	Individual units	Species, in relation to function (masticatory)	Variables, such as those with masticatory significance
	All regions combined	Sexes and species, in relation to evolution	Variables, such as genetic and developmental processes
Niche	Individual units	Species, in relation to the niche	Variables, reflecting ecology
	All units combined	Families, in relation to evolution	Variables, by removing convergences

6. SUMMARY

As a result of these investigations, therefore, a summary plan can now be drawn up about the information content of the various analyses (Table 5).

In terms of regional form and function, Table 5 documents that separations of species relate mostly to the functional milieux within which individual anatomical regions operate. Additional proof that this is so is that the clustering of variables seems to relate to the anatomical adaptations pertinent to those functional milieux.

In contrast, in terms of the whole organism (the same sets of variables but combined) Table 5 demonstrates that the separations of species are most closely linked to what is known about evolution (and provides information useful for evolutionary controversies such as the positions of tarsiers and aye-ayes). Additional proof that this is so is that the clusterings of the variables are most closely allied with the genetic and developmental processes that underpin whole organism structure and its diversity.

These investigations also speak especially to the situation of humans among the primates. In terms of form in individual regions, humans are generally quite different from other primates (e.g., in the upper and lower limbs, figures 7.1 and 7.2 in Oxnard, 1983/84). This seems to be because of regional anatomical adaptations to the totally new functional and behavioral milieux that humans have come to inhabit.

But in terms of overall form, the functional uniqueness of humans comes to be appropriately buried in close overall similarities with the African apes (e.g., figure 8.19 of Oxnard, 1983/84). This picture reflects those common molecular, genetic and developmental phenomena that apply as much to humans as they do to African apes. Perhaps this is as close as we can get, with whole-organism investigations, in resolving the paradox of the human position. The studies require us to accept and meld two different views of humans at one and the same time.

ACKNOWLEDGMENTS

I am grateful to many colleagues and students who, over the years, have participated in these investigations. In addition, I am indebted to my graduate students for permission to describe the findings in their doctoral dissertations. These individuals are all cited in the text.

The studies could not have been carried out without the help of a number of museums on three continents and the different curators of the collections that were used (Powell Cotton Museum, Birchington, UK, British Museum of Natural History, London, UK, Field Museum, Chicago, USA, Los Angeles County Museum, Los Angeles, USA, Western Australian Museum, Perth, Australia).

I am especially indebted to Professor F. P. Lisowski and Dr. Len Freedman for comments upon this and related scripts, and for broader discussions of primate morphology and evolution. I also thank three anonymous reviewers and Elizabeth Strasser for their helpful comments on the manuscript.

Most of all I wish to document that these ideas have arisen through a series of discussions with various morphological, genetic and developmental scientists: especially Lewis Wolpert, Brian Hall and Paul O'Higgins. The ideas first arose in the preparation of *The Order of Man* (1983/1984). They were especially stimulated by an Australian Academy of Science Discussion Meeting held in Sydney, 1992, in commemoration of the late Professor N. W. G. Macintosh. They were further explored in preparation of the Key Note Lecture for the Primate Locomotion, 1995, Symposium in commemoration of the late Professor Warren Kinzey. They are currently the topic of discussions and collaborations with colleagues in the Department of Anatomy and Developmental Biology, University College, London.

The overall research program is supported by several grants from the Australian Research Council and by the Centre for Human Biology, UWA. The final stages were been greatly aided by my appointment in 1995 as Australian Academy of Science Visiting Scholar to the Academia Sinica, Kunming Institute of Zoology, PRC, and in 1996 as Visiting Professor in the Department of Anatomy, University of Hong Kong. Professor Y. C. Wong, Head of Anatomy in Hong Kong, is especially thanked for his support during that extended period.

REFERENCES

Atchley WR (1993) Genetic and developmental aspects of variability in the mammalian mandible. In J Hanken and BJ Hall (eds.): The Skull, Volume 1: Development. Chicago: The University of Chicago Press, pp 207–247.

Clark RJ, and Tobias PV (1995) Sterkfontein Member 2 foot bones of the oldest South African hominid. Science *269*:521–524.

Corrucini RS, and Ciochon RL (1978) Morphometric affinities of the human shoulder. Am. J. Phys. Anthropol. *45*:19–38.

Crompton RH, Lieberman SS, and Oxnard CE (1987) Morphometrics and nichemetrics in prosimian locomotion. An approach to measuring locomotion, habitat and diet. Am. J. Phys. Anthropol. *73*:149–177.

Dolle P, Izpisua-Belmonte J-C, Falkenstein H, Rennucci A, and Douboule D (1989) Coordinate expression of the murine *Hox-5* complex homeobox-containing genes during limb pattern formation. Nature *342*:767–769.

Feldesman M (1976) The primate forelimb: A morphometric study of locomotor diversity. University of Oregon Anthropological Papers *10*:1–154.

Feldesman M (1979) Further morphometric studies of the ulna from the Omo Basin, Ethiopia. Am. J. Phys. Anthropol. *51*:409–416.

Feldesman M (1982) Morphometric analysis of the distal humerus of some Cenozoic catarrhines: The late divergence hypothesis revisited. Am. J. Phys. Anthropol. *59*:73–76.

Hanken J, and Hall BJ, eds. (1993) The Skull, Vol. 1, Development. Chicago: University of Chicago Press.

Hayes V (1994) Sexual dimorphism in the dentition of African Old World monkeys. Ph.D. Dissertation, University of Western Australia.

Hayes V, Freedman L, and Oxnard CE (1990) Taxonomy of savannah baboons: An odontomorphometric approach. Am. J. Primatol. *22*:171–190.

Hayes V, Freedman L, and Oxnard CE (1995) The differential expression of dental sexual dimorphism in subspecies of *Colobus guereza*. Int. J. Primatol. *17*:971–996.

Hayes V, Freedman L, and Oxnard CE (1996) Dental sexual dimorphism and morphology in African colobus monkeys. Int. J. Primatol. *17*:725–757.

Jacobson AG (1993) Somitomeres: Mesodermal segments in the head and trunk. In J Hanken and BJ Hall (eds.): The Skull, Volume 1: Development. Chicago: The University of Chicago Press, pp 42–76.

Jungers WL, and Stern JT Jr. (1983) Telemetered electromyography of forelimb muscle chains in gibbons (*Hylobates lar*). Science *206*:617–619.

Kidd RS (1995) An investigation into the patterns of morphological separation in the proximal tarsus of selected human groups, apes and fossils: A morphometric analysis. Ph.D. Dissertation, University of Western Australia.

Kidd RS, O'Higgins P, and Oxnard CE (1996) The OH8 foot: A reappraisal of the functional morphology of the hindfoot using a multivariate analysis. J. Hum. Evol. *31*:269–291.

Kidd RS, O'Higgins P, and Oxnard CE (1997) Patterns of morphological discrimination in the human talus: A consideration of the case for negative function. Perspect. Human Biol. *3*:57–69.

Lieberman SS, Gelvin BR, and Oxnard CE (1985) Dental sexual dimorphisms in some extant hominoids and ramapithecines from China: A quantitative approach. Am. J. Primatol. *9*:305–326.

McHenry HM, and Corrucini RL (1975) Multivariate analysis of early hominoid pelvic bones. Am. J. Phys. Anthropol. *46*:263–270.

Manaster BJM (1975) Locomotor adaptations within the *Cercopithecus, Cercocebus* and *Presbytis* genera: A multivariate approach. Ph.D. Dissertation, The University of Chicago.

Manaster BJM (1979) Locomotor adaptations within the *Cercopithecus* genus: A multivariate approach. Am. J. Phys. Anthropol. *50*:169–182.

Milne N, O'Higgins P, and Oxnard CE (1996) Metameric variation in the vertebral column of hominoids. Am. J. Phys. Anthropol., Suppl. *22*:170.

Oxnard CE (1967) The functional morphology of the primate shoulder as revealed by comparative anatomical, osteometric and discriminant function techniques. Am. J. Phys. Anthropol. *26*:219–240.

Oxnard CE (1968) The architecture of the shoulder in some mammals. J. Morph. *126*:249–290.

Oxnard CE (1973) Form and Pattern in Human Evolution: Some mathematical, physical and engineering approaches. Chicago: The University of Chicago Press.

Oxnard CE (1975) Uniqueness and Diversity in Human Evolution: Morphometric studies of australopithecines. Chicago: The University of Chicago Press.

Oxnard CE (1978) The problem of convergence and the place of *Tarsius* in primate phylogeny. In DJ Chivers and KA Joysey (eds.): Recent Advances in Primatology, Vol. 3, Evolution. London: Academic Press, pp 239–247.

Oxnard CE (1983a) Anatomical, biomolecular and morphometric views of the living primates. In RJ Harrison and V Navaratnam (eds.): Progress in Anatomy. Cambridge: Cambridge University Press, pp. 113–142.

Oxnard CE (1983b) Sexual dimorphisms in the overall proportions of primates. Am. J. Primatol. *4*:1–22.

Oxnard CE (1983/84) The Order of Man: A biomathematical anatomy of the primates. Hong Kong: Hong Kong University Press (1983), New Haven: Yale University Press (1984).

Oxnard CE (1987) Fossils, Teeth and Sex: New perspectives on human evolution. Hong Kong: Hong Kong University Press, Seattle: Washington University Press.

Oxnard CE (1991) Anatomies and lifestyles, morphometrics and niche metrics: Tools for studying primate evolution. J. Hum. Evol. *6*:97–115.

Oxnard CE (1992) Developmental processes and evolutionary diversity: Some factors underlying form in primates. Archaeol. in Oceania, *27*:95–104.

Oxnard CE, Crompton RH, and Lieberman SS (1990) Animal Lifestyles and Anatomies: The case of the prosimian primates. Seattle: Washington University Press.

Oxnard CE, and Hoyland-Wilkes C (1993) Hominid bipedalism: The pelvic evidence. Persp. Hum. Biol. *No. 4*:13–34.

Oxnard CE, Lieberman SS, and Gelvin B (1985) Sexual dimorphism in dental dimensions of higher primates. Am. J. Primatol. *8*:127–152.

Pan R-L (1998) A morphometric approach to the skull of macaques: Implications for *Macaca arctoides* and *M. thibetana*. Ph.D. Dissertation, University of Western Australia, Perth.

Pan R-L, Jablonski NG, and Oxnard CE (in press) Morphometric analysis of *Macaca arctoides* and *M. thibetana* in relation to other macaque species. Primates.

Prost J (1980) The origin of bipedalism. Am. J. Phys. Anthropol. *52*:175–190.

Richardson MK, Hornbruch A, and Wolpert L (1990) Mechanisms of pigment pattern formation in the quail embryo. Development *109*:81–89.

Ripley S (1967) The leaping of langurs: A problem in the study of locomotor adaptation. Am. J. Phys. Anthropol. *26*:149–170.

Ruizi IA, and Melton D (1990) Axial patterning and the establishment of polarity in the frog embryo. Trends Genet. *6:*57–67.

Schmidt P (1984) Eine Rekonstruktion des Skellettes von AL 288–1 und deren Konsequenzen. Folia Primatol. *40*:283–306.

Senut B (1981) Humeral outlines in some hominoid primates and in Plio-Pleistocene hominids. Am. J. Phys. Anthropol. *56*:275–284.

Stern JT Jr., and Sussman RL (1983a) The locomotor anatomy of *Australopithecus afarensis*. Am. J. Phys. Anthropol. *60*:279–318.

Stern JT Jr., and Sussman RL (1983b) Functions of peroneus longus and brevis during locomotion in apes and humans. Am. J. Phys. Anthropol. *60*:256–257.

Susman RL, Stern JT Jr., and Rose MD (1983) Morphology of KNM-ER 3228 and OH 28 innominates from East Africa. Am. J. Phys. Anthropol. *60*:259–260.

Tardieu C (1981) Morpho-functional analysis of the articular surface of the knee joint in primates. In AB Chiarelli and RS Corrucini (eds.): Primate Evolutionary Biology. Berlin: Springer Verlag, pp. 68–80.

Tattersall I, and Schwartz JH (1975) Relationships among the Malagasy lemurs. The craniodental evidence. In WP Luckett and FS Szalay (eds.): Phylogeny of the Primates: A Multidisciplinary Approach. New York: Plenum Press, pp. 299–312.

Wolpert L, and Hornbruch A (1992) Double anterior chick limb buds and models for cartilage rudiment specification. Development *109*:961–966.

Wolpert L, Lewis J, and Summerbell D (1975) Morphogenesis of the vertebrate limb. Ciba Foundation Symp. *29*:95–129.

Zonnenfeldt FW, and Wind J (1985) A new method for high resolution computed tomography of hominid fossils. Proc. Taung Diamond Jubilee Symposium, Johannesburg, pp. 427–436.

16

HETEROCHRONIC APPROACHES TO THE STUDY OF LOCOMOTION

Laurie R. Godfrey,[1] Stephen J. King,[1] and Michael R. Sutherland[2]

[1]Department of Anthropology
Machmer Hall, Box 34805
[2]Statistical Consulting Center
Lederle Tower, Box 34535
University of Massachusetts
Amherst, Massachusetts 01003

1. INTRODUCTION

Despite the advances that have been made in documenting ontogenetic changes in primate skeletal morphology, relatively little is known about the ontogeny of locomotor behavior. Even less is known about differences among behavioral developmental trajectories of closely related species. The study of evolutionary changes in developmental trajectories is the purview of heterochrony. Heterochrony is generally defined as the study of perturbations and displacements in existing ontogenetic pathways caused by changes in developmental timing and rates. As such, it has been conceived by some researchers (e.g., Gould, 1977; Raff et al., 1990; Zelditch and Fink, 1996; Rice, 1997) as narrowly encompassing only a subset of possible evolutionary shifts in ontogeny, and by others (e.g., McKinney and McNamara, 1991) as all-inclusive. The goal of heterochrony is to understand the proximate causes of differences among adult phenotypes of closely related species. This entails understanding how, *through development*, morphology and behavior emerge, and how developmental trajectories themselves evolve. Applied to the study of locomotion, it means understanding the evolutionary basis for variation, if any, in the ontogeny of positional behavior, and its relation to the ontogeny of form.

There is today a standard vocabulary of heterochronic "processes," derived largely from the work of Gould (1977) with modifications by others (e.g., Alberch et al., 1979; Shea, 1981, 1983, 1986; McKinney, 1986, 1988; McKinney and McNamara, 1991). Gould (1977) sought to explain how simple developmental perturbations might lead to "juvenilization" ("paedomorphosis") or "overdevelopment" (Gould's "recapitulation," later called "peramorphosis") of descendant adults. Descendant adults are juvenilized (or "paedomorphic") if they closely resemble ancestral juveniles. They are overdeveloped (or "peramor-phic") if they compare favorably to what might be expected of an ancestor that developed *beyond* its normal endpoint—an ancestral "hyperadult."

Strongly influenced by Huxley's (1932) *Problems of Relative Growth*, Gould also linked the study of heterochrony with the study of allometry, or size-correlated shape change. To Gould, a fundamental task of heterochrony was the diagnosis of the preservation or disruption, in descendants, of ancestral ontogenetic linkages between size (metric dimensions) and shape (dimensionless proportions). Gould called ontogenetic change in size "growth," and ontogenetic change in shape "development." He built a three-parameter clock model showing possible evolutionary outcomes of alterations in 1) rates of growth (or change in size), 2) rates of development (or change in proportions), and/or 3) age at reproductive maturation (or some other maturational stage). Preservation, in the descendant, of ancestral size/shape linkages would result in ancestor/descendant allometric commonality. Changes in ontogenetic size/shape linkages ("dissociation" of size and shape) could also be diagnosed using the standard tools of bivariate allometry (see Gould, 1977: 239, Fig. 30; see Figure 1). Ancestral growth allometries could be used, following Huxley (1932), as a criterion of subtraction from which shifts in descendant ontogenetic pathways could be assessed (Gould, 1966, 1975a,b, 1977).

Over the past two decades, primatologists have increasingly used heterochrony to understand locomotor form and function. Much of the literature on heterochrony and primate locomotion takes Gould's ontogenetic criterion of subtraction, and his treatment of the relationship between heterochrony and allometry, as its point of departure (Shea, 1981, 1983, 1986, 1988, 1992; Gomez, 1992; Inouye, 1992; Ravosa, 1992; Ravosa et al., 1993; Taylor, 1995, 1997). The strategy is to compare species' differences in *growth allometries*.

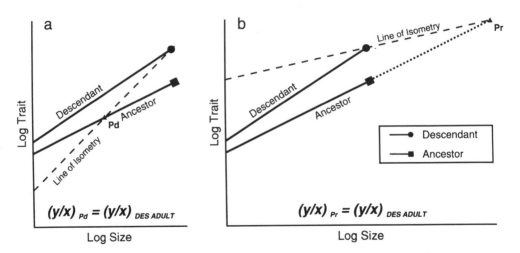

Figure 1. Gould's test of dissociation of size and shape: Ascertain where the line of descendant adult isometry (dashed) intersects the ancestor's growth allometry. (*a*) Under paedomorphosis via neoteny, the line of descendant adult isometry intersects the juvenile portion of the ancestral growth allometry (Pd) and the descendant adult is a paedomorph. (*b*) Under peramorphosis via acceleration, the line of descendant adult isometry intersects the hyperadult or extrapolated (dotted) portion of the ancestor's growth allometry (Pr) and the descendant adult is a peramorph. Note that the results of this test depend not solely on whether the descendant's growth allometry is steeper or shallower than that of the ancestor, but also on whether the ancestral and descendant growth allometries are shallower (as in case *a*, where both are negative) or steeper (as in case *b*, where both are positive) than the line of descendant adult isometry.

This is usually accomplished by examining the linear approximations of the ontogenetic relationships between pairs of variables plotted on logarithmic scales. Growth allometries are identified as "ontogenetically scaled" or "allometrically dissociated" based on the statistical significance of differences between their allometric coefficients (or slopes) and intercepts. Inferences are sometimes drawn regarding *heterochronic processes* (e.g., rate or time hypo- or hypermorphosis, pre- or postdisplacement, neoteny, acceleration, etc.) and the *functional significance* of the evolutionary perturbations. Ontogenetic scaling is sometimes taken to signal *evolutionary conservatism* (in that shape changes may be "size-required" and not the specific objects of selective pressure) whereas allometric dissociation is taken to signal possible *evolutionary innovation* (Inouye, 1992; Shea, 1992). In addition, *behavioral inferences* are sometimes drawn, based on a postulated correspondence between ontogenetic behavioral and morphological trajectories, or between paedomorphosis (or peramorphosis) in form and juvenile (or hyperadult) behaviors (Gould, 1977; Doran, 1992). There is a strong conviction that one can learn a great deal about locomotor form and function by employing an ontogenetic criterion of subtraction (see Ravosa et al., 1993), and that, in general, "...studies attempting to relate form to function by utilizing the comparative approach among closely-related adults of differing body size would be on much firmer ground if ontogenetic allometric data were incorporated in order to examine scaling patterns and utilize the appropriate criterion of subtraction for such comparisons" (Shea, 1992: 299).

This chapter focuses on heterochronic analyses of primate locomotion. It addresses the following questions: Do growth allometries reveal heterochronic process? To what extent does ontogenetic scaling imply functional conservatism, and allometric dissociation imply the converse—i.e., functional novelty? Does allometric commonality imply commonality of behavioral ontogenies? Does paedomorphosis or peramorphosis of *form* imply juvenilization or adultification of positional *behavior*? Finally, how might heterochrony contribute to our understanding of the evolution of positional behavior in primates?

2. DO GROWTH ALLOMETRIES REVEAL HETEROCHRONIC PROCESS?

What is the relationship between heterochrony and allometry? Neither Gould (1977) nor Alberch et al. (1979) treated this subject thoroughly, and recent treatments have been inconsistent and sometimes contradictory (see Shea, 1983, 1985, 1988, 1989; McKinney and McNamara, 1991; Godfrey and Sutherland, 1995b, 1996). It is actually a simple matter to model the allometric expectations of evolutionary perturbations in any of Gould's three parameters of heterochronic change, as long as 1) the descendant follows exactly its ancestor's path of *shape* change (Gould's "development"), and 2) the traits whose allometric relationships are being examined are simple power functions of one another. These conditions are rarely met. Because, even under these ideal conditions, the allometric expectations of heterochronic perturbations have been and continue to be misunderstood, they warrant brief clarification here (see Table 1).

2.1. Rate Hypo- and Rate Hypermorphosis

As noted by Shea (1983, 1988), as long as relationships between Y and X are simple power functions and are thus linear on logarithmic scales, several heterochronic processes

Table 1. Allometry and heterochrony under restricted conditions[1]

Ontogenetic perturbation	What happens to growth allometries in the descendant, which in the ancestor are:			Ancestral size/shape linkages?	Descendant size at maturation	Descendant age at maturation	Product
	Positive	Negative	Isometric				
Rate hypomorphosis	No change	No change	No change	Preserved	Smaller	Same	Paedomorph
Rate hypermorphosis	No change	No change	No change	Preserved	Bigger	Same	Peramorph
Time hypomorphosis	No change	No change	No change	Preserved	Smaller	Younger	Paedomorph
Time hypermorphosis	No change	No change	No change	Preserved	Bigger	Older	Peramorph
Pre-displacement	No change	No change	No change	Preserved	Bigger	Same	Peramorph
Post-displacement	No change	No change	No change	Preserved	Smaller	Same	Paedomorph
Neoteny	Slopes decrease	Slopes increase	No change	Disrupted	Same	Same	Paedomorph
Acceleration	Slopes increase	Slopes decrease	No change	Disrupted	Same	Same	Peramorph
Proportioned giantism	Slopes decrease	Slopes increase	No change	Disrupted	Bigger	Same	Neither
Proportioned dwarfism	Slopes increase	Slopes decrease	No change	Disrupted	Smaller	Same	Neither

[1]All growth allometries are linear on logarithmic scales. All ancestral shape paths are conserved.

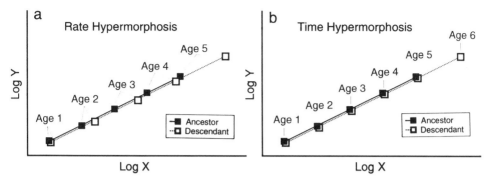

Figure 2. Two examples of hypermorphosis. (*a*) Under rate hypermorphosis, the ancestor and descendant follow common growth allometries at different rates. (*b*) Under time hypermorphosis, the descendant follows ancestral growth allometries at the same rate, but for a longer period of time. After Ravosa et al. (1993).

predict coincidence of ancestral and descendant linear regression parameters, or "ontogenetic scaling."

These include "rate hypo-" and "rate hypermorphosis," first defined by Shea in 1983. Under rate hypo- or hypermorphosis, ancestors and descendants follow common growth allometries, but do so at different rates (Figure 2a). In other words, ancestors and descendants exhibit similar shapes at similar sizes but different ages. The descendant grows and develops more slowly (rate hypomorphosis) or more quickly (rate hypermorphosis) than its ancestor. Ancestral size/shape linkages are preserved.

2.2. Progenesis and Hypermorphosis (= Time Hypo- and Time Hypermorphosis)

Gould's "progenesis" and "hypermorphosis" (renamed by Shea, 1983, "time hypo-" and "time hypermorphosis," respectively), also yield coincident ancestral and descendant allometric slopes and intercepts on logarithmic scales (as long as the relationships between Y and X are simple power functions; see Figure 2b). Ancestors and descendants follow common allometries (i.e., they exhibit similar shapes [values of Y/X] at similar sizes [values of X]), and ancestral rates of trait growth remain unchanged, but the duration of growth decreases (time hypomorphosis) or increases (time hypermorphosis).

2.3. Predisplacement and Postdisplacement

Alberch et al. (1979) defined predisplacement and postdisplacement as positive vs. negative perturbations in the timing of the *onset* of the development (or change in shape) of some body part. Represented on development (i.e., shape, or Y/X) vs. age scales, predisplacement and postdisplacement yield descendant trajectories that parallel those of the ancestor, as depicted by Alberch et al. (1979:309, Fig. 20).

Plotted on logarithmic scales (log Y vs. log X), predisplaced or postdisplaced descendant trajectories thus defined will coincide with those of the ancestor. Predisplacement and postdisplacement of the development of some body part entails no change in the relationships *among Y and X components of that part*—only a change in the relationship of those components to other body parts. The allometric prediction of such a change, whenever ancestral and descendant relationships between log Y and log X are compared, is the coincidence of allometric slopes and intercepts, or "ontogenetic scaling." Ancestors and descendants will exhibit similar shapes (Y/X) at similar sizes (values of X), but at different ages. The age difference results from a change in developmental onset age and not from a change in developmental rate.[*]

2.4. Neoteny and Acceleration

Neoteny and acceleration are two of many ontogenetic perturbations that result in the "allometric dissociation," or non-coincidence, of *at least some* allometric slopes and intercepts. Gould (1977) defined "neoteny" as paedomorphosis acquired via a retardation in the rate of development (or shape change) with no concomitant change in overall size or age at maturation. Thus defined, neoteny requires a weakening, from ancestor to descendant, of growth allometries (Shea, 1989; Godfrey and Sutherland, 1996). This means that all allometries will converge toward isometry (the slopes of ancestral positive allometries will decrease, thereby becoming more weakly positive; the slopes of ancestral negative allometries will increase, thereby becoming more weakly negative; and ancestral isometries will not change). Neoteny contrasts with "acceleration," or peramorphosis acquired via an increase in the rate of development (or shape change) with no concomitant change in overall size or age at maturation. Acceleration requires a strengthening of growth allometries (the slopes of ancestral positive allometries increase, thereby becoming more strongly positive; the slopes of ancestral negative allometries decrease, thereby becoming more strongly negative; and the slopes of ancestral isometries remain unperturbed). These allometric predictions have been poorly understood (e.g., McKinney and McNamara, 1991; Klingenberg and Spence, 1993; Vrba et al., 1994; Rice 1997; see Godfrey and Sutherland, 1995b, 1996 for critiques), despite their having been explicitly described by Shea (1989, see pp. 73 and 96)[†] and others (Godfrey and Sutherland, 1996, see pp. 33–34).

Whenever ancestral slopes are at or near isometry, little or no change in allometric slopes and intercepts is the expectation of heterochronic processes such as neoteny and acceleration. For example, under neoteny, a further weakening (as expected) of already weak ancestral growth allometries may not effect a statistically significant change in slope. Ancestors and descendants may well appear "ontogenetically scaled" for the relationships under consideration. Thus, whenever *either* the ancestor or descendant shows isometry,

---

[*] Many authors claim that predisplacement and postdisplacement predict simple upward or downward transpositions of descendant growth trajectories on log Y vs. log X scales (e.g., McKinney and McNamara, 1991). This is not merely incorrect for predisplacement and postdisplacement as originally defined by Alberch et al. (1979), but also for other commonly accepted meanings of these terms (e.g., as perturbations in the onset of the growth of one trait, Y, relative to another, X). The allometric prediction of a perturbation in the onset of the growth of one trait relative to another is neither "ontogenetic scaling" nor simple upward or downward transposition of ancestral growth allometries. Godfrey and Sutherland (1995a) show that a shift in the onset timing of the growth of Y (relative to X) will generally result in convergence of ancestral and descendant growth trajectories on logarithmic scales, because the trajectories cannot share the same origin on original measurement scales. In contrast, simple transpositions of growth allometries on logarithmic scales describe changes in the scalar rate component of simple power relationships between Y and X.

[†] Shea (1989:91) defined neoteny as "allometric dissociation and retardation of shape change." Under neoteny, "trajectories of positive allometry require slope decreases and/or downward transpositions to yield paedomorphosis, while trajectories of negative allometry require slope increases and/or upward transpositions" (Shea, 1989:96)

"ontogenetic scaling" is non-diagnostic of the maintenance of ancestral size/shape linkages.

Other ontogenetic perturbations predict changes in growth allometries similar to those expected under neoteny and acceleration. In fact, any case of retardation of development (change in Y/X) relative to growth (change in X) yields weakened growth allometries, whereas any case of acceleration of development relative to growth yields strengthened growth allometries. Thus, for example, proportioned giantism involves a retardation of shape change *relative to growth*, and it resembles neoteny in its allometric expectations, whereas proportioned dwarfism involves an acceleration of shape change relative to growth, and it resembles acceleration (see Godfrey and Sutherland, 1996).

2.5. Complex Allometry, Neomorphosis, and Mosaic Perturbations

Disjunction of ancestral and descendant growth allometries can also result from a variety of ontogenetic perturbations when Y and X are not related to one another as simple power functions and are thus not linear on logarithmic scales (see Shea, 1986; Shea et al., 1990; Sutherland and Godfrey, in prep.; Table 2). When ancestral allometries are "complex" (i.e., non-linear on logarithmic scales), heterochronic processes such as time or rate hypo- or hypermorphosis, as well as predisplacement or postdisplacement as originally defined by Alberch et al. (1979), can yield apparent "allometric dissociation" (Figure 3a; see Shea et al., 1990). The descendant follows the same trait-trait trajectory as its ancestor, but to a different terminus. Ancestral size/shape linkages are preserved, but disjunct slopes or

Table 2. Changes in descendant growth allometries under specified conditions

Perturbation	What happens to descendant growth allometries (vis-à-vis those of the ancestor)	What happens to ancestral size/shape linkages	Product
Truncation of ancestral complex allometries[1]	?	Preserved	Paedomorph
Extrapolation of ancestral complex allometries[1]	?	Preserved	Peramorph
Deviation from ancestral shape path	?	Disrupted	Neomorph
Dissociation of growth fields, with local retardation of size and shape	When traits within affected fields are plotted against size surrogates outside of the affected field, descendant growth allometries will show a decrease in slope and/or downward transposition via a decrease in the Y-intercept.	Local linkages preserved; global linkages disrupted	Affected growth fields will be paedomorphic; descendant as a whole will be neither paedomorphic nor peramorphic.
Dissociation of growth fields, with local acceleration of size and shape	When traits within affected fields are plotted against size surrogates outside of the affected field, descendant growth allometries will show an increase in slope and/or upward transposition via an increase in the Y-intercept.	Local linkages preserved; global linkages disrupted	Affected growth fields will be peramorphic; descendant as a whole will be neither paedomorphic nor peramorphic.

[1]We assume that the data are treated as linear on double-logarithmic scales, whether or not they show some curvature that may indicate the existence of complex (nonlinear) allometry.

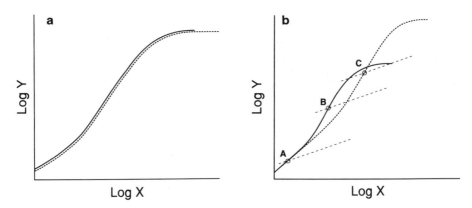

Figure 3. Two examples of extrapolation of ancestral complex allometry. (*a*) The descendant (dashed curve) follows its ancestor's (solid curve) complex allometric trajectory to a new terminus. (*b*) The descendant (dashed curve) extends each phase of its ancestor's allometric pattern (solid curve) sequentially. The parallel dashed lines represent hypothetical lines of isometry. At **A**, the ancestor and descendant are the same size and shape. They are not the same size and shape midway through the steepest portion of their allometric trajectories (**B** for the ancestor, and **C** for the descendant). The shape pathways have diverged. Modified from Shea et al. (1990).

intercepts can occur because neither trajectory is correctly described by a linear regression. In cases of nonlinear allometry, standard tests of the statistical significance of differences between ancestral and descendant linear regression parameters on logarithmic scales will not necessarily diagnose allometric commonality as "ontogenetic scaling." Because of this problem, Shea et al. (1990) suggest visual examination of the coincidence of ancestral and descendant allometric patterns.

An interesting question raised by Shea et al. (1990) is what happens when changes in complex allometric slopes are age- rather than size-dependent (Figure 3b). Suppose, for example, a descendant extends each phase of its ancestor's allometric pattern sequentially, as shown in Figure 3b—i.e., maintaining the overall "shape" of the ancestor's ontogenetic trajectory on logarithmically transformed trait-trait scales. This sort of evolutionary change has been called "sequential hypermorphosis" (McNamara, 1983; McKinney and McNamara, 1991; Rice, 1997). The resulting dissociation of ancestral and descendant allometric slopes and intercepts is not an artifact of fitting linear regressions to nonlinear data. Ancestral linkages between size and shape *are* actually disrupted. There is, in addition, a change, from ancestor to descendant, in the sequence of proportions. The descendant is no longer on its ancestor's *shape path* (Figure 4). This is one of many ways in which *neomorphosis* (or entirely novel shapes) can be achieved (Table 2). Unless the descendant follows its ancestor's shape pathway, concepts such as *paedomorphosis* and *peramorphosis* have little meaning.

Upward or downward transpositions of growth allometries (or changes in their Y-intercepts) are likely to result from mosaic evolution, or changes in the rates of growth and development of parts, when perturbed traits are plotted against unaffected traits (see Table 2). This is true whether or not ancestral size/shape linkages are retained within affected parts. Other perturbations can produce upward or downward transposition of growth allometries. For example, the nonparallel growth allometries produced by shifts in the onset of the development of some trait plotted against some unaffected trait may nevertheless appear to be parallel if the shift in onset timing occurs well before the age span covered by the recorded data (see Godfrey and Sutherland, 1995a).

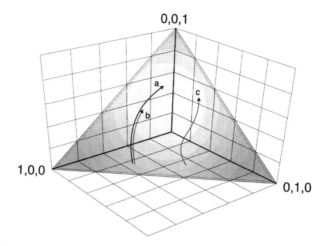

Figure 4. A shape space can be constructed by converting trait measurements to proportions (such that they always sum to 1), and then plotting these proportions (from 0 to 1) on orthogonal axes. Shown here is a 3-dimensional shape space representing traits X (from the origin, 0,0,0 to 1,0,0), Y (from the origin to 0,1,0), and Z (from the origin to 0,0,1). In 3-dimensional shape space, all shape paths are constrained to lie on a single plane. Shape paths of three species, labeled "**a**" (ancestor), "**b**" and "**c**" (two hypothetical descendants), are shown. Arrows indicate directions of ontogenetic change. Descendant **b** follows its ancestor's shape path, whereas descendant **c** does not. Note that the shape paths do not reveal the sizes of species at any shape or their rates of development (age at any shape). Bivariate comparisons of growth allometries of species **a** and **b**, as well as species **a** and **c**, may reveal a combination of allometrically dissociated and ontogenetically scaled relationships. Only species **b**, however, is a true paedomorph for the growth field represented by these three traits.

It is clear from the above discussion that inferences regarding heterochronic process are difficult to draw from an examination of changes in slopes and intercepts of growth allometries. The *coincidence* of allometric slopes and intercepts may imply: 1) rate hypo- or hypermorphosis (i.e., a negative or positive perturbation in the rates of growth and development); 2) time hypo- or hypermorphosis (i.e., a negative or positive perturbation in the age of maturation, or cessation of growth and development); 3) predisplacement or postdisplacement of onset timing (i.e., a negative or positive perturbation in the timing of the initiation, or onset, of growth and development of some part); or 4) under ancestral or descendant growth isometry, some process (such as neoteny) that results in the disruption of ancestral size/shape linkages.

"Allometric dissociation" is similarly a catch-all category comprising many kinds of heterochronic perturbations. The *non-coincidence* of allometric slopes and/or intercepts may imply: 1) neoteny, acceleration or any other simple disruption of ancestral linkages between size and shape, but with preservation of ancestral shape pathways; 2) inappropriate use of linear regression to describe truncated or extrapolated ancestral complex (or nonlinear) allometries (preservation of ancestral size/shape linkages not withstanding); 3) neomorphosis via a deviation of the descendant from the ancestor's shape path; 4) a mosaic change in the rate of growth and development of a part but not the whole, or in the onset timing of a part, but not the whole; or 5) some combination of the above.

Given the above considerations, we can specify conditions under which *coincidence of allometric slopes and intercepts must be considered trivial* (at least in the sense that it is *non-diagnostic* of the preservation of ancestral size/shape linkages). "Ontogenetic scaling" must be considered trivial whenever either of the two species being compared shows an

Table 3. Assessment of preservation of ancestral size/shape linkages, with and without consideration of relationships that are non-diagnostic, or "trivial"

Source	Database	Taxa Compared	Relationships reported to be ontogenetically scaled	Conclusion drawn	Relationships that are non-trivially ontogenetically scaled
Inouye, 1992	Manual rays	*Pan* and *Gorilla*	15 of 22	Pervasive ontogenetic scaling	3 of 22
Shea, 1992	Limbs and trunk	*Cercopithecus talapoin* and *C. cephus*	9 of 9	Pervasive ontogenetic scaling	2 of 9
Ravosa et al., 1993	Limbs and trunk	*Propithecus diadema edwardsi* and *P. tattersalli*	17 of 19	Pervasive ontogenetic scaling	3 of 19
Taylor, 1995	Scapula	Male and female *Gorilla gorilla beringei*	6 of 6	Pervasive ontogenetic scaling	1 of 6

isometric relationship between Y and X. This may well affect assessments of the prevalence of size/shape conservation in the primate postcranial skeleton (see Table 3). Similarly, *allometric dissociation must be considered trivial* (or non-diagnostic of the disruption of ancestral size/shape linkages) whenever the bivariate relationship being examined is not a simple power function. If ancestral and descendant growth allometries are both significantly positive or negative, and if both are linear on logarithmic scales, then coincidence of allometric slopes and intercepts is not trivial. The strengths and shapes of allometric trajectories must be considered.

The upshot is that cataloguing differences between ancestral and descendant allometric slopes and intercepts does not reveal heterochronic process in a straightforward manner. "Ontogenetic scaling" (defined as the coincidence of ancestral and descendant slopes and intercepts of linear ontogenetic regressions plotted on logarithmic scales) can be produced by many different heterochronic processes, including some that *disrupt* ancestral size/shape linkages. "Allometric dissociation" (defined as the non-coincidence of ancestral and descendant slopes or intercepts of linear regressions plotted on logarithmic scales) is, similarly, an expectation of many different heterochronic processes, including some that *preserve* ancestral size/shape linkages. Multiple heterochronic processes yield the same allometric clues. If heterochronic process is to be inferred, and if growth allometries are to be used as diagnostic tools, then the full set of allometric implications of different heterochronic processes must be understood. These distinctions must be embraced if the functional importance of *neomorphosis*, as opposed to upwardly or downwardly *scaled* paedomorphosis or peramorphosis, is to be addressed.

3. FUNCTIONAL CONSERVATISM AND NOVELTY

Despite the nebulous connection between "ontogenetic scaling," "allometric dissociation," and heterochronic processes, the distinction between ontogenetic scaling and allometric dissociation may be theoretically important and useful for heterochronic analysis.

Huxley (1932) studied ontogenetic series of termites, ants, and other organisms; he found that many organs grow according to a "simple heterogony formula" (that is, Y increases as a power function of size, or X). Thus, for example, differences in the shape of soldiers and workers are achieved via feeding discrimination; they are consequential effects of heterogonic growth to different sizes. Huxley suggested that shape differences produced by extrapolation or truncation of the same simple power functions should be adaptively neutral and of no taxonomic importance, whereas shape differences produced by different growth allometries should be functionally and taxonomically important. This notion had a strong influence on the development of Gould's (1966, 1975a,b, 1977) thinking about ontogeny and phylogeny, as is reflected in Pilbeam and Gould's (1974: 892) reference to ontogenetic scaling as the generation of "the same animals at different sizes."

This concept has also had a profound influence on the literature on heterochrony and primate locomotion. Shea (1988: 242) states, "The primates present many cases where ontogenetic allometries are dissociated, presumably in response to selection for specific novel proportions (as opposed to simple size change or growth duration)." Inouye (1992) uses the distinction between ontogenetic scaling and allometric dissociation as a vehicle to identify those shape differences in the manual rays of chimpanzees and gorillas that do not require adaptive explanation vs. those that do. Traits (such as metacarpal head depth) that are ontogenetically scaled vis-à-vis several size surrogates have "equal expression" in chimpanzees and gorillas at common sizes (Inouye, 1992:127), and their differences among adult chimpanzees and gorillas need no special functional interpretation. In contrast, locomotor functional explanations (increased terrestriality) are sought for shape differences (such as the relative shortening of gorilla metacarpals and phalanges) that result from allometric dissociation. Ravosa et al. (1993) use the distinction between ontogenetic scaling and allometric dissociation in a similar manner: Proportional differences that arise through differential extension of common growth allometries in diademed and Verreaux' sifakas are interpreted as by-products of selection for differences in size (in different ecological regimes), whereas traits that show allometric dissociation are treated as requiring independent functional explanation (possibly related to differences in locomotion). Finally, on the basis of "predominant" ontogenetic scaling of the scapulae of male and female mountain gorillas, Taylor (1995: 442) concludes that sexual differences in gorilla locomotor behavior do not depend on "unique morphological adaptations to different ecological niches."

If systemic size perturbations produce descendants that are ontogenetically scaled, then ontogenetic scaling can be viewed as a kind of null hypothesis for the expected degree of *shape* change that might occur without selection for change in shape. In other words, it might reveal the shape changes that can be expected if selection operates not to change shape per se, but only to increase or decrease overall size. Thus, ontogenetic scaling may be a powerful analytical tool, providing an interesting and viable alternative to the common premise that interspecific shape differences are functionally meaningful and selected. It may also provide a "criterion of subtraction" from which deviations from (apparently) passive (or, at least, ontogenetically size-correlated) shape changes can be quantified. Allometric truncation or extension via ontogenetic scaling might be used to predict proportional differences among species that differ in given amounts of size. Essentially, ontogenetic scaling might be taken as 1) the appropriate test of the hypothesis that interspecific differences in adult proportions are merely allometric consequences of overall size changes, and 2) "the proper 'criterion of subtraction' with which to assess deviations from expected allometric baselines" (Shea, 1992: 284). Ontogenetic scaling might signal evolutionary conservatism and allometric dissociation evolutionary innovation.

Lande (1985) shows, however, that ontogenetic scaling is neither an obvious nor generally appropriate null hypothesis for passive, size-correlated, evolutionary shape change (also see Lande, 1979). Arguing from a quantitative genetics perspective, he observes that selection on body size alone will effect shape change in a direction determined by the complex *genetic* variance and covariance structure of the ancestor, and not by its *phenotypic* variance and covariance structure. For particular sets of variables, there may be a strong correlation between genetic and phenotypic variance/covariance matrices, but such a relationship cannot be assumed. Thus, according to Lande (1985: 26), the null hypothesis that selection on body size alone will extrapolate ontogenetic allometries "is incorrect in theory and may be seriously misleading in practice." There is also no guarantee that pre-adult genetic variances and covariances will be strongly correlated with adult variances and covariances; thus, if selection operates to alter adult morphologies, pre-adult variances and covariances may have little bearing on evolutionary change in shape (Lande, 1982). Cheverud (1982) gives further reasons to doubt that, under selection for change in size, evolution will extrapolate ontogenetic allometries. If the slopes and lengths of ontogenetic allometries are genetically correlated, then ontogenetic scaling is an unlikely outcome. Finally, as Huxley recognized, many traits do not scale ontogenetically as simple power functions of size. When they do not, there can be no theoretical reason to expect size perturbations to maintain ancestral linear allometries on logarithmic scales.

Ultimately, the degree to which systemic size perturbations preserve ancestral allometries must be assessed empirically. Growth allometries of transgenic and normal mice have been studied specifically to address this question (Shea et al., 1987, 1990; Shea, 1988); transgenic mice differ from normal mice in their systemic levels of growth hormone (GH) and insulin-like growth factor I (IGF-I). The results, however, are equivocal. Of 11 postcranial skeletal comparisons (Shea et al., 1990:29), six show ontogenetic scaling on the basis of coincidence (on logarithmic scales) of their linear regression parameters and a P value for *rejecting* ontogenetic scaling set at <0.01. If we set P at <0.05, then only 3 of the 11 selected postcranial comparisons exhibit ontogenetic scaling. For cranial comparisons (Shea et al., 1990:29), 15 of 19 comparisons show ontogenetic scaling at P <0.01. Six of these 15 cases of ontogenetic scaling are trivial, and a 7th is not ontogenetically scaled when P is set at <0.05. Thus only 8 or 9 of 19 selected cranial comparisons show non-trivial ontogenetic scaling. Whereas some cases of "allometric dissociation" may involve extrapolation of nonlinear allometries, this cannot be assessed from the information offered the reader.

Empirical support for functional neutrality under ontogenetic scaling is also mixed. Differences in postcranial shape among some species of guenons appear to be a function of passive shape change under selection for change in size (Shea, 1992). Talapoin monkeys are smaller, but leap more frequently, than moustached monkeys, yet they have higher intermembral indices, higher humerofemoral indices, and higher crural indices. According to Shea (1992: 299), these counterintuitive inter- and intralimb proportions could not be anticipated outside the context of the pervasive ontogenetic scaling of talapoin and moustached monkeys: "...the particular body proportions of the talapoin monkey owe more to its evolutionary size decrease via allometric truncation, than they do to the results of concerted selective efforts to alter individual skeletal elements in relation to specific locomotor kinematics and frequencies."

Other studies of postnatal scaling patterns in primates do not support the functional neutrality of ontogenetic scaling. Thus, for example, Ravosa et al. (1993) observe that the positive allometry of the intermembral indices of ontogenetically-scaled sifaka species fits Cartmill's (1974) biomechanical model of vertical climbing. Critics of Gould's use of bivariate allometry as a "criterion of subtraction" (for example, Clutton-Brock and Harvey, 1979; Smith, 1980; Jungers, 1984) have offered numerous examples of cross-species differences in form and function that correlate strongly with size. Most investigators today, including proponents of the use of ontogenetic scaling as a criterion of subtraction (e.g., Shea, 1988, 1992; Gomez, 1992), would agree that extrapolation of ancestral growth allometries does not necessarily imply functional conservatism. Function must be assessed independently of ontogenetic patterns of growth and development.

There is, of course, no necessary connection between the degree of interspecific shape differentiation and the mechanism through which those differences are achieved. Both ontogenetic scaling and allometric dissociation can effect *any* amount of shape difference between ancestor and descendant. Under ontogenetic scaling, differences in adult proportions depend on the strength of the allometric coefficients and on the degree of size differentiation between ancestral and descendant adults. Whenever traits scale isometrically, there will be no correlation between size (X) and shape (Y/X), regardless of the strength of the correlation between traits Y and X. Only if the scaling coefficients themselves have no adaptive significance will shape differences generated through ontogenetic scaling be functionally neutral. It may well be that allometric dissociation (rather than ontogenetic scaling) is a prediction of functional conservatism. For example, allometric dissociation may be a means to maintain similar shapes at different sizes. If shape is important (regardless of size), then allometric dissociation may signal behavioral conservatism.

Not all studies of ontogeny and allometry in primate locomotion focus on the distinction between ontogenetic scaling and allometric dissociation. Another approach uses ontogenetic allometries to ascertain the direction and degree to which shape changes postnatally (Jungers and Fleagle, 1980; Jungers and Hartman, 1988; Jungers and Cole, 1992; Falsetti and Cole, 1992). For each species, the critical questions are: 1) the degree of departure from postnatal isometry; and, 2) the adaptive significance of species-specific patterns of nonisometric scaling. Shape differences that are present at birth are contrasted with those that arise postnatally. Both are evaluated adaptively. The degree to which postnatal scaling patterns preserve or exaggerate shape differences that are present at birth is also assessed. Ontogenetic allometric extrapolation is viewed as but one of a number of means to achieve adult shape differences, and is not imbued with functional neutrality.

A major finding of this research is that patterns of postnatal nonisometric scaling are generally adaptive, and may well preserve or exaggerate interspecific shape differences that are already manifested at birth. In other words, prenatal shape divergence and postnatal patterns of nonisometric scaling can carry the same functional signals. Thus, Jungers and Cole (1992) showed that the larger-bodied siamangs (*Hylobates syndactylus*) have relatively longer forearms and forelimbs at birth than do smaller gibbons (*H. lar*), and that, in both species, brachial and humerofemoral indices scale with postnatal positive allometry. These differences serve to enlarge the feeding sphere and modify the frictional constraints on climbing in the larger-bodied animals. Siamangs, however, are not overgrown (or peramorphic) gibbons. For many postcranial characteristics, "bivariate scaling patterns were predominantly allometric and were similar in both species, but postnatal starting points were usually quite different" (Jungers and Cole, 1992: 98). Jungers and Hartman (1988) assess the functional and phylogenetic significance of differences in postnatal nonisometric scaling patterns and prenatal shape divergence among hominoids in general. Both function and phylogeny play a role in establishing shape differences among species, but the phylogenetic signal tends to be weak. The human pattern of postnatal growth allometry is very different from those of other hominoids; it exaggerates differences already manifested at birth, and is clearly related to bipedalism. In general, postnatal allometries effect proportional changes that are compatible with biomechanical expectations; thus, for example, humerofemoral indices increase with increasing size among large-bodied climbers, but not among bipeds.

4. ALLOMETRIC COMMONALITY, BEHAVIORAL ONTOGENY, AND THE EVOLUTION OF BEHAVIOR

Does allometric commonality imply commonality of behavioral ontogenies? Does paedomorphosis or peramorphosis of *form* imply juvenilization or adultification of positional *behavior*? These two questions are, of course, not the same. Paedomorphosis or peramorphosis can be generated without allometric commonality—i.e., through the weakening or strengthening of ancestral growth allometries. Ontogenetically scaled paedomorphs and peramorphs will resemble their ancestor's juvenile or hyperadult stage in both size and shape, whereas paedomorphs or peramorphs produced under allometric dissociation will resemble their ancestor's juvenile or hyperadult stage in shape but not size.

The idea that paedomorphosis or peramorphosis in *form* has direct implications for behavior (regardless of whether or not ancestral size/shape linkages are preserved) has deep roots outside the primate literature (see, for example, Olson, 1973; Gould, 1977; Feduccia, 1980; Coppinger and Coppinger, 1982; Coppinger et al., 1987; Lawton and Lawton, 1986; Hafner and Hafner, 1988; MacDonald and Smith, 1994; Livezey, 1995). This idea has not gained a strong foothold in primatology; primatologists have tended to show more interest in the mechanical and functional *dissimilarities* of similarly-shaped organisms that differ in size (Demes and Gunther, 1989; Strasser, 1992). It is widely appreciated that, to maintain mechanical equivalence at different sizes, linear proportions must change. The meaning of paedomorphosis and peramorphosis is also somewhat ambiguous under size/shape dissociation, since such "paedomorphs" and "peramorphs" can never be paedomorphic or peramorphic for all characters. This is because tissues, cells, and cell structures have size constraints that are independent of body mass; they do not scale up or down proportionally as organisms increase or decrease in overall mass. Developing a set of behavioral expectations for such organisms is not straightforward.

In contrast, there is no *a priori* reason why paedomorphs or peramorphs that have evolved via rate or time hypo- or hypermorphosis cannot be paedomorphic or peramorphic for all characteristics. Allometric commonality may thus provide a null hypothesis for *behavioral or functional* differences among closely related species that differ in size, provided that there is some behavioral or functional significance to the ontogenetic changes in ancestral morphology. Thus, for example, if a descendant is juvenilized in size and shape, one might posit that the behavioral differences between descendant and ancestral adults will resemble, in nature and degree, those that distinguish ancestral juveniles from ancestral adults (Shea, 1986).

Evidence collected to date, however, reveals a nebulous correspondence between allometric commonality and the commonality of behavioral trajectories—at least for primates. For example, consider the locomotor behavioral ontogenies of pygmy and common chimpanzees. Doran (1992) demonstrated that *Pan paniscus* and *Pan troglodytes* follow similar locomotor ontogenies. With increasing age, there is a concomitant decrease in suspensory behavior and an increase in quadrupedalism. Doran also showed that the two species do not follow this developmental trajectory to the same terminus. Instead, there is a close match between the locomotor behavioral profiles of adult *P. paniscus* and "Stage 3 infant" *P. troglodytes*. In *P. troglodytes*, Stage 3 infants are older than 2 years of age, but they do not yet travel independently of their mothers. This stage is characterized by relatively greater amounts of suspensory behavior and quadrupedalism and less quadrumanous climbing and scrambling than is common in adult *P. troglodytes*. Adult *P. paniscus* exhibit more suspensory and quadrupedal behaviors than do adult common chimpanzees; they more closely resemble much younger common chimpanzees. Doran (1992) suggests that this behavioral difference results from juvenilization via ontogenetic scaling.

Doran's (1992) argument is appealing, but breaks down in some of its details. The argument has two facets: The first is the proposition (derived from the work of Shea, 1986) that, in many aspects of its postcranial anatomy, *P. paniscus* is paedomorphic vis-à-vis *P. troglodytes* due to a truncation of common scapular allometries. The second is the proposition that the behavioral resemblance of adult *P. paniscus* to Stage 3 infant *P. troglodytes* is a by-product of ontogenetic scaling.

We repeated Shea's (1986) test of the significance of differences among the slopes and intercepts of six scapular growth allometries in *P. paniscus* and *P. troglodytes* (Tables 4 and 5). Five of the six show ontogenetic scaling. This result differs slightly from that of Shea (1986) for the same set of six comparisons; Shea found ontogenetic scaling in all six relationships. However, two of these five (II and VI) show isometry in *P. troglodytes*. There are no significant differences between the adult ratios for these two relationships in *P. paniscus* and *P. troglodytes*—that is, adults of the two species exhibit the same "shape." The other three relationships show significant negative allometry in *P. troglodytes* (e.g., Figure 5a). Given the truncated allometries and the smaller adult scapulae of *P. paniscus*, one might expect the latter species to be paedomorphic for these three relationships. Only one, however, reveals a significant difference between adult ratios in the predicted direction (see Table 5, ratio III). This is the relationship between the morphological length of the scapula and infraspinous breadth. It is this relationship that argues most strongly for paedomorphosis in *P. paniscus*.

But even this last example is problematic (Figure 5b). As Shea (1986) pointed out, the relationship between the morphological length of the scapula and infraspinous breadth in *P. paniscus* and *P. troglodytes* is not linear on logarithmic scales. The problem here is that, due to this curvilinearity, shape changes rather little until close to the end of the common trajectory, where the curve flattens. It is the greater extension of the flattened part of the curve in *P. troglodytes* that is responsible for the significant difference between the adult ratios of morphological length to infraspinous breadth in *P. paniscus* and *P. troglodytes*.[*] Whereas it is true that the values of this ratio in adult *P. paniscus* are typical for young *P. troglodytes* (including Infant Stage 3), they are also typical for much older juvenile and subadult, and even many adult, *P. troglodytes* (Figure 5c).[†] Furthermore, the size of adult *P. paniscus* scapulae is hardly similar to that of Stage 3 infant *P. troglodytes*. Thus, *P. paniscus* adult scapulae are *not* most similar *in size and shape* to those "ancestral" individuals with similar behavioral profiles (Figure 6).

---

[*] Shea (1986) reports significant differences between adult ratios for three relationships in *Pan troglodytes* and *P. paniscus* – I and IV (in addition to III). We found no significant difference for Ratio I (the relationship between morphological length and total breadth); in any case, Ratio I suffers the same problem that we describe for Ratio III. Ratio IV is the relationship between infraspinous breadth and supraspinous breadth. For the two species of *Pan*, we found the slopes of the regressions of log infraspinous breadth to be (barely) significantly different; therefore, we do not treat the observed difference between adult means for Ratio IV as a product of ontogenetic scaling.

[†] Shea (1986:486) makes this same point when he states that "adult pygmy chimpanzees have scapulae of the same size and shape as those [of] *subadult and small adult* common chimpanzees" (emphasis ours).

Table 4. Comparison of six scapular growth allometries in *Pan paniscus* and *Pan troglodytes*[1]

Regression of	Parameters	*Pan paniscus*		*Pan troglodytes*		Significance[2]
I. Log morphological length (y) on log total breadth (x)	N	25		75		
	k (SE)	0.920 (0.047)	**0.936 (0.044)**	0.855 (0.025)	**0.879 (0.026)**	NS
	Allometry?[3]	Isometry		Negative		
	y-intercept (SE)	0.018 (0.095)	**−0.016 (0.091)**	0.134 (0.051)	**0.087 (0.052)**	NS
II. Log morphological length (y) on log supraspinous breadth (x)	N	27		85		
	k (SE)	0.905 (0.052)	**0.934 (0.048)**	0.971 (0.026)	**1.00 (0.027)**	NS
	Allometry?	Isometry		Isometry		
	y-intercept (SE)	0.285 (0.092)	**0.234 (0.086)**	0.179 (0.046)	**0.128 (0.047)**	NS
III. Log morphological length (y) on log infraspinous breadth (x)	N	29		88		
	k (SE)	0.803 (0.072)	**0.856 (0.076)**	0.748 (0.033)	**0.793 (0.034)**	NS
	Allometry?	Negative	**Isometry**	Negative		
	y-intercept (SE)	0.503 (0.125)	**0.411 (0.134)**	0.568 (0.057)	**0.488 (0.060)**	NS
IV. Log infraspinous breadth (y) on log supraspinous breadth (x)	N	30		95		
	k (SE)	0.884 (0.106)	**0.929 (0.110)**	1.122 (0.055)	**1.267 (0.061)**	Yes
	Allometry?	Isometry		Positive		
	y-intercept (SE)	0.159 (0.190)	**0.171 (0.194)**	−0.215 (0.096)	**−0.469 (0.107)**	NS (**Yes**)
V. Log morphological length (y) on log spine length (x)	N	26		71		
	k (SE)	0.929 (0.03)	**0.930 (0.028)**	0.927 (0.014)	**0.933 (0.014)**	NS
	Allometry?	Negative		Negative		
	y-intercept (SE)	0.034 (0.06)	**0.032 (0.058)**	0.040 (0.027)	**0.028 (0.027)**	NS
VI. Log total breadth (y) on log spine length (x)	N	29		90		
	k (SE)	0.906 (0.049)	**0.946 (0.047)**	0.977 (0.025)	**1.006 (0.026)**	NS
	Allometry?	Isometry		Isometry		
	y-intercept (SE)	0.222 (0.100)	**0.142 (0.096)**	0.089 (0.050)	**0.033 (0.051)**	NS

[1] Results of least squares analysis (Model I) in normal font. Results of maximum likelihood analysis (Model II) in bold font.

[2] Significance of difference in slope (k) or in y-intercept between *Pan paniscus* and *Pan troglodytes*. Yes, $P<0.05$; NS, not significant, $P>0.05$.

[3] Judged to be negative, isometric or positive, depending on significance of difference of slope (k) from 1.00. Negative allometry, k significantly <1.00 ($P<0.05$); Isometry, k insignificantly different from 1.00 ($P>0.05$); Positive allometry, k significantly > 1.00 ($P<0.05$). When not otherwise indicated, Models I and II yield identical allometric conclusions.

Table 5. Test of hypothesis that *Pan paniscus* is paedomorphic (vis-à-vis *Pan troglodytes*) due to truncation of common allometries

Ratio[1]	Observed allometries	Ontogenetically scaled?	Theoretical expectation (comparison of adult ratios)	Observed (comparison of adult ratios), Significance	Hypothesis supported?
I	Isometric (*P. p.*) or Negative (*P.t.*)	Yes	*P. paniscus* > *P. troglodytes*	*P. paniscus* = *P. troglodytes* NS	No. *Pan paniscus* adults are not paedomorphic.
II	Consistently isometric	Yes	*P. paniscus* = *P. troglodytes*	*P. paniscus* = *P. troglodytes* NS	Yes, but trivial. No difference in adult shape.
III	Consistently negative[2]	Yes	*P. paniscus* > *P. troglodytes*	*P. paniscus* > *P. troglodytes* $P<0.05$	Yes. *Pan paniscus* adults are paedomorphic through ontogenetic scaling.
IV	Isometric (*P. p.*) or Positive (*P.t.*)	No	–	–	No. There is no ontogenetic scaling.
V	Consistently negative	Yes	*P. paniscus* > *P. troglodytes*	*P. paniscus* = *P. troglodytes* NS	No. *Pan paniscus* adults are not paedomorphic.
VI	Consistently isometric	Yes	*P. paniscus* = *P. troglodytes*	*P. paniscus* = *P. troglodytes* NS	Yes, but trivial. No difference in adult shape.

[1]Ratio I (morphological length X 100/total breadth); Ratio II (morphological length X 100/supraspinous breadth); Ratio III (morphological length X 100/infraspinous breadth); Ratio IV (infraspinous breadth X 100/supraspinous breadth); Ratio V (morphological length X 100/spine length); Ratio VI (total breadth X 100/spine length). On the basis of maximum likelihood analysis, following Shea (1986). On the basis of maximum likelihood analysis, this observation changes to Isometric (in *P. paniscus*) and Negative (in *P. troglodytes*). All other results are identical for least squares (Model I) and maximum likelihood (Model II) analyses.

[2]These results are based on least squares analysis, following Shea (1986). On the basis of maximum likelihood analysis, this observation changes to Isometric (in *P. paniscus*) and Negative (in *P. troglodytes*). All other results are identical for least squares (Model I) and maximum likelihood (Model II) analyses.

An even more compelling example of discordance between allometric and behavioral ontogenies is that of talapoin and moustached monkeys. Shea (1992) found these guenons to be predominantly ontogenetically scaled in their postcranial anatomy (albeit largely via growth *isometries*; see Table 3). Adult talapoins are closest, in postcranial size and shape, to infant moustached monkeys. Whereas talapoin morphology may have evolved via allometric truncation from a guenon similar to a moustached monkey, adult talapoin locomotor behavior certainly did not evolve via truncation of the locomotor behavioral ontogeny of such an ancestor.

In summary, the above examples offer little support for the notion that "paedomorphosis via ontogenetic scaling" preserves the behaviors of "ancestors" of corresponding size and shape. Because behavior is potentially influenced by so many variables, we must devise rigorous tests for our heterochronic hypotheses. Furthermore, many reported examples of "pervasive" or "predominant" ontogenetic scaling are based largely on traits that are isometric or nearly so; the little change in proportions that these relationships imply contrasts markedly with the dramatic ontogenetic changes in positional behavior that so often occur. Growth *isometries* are decidedly unhelpful in elucidating ontogenetic *changes* in behavior.

5. ONTOGENETIC SCALING, DEVELOPMENTAL CONSERVATISM, AND BEHAVIORAL COMMONALITY: AN ILLUSTRATIVE EXAMPLE

Thus far we have examined patterns of postnatal scaling among closely related species. In this section we extend the analysis to distantly related species. If the distinction between ontogenetic scaling and allometric dissociation is meaningful, then one might reasonably expect distantly-related species that differ in morphology and behavior throughout their life cycles to exhibit predominant allometric dissociation. Ontogenetic scaling in behaviorally distinct and phylogenetically distant species might be expected to reflect developmental conservatism.

We examined Shea's (1986) set of six scapular growth allometries in *Macaca fascicularis* and *Hylobates lar* (Tables 6 and 7; Figure 7). We anticipated that these species would exhibit significantly greater dissociation of growth allometries than the two species of *Pan*. For each pair of regressions, we tested the significance of differences between regression parameters (Y-intercepts and slopes). Following Shea (1986), ontogenetic scaling was diagnosed whenever neither showed a significant difference. Mean adult ratios were calculated and compared. Adults of both species were identified by full adult dental development and fusion of the basioccipital suture. Table 7 reports the significance of the dif-

───▶

Figure 5. (*a*) The relationship between log morphological length and log infraspinous breadth in two species of *Pan*. A comparison of growth allometries reveals "ontogenetic scaling" with paedomorphosis of *Pan paniscus*. Original measurements are in mm. (*b*) The same relationship showing the inherent curvilinearity (and flattening in the adult portions of the two species' ranges) on logarithmic scales. (*c*) The same example on original measurement scales (in mm.). Ellipses show the ranges of adults of both species, and of older infant *Pan troglodytes* (including Infant Stage 3, aged by dental developmental stage). Note that, on original measurement scales, lines of shape identity (or isometry) are obtained by connecting individual points to the origin (0, 0). The ellipses indicate that adult *Pan paniscus* are the same shape as infant and small adult *Pan troglodytes*. They are not the same *size* and *shape* as older infant *Pan troglodytes*.

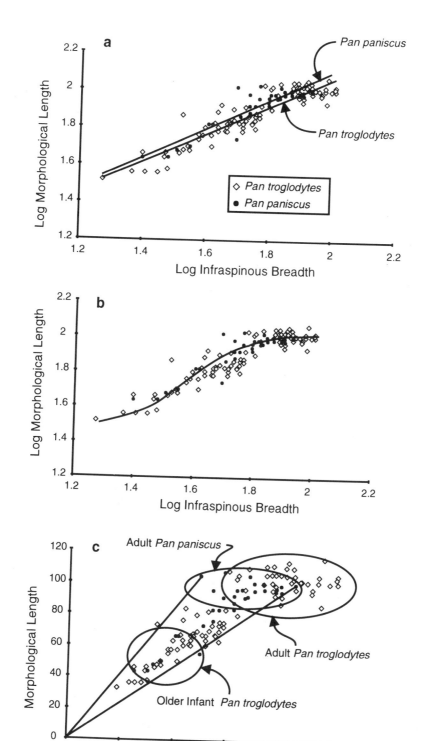

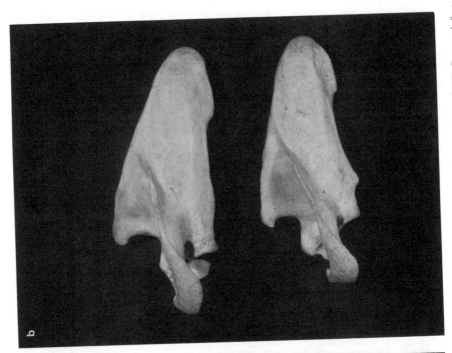

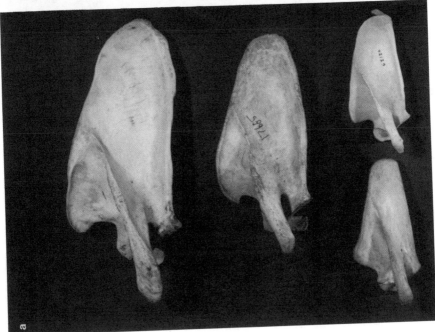

Figure 6. (*a*) Ontogenetic series of scapulae of *Pan troglodytes*. Top, adult male (MCZ 48686). Middle, juvenile of approximately 8 years (MCZ 17685). Bottom, infants of approximately 3–4 years (MCZ 42129 and 34101). (*b*) Adult male *Pan paniscus* (MCZ 38020, top, and MCZ 38018, bottom).

Table 6. Comparison of six scapular growth allometries in *Macaca fascicularis* and *Hylobates lar*[1]

Regression of	Parameters	*Macaca fascicularis* (Model I)	*Macaca fascicularis* (Model II)	*Hylobates lar* (Model I)	*Hylobates lar* (Model II)	Significance[2]
I. Log morphological length (y) on log total breadth (x)	N	49		54		
	k (SE)	0.836 (0.029)	**0.832 (0.028)**	0.796 (0.028)	**0.816 (0.028)**	NS
	Allometry?[3]	Negative		Negative		
	y-intercept (SE)	0.380 (0.044)	**0.360 (0.045)**	0.410 (0.047)	**0.377 (0.047)**	NS
II. Log morphological length (y) on log supraspinous breadth (x)	N	49		55		
	k (SE)	0.721 (0.052)	**0.652 (0.033)**	0.730 (0.034)	**0.760 (0.035)**	NS
	Allometry?	Negative		Negative		
	y-intercept (SE)	1.018 (0.048)	**1.034 (0.032)**	0.625 (0.053)	**0.578 (0.054)**	Yes
III. Log morphological length (y) on log infraspinous breadth (x)	N	49		54		
	k (SE)	0.826 (0.032)	**0.856 (0.031)**	0.863 (0.056)	**0.949 (0.061)**	NS
	Allometry?	Negative		Negative	**Isometry**	
	y-intercept (SE)	0.461 (0.047)	**0.440 (0.045)**	0.625 (0.074)	**0.512 (0.080)**	NS
IV. Log infraspinous breadth (y) on log supraspinous breadth (x)	N	49		54		
	k (SE)	0.825 (0.067)	**0.764 (0.039)**	0.699 (0.054)	**0.776 (0.059)**	NS
	Allometry?	Negative		Negative		
	y-intercept (SE)	0.717 (0.061)	**0.691 (0.038)**	0.227 (0.085)	**0.107 (0.092)**	Yes
V. Log morphological length (y) on log spine length (x)	N	49		55		
	k (SE)	1.021 (0.010)	**1.023 (0.010)**	0.991 (0.013)	**0.995 (0.012)**	NS
	Allometry?	Positive (barely)		Isometry		
	y-intercept (SE)	−0.104 (0.017)	**−0.108 (0.017)**	−0.085 (0.023)	**−0.093 (0.023)**	NS
VI. Log total breadth (y) on log spine length (x)	N	49		54		
	k (SE)	1.162 (0.038)	**1.230 (0.040)**	1.175 (0.040)	**1.216 (0.041)**	NS
	Allometry?	Positive		Positive		
	y-intercept (SE)	−0.477 (0.066)	**−0.563 (0.070)**	−0.492 (0.074)	**−0.570 (0.076)**	NS

[1] Results of least squares analysis (Model I) in normal font. Results of maximum likelihood analysis (Model II) in bold font.

[2] Significance of difference in slope (k) or in y-intercept between *Macaca fascicularis* and *Hylobates lar*. Yes, $P<0.05$; NS, not significant, $P>0.05$.

[3] Judged to be negative, isometric or positive, depending on significance of difference of slope (k) from 1.00. Negative allometry, k significantly < 1.00 ($P<0.05$); Isometry, k insignificantly different from 1.00 ($P>0.05$); Positive allometry, k significantly > 1.00 ($P<0.05$). When not otherwise indicated, Models I and II yield identical allometric conclusions.

Table 7. Test of hypothesis that *Hylobates lar* is peramorphic (vis-à-vis *Macaca fascicularis*) due to an extension of common allometries

Ratio[1]	Observed allometries	Ontogenetically scaled?	Theoretical expectation (comparison of adult ratios)	Observed (comparison of adult ratios), significance	Hypothesis supported?
I	Consistently negative	Yes	*H. lar* < *M. fascicularis*	*H. lar* < *M. fascicularis* P<0.05	Yes. *Hylobates lar* adults are peramorphic.
II	Consistently negative	No	–	–	No. There is no ontogenetic scaling.
III	Consistently negative [2]	Yes	*H. lar* < *M. fascicularis*	*H. lar* > *M. fascicularis* P<0.05	No. *Hylobates lar* adults are paedomorphic.
IV	Consistently negative	No	–	–	No. There is no ontogenetic scaling.
V	Isometric (*H. l.*) or Positive (*M. f.*)	Yes	*H. lar* > *M. fascicularis*	*H. lar* < *M. fascicularis* P<0.05	No. *Hylobates lar* adults are paedomorphic.
VI	Consistently positive	Yes	*H. lar* > *M. fascicularis*	*H. lar* > *M. fascicularis* P<0.05	Yes. *Hylobates lar* adults are peramorphic.

[1] Ratio I (Morphological length X 100/total breadth); Ratio II (Morphological length X 100/supraspinous breadth); Ratio III (Morphological length X 100/infraspinous breadth); Ratio IV (Infraspinous breadth X 100/supraspinous breadth); Ratio V (Morphological length X 100/spine length); Ratio VI (Total breadth X 100/spine length).

[2] These results are based on least squares analysis, following Shea (1986). On the basis of maximum likelihood analysis, this observation changes to Negative (in *M. fascicularis*) and Isometric (in *H. lar*). All other results are identical for least squares (Model I) and maximum likelihood (Model II) analyses.

ferences between adult ratios, and, for those instances for which significant differences exist, the directionality of the difference.

Four of the six bivariate comparisons of *M. fascicularis* and *H. lar* show "ontogenetic scaling" (e.g., Figure 8; Table 7). Given the striking differences in locomotor ontogeny and adult morphology and behavior between these two species, this result seems counterintuitive. Surely, nobody would claim that the morphology of the scapula of the slightly larger-bodied *H. lar* (see Smith and Jungers, 1997) is similar to that of a hyperadult, or overdeveloped, *M. fascicularis*. Nor would anyone seek a behavioral corollary for the signal of predominant ontogenetic scaling. No extrapolation of the locomotor behavioral ontogeny of an "ancestor" such as *M. fascicularis* would produce a species resembling a "descendant" such as *H. lar*.

Does the predominant ontogenetic scaling of the scapulae of *M. fascicularis* and *H. lar* imply that scapular development in Catarrhini is highly conservative? Is it possible that the four ontogenetically scaled relationships reflect conservatism of *certain aspects* of scapular development, whereas the two that show allometric dissociation carry all of the information of functional import?

Upon scrutiny of the four relationships that show "ontogenetic scaling," it becomes clear that the answer is "no." If scapular development were largely conservative in the above-described manner, then we would expect adult shape differences to be passive consequences of allometric extrapolation. The larger-bodied species (in this case, *H. lar*) should be peramorphic for all shape differences that arise via ontogenetic scaling. Peramorphosis requires the larger-bodied species to manifest the *smaller* adult ratios whenever

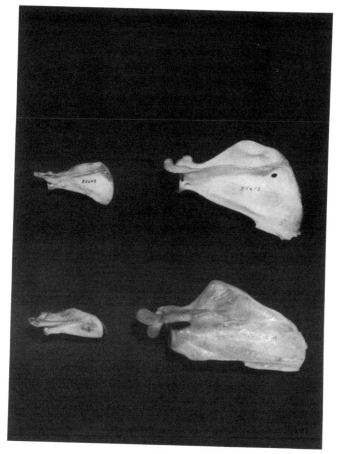

Figure 7. *Top,* infant (MCZ 35643, male) and adult (MCZ 35613, male) *Macaca fascicularis. Bottom,* infant (MCZ 35456, male) and adult (MCZ 41415, male) *Hylobates lar.*

the common allometries are negative, and the *larger* adult ratios whenever the common allometries are positive. In only two of these four examples do the observed adult ratios differ in the "expected" direction (Table 7). Thus, the four examples, taken together, fail to support a hypothesis of developmental conservatism.

The two examples that do support peramorphosis via ontogenetic scaling in the larger-bodied *H. lar* are flip sides of the same coin. These are relationships I (between log morphological length and log total breadth) and VI (between log total breadth and log spine length). Because morphological length and spine length are highly redundant (see relationship V; Figure 9), the negative allometry of relationship I (where log total breadth is taken as the X variable) and positive allometry of relationship VI (where log total breadth is taken as the Y variable) are also redundant. Because total scapular breadth is a reflection of both infraspinous breadth and supraspinous breadth, true ontogenetic conservatism should imply conservatism in the relationships between each of the latter two variables and morphological (or spine) length. In *M. fascicularis* and *H. lar*, these expectations are not met. Indeed the two examples of "allometric dissociation" (relationships II and IV) involve supraspinous and infraspinous breadth (as related to each other or to morphological length).

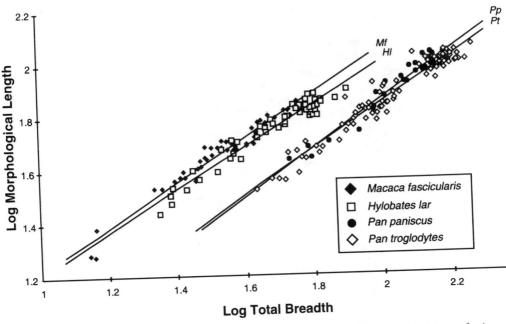

Figure 8. The relationship between log morphological length and log total scapular breadth in *Macaca fascicularis* and *Hylobates lar*, as well as *Pan paniscus* and *Pan troglodytes*. Both species pairs show "ontogenetic scaling" and all four species exhibit weak negative allometry for this relationship. The *Hylobates/Macaca* example supports the "expectation" that differential extension of common growth allometries will reveal differences in adult proportions. The *Pan* comparison does not.

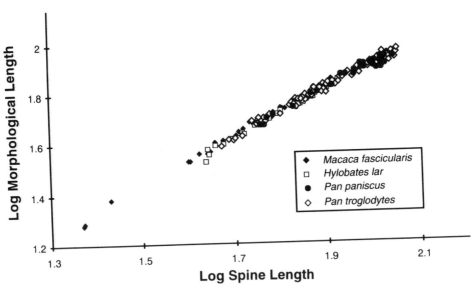

Figure 9. The relationship between log morphological length and log spine length shows ontogenetic scaling in *Macaca fascicularis* and *Hylobates lar*. However, these variables are strongly redundant. The relationship between them is virtually identical in other primate species, including *Pan troglodytes* and *Pan paniscus* (which also show ontogenetic scaling for this relationship).

Macaca fascicularis and *Hylobates lar* do differ from one another in their scapular ontogenies more than the two species of *Pan*. However, a tabulation of the degree to which the same six scapular bivariate relationships show ontogenetic scaling does not make this exceedingly obvious. Although *H. lar* is larger in body mass (on average) than *M. fascicularis*, all scapular dimensions are not larger. Shape differences, even for suites of traits that show "ontogenetic scaling," do not fall along a simple spectrum from paedomorphosis to peramorphosis. For regressions that show "allometric dissociation," *Macaca* and *Hylobates* do not exhibit the same shapes at different sizes; rather, they exhibit different shapes throughout their ontogenies (Figure 10). Within a multivariate context (Table 8, Sutherland and Godfrey, in prep.), it is easy to document the degree of separation of shape pathways and the relative contributions of different trait complexes to that separation. In

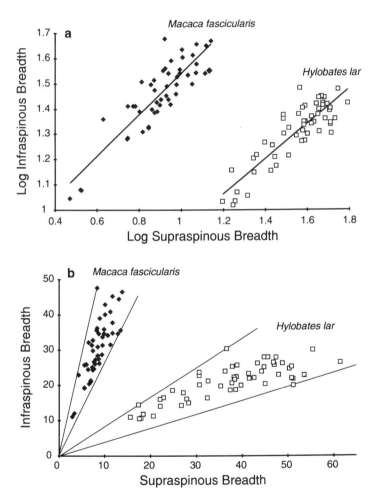

Figure 10. (*a*) The relationship between log infraspinous breadth and log supraspinous breadth in *Macaca fascicularis* and *Hylobates lar*. (*b*) The same relationship on original measurement scales. Shown here are lines of isometry for individuals situated at either end of each species' range. Note that there is little ontogenetic shape change (for this relationship) in either species (i.e., adults of both species exhibit proportions that occur in very young individuals). Furthermore, there is no shape commonality (for this relationship) at any point in the species' ontogenies. Under allometric dissociation, shapes may be similar at *different* sizes, or shapes may differ throughout ontogeny. This example conforms to the latter situation.

Table 8. Euclidean distances between stage centroids in 5-dimensional shape space[1]

Distance separating stage centroids of *Macaca fascicularis* and *Hylobates lar*				Distance separating centroids of infants and adults		Distance separating *Pan* species' centroids
Infants	Juveniles	Subadults	Adults	*Macaca*	*Hylobates*	Adults
0.117	0.128	0.132	0.138	0.040	0.032	0.028

[1] The five axes are: the morphological length of the scapula, the morphological breadth of the scapula, infraspinous breadth, supraspinous breadth, and spine length—each is scaled as a proportion of the whole.

other words, paths in shape space can be used to measure neomorphosis and to isolate the ways in which developmental pathways differ throughout ontogeny. In this case, it is not scapular breadth, but the combination of infraspinous and (especially) supraspinous breadth that distinguishes the two species (cf. Figure 11a,b).

Table 8 documents the separation (or Euclidean distances in shape space) between developmental stage centroids for *Macaca* and *Hylobates*. We constructed a 5-dimensional shape space from the five scapular traits that were used to assess ontogenetic scaling vs. allometric dissociation in *M. fascicularis*, *H. lar*, and the two species of *Pan*. The separation of stage centroids for *Macaca* and *Hylobates* is nearly as great at infancy as at adulthood, and is more than four times the distance between *adult* centroids of the two species of *Pan*. The total distances that *Macaca* and *Hylobates* travel through shape space, from infancy to adulthood, are small. Each traverses about a quarter of the total distance separating the two species' adult centroids. *Macaca* and *Hylobates* occupy their own "niches" in shape space, and maintain those niches throughout life. Prenatal development is critical to establishing species-specific shape differences. If there is common directionality to postnatal shape change, it is a slight decrease in scapular length relative to scapular breadth (Figure 11b)—a phenomenon that appears to relate to the mechanical efficiency of forelimb protraction and retraction in individuals of different body size.

6. HETEROCHRONY IN PRIMATOLOGY: WHERE NOW?

There are numerous ways in which ontogenetic research can contribute to our understanding of primate postcranial skeletons and locomotor behavior. One approach has largely dominated the literature on heterochrony in primate locomotion. It values ontogenetic studies for their potential to reveal that an observed interspecific shape difference is correlated with (and possibly caused by) factors that generate evolutionary changes in size. Certainly, passive (size-correlated) shape change represents a viable and interesting alternative to the premise that interspecific shape differences are functionally meaningful and "deliberately" selected. Many primatologists use "ontogenetic scaling" (or the coincidence of ontogenetic linear regression parameters on logarithmic scales) both to test the

Figure 11. Scapular ontogeny in *Macaca fascicularis* and *Hylobates lar* as glimpsed in two 3-dimensional constructions of shape space. Plotted here are the relative proportions for sets of three variables, averaged over four dental developmental stages (infants, juveniles, subadults, and adults). Mean values for these proportions are given for each stage, in the order X, Y, and Z, top to bottom. Arrows show the direction of ontogenetic change in shape (from infants to adults). (*a*) The three variables shown here are "allometrically dissociated" on bivariate plots. (*b*) The three variables shown here are "ontogenetically scaled" on bivariate plots. Even traits that are "ontogenetically scaled" when examined as pairs on bivariate plots may not follow the same trajectories when examined as multivariate sets in shape space.

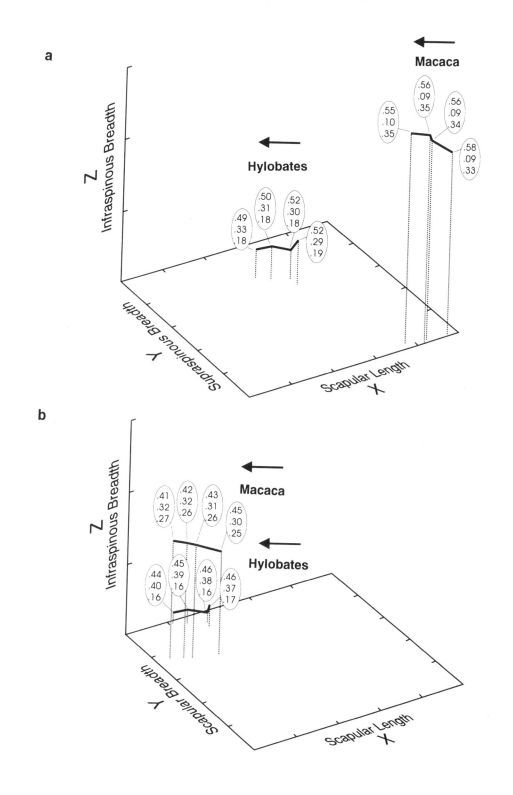

prevalence of size-correlated shape change among closely related species, and to construct a "criterion of subtraction" from which deviations from size-correlated shape changes can be quantified. It is also sometimes used to construct a null hypothesis for size-correlated changes in function.

The efficacy of this approach depends on 1) trait-trait ontogenetic relationships being linear on logarithmic scales, and 2) passive (size-correlated) evolutionary shape change extrapolating or truncating ancestral ontogenetic allometries. Given the caveats that we have discussed in this paper, we would suggest that "ontogenetic scaling" be used as the null hypothesis for passive (size-correlated) evolutionary changes in shape or in behavior only with extreme caution. We offer the following recommendations:

1. It should be recognized that selection for change in size alone can result in dissociated growth allometries. Indeed, extrapolation or truncation of linear or nonlinear ontogenetic allometries is a prediction of selection for change in size (and not shape) *only under a limited set of conditions*. One such condition (i.e., strong correlation between ontogenetic phenotypic and genetic variance/covariance matrices) can be tested by examining the heritability of traits.

2. The linearity of trait-trait relationships on logarithmic scales should be assessed. When trait-trait relationships are not linear on logarithmic scales, linear regression on logarithmic scales should not be used to test the hypothesis that shape differences are correlated with (and possibly caused by) differences in size. Multivariate tools that do not depend on *a priori* assumptions regarding the nature of trait-trait relationships can be employed to test the preservation or disruption of ancestral size/shape linkages (Sutherland and Godfrey, in prep.). They can be used to assess the commonality of patterns of change in allometric slopes, including the sort of heterochronic process that has been called, by some, *sequential* hypo- or hypermorphosis.

3. Growth isometries should be documented and treated with reservation. Commonality of growth isometries cannot demonstrate that shape changes passively with size. To the contrary, isometric growth demonstrates no relationship between size and shape. Furthermore, under ancestral growth isometry, "ontogenetic scaling" is the prediction of some heterochronic processes that disrupt as well as those that preserve ancestral size/shape linkages (and is therefore diagnostically trivial).

4. Adult ratios must be compared to verify that shape does indeed differ in the direction or directions predicted by extrapolation or truncation of ancestral ontogenetic allometries.

5. Function must be assessed independently of the ontogenetic trajectories through which changes in skeletal form are achieved.

Ontogenetic studies can contribute to our understanding of locomotor skeletal anatomy and locomotor behavior in other ways. Only through ontogenetic research can we appreciate the degree to which interspecific shape differences are manifested at birth. Only through ontogenetic research can we discover the interplay between size and age as triggers of change in form, and the impacts of variation in absolute rates of growth and development on adult morphology and behavior. Ontogenetic research can reveal developmental dissociation of parts. Dissociation of parts is not the same as allometric dissociation. Allometric dissociation can be generated by processes that preserve ancestral developmental pathways (but not size/shape linkages). Under nonlinear allometry, allometric dissociation can even be produced by processes that preserve ancestral size/shape linkages. By distinguishing dissociation of parts (or developing modularity) from allometric dissociation, investigators may be able to identify sites of locally targeted selection. Distinguishing allometric dissociation from ontogenetic scaling does not address this problem.

In this paper, we have reviewed the relationship between heterochrony and allometry. We have shown that tests of the coincidence of allometric slopes and intercepts are only partially successful in revealing heterochronic processes. They have uncertain utility in revealing size-correlated (and possibly size-required) evolutionary shape change. They do not diagnose developmental or functional conservatism. Allometric commonality does not reveal commonality of behavioral ontogenies. Testing any heterochronic processual model means understanding the model's full set of predictions and how those predictions differ from those of other models. The utility of diagnostic tools depends on their ability to identify features that distinguish alternative models. Progress in understanding those evolutionary processes that underlie variation in skeletal form and function means being willing to modify our processual models, and the tools we use to test them, when in fact those models or those diagnostic tools do not work.

ACKNOWLEDGMENTS

Research for this paper was supported in part by NSF grant GER 9450175 to LRG and by the Statistical Consulting Center of the University of Massachusetts. We thank Elizabeth Strasser for inviting us to submit this paper, and for her patience and encouragement along the way; we thank William Jungers, Brian Shea, Richard Smith, and an anonymous reviewer for their insightful comments on our first draft. We are grateful to Darren Godfrey and Emily Heinisch for their skillful preparation of the figures. Data analyzed here were collected by Stephen J. King or drawn from the literature. We thank Maria Rutzmoser and the curatorial staff of the Museum of Comparative Zoology at Harvard University for access to scapulae of *Macaca fascicularis*, *Hylobates lar*, *Pan troglodytes*, and *Pan paniscus*, and for permission to photograph individuals belonging to this collection. We also thank AW Crompton for the use of photographic copy equipment in his laboratory.

REFERENCES

Alberch P, Gould SJ, Oster GF, and Wake DB (1979) Size and shape in ontogeny and phylogeny. Paleobiology 5:296–317.

Cartmill M (1974) Pads and claws in arboreal locomotion. In FA Jenkins (ed.): Primate Locomotion. New York: Academic Press, pp. 45–83.

Cheverud J (1982) Relationships among ontogenetic, static, and evolutionary allometry. Am. J. Phys. Anthropol. 59:139–149.

Clutton-Brock TH, and Harvey PH (1979) Comparison and adaptation. Proc. R. Soc. Lond. B. 205:547–565.

Coppinger L, and Coppinger R (1982) Livestock-guarding dogs that wear sheep's clothing. Smithsonian 13:65–73.

Coppinger R, Glendinning J, Matthay C, Smith C, Sutherland M, and Torop E (1987) Degree of behavioral neoteny differentiating canid polymorphs. Ethology 75:89–108.

Demes B, and Gunther MM (1989) Biomechanics and allometric scaling in primate locomotion and morphology. Folia Primatol. 53:125–141.

Doran DM (1992) The ontogeny of chimpanzee and pygmy chimpanzee locomotor behavior: A case study of paedomorphism and its behavioral correlates. J. Hum. Evol. 23:139–157.

Falsetti AB, and Cole TM III (1992) Relative growth of the postcranial skeleton in callitrichines. J. Hum. Evol. 23:79–92.

Feduccia A (1980) The Age of Birds. Cambridge, Ma.: Harvard University Press.

Godfrey LR, and Sutherland MR (1995a) Flawed inference: Why size-based tests of heterochronic process do not work. J. Theor. Biol. *172*:43–61.

Godfrey LR, and Sutherland MR (1995b) What's growth got to do with it? Process and product in the evolution of ontogeny. J. Hum. Evol. *29*:405–431.

Godfrey LR, and Sutherland MR (1996) Paradox of peramorphic paedomorphosis: Heterochrony and human evolution. Am. J. Phys. Anthropol. *99*:17–42.

Gomez AM (1992) Primitive and derived patterns of relative growth among species of Lorisidae. J. Hum. Evol. *23*:219–233.

Gould SJ (1966) Allometry and size in ontogeny and phylogeny. Biol. Rev. 41: 587–640.

Gould SJ (1975a) Allometry in primates, with emphasis on scaling and the evolution of the brain. In FS Szalay (ed.): Approaches to Primate Paleobiology. Karger: Basel. Contr. Primatol. 5:244–292.

Gould SJ (1975b) On the scaling of tooth size in mammals. Am. Zool. *15*:351–362.

Gould SJ (1977) Ontogeny and Phylogeny. Cambridge, Ma.: Harvard University Press.

Hafner JC, and Hafner MS (1988) Heterochrony in rodents. In ML McKinney (ed.): Heterochrony in Evolution: A Multidisciplinary Approach. New York: Plenum Press, pp. 217–235.

Huxley JS (1932) Problems of Relative Growth. London: Methuen & Co.

Inouye SE (1992) Ontogeny and allometry of African ape manual rays. J. Hum. Evol. *23*:107–138.

Jungers WL (1984) Aspects of size and scaling in primate biology with special reference to the locomotor skeleton. Yrbk. Phys. Anthropol. *27*:73–97.

Jungers WL, and Cole MS (1992) Relative growth and shape of the locomotor skeleton in lesser apes. J. Hum. Evol. *23*:93–105.

Jungers WL, and Fleagle JG (1980) Postnatal growth allometry of the extremities in *Cebus albifrons* and *Cebus apella*: A longitudinal and comparative study. Am. J. Phys. Anthropol. *53*:471–478.

Jungers WL, and Hartman SE (1988) Relative growth of the locomotor skeleton in orang-utans and other large-bodied hominoids. In JH Schwartz (ed.): Orang-utan Biology. Oxford: Oxford University Press, pp. 347–359.

Klingenberg CP, and Spence JR (1993) Heterochrony and allometry: Lessons from the water strider genus *Limnoporus*. Evol. *47*: 1834–1853.

Lande R (1979) Quantitative genetic analysis of multivariate evolution applied to brain:body size allometry. Evol. *33*:402–416.

Lande R (1982) A quantitative genetic theory of life history evolution. Ecology *63*:607–615.

Lande R (1985) Genetic and evolutionary aspects of allometry. In WL Jungers (ed.): Size and Scaling in Primate Biology. New York: Plenum Press, pp. 21–32.

Lawton MF, and Lawton RO (1986) Heterochrony, deferred breeding, and avian sociality. In RF Johnston (ed.): Current Ornithology, Volume 3. New York: Plenum Press, pp. 187–221.

Livezey BC (1995) Heterochrony and avian flightlessness. In KJ McNamara (ed.): Evolutionary Change and Heterochrony. New York: John Wiley & Sons, pp. 169–193.

McDonald MA, and Smith MH (1994) Behavioral and morphological correlates of heterochrony in Hispaniolan palm-tanagers. The Condor *96*:433–446.

McKinney ML (1986) Ecological causation of heterochrony: A test and implications for evolutionary theory. Paleobiology *12*:282–289.

McKinney ML (1988) Classifying heterochrony: Allometry, size, and time. In ML McKinney (ed.): Heterochrony in Evolution, A Multidisciplinary Approach. New York: Plenum Press, pp. 17–34.

McKinney ML, and McNamara KJ (1991) Heterochrony: The Evolution of Ontogeny. New York: Plenum Press.

McNamara KJ (1983) Progenesis in trilobites. In DEG Briggs and PD Lane (eds.): Trilobites and Other Early Arthropoda: Papers in Honour of Professor H. B. Whittington, FRS. Special Papers in Paleontology 30. London: Paleontological Association, pp. 59–68.

Olson SL (1973) Evolution of the rails of the South Atlantic islands (Aves: Rallidae). Smithson. Contr. Zool. *152*:1–53.

Pilbeam DR, and Gould SJ (1974) Size and scaling in human evolution. Science 186: 892–901.

Raff RA, Parr BA, Parks AL, and Wray GA (1990) Heterochrony and other mechanisms of radical evolutionary change in early development. In MH Nitecki (ed.): Evolutionary Innovations. Chicago: University of Chicago, pp. 71–98.

Ravosa MJ (1992) Allometry and heterochrony in extant and extinct Malagasy primates. J. Hum. Evol. *23*:197–217.

Ravosa MJ, Meyers DM, and Glander KE (1993) Relative growth of the limbs and trunk in sifakas: Heterochronic, ecological and functional considerations. Am. J. Phys. Anthropol. *92*:499–520.

Rice SH (1997) The analysis of ontogenetic trajectories: When a change in size or shape is not heterochrony. Proc. Natl. Acad. Sci. USA *94*:907–912.

Shea BT (1981) Relative growth of the limbs and trunk in the African apes. Am. J. Phys. Anthropol. *56*:179–201.

Shea BT (1983) Allometry and heterochrony in the African apes. Am. J. Phys. Anthropol. *62*:275–289.

Shea BT (1985) Bivariate and multivariate growth allometry: Statistical and biological considerations. J. Zool., Lond. *206*:267–290.

Shea BT (1986) Scapula form and locomotion in chimpanzee evolution. Am. J. Phys. Anthropol. *70*:475–488.

Shea BT (1988) Heterochrony in primates. In ML McKinney (ed.): Heterochrony in Evolution. New York: Plenum Press, pp. 237–266.

Shea BT (1989) Heterochrony in human evolution: The case for neoteny reconsidered. Yrbk. Phys. Anthropol. *32*:69–101.

Shea BT (1992) Ontogenetic scaling of skeletal proportions in the talapoin monkey. J. Hum. Evol. *23*:283–307.

Shea BT, Hammer RE, and Brinster RL (1987) Growth allometry of the organs in giant transgenic mice. Endocrin. *121*:1924–1930.

Shea BT, Hammer RE, Brinster RL, and Ravosa MJ (1990) Relative growth of the skull and postcranium in giant transgenic mice. Genet. Res. Camb. *56*:21–34.

Smith RJ (1980) Rethinking allometry. J. Theor. Biol. *87*:97–111.

Smith RJ, and Jungers WL (1997) Body mass in comparative primatology. J. Hum. Evol. *32*:523–559.

Strasser E (1992) Hindlimb proportions, allometry, and biomechanics in old world monkeys (Primates, Cercopithecidae). Am. J. Phys. Anthropol. *87*:187–213.

Sutherland MR, and Godfrey LR (in prep.) A multivariate matrix model for heterochronic analysis.

Taylor AB (1995) Effects of ontogeny and sexual dimorphism on scapula morphology in the mountain gorilla (*Gorilla gorilla beringei*). Am. J. Phys. Anthropol. *98*:431–445.

Taylor AB (1997) Relative growth, ontogeny, and sexual dimorphism in *Gorilla* (*Gorilla gorilla gorilla* and *G. g. beringei*): Evolutionary and ecological considerations. Am. J. Primatol. *43*:1–31.

Vrba ES, Vainys JR, Gatesy JE, De Salle R, and Wei K-Y (1994) Analysis of paedomorphosis using allometric characters: The example of the Reduncini antelopes. Syst. Biol. *43*:92–116.

Zelditch ML, and Fink WL (1996) Heterochrony and heterotopy: Stability and innovation in the evolution of form. Paleobiology *22*:241–254.

17

BODY SIZE AND SCALING OF LONG BONE GEOMETRY, BONE STRENGTH, AND POSITIONAL BEHAVIOR IN CERCOPITHECOID PRIMATES

William L. Jungers,[1] David B. Burr,[2] and Maria S. Cole[3]

[1]Department of Anatomical Sciences
School of Medicine
SUNY at Stony Brook
Stony Brook, New York 11794-8081
[2]Departments of Anatomy and Orthopedic Surgery
Biomechanics and Biomaterials Research Center
Indiana University Medical Center at Indianapolis
Indianapolis, Indiana 46202-5120
[3]Department of Anatomy
University of Health Sciences
College of Osteopathic Medicine
Kansas City, Missouri 64106-1453

1. INTRODUCTION

Allometry in the strictest biometrical sense—size-correlated differences in shape - explains nothing. It is also not a biological "principle" (Smith, 1980; Jungers, 1984; Jungers et al., 1995; contra Gould, 1975; contra Martin, 1993). Rather, allometry is merely a quantitative description or signal that may or may not serve to test an explicit hypothesis. Without explicit hypotheses of how and why things should change as a function of body size (i.e., **similarity criteria**), allometry cannot be diagnosed except with respect to the statistical, dimensional null hypothesis of "isometry" or geometric similarity. In special circumstances, isometry can itself be a hypothetical criterion of biological similarity (Alexander et al., 1979; Biewener, 1990; Prothero, 1992). If such criteria cannot be specified and justified *a priori*, it also follows that even when allometry is discovered, it cannot be assumed that the observed **size-correlated** differences are evidence of **size-required** changes sufficient to insure "functional equivalence" (Smith, 1980). Empirical lines used

Primate Locomotion, edited by Strasser *et al.*
Plenum Press, New York, 1998

to describe allometric patterns of interspecific scaling can rarely, if ever, be rationalized into meaningful, adaptive "criteria of subtraction" for the subsequent analysis of residuals (Smith, 1984; Jungers et al., 1995). The scaling of mammalian long-bone dimensions makes these points clearly and unequivocally: although long bone robusticity is expected to increase with body size according to most biomechanical theories, positively allometric distortions in the shape of the long bones of larger vertebrates do not produce functional equivalence in any mechanical or behavioral sense. To the contrary, further behavioral and structural modifications are still required to maintain adequate safety factors at larger body sizes (Biewener, 1982, 1990; Rubin and Lanyon, 1984; Selker and Carter, 1989; Bertram and Biewener, 1990; Demes and Jungers, 1993; Jungers and Burr, 1994).

One of the very few explicit similarity criteria in organismal biology was developed by McMahon (1973, 1975) and applied widely to long bone scaling, "elastic similarity", whereby diameters were predicted to increase disproportionately relative to lengths in order to prevent Euler buckling; i.e., strong positive allometry of diameters was the specified expectation such that length scales to diameter to the two-thirds power. This criterion was qualified (Economos, 1983), modified (Hokkanen, 1986), generalized to other loading regimes (Prange, 1977; McMahon, 1984; Alexander, 1988), and frequently rejected for a wide array of mammals, including primates (e.g., Alexander et al., 1979; Aiello, 1981; Jungers, 1984; Casinos et al., 1986; Ruff, 1987; Bertram and Biewener, 1990; Demes et al., 1991; Biknevicius, 1993; Demes and Jungers, 1993). It is now recognized that no universal allometric similarity criterion exists for vertebrate long bones (Rubin and Lanyon, 1984). Instead, the scaling criteria themselves appear to change as a function of body size (Biewener, 1989, 1990, 1991). Biewener's valuable synthesis of theory and data predicts that for terrestrial mammals in the size range from 100 g to 300 kg (i.e., encompassing almost the entire range of extant primates), only slightly positive allometry of limb bone diameters is likely to occur in concert with changes in limb posture (more extended limbs in larger species) and muscular mechanical advantages (better mechanical advantages in larger animals). In terms of the dynamics of positional behavior, this multifactorial response probably implies somewhat reduced agility and compromised "athletic ability" in larger species (Alexander, 1989). We propose to test aspects of this general hypothesis here in two different ways for **cercopithecoid monkeys**: (1) via analysis of external long bone diameters for a large data set of wild-collected individuals, and (2) by similar analysis of a subset of these specimens using a cross-sectional methodology based on photon absorptiometry and linked closely to bone strength (Martin and Burr, 1984; Schaffler et al., 1985; Martin, 1991; Jungers and Burr, 1994). Note that Biewener's comprehensive model, unlike that of McMahon and others, provides no exact magic number of how diameters and cross-sectional dimensions need to scale allometrically for long bones to cope with gravity-related strains and stresses. Indeed, precise "equivalence" remains an elusive and quixotic concept in scaling, even in the realm of bone biomechanics.

As Strasser (1992) has observed, cercopithecoids are an excellent "natural experiment" for the study of size and postcranial scaling. There is a similar Bauplan for the entire group (Schultz, 1970), and the two major clades — cercopithecines and colobines — are clearly monophyletic (Strasser and Delson, 1987). Adult body mass of cercopithecines ranges from just over 1 kg (talapoin monkeys) to over 30 kg (mandrill males); colobines range from under 5 kg (olive colobus) to over 20 kg (proboscis monkey males) (Smith and Jungers, 1997). At the same time, there are noteworthy clade-specific differences in locomotor behaviors (e.g., more leaping in colobines) and substrate preferences (more terrestriality in cercopithecines) that should have mechanical and structural consequences (reviewed in Strasser, 1992). We also place our findings about cercopithecoid scaling into a broader anthropoid context and discuss several eco-behavioral implications of our results.

2. SCALING OF HUMERAL AND FEMORAL EXTERNAL DIAMETERS AT MIDSHAFT

2.1. Measurements and Skeletal Sample

It is relatively easy to capture data on the external dimensions of long bones. Here we use sliding calipers to measure inter-articular lengths of the humerus and femur as well as anteroposterior (a-p) and mediolateral (m-l) diameters at midshaft. Our skeletal sample is listed in Table 1 along with sex-specific averages for body mass. Our colobine sample includes 78 adult individuals of known body mass (data taken from field notes, museum records and specimen labels); this includes data on 15 colobine species that range in body mass here from 4.57 kg (female *Procolobus verus*) to 21.4 kg (male *Nasalis larvatus*). The cercopithecine sample is comprised of 131 individuals, all of which have associated body masses except for *Miopithecus talapoin*; we used the sex-specific averages reported by Gautier-Hion (1975) for talapoins. The body mass range for cercopithecines used here is 1.12 kg for female talapoins to 29 kg (chacma baboon males) and a single giant mandrill male at 45 kg. Trimming the size distribution of cercopithecines to exclude the talapoins and the mandrill had no significant effect on any of our results or inferences.

2.2. Statistical Methods

We have estimated the relevant scaling parameters—slope and intercept—of lengths and diameters on body mass in log-log space (base *e*) using the Model II reduced major axis (RMA) line-fitting method (Ricker, 1984; Rayner, 1985; McArdle, 1988). Ordinary least squares regression is clearly inappropriate because it assumes that there is no error term associated with the X (mass) variable. Sex-specific means of all variables were calculated for each species; these are then used as our raw data. Despite sexual dimorphism in body size in most species and unbalanced sample sizes, preliminary Model II comparisons of slopes and elevations (see below) disclosed no significant differences between sex-specific samples (all P values were > 0.5). We have adjusted degrees of freedom downward, however, due to significant phylogenetic inertia by using the variance components approach developed by Smith (1994), as implemented by a maximum likelihood procedure in the mainframe version of SAS (1985). Our reduced, "effective" sample size used for calculating confidence limits and statistical testing is therefore 22 rather than either 209 (individuals) or 57 (sex-specific averages). Colobines and cercopithecines were also examined separately and compared via Clarke's (1980) t-test for RMA slopes and Tsutakawa and Hewett's (1977) "quick test" for elevations. Rejection of statistical null hypotheses requires a probability value less than 5 percent (i.e., P<0.05). Ninety-five percent confidence intervals for the RMA slopes are computed following the recommendations of Jolicoeur and Mosimann (1968). If the 95% interval of a slope does not include the value for geometric similarity, the slope is said to describe **significant** allometry. Note that significant positive allometry of diameters (Biewener's expectation) need not imply "strong" positive allometry of the magnitude predicted by elastic or static stress similarity criteria (McMahon, 1975).

Table 1. Taxa, samples and average (sex-specific) body masses for the two analyses (external diameters and bone mineral analyzer geometry)

Cercopithecine species	External dimensions		"Mineral" geometry	
	N	Mass (kg)	N	Mass (kg)
Allenopithecus nigroviridis- m	2	5.5		
Cercopithecus aethiops- m	4	5.62	1	5.9
C. aethiops- f	6	3.57	1	3.6
C. ascanius- m	4	5.45		
C. ascanius- f	1	2.48		
C. cephus- m	3	3		
C. cephus- f	3	2.67		
C. mitis- m	7	7.98		
C. mitis- f	6	3.89		
C. neglectus- m	2	6.9		
C. neglectus- f	3	4.25		
C. diana- m	1	5.4		
C. petaurista- m	1	6		
Miopithecus talapoin- m	5	1.38		
M. talapoin- f	6	1.12		
Erythrocebus patas- f	1	4.9		
Lophocebus albigena- m	4	8.89	1	8.5
L. albigena- f	2	6.77	1	6.9
Macaca arctoides- m	2	10.1		
M. arctoides- f	1	6.02		
M. fascicularis- m	10	5.14	8	4.91
M. fascicularis- f	8	3.18	8	3.18
M. nemestrina- m	2	9.25	2	9.25
M. nemestrina- f	5	5.78	7	5.95
M. mulatta- m	2	9.46		
Mandrillus sphinx- m	1	45		
Papio h. hamadryas- m	4	20.5		
P. h. cynocephalus- m	17	24.4	12	23.6
P. h. cynocephalus- f	14	13.4	10	12.3
P. h. ursinus- m	2	29		
P. h. ursinus- f	2	16		
Theropithecus gelada- f	1	13.8		

Colobine species	External dimensions		"Mineral" geometry	
	N	Mass (kg)	N	Mass (kg)
Colobus angolensis- m	2	9.66		
C. angolensis- f	1	9.1		
C. guereza- m	7	10.4	5	10.5
C. guereza- f	10	8.04	5	8.84
C. polykomos- f	2	7.91		
Piliocolobus badius- m	4	8.91		
P. badius- f	1	7.15		
Procolobus verus- m	5	4.77		
P. verus- f	2	4.57	1	3.64
Presbytis entellus- m	1	23.6		
P. frontata- m	2	5.57		
P. hosei- f	1	5.57		
P. rubicunda- m	3	5.68		
P. rubicunda- f	2	6.14	1	6.86
P. melalophos- m	2	6.69	1	6.88
P.melalophos- f	3	6.88	8	6.78
Trachypithecus cristata- m	7	7.09	7	5.88
T. cristata- f	6	5.95		
T. obscura- m	2	7.58		
T. obscura- f	2	6.6		
T. phayrei- m	1	7.05		
T. phayrei- f	1	7.05		
Pygathrix nemaeus- m	2	10.9	2	10.9
Nasalis larvatus- m	4	21.4	4	21.4
N. larvatus- f	5	10.6	5	10.6

¹All body masses are sample-specific, wild-collected values except for *Miopithecus talapoin*. Body mass data for *M. talapoin* are taken form Gautier-Hion (1975).

2.3. Results

As a group, cercopithecoids exhibit slight, but not significant, positive allometry of humeral and femoral lengths (Table 2, Figure 1) and all external diameters except for the humeral a-p dimension (Table 2, Figure 2). The a-p humeral diameter scales with significant positive allometry (RMA slope = 0.405), but the five other variables have RMA slopes for which the 95% confidence intervals include 0.333. Most of this pattern of scaling is preserved when cercopithecines and colobines are analyzed separately; humeral a-p diameter remains the only significant departure from isometry in both clades. Femoral length slopes are now less than 0.333 in both, but not significantly so. None of the slopes are significantly different between colobines and cercopithecines, but there are two significant differences in elevation. At any given body mass, cercopithecines tend to have predictably greater m-l humeral diameters, and colobines possess predictably longer femora (also see Strasser, 1992).

3. SCALING OF HUMERAL AND FEMORAL CROSS-SECTIONAL GEOMETRY (BONE MINERAL)

3.1. Measurements and (Sub)Sample

Logistical constraints (e.g., availability and portability of bone mineral analyzers; restrictions on the transport of radioactive materials) prevented us from extending our analysis from external diameters to cross-sectional geometry for our entire sample, but it is worth noting that important research in this area has often been based on relatively small samples (e.g., McMahon, 1975; Alexander et al., 1979; Biewener, 1982; Selker and Carter, 1989). Studies such as this one focusing on specimens of known body mass are extremely rare. The subset of species and specimens that were used for this analysis is also listed in Table 1. Sex-specific mean masses are again reported by species, and it can be seen that the ranges (total, cercopithecine, and colobine) are comparable to the full cercopithecoid sample used above. The total range is from 3.6 kg (female vervet) to 23.6 kg (male Darajani baboons); the colobines range from 3.64 kg (olive colobus female) to 21.4 kg (proboscis monkey males). A total of 90 adult individuals were included in this part of the study, 51 cercopithecines and 39 colobines.

We have measured the cross-sectional, midshaft geometry of the humeri and femora using a noninvasive method based on single photon-absorptiometry (Martin and Burr, 1984). A Norland-Cameron Bone Mineral Analyzer is depicted schematically at the top of Figure 3 with an elliptical bone section. Each long bone was wrapped in a water-filled, tissue equivalency bag and scanned in the a-p and m-l anatomical planes; an average of three scans in each direction for each individual was used in all subsequent calculations to reduce the possibility of measurement error (Yezerinac et al., 1992). A hypothetical absorption curve is seen at the bottom of Figure 3; this curve is integrated to calculate bone mineral content (BMC), cortical area and second moment of area (SMA). Figure 4 is a sample a-p scan of a female Darajani baboon that reveals the peaks associated with the cortices. This method adjusts all cross-sectional geometrical measurements for differential mineralization and microscopic voids in cortical bone (Schaffler et al., 1985). It has also been demonstrated that estimates of long bone strength computed this way are highly correlated with actual bending strength (to failure) in primate long bones (r = 0.94; Martin, 1991).

Table 2. Scaling of external dimensions of the humerus and femur in cercopithecoid primates[1]

Variable	N	Cercopithecoidea			Cercopithecinae			Colobinae			Comparisons[2]	
		Slope	ln intercept	r	Slope	ln intercept	r	Slope	ln intercept	r	H_0: = slope	H_0: = elevation
Humeral A-P Diameter	22	0.405*	1.4641	0.94	0.401*	1.4891	0.96	0.444*	1.3609	0.91	accept	accept
Humeral M-L Diameter	22	0.354	1.6217	0.94	0.351	1.6672	0.97	0.381	1.5141	0.94	accept	Reject cerc > colo
Femoral A-P Diameter	22	0.359	1.6675	0.97	0.353	1.6749	0.97	0.385	1.6191	0.94	accept	Accept
Femoral M-L Diameter	22	0.347	1.7133	0.97	0.338	1.7148	0.98	0.365	1.6931	0.94	accept	Accept
Humeral Length	22	0.349	4.2947	0.97	0.345	4.3084	0.98	0.380	4.2243	0.92	accept	Accept
Femoral Length	22	0.338	4.5096	0.93	0.323	4.4962	0.96	0.306	4.6283	0.87	accept	Reject colo > cerc

[1]Reduced major axis line-fitting of ln-ln data.
[2]Slope comparisons based on Clarke's (1980) t-test; elevation comparisons based on "quick test" (Tsutakawa and Hewett, 1977).
*significant positive allometry.
N, "effective sample size" based on variance components (Smith, 1994); total N is 57 sex-specific averages. r, Pearson's correlation coefficient.

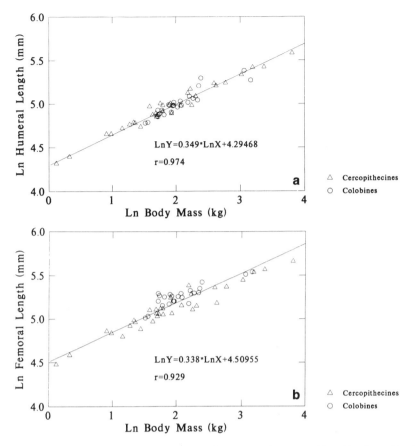

Figure 1. Bivariate scaling of (a) humerus length and (b) femoral length in cercopithecoids. Cercopithecines are triangles; colobines are circles. The reduced major axis (RMA) slope of the ln-ln relationship is provided along with the equation for the line and the parametric correlation coefficient. Neither relationship departs significantly from isometry. The separate slopes for cercopithecines and colobines are not significantly different, but colobines possess predictably longer femora.

The relevant measurements derived from the bone mineral analyzer include a-p and m-l bone widths (BW), cortical areas (CA), and second or cross-sectional moments of inertia (SMA) in both a-p and m-l planes. Section modulus (Z) was calculated in each plane as $[SMA/(0.5 \times BW)]$; Z is proportional to bending strength. Torsional rigidity is estimated by the polar moment of area (J), computed for nearly circular sections as the sum of a-p and m-l SMAs. Theoretical indices of compressive and bending strength can also be calculated (Alexander, 1983):

$$\text{Compressive Strength} = CA/\text{Body Mass} \; [cm^2/kg]$$

$$\text{Bending Strength} = Z/(0.5 \times \text{Length} \times \text{Body Mass}) \; [cm^2/kg]$$

Mass was used here as a surrogate for force because the two differ only by the gravitational constant. Theoretical bending strength has also been called an index of athletic abil-

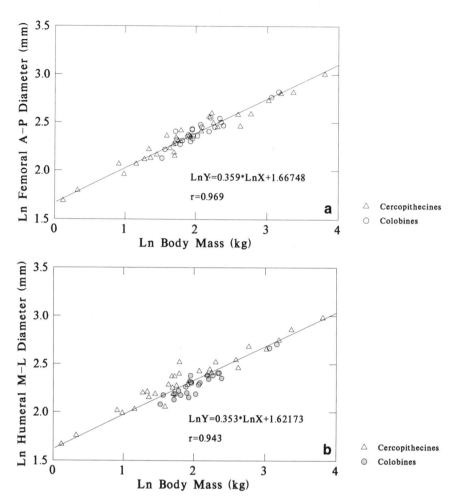

Figure 2. Bivariate scaling of external diameters in cercopithecoids: (a) the RMA slope for a-p midshaft femoral diameter is slightly positively allometric (i.e., > 0.333) but not statistically significantly so; (b) the RMA slope for m-l humeral midshaft diameter is also not significantly different from isometry, but cercopithecines have predictably larger diameters than colobines.

ity or maneuverability by Alexander (1989). Dimensionality alone indicates that both indices should decrease with increasing body size unless there is very strong positive allometry of the cross-sectional variables in the numerators (Jungers and Burr, 1994). Although these indices are static formulations of strength, they can be made more "dynamic" by acknowledging that forces within and acting upon animals may scale not in proportion to body mass, but to body mass raised to some fractional power; if Alexander's (1980, 1985a; also see Biewener, 1982) findings can be generalized, loads acting on bones may be approximately proportional to mass$^{0.67}$. This is probably accomplished in part through the size-related postural and behavioral modifications inherent in Biewener's model of terrestrial biomechanics (e.g., higher duty factors and relatively lower peak substrate forces in larger animals). Accordingly, a second set of indices of theoretical strength were calculated with mass$^{0.67}$ in addition to those using mass$^{1.0}$.

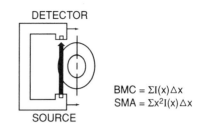

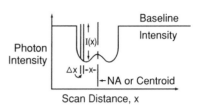

Figure 3. Schematic illustration of the bone mineral analyzer and an elliptical bone section is seen above. The hypothetical absorption curve figured below is integrated to provide bone mineral content (BMC), cortical areas and second moments of area (SMA) adjusted for mineral content. "I" refers here to photon intensity; "NA" is the neutral axis or centroid. Less intensity (the downward deflection of the curve) implies more bone mineral (see Martin and Burr [1984] for a full explanation of the method).

3.2. Statistical Methods

Slopes and intercepts of log-log bivariate relationships were again calculated using RMA line-fitting methods. Cercopithecoids are first analyzed as a group followed by separate analyses for colobines and cercopithecines. Due to very uneven sample sizes for different species and sexes, and for maximal comparability to the analysis of external diameters, sex-specific species averages were again used throughout this part of the study. In order to control partially for phylogenetic inertia and to reduce inflated degrees of freedom, effective sample sizes were estimated by the number of species rather than the number of sex-specific groups (i.e., 12 rather than 20; see Smith and Jungers, 1997; also see Felsenstein, 1985). Colobines and cercopithecines are compared again by Clarke's t-test for RMA slopes and by the "quick test" for elevations.

Spearman rank order correlations (rho) were computed between body size and the indices of strength to test for "allometry" (sensu Mosimann and James, 1979); a significant negative correlation, for example, would indicate that strength decreases predictably with increasing body size (i.e., "negative allometry"). Locally weighted smoothing or LOWESS regressions (Cleveland, 1981; Hardle, 1990) were fit to each of these relation-

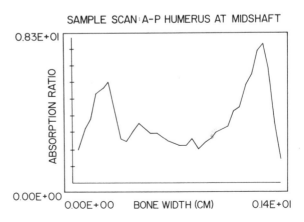

Figure 4. Sample bone mineral scan of a female Darajani baboon humerus. The absorption peaks indicate the locations of the anterior and posterior cortices.

ships ("tension" or window width was set at 0.7, a relatively stiff factor). This nonparametric approach for detecting trends within scatter plots assumes nothing about monotonicity or probable shape of the underlying distribution. A 2-by-2 matrix was created by tabulating the number of colobines and cercopithecines above and below each LOWESS curve, and the significance of clade differences was assessed by the G-statistic for independence in a row-by-column contingency table.

3.3. Results

Table 3 summarizes the information about scaling of cross-sectional geometric variables and bone lengths in our subset of Old World monkeys. Although slope values for humeral and femoral lengths in cercopithecoids as a group are again all greater than the null isometric expectation of 0.333, they are not significantly greater. The colobine results for these two variables are very similar to the larger sample considered above, but the cercopithecine slopes are higher within the subset, and the femoral length slope is greater than the humeral (although neither are significantly different from 0.333). Slopes for humeral cortical area are greater than the isometric value of 0.67 regardless of grouping (lowest in colobines), but this positive allometry is not statistically significant (Figure 5a). Femoral cortical area (Figure 5b), however, scales in a significantly positively allometric fashion for cercopithecoids and for cercopithecines (but not for colobines considered alone despite the 0.738 slope value). Humeral J (Figure 6a) is much like humeral CA (positive allometry with slopes greater than the isometric expectation of 1.333, but with confidence limits that include this isometric value), and femoral J (Figure 6b) scaling recalls that seen for femoral CA (significant positive allometry in cercopithecoids and cercopithecines, but not in colobines). Humeral SMAs and Zs are slightly positively allometric in both a-p and m-l planes for cercopithecoids as a group (i.e., slope values exceed isometric values, but not significantly so). Colobines exhibit significant positive allometry for humeral a-p SMA, and cercopithecines have humeral Zs that are significantly allometric in both anatomical planes. Femoral SMAs and Zs are all significantly allometric for cercopithecoids and cercopithecines, but not for colobines.

Despite these differences between cercopithecines and colobines, with the former exhibiting more frequent significant departures from isometry in the direction expected by Biewener's model, none of the slopes are significantly different from each other (i.e., the null hypothesis of slope equality is accepted for all 14 variables in Table 3; P>0.05). Nevertheless, cercopithecines have predictably greater slope values for all 12 geometric variables. There are also several significant differences in elevations between the two clades. Across the range of body sizes, cercopithecines have predictably larger humeral cortical areas, humeral polar moments of area, humeral second moments of area, and humeral section moduli; colobines have predictably longer femora at any given body size.

Despite the pervasive trend for geometrical properties to exceed isometric values, and often significantly so, **this positive allometry alone is clearly not sufficient to maintain comparable levels of theoretical strength as body size increases**. If force acting on bones is indeed proportional to body mass, then there is a predictable decrement in all indices of bone strength at larger body sizes (Table 4). All six rank order correlations are significantly negative. Figures 7 and 8 illustrate this finding for both compressive and bending indices of strength. However, if forces acting on bones are really proportional to mass$^{0.67}$, then a rather different picture emerges. None of the indices now have a significant negative relationship. Rather, the three humeral indices show no significant relationship with size at all, and femoral strengths actually increase with increasing body size.

Table 3. Scaling of cross-sectional geometry (mineral) of the humerus and femur in cercopithecoid primates[1]

Variable	N	Cercopithecoidea			Cercopithecinae			Colobinae			Comparisons	
		Slope	ln intercept	r	Slope	ln intercept	r	Slope	ln intercept	r	H_o: = slope	H_o: = elevation
Humeral Cortical Area	12	0.732	−3.1707	0.93	0.767	−3.1322	0.98	0.683	−3.1724	0.96	accept	Reject cerc >colo
Femoral Cortical Area	12	0.776*	−2.9767	0.98	0.800*	−3.0221	0.99	0.738	−2.8963	0.97	accept	Accept
Humeral J	12	1.491	−6.5439	0.94	1.532	−6.4020	0.98	1.438	−6.6531	0.99	accept	Reject cerc >colo
Femoral J	12	1.536*	−6.1505	0.99	1.590*	−6.2218	0.99	1.472	−6.0491	0.99	accept	Accept
Humeral A-P SMA	12	1.551	−7.3865	0.94	1.566	−7.1797	0.98	1.536*	−7.5933	0.99	accept	Reject cerc >colo
Humeral M-L SMA	12	1.456	−7.1460	0.94	1.524	−7.0707	0.98	1.362	−7.1554	0.98	accept	Reject cerc >colo
Femoral A-P SMA	12	1.547*	−6.8953	0.98	1.618*	−6.9846	0.99	1.461	−6.7658	0.99	accept	Accept
Femoral M-L SMA	12	1.531*	−6.8072	0.99	1.567*	−6.8591	0.99	1.487	−6.7318	0.99	accept	Accept
Humeral A-P Z	12	1.161	−5.8731	0.94	1.191*	−5.7610	0.98	1.127	−5.9694	0.99	accept	Reject cerc >colo
Humeral M-L Z	12	1.110	−5.7340	0.94	1.174*	−5.7011	0.98	1.018	−5.6969	0.98	accept	Reject cerc >colo
Femoral A-P Z	12	1.175*	−5.5254	0.99	1.238*	−5.6191	0.99	1.090	−5.3749	0.99	accept	Accept
Femoral M-L Z	12	1.166*	−5.4670	0.99	1.210*	−5.5373	0.99	1.105	−5.3550	0.99	accept	Accept
Humeral Length	12	0.372	1.9711	0.95	0.365	1.9905	0.98	0.386	1.9335	0.91	accept	Accept
Femoral Length	12	0.363	2.1850	0.94	0.377	2.1309	0.97	0.303	2.3412	0.93	accept	Reject colo >cerc

[1] Reduced major axis line-fitting of ln-ln data. (Same conventions as in Table 2.)

N, "effective sample size" based on number of species rather than number of sex-specific averages (20).

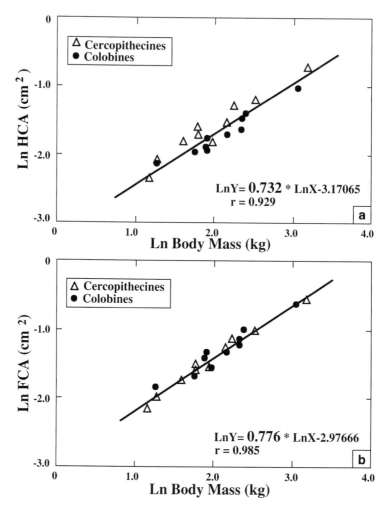

Figure 5. Bivariate scaling of (a) humeral cortical area and (b) femoral cortical area in the subset of cercopithe-coid monkeys. Cercopithecines are triangles; colobines are circles. The RMA slope for the humerus is slightly al-lometric (>0.67), but not significantly so; however, cercopithecines have predictably more cortical area than colobines. The RMA slope for the femur is significantly positively allometric, and there are no predictable clade-specific differences in cortical area (but recall that colobines have longer femora).

Clearly, how loads acting on bones scale is of critical importance, and if animals can somehow reduce peak forces at larger body sizes, then decreases in strength can be avoided with only the modest structural allometries documented here.

The scaling trends seen above, especially the series of significant elevational differences between the two clades, result in some consistent differences in theoretical strengths between cercopithecines and colobines. More specifically, G-tests of the distribution of taxa above and below LOWESS curves disclose significant differences between clades for all indices of strength except femoral compressive strength. In all other comparisons, cercopithecines possess predictably stronger bones than their colobine counterparts. These trends are easily discernible in Figures 7 and 8. This difference holds no matter which estimate of force is used.

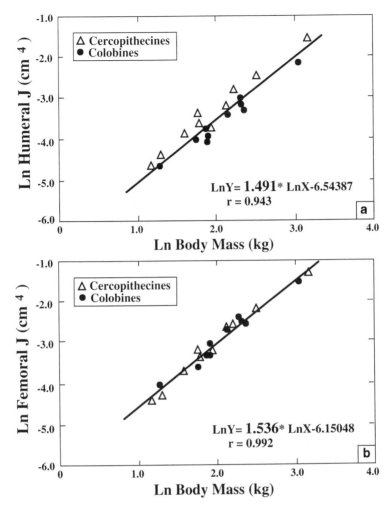

Figure 6. Bivariate scaling of (a) humeral polar moment of area and (b) femoral polar moment of area in Old World monkeys. The RMA slope for the humerus is not significantly different from isometry, but cercopithecines possess predictably larger polar moments than colobines. The femoral slope is significantly positively allometric (>1.33), but there are no clade-specific differences in elevations.

4. DISCUSSION

4.1. General Trends and Evaluation of Model Expectations

Although there are differences in the details of how external diameters and cross-sectional geometry scale within cercopithecoid long bones (see below), the results of both types of analyses are consistent with and corroborate the structural component of Biewener's multifactorial model for "terrestrial" locomotion and long bone allometry. Primates, including cercopithecoids, are different from most other mammals in various aspects of their anatomy and quadrupedalism: they have longer and more robust limb bones (Alexander, 1985b; Kimura, 1991; Polk et al., 1997), their long bones tend to be less

Table 4. Spearman rank order correlations (rho) between body mass and theoretical bone
strength in cercopithecoid primates[1]

Variable	Force ∝ Mass	Sig.	Force ∝ Mass$^{0.67}$	Sig.
Humeral compressive strength	−0.71	P<0.001	−0.03	ns, P>0.9
Femoral compressive strength	−0.82	P<0.001	0.52	P<0.01
Humeral a-p bending strength	−0.49	P<0.03	0.11	ns, P>0.6
Humeral m-l bending strength	−0.69	P<0.001	−0.01	ns, P>0.9
Femoral a-p bending strength	−0.63	P<0.005	0.52	P<0.05
Femoral m-l bending strength	−0.62	P<0.005	0.62	P<0.005

[1]Mass was used as "force" here without multiplying by the gravitational constant; load-sharing by the upper
and lower limb was assumed to be roughly equal under both force regimes.
Sig., significance; ns, non-significant.

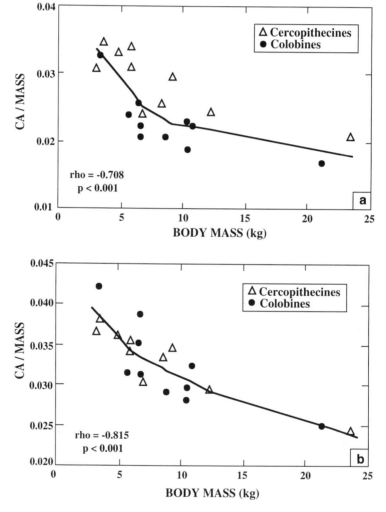

Figure 7. Theoretical compressive strength of the (a) midshaft humerus and (b) midshaft femur as a function of
body mass in cercopithecoids. Strength decreases in both elements as size increases; the LOWESS curve follows
this decline. Cercopithecines tend to have stronger humeri at any given body size than do colobines. Force acting
on the bones is assumed to be proportional to body mass, but changes in posture and locomotor behavior probably
also change allometrically to preserve adequate safety factors.

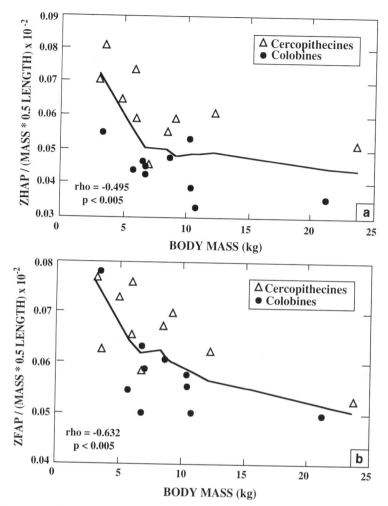

Figure 8. Theoretical a-p bending strength of the (a) humeral midshaft and (b) femoral midshaft in Old World monkeys. Strength decreases as a function of body size; the LOWESS curves track this decline in bending strength. Behavioral changes presumably compensate for this apparent lack of functional equivalence. Note that cercopithecines have predictably stronger bones than colobines, and this may reflect clade-specific differences in the compliance of their habitual substrates (i.e., stiffer terrestrial substrates are experienced by many cercopithecines).

curved (Swartz, 1990), they tend to take longer strides with lower stride frequencies (Alexander and Maloiy, 1984), they support more of their body weight on their hindlimbs (Kimura et al., 1979; Reynolds, 1985; Demes et al., 1994), and their footfall sequences are unusual for mammals (Hildebrand, 1967; Shapiro et al., 1997). Many of the cercopithecoids are also better characterized as arboreal quadrupeds rather than as terrestrial quadrupeds (Rowe, 1996), and the mechanical properties of an arboreal substrate differ considerably from those experienced on the ground (Alexander, 1991; Cant, 1994; Schmitt, 1994). Nevertheless, the pervasive finding here for cercopithecoids was slightly positive allometric scaling of supporting dimensions of the long bones, only some of which were statistically significant departures from isometry. This closely matches

Biewener's general conclusions for quadrupedal species within the size range of Old World monkeys (Biewener, 1990).

The uniformly negative allometry of theoretical bone strength (with forces assumed to be proportional to body mass or weight) in cercopithecoids indicates that this slight degree of allometric reinforcement is insufficient to maintain mechanical equivalence at larger body sizes. In other words, long bone allometry alone is an inadequate solution to size-related increases in loads and stresses in this group; hence, it would be illogical and simply wrong to view these allometries as evidence of "functional equivalence". Recall that Biewener's model predicts other types of size-required changes to maintain reasonable safety factors, and these include changes in limb posture and muscular mechanical advantages. We currently possess little more than anecdotal evidence for primates, but there is reason to believe that larger primate quadrupeds do move with more extended limb joints (see Figure 9, adapted from Fleagle, 1988); this could rotate the substrate reaction force acting on each limb more into line with the limb's long axis and thereby reduce its flexing moment arm. Bending moments would be reduced as a consequence, and extended limbs could then support weight-related loads with less muscular effort, especially if bony lever arms of the extensor musculature are also longer and/or more favorably oriented in larger species (e.g., re-oriented and relatively longer olecranon processes for the triceps brachii muscle mass; see Jolly, 1972; Rose, 1993). Although much more data are needed, some of these suspicions have been confirmed recently for cercopithecine forelimbs by Schmitt (1995), and cercopithecoids therefore appear to present the full package of expectations developed by Biewener and others for truly terrestrial mammals. If peak forces per unit body mass are somehow reduced at larger cercopithecoid body sizes, then the more "dynamic" version of the strength indices (the third column in Table 4) might be more realistic; as such, they would predict that bony stresses and strains do not increase much, if at all, with increasing body size due to a combination of structural and behavioral modifications (Biewener, 1982, 1993; Rubin, 1984; Rubin and Lanyon, 1984).

Limb postures may differ between small and large monkeys, but it is not clear if overall "athletic ability", agility and/or dynamic maneuverability have also been sacrificed to any great degree by larger cercopithecoids. For example, in McGraw's (1996) study of Tai Forest (Ivory Coast) cercopithecoids, body size alone did not predict well the observed frequencies of climbing versus leaping behaviors; e.g., the most frequent climber was one of the smallest monkeys and leaping was correlated more with phylogeny than size (colo-

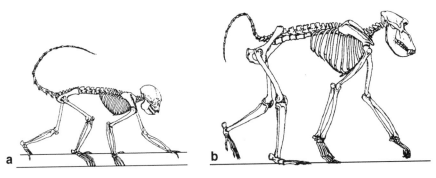

Figure 9. Cartoon contrast between a smaller, more flexed monkey on the left versus a larger, more extended monkey on the right. These are some the predicted size-related differences in behavior that are believed to compensate for only slightly positive allometry of supporting dimensions. Adapted from Fleagle (1988).

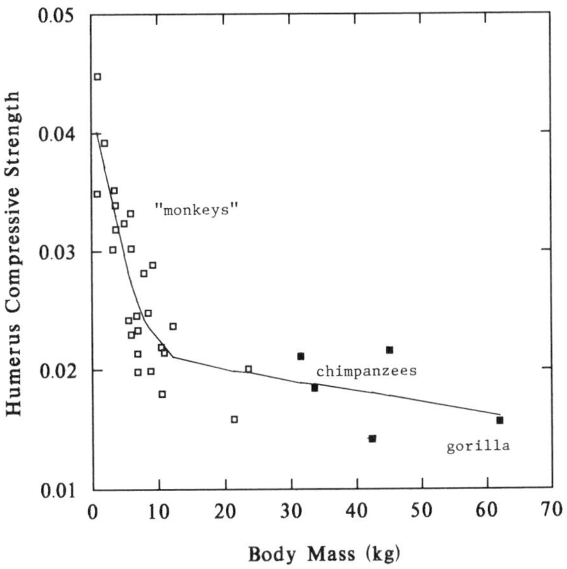

Figure 10. Theoretical compressive strength of the midshaft humerus in Neotropical and Old World monkeys (open squares) and African apes (filled squares). The addition of smaller monkeys such as the squirrel monkey (extreme left open squares) and the larger apes has little impact on the size-related trends documented for cercopithecoids alone. The rapid decline in theoretical strength levels off by approximately 20 kg; the LOWESS curve follows this decay graphically.

bines were the most frequent leapers at Tai). Comparable data on the largest cercopithecoids would be very interesting, but are currently lacking.

The negatively allometric pattern of theoretical strengths documented for cercopithecoids remains intact if New World monkeys, including small squirrel monkeys, are added to the sample (see Jungers and Burr, 1994, for details of the platyrrhine sample). The largest quadrupedal anthropoids are African apes, and even if chimpanzees and gorillas are added to the overall picture with platyrrhines and cercopithecoids (Figure 10), this negatively allometric pattern for strength still persists, although it does appear to level off quickly after approximately 20 kg. Comparable size-related decrements in the theoretical bone strength of prosimian primates have also been documented (Demes and Jungers, 1993), and Biewener's model probably obtains for all primates whether they are quadrupedal or not. In fact, if Asian apes (gibbons, siamang and orang-utans) are considered together as a size-graded group of suspensory primates (Figure 11), their theoretical strengths also decrease and can be superimposed onto the quadrupedal trends. It is also the case that orang-utan locomotion is considerably less dynamic than that practiced by lesser apes, so functional safety factors may differ little among them despite dramatic differences in body size.

4.2. External Dimensions versus Cross-Sectional Geometry

Only the humeral a-p diameter was significantly positively allometric in the analysis of external dimensions, whereas significant departures from isometry were more common for the cross-sectional data collected with the bone mineral analyzer. There are several possible, and not mutually exclusive, reasons for this difference. Although information drawn from external diameters can be and often is correlated with geometrical data extracted from cross-sections (e.g., Jungers and Minns, 1979; Ruff, 1989), a more complete picture of bone distribution is provided by approaches that take into account the internal boundaries of cortical bone (Biewener, 1982; Ruff, 1989; Ruff and Runestad, 1992). In human evolution, for example (Ruff et al., 1993; Grine et al., 1995), differences in the amount of cortical bone and corresponding diaphyseal strengths are profoundly affected

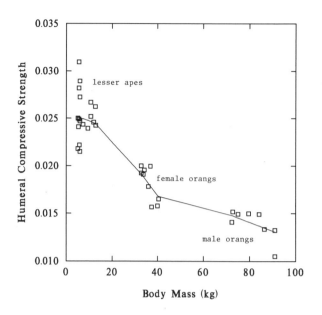

Figure 11. Theoretical compressive strength of the midshaft humerus in lesser apes and orang-utans. The size-related decrease in strength seen in these suspensory primates mirrors that seen in more pronograde quadrupeds.

by significant variation in medullary diameters—information simply unavailable to the most precise calipers. Theoretical and empirical measures of bone strength are clearly more closely related to sectional moduli and cortical areas than they are to external diameters alone. As such, the allometric trends in cross-sectional geometry are probably tracking the most relevant biomechanical signals (i.e., cross-sectional variables could be allometric even if external dimensions are consistently closer to isometry). It should also be noted that the method used here that weights cortical bone distribution by mineral content (Martin and Burr, 1984; Gilbert et al., 1989; Ruff, 1989) could also contribute to observed differences from external diameter scaling. Finally, the cross-sectional analysis presented here is based only on a subset of the sample used for external diameters, and some of the observed differences could simply reflect these different sampling strategies (e.g., the absolute ranges of body mass are slightly different). Data and inferences drawn from external diameters are certainly valuable, and many of the general conclusions offered here based on midshaft diameters are similar to those derived from more detailed cross-sectional geometry. Given a choice, however, we would still opt for the cross-sectional data whenever a more complete picture of long bone biomechanics is desired.

4.3. Clade-Specific Differences and Substrate Influences

For both cross-sectional variables and external diameters, no significant differences were found between colobines and cercopithecines in slope values. Oddly, however, cercopithecines had greater slopes for all geometric variables, but colobines had greater slopes for all external diameters. Significant differences were found for several elevations, with cercopithecines having more bone in cross-section than colobines of comparable body size for all humeral variables. The finding that colobines had predictably longer femora than cercopithecines is perhaps not surprising given the greater incidence of leaping in colobines (Fleagle, 1988; Strasser, 1992; McGraw, 1996). With the exception of their femoral compressive strength, cercopithecines had predictably stronger long bones than colobines. One might argue that the longer femora of colobines contributes to their lower bending strengths for this element (since the load arm is proportional to bone length), but this argument would fail to account for the same pattern in humeral compressive and bending strengths. Why, then, are cercopithecine long bones structurally reinforced in comparison to those of colobines?

Arboreal and terrestrial substrates have quite different mechanical properties. In particular, branches and trees are usually more compliant than the ground, and this difference should impact on the mechanics of locomotion (Alexander, 1991; Cant, 1994). In fact, Schmitt (1994, 1995, this volume) has documented reductions in peak forces acting on primates moving on branches compared to their locomotion on the ground; substrate reaction force directions are also predictably different in the mediolateral plane. Cercopithecines as a group are more terrestrial than colobines, and it seems reasonable to connect differences in habitual substrate stiffness to differences in long bone geometry and strength. In other words, despite their propensity for leaping from and to arboreal supports, quadrupedal colobines may experience relatively lower peak stresses because of greater substrate compliance, and hence require less bone in cross-section to resist bending moments and axial loads. The greater commitment to terrestriality seen in cercopithecines requires them to move more frequently on stiffer substrates; higher peak forces may be one of the prices of this substrate preference, and stronger long bones may be a necessary corollary of this habitat difference (also see Burr et al., 1989).

We now know much about the scaling of bone geometry (although its developmental basis remains highly theoretical [van der Meulen and Carter, 1995]), but the study of how substrate mechanics are connected to primate locomotor biomechanics and morphology is still very much in its infancy (Alexander, 1991; Crompton et al., 1993; Cant, 1994; Schmitt, 1995; but also see McMahon [1984] on compliant running tracks, and Demes et al. [1995] on compliant supports in leaping prosimians). We also need a great deal more and better focused information about the allometry of primate locomotion itself, especially quantitative data on size-correlated differences in kinematics and kinetics.

ACKNOWLEDGMENTS

We thank the editors for inviting us to contribute to this volume and for the opportunity to celebrate the memory of Warren Kinzey. We gratefully acknowledge the assistance of the curators of valuable skeletal material for giving us access to specimens in their charge (at the Smithsonian Institution, the American Museum of Natural History, the Museum of Comparative Zoology, the Field Museum of Natural History, the University of Texas, the Tervuren Museum, the Powell-Cotton Museum, the British Museum of Natural History, and the University of Gottingen). R. Susman and S. McGraw kindly allowed us to measure specimens in their private collections. We also thank Luci Betti-Nash for her skillful artwork. Discussions with Brigitte Demes about bone biomechanics and with Eric Delson about cercopithecoid body size were very helpful. We also offer our sincere thanks to Beth Strasser, Steve Churchill, and two of the three anonymous reviewers for their careful reading of our paper and their valuable suggestions. This research was supported by NSF Grants BNS 8606781 and SBR 9507078, the Leakey Foundation, and the Boise Fund.

REFERENCES

Aiello LC (1981) The allometry of primate body proportions. Symp. Zool. Soc. Lond. *48*:331–358.

Alexander R McN (1980) Forces in animal joints. Engineering in Medicine *9*:93–97.

Alexander R McN (1983) On the massive legs of a Moa (*Pachyornis elephantopus*, Dinornithes). J. Zool., Lond. *201*:363–376.

Alexander R McN (1985a) The maximum forces exerted by animals. J. exp. Biol. *115*:231–238.

Alexander R McN (1985b) Body size and limb design in primates and other mammals. In WL Jungers (ed.): Size and Scaling in Primate Biology. New York: Plenum Press, pp. 337–343.

Alexander R McN (1988) Elastic Mechanisms in Animal Movement. Cambridge: Cambridge University Press.

Alexander R McN (1989) Mechanics of fossil vertebrates. J. Geol. Soc. Lond. *146*:41–52.

Alexander R McN (1991) Elastic mechanisms in primate locomotion. Z. Morph. Anthropol. *78*:315–320.

Alexander R McN, Jayes AS, Maloiy GMO, and Wathuta EM (1979) Allometry of the limb bones of mammals from shrews (*Sorex*) to elephant (*Loxodonta*). J. Zool., Lond. *189*:305–314.

Alexander R McN, and Maloiy GMO (1984) Stride lengths and stride frequencies of primates. J. Zool., Lond. *202*:577–582.

Bertram JEA, and Biewener AA (1990) Differential scaling of the long bones in the terrestrial Carnivora and other mammals. J. Morph. *204*:157–169.

Biewener AA (1982) Bone strength in small mammals and bipedal birds: Do safety factors change with body size? J. exp. Biol. *98*:289–301.

Biewener AA (1989) Mammalian terrestrial locomotion and size. BioSci. *39*:776–783.

Biewener AA (1990) Biomechanics of mammalian terrestrial locomotion. Science *250*:1097–1103.

Biewener AA (1991) Musculoskeletal design in relation to body size. J. Biomech. *24* (suppl. 1):19–29.

Biewener AA (1993) Safety factors in bone strength. Calcif. Tissue Int. *53* (suppl.):S68-S74.

Biknevicius AR (1993) Biomechanical scaling of limb bones and differential limb use in caviomorph rodents. J. Mammal. *74*:95–107.

Burr DB, Ruff CB, and Johnson C (1989) Structural adaptations of the femur and humerus to arboreal and terrestrial environments in three species of macaque. Am. J. Phys. Anthropol. *79*:357–368.

Cant JGH (1994) Positional behavior of arboreal primates and habitat compliance. In B Thierry (ed.): Current Primatology, Vol. 1. Ecology and Evolution. Strasbourg: Université Louis Pasteur, pp. 187–193.

Casinos A, Bou J, Castiella MJ, and Viladiu C (1986) On the allometry of long bones in dogs. J Morph. *190*:73–79.

Clarke MRB (1980) The reduced major axis of a bivariate sample. Biometrika *67*:441–446.

Cleveland WS (1981) LOWESS: A program for smoothing scatterplots by robust locally weighted regression. Amer. Statistician *35*:54.

Crompton RH, Sellers WI, and Gunther MM (1993) Energetic efficiency and ecology as selective forces in the saltatory adaptation of prosimian primates. Proc. Roy. Soc. Lond.(B) 254:41–45.

Demes B, and Jungers WL (1993) Long bone cross-sectional dimensions, locomotor adaptations and body size in prosimian primates. J Hum. Evol. *25*:57–74.

Demes B, Jungers WL, Gross TS, and Fleagle JG (1995) Kinetics of leaping primates: Influence of substrate orientation and compliance. Am. J. Phys. Anthropol. *96*:419–430.

Demes B, Larson SG, Stern JT Jr., Jungers WL, Biknevicius AR, and Schmitt D (1994) The kinetics of primate quadrupedalism: "hindlimb drive" reconsidered. J. Hum. Evol. *26*:353–374.

Demes B, Jungers WL, and Selpien K (1991) Body size, locomotion, and long bone cross-sectional geometry in indriid primates. Am. J. Phys. Anthropol. *86*:537–547.

Economos AC (1983) Elastic and/or geometric similarity in mammalian design? J. Theor. Biol. *103*:167–172.

Felsenstein J (1985) Phylogenies and the comparative method. Am. Natur. *125*:1–15.

Fleagle JG (1988) Primate Adaptation and Evolution. New York: Academic Press.

Gautier-Hion A (1975) Dimorphisme sexuel et organisation sociale chez les cercopithecines forestiers Africains. Mammalia *39*:365–374.

Gilbert JA, Skrzynski MC, and Lester GE (1989) Cross-sectional moment of inertia of the distal radius from absorptiometric data. J. Biomech. *22*:751–754.

Gould SJ (1975) On the scaling of tooth size in mammals. Am. Zool. *15*:351–362.

Grine FE, Jungers WL, Tobias PV, and Pearson OM (1995) Fossil *Homo* femur from Berg Aukas, Northern Namibia. Am. J. Phys. Anthropol. *97*:151–185.

Hardle W (1990) Applied Nonparametric Regression. Cambridge: Cambridge University Press.

Hildebrand M (1967) Symmetrical gaits of primates. Am. J. Phys. Anthropol. *26*:119–130.

Hokkanen JEI (1986) Notes concerning elastic similarity. J. Theor. Biol. *120*:499–501.

Jolicoeur P, and Mosimann JE (1968) Intervalles de confiance pour la pente de l'axe majeur d'une distribution normale bidimensionnelle. Biom.-Praxim. *9*:121–140.

Jolly CJ (1972) The classification and natural history of *Theropithecus* (*Simopithecus*) (Andrews, 1916), baboons of the African Plio-Pleistocene. Bull. Brit. Mus. (Nat. Hist.) Geol. Ser. *22*:1–123.

Jungers WL (1984) Aspects of size and scaling in primate biology with special reference to the locomotor skeleton. Yrbk. Phys. Anthropol. *27*:73–97.

Jungers WL, and Burr DB (1994) Body size, long bone geometry and locomotion in quadrupedal monkeys. Z. Morph. Anthropol. *80*:89–97.

Jungers WL, Falsetti AB, and Wall CE (1995) Shape, relative size, and size-adjustments in morphometrics. Yrbk. Phys. Anthropol. *38*:137–161.

Jungers WL and Minns RJ (1979) Computed tomography and biomechanical analysis of fossil long bones. Am. J. Phys. Anthropol. *50*:285–290.

Kimura T (1991) Long and robust limb bones of primates. In A Ehara (ed.): Primatology Today. Amsterdam: Elsevier Science Publishers, pp. 495–498.

Kimura T, Okada M, and Ishida H (1979) Kinesiological characteristics of primate walking: its significance in human walking. In M Morbeck, H Preuschoft, and N Gomberg (eds.): Environment, Behavior and Morphology: Dynamic Interactions in Primates. New York: Fischer, pp. 297–311.

Martin RB (1991) Determinants of the mechanical properties of bones. J. Biomech. *24* (suppl. 1):79–88.

Martin RB, and Burr DB (1984) Non-invasive measurement of long bone cross-sectional moments of inertia by photon absorptiometry. J. Biomech. *17*:195–201.

Martin RD (1993) Allometric aspects of skull morphology in *Theropithecus*. In NG Jablonski (ed.): *Theropithecus*. The Rise and Fall of a Primate Genus. Cambridge: Cambridge University Press, pp. 273–298.

McArdle BH (1988) The structural relationship: regression in biology. Can. J. Zool. 66:2329–2339.

McGraw WS (1996) The positional behavior and habitat use of six sympatric monkeys in the Tai Forest, Ivory Coast. Ph.D Dissertation, State University of New York at Stony Brook.

McMahon TA (1973) Size and shape in biology. Science *179*:1201–1204.

McMahon TA (1975) Using body size to understand the structural design of animals: quadrupedal locomotion. J. Appl. Physiol. *39*:619–627.

McMahon TA (1984) Muscles, Reflexes, and Locomotion. Princeton: Princeton University Press.

Mosimann JE, and James FC (1979) New statistical methods for allometry with applications to Florida red-winged blackbirds. Evolution 23:444–459.

Polk JD, Demes B, Jungers WL, Heinrich RE, Biknevicius AR, and Runestad JA (1997) Cross-sectional properties of primate and nonprimate limb bones. Am. J. Phys. Anthropol., suppl. *24*:188.

Prange HD (1977) The scaling and mechanics of arthropod exoskeletons. In TJ Pedley (ed.): Scale Effects in Animal Locomotion. London: Academic Press, pp. 169–181.

Prothero J (1992) Scaling of bodily proportions in adult terrestrial mammals. Am. J. Physiol. *262*:R492-R503.

Rayner JMV (1985) Linear relationships in biomechanics: the statistics of scaling functions. J. Zool., Lond. *206*:415–439.

Reynolds TR (1985) Stress on the limbs of quadrupedal primates. Am. J. Phys. Anthropol. 67:351–362.

Ricker WE (1984) Computation and uses of central trend lines. Can. J. Zool. *62*:1897–1905.

Rose MD (1993) Functional anatomy of the elbow and forearm in primates. In DL Gebo (ed.): Postcranial Adaptation in Nonhuman Primates. DeKalb: Northern Illinois University Press, pp. 70–95.

Rowe N (1996) The Pictorial Guide to the Living Primates. East Hampton: Pogonias Press.

Rubin CT (1984) Skeletal strain and the functional significance of bone architecture. Calcif. Tissue Int. *36*:S11-S18.

Rubin CT, and Lanyon LE (1984) Dynamic strain similarity in vertebrates: an alternative to allometric limb bone scaling. J. Theor. Biol. *107*:321–327.

Ruff CB (1987) Structural allometry of the femur and tibia in Hominoidea and *Macaca*. Folia Primatol. *48*:9–49.

Ruff CB (1989) New approaches to structural evolution of limb bones in primates. Folia Primatol. *53*:142–159.

Ruff CB and Runestad JA (1992) Primate limb bone structural adaptations. Ann. Rev. Anthropol. *21*:407–433.

Ruff CB, Trinkaus E, Walker A, and Larsen CP (1993) Postcranial robusticity in *Homo*. I: Temporal trends and mechanical interpretation. Am. J. Phys. Anthropol. *91*:21–53.

SAS (1985) SAS Users's Guide: Statistics, Version 5. Cary: SAS Institute Inc.

Schaffler MB, Burr DB, Jungers WL, and Ruff CB (1985) Structural and mechanical indicators of limb specialization in primates. Folia Primatol. *45*:61–75.

Schmitt D (1994) Forelimb mechanics as a function of substrate type during quadrupedalism in two anthropoid primates. J. Hum. Evol. 26:441–458.

Schmitt D (1995) A kinematic and kinetic analysis of forelimb use during arboreal and terrestrial quadrupedalism in Old World monkeys. Ph.D Dissertation, State University of New York at Stony Brook.

Schmitt D (this volume) Forelimb mechanics during arboreal and terrestrial quadrupedalism in primates. In E Strasser, JG Fleagle, AL Rosenberger, and HM McHenry (eds.) Primate Locomotion: Recent Advances. New York: Plenum Press, pp. 175–200.

Schultz AH (1970) The comparative uniformity of the Cercopithecoidea. In J Napier and P Napier (eds.): Old World Monkeys. New York: Academic Press, pp. 39–51.

Selker F, and Carter DR (1989) Scaling of long bone fracture strength with animal mass. J. Biomech. *22*:1175–1183.

Shapiro LJ, Anapol FC, and Jungers WL (1997) Interlimb coordination, gait, and neural control of quadrupedalism in chimpanzees. Am. J. Phys. Anthropol. *102*:177–186.

Smith RJ (1980) Rethinking allometry. J. Theor. Bio. *87*:97–111.

Smith RJ (1984) Determination of relative size: the "criterion of subtraction" problem in allometry. J. Theor. Bio. *108*:131–142.

Smith RJ (1994) Degrees of freedom in interspecific allometry: an adjustment for the effects of phylogenetic constraint. Am. J. Phys. Anthropol. *93*:95–107.

Smith RJ, and Jungers WL (1997) Body mass in comparative primatology. J. Hum. Evol. *32*:523–559.

Strasser E (1992) Hindlimb proportions, allometry, and biomechanics in Old World monkeys (Primates, Cercopithecidae). Am. J. Phys. Anthropol. *87*:187–213.

Strasser E and Delson E (1987) Cladistic analysis of cercopithecid relationships. J. Hum. Evol. *16*:81–99.

Swartz SM (1990) Curvature of the forelimb bones of anthropoid primates: overall patterns and specializations in suspensory species. Am. J. Phys. Anthropol. *83*:477–498.

Tsutakawa RK, and Hewett JE (1977) Quick test for comparing two populations with bivariate data. Biometrics *33*:215–219.

Van der Meulen MCH, and Carter DR (1995) Developmental mechanics determine long bone allometry. J. Theor. Biol. *172*:323–327.

Yezerinac SM, Lougheed SC, and Handford P (1992) Measurement error and morphometric studies: statistical power and observer experience. Syst. Bio. *41*:471–482.

IV

FOSSILS AND RECONSTRUCTING THE ORIGINS AND EVOLUTION OF TAXA

INTRODUCTION TO PART IV

Henry M. McHenry

There has been a spectacular enrichment of the fossil record of monkeys, apes and people in the last two decades. There has been a concomitant improvement of methodology to interpret this record. The final six chapters reflect this progress.

As Carol Ward points out in Chapter 18, the two best known early Miocene hominoid genera, *Proconsul* and *Afropithecus*, are very different in their craniodental morphology. In a previous study, Ward and co-authors (Begun et al., 1997) found that the most parsimonious cladogram defined by 240 characters separates these genera into two widely separate clades. The face and anterior dentition of *Afropithecus* appears to be specialized for sclerocarp foraging similar to that seen in *Cacajao* and other members of the Pithecinae of South America as described by Kinzey (1992). In this respect *Afropithecus* resembles the middle Miocene genus, *Kenyapithecus* (McCrossin et al., Chapter 19). The craniodental morphology of *Proconsul*, on the other hand, is adapted to a more generalized frugivorous diet. But the locomotor skeletons of the two genera are remarkably similar. Both have some peculiarities seen in modern Hominoidea such as a well-defined *zona conoidea* of the distal humerus associated with use of the forelimb in a wide variety of positions of pronation and supination. Both share a suite of traits associated with arboreal quadrupedalism without the forelimb-dominated climbing and suspensory features characteristic of modern hominoids. "The disparity in craniofacial and dental form between *Afropithecus* and *Proconsul*" Ward (pg. 350) points out, "suggests that early hominoid locomotor adaptations did not limit or constrain ecological diversification, and may even have facilitated the initial hominoid radiation."

By Middle Miocene times (16–14 m.y.a.) terrestrial adaptations appear in both Cercopithecoidea (*Victoriapithecus*) and Hominoidea (*Kenyapithecus*) as Monte McCrossin and his collaborators present in Chapter 19. This is a remarkable new insight made possible especially by the authors and their crews' heroic efforts on Maboko Island of Kenya. Both fossil genera have features of the shoulder, elbow, and foot that are associated with the more terrestrially adapted modern cercopithecoids. The paleoenvironment is woodland, not open-country savanna. The morphology of the face and anterior dentition of *Kenyapithecus* shows the key traits associated with sclerocarp foraging as seen in *Afropithecus, Chiropotes, Pithecia* and *Cacajao*. Perhaps, the authors speculate, terrestriality in Hominoidea arose as an adaptation to this type of foraging.

In what context did hominid bipedalism arise? As McCrossin et al. point out, the terrestrial adaptation of *Kenyapithecus* is one clue: it is likely that terrestrial quadrupedalism preceded bipedalism. Alternative views still hold strong appeal such as that articulated by Tuttle (1974) and others that the immediate ancestors of the first bipeds were arboreal, small-bodied apes whose terrestrial traverses were bipedal similar to those seen in modern species of *Hylobates*. But the next two chapters (20 and 21) lend support to the view that the immediate ancestors of hominid bipeds were terrestrial quadrupeds not arboreal hylobatians.

Kevin Hunt (Chapter 20) approaches bipedal origins from his experience with observing free-ranging chimps in East Africa. With 701 hours of observations behind what he says, the fact that 80% of chimp bipedalism involves feeding carries authority. He points out that the morphology of *A. afarensis* and other early hominids corresponds to what one would expect from an animal that, like the modern chimps, stands bipedally to feed on plant material either on the ground (mostly) or in trees. Robert Foley and Sarah Elton (Chapter 21) follow by explaining the energetics of hominoid locomotion with models predicting the costs of different forms of locomotion in various contexts. The results show that it is likely that bipedalism evolved from an ancestor who spent a majority of its daylight hours on the ground and only a smaller part of its activity in the trees.

In Chapter 22 Russell Tuttle and collaborators reflect on the key evidence for early hominid bipedalism and particularly the footprints of Laetoli and the foot morphology of the Hadar australopithecines. By their extensive analyses the footprints are indistinguishable from modern people. The Hadar fossils, however, retain many primitive features including long and curved toes. From this it is reasonable to propose that there is more than one kind of bipedal hominoid in the Pliocene. It is also likely, they argue, that squatting played a significant role in the early stages of bipedalism.

The best known early hominid biped, *Australopithecus afarensis*, is assessed by its original describers as having "...adaptation to full and complete bipedality" (Johanson et al., 1982:386) and a forelimb "...not primarily involved in locomotor behavior" (Johanson et al., 1982:385). But as Hunt, Tuttle et al., and many others point out, there are many peculiarities of its postcranial skeleton that need to be explained. In Chapter 23 Christopher Ruff provides one explanation by careful application of biomechanical principles and particularly his own analyses of the mechanical properties of the femoral shaft. In the mediolateral plane, the hip of A.L. 288-1 (Lucy) is unlike modern humans in its relatively short femur, long femoral neck, wide biacetabular diameter, cortical thickness of the femoral shaft and other features. Some of these peculiarities are consistent with the model describing modern human hips, but not all. If the pelvis is tilted up slightly on the unsupported side during stance phase, however, the predictions from the model fit much more reasonably. This also fits with the elongated foot observed in *A. afarensis*.

Unanimity of opinion on all aspects of interpretation is not achieved here. Careful reading reveals many differences of method and conclusion. This is a measure of health and vitality of the field, of course. Add the fact that the fossil record samples only a tiny portion of our family's history to appreciate the open field before us. We can delight in what new discoveries and approaches will bring.

REFERENCES

Begun DR, Ward CV, and Rose MD (1997) Events in hominoid evolution. In DR Begun, CV Ward, and MD Rose (eds.): Function, Phylogeny, and Fossils: Miocene Hominoid Evolution and Adaptation. New York: Plenum Press, pp. 389–415.

Johanson DC, Taieb M, and Coppens Y (1982) Pliocene hominids from the Hadar Formation, Ethiopia (1973–1977): Stratigraphic, chronologic, and paleoenvironmental contests, with notes on hominid morphology and systematics. Am. J. Phys. Anthropol. *57*:373–420.

Kinzey WG (1992) Dietary and dental adaptations in the Pithecinae. Am. J. Phys. Anthropol. *88*:499–514.

Tuttle RH (1974) Darwin's apes, dental apes, and descent of man - normal science in evolutionary anthropology. Curr. Anthropol. *15*:389–398.

18

AFROPITHECUS, PROCONSUL, AND THE PRIMITIVE HOMINOID SKELETON

Carol V. Ward

Departments of Anthropology and Pathology and Anatomical Sciences
107 Swallow Hall
University of Missouri
Columbia, Missouri 65211

1. INTRODUCTION

Many distinctive synapomorphies of modern apes (and humans) are found in the postcranial skeleton. These characters reflect a basic adaptation, variably developed and practiced among modern species, to forelimb-dominated arboreal locomotion, including climbing, brachiation, and/or forelimb suspension (e.g., Cartmill and Milton, 1977; Fleagle et al., 1981; Hunt, 1992; Keith, 1923; Stern, 1971; Stern et al., 1977). The morphological pattern shared by modern hominoids has led to the general assumption that locomotor divergence was an initial hallmark of the hominoid lineage, setting them apart from their monkey-like forbears. As more is learned about the earliest hominoids, however, paleontologists realize that not all apes share a similar pattern of postcranial anatomy and locomotor behavior, and that the suite of features seen in extant apes evolved in a mosaic fashion over the course of hominoid evolutionary history (reviews and references in Begun et al., 1997a).

Hundreds of hominoid fossils have been discovered from the Miocene of Africa, Asia and Europe. The earliest of these occur in East Africa, and comprise several genera. Many of these taxa are known from postcranial elements, allowing us to assess their morphological similarities to and differences from their modern relatives. A recent cladistic analysis of 240 cranial and postcranial characters of nine fossil and four extant hominoid taxa (Begun et al., 1997b; Figure 1) suggests that the early part of the hominoid lineage was characterized by taxa that lacked many derived traits of the postcranial skeleton found in extant apes (reviews and references in Rose, 1997; Ward, 1997). The combination of features typical for many early Miocene genera consists of hominoid synapomorphies, primitive retentions, and unique characters. This combination is unique among anthropoids (Rose, 1983, 1993, 1994), and reflects a generalized arboreal locomotor adaptation,

Primate Locomotion, edited by Strasser *et al.*
Plenum Press, New York, 1998

characterized by more frequent pronograde postures and a lesser emphasis on suspensory activities than in modern apes.

The generalized pattern of postcranial anatomy and inferred locomotor behavior characterized most early to middle Miocene hominoids, except perhaps for the enigmatic *Morotopithecus* (Gebo et al., 1997), for which comparatively little fossil evidence is known. Only later do the beginnings of extant hominoid-like forelimb-dominated arboreal adaptations begin to appear, and only in the later Miocene did apes begin to take on their modern postcranial form.

Despite the postcranial similarity among the earliest hominoids, these apes appear to have had an array of dietary and ecological craniofacial adaptations (Harrison, 1982; Kay, 1977; Kay and Ungar, 1997; Walker et al., 1994). The degree of postcranial similarity coupled with craniofacial difference among early Miocene genera equals or exceeds that found in any subsequent period of hominoid evolutionary history. In no two taxa among early apes is this pattern of intergeneric postcranial similarity and craniofacial diversity more apparent than *Afropithecus* and *Proconsul*.

Because they are so well known, *Afropithecus* and *Proconsul* provide an excellent opportunity to compare cranial and postcranial adaptations in two early Miocene apes. This paper reviews the postcranial similarities between these genera, and contrasts this with their very different craniodental adaptations. This contrast between cranial and postcranial diversity in *Afropithecus* and *Proconsul* illustrates the complex nature of mosaic evolution in primates, and provide a basis for understanding the early evolution of the hominoid lineage.

2. THE FOSSILS

For many years, *Proconsul* has been the best-known genus of Miocene ape, known from virtually all body parts, and for multiple individuals of four species: *P. africanus* from Koru and other Tinderet localities (Hopwood, 1933), *P. heseloni* from site R114 on Rusinga Island (Walker et al., 1993) and other Kisingiri localities, *P. nyanzae* from type locality R1-3 (Clark and Leakey, 1951) and other sites on Rusinga Island and from Mfangano Island, and *P. major* from Songhor (Clark and Leakey, 1951) and other Tinderet localities (review and references in Walker, 1997). All of these sites date to the early Miocene, around 18 million years ago.

Proconsul is known from a nearly complete skull, as well as many maxillary, mandibular and dental specimens. As a result, its cranial and postcranial adaptations are fairly well-understood. *Proconsul* species show little postcranial variability, despite their size differences (but see Nengo and Rae, 1992), so their functional morphology is considered here at the generic level.

In 1986, Leakey and Leakey published the discovery of new *Afropithecus turkanensis* fossils from Kalodirr in northern Kenya. Although *Afropithecus* is known from four sites, Kalodirr, Buluk, Moruorot and Locherangan, postcrania are only known from Kalodirr and Buluk (Leakey et al., 1988). All *Afropithecus* sites date to the early Miocene, 16–18 million years ago, so they were broadly contemporaneous with *Proconsul* sites. Among the Kalodirr fossils is a nearly complete facial and anterior cranial skeleton (Leakey and Leakey, 1986; Leakey et al., 1988). It is probably from a male, and was about the size of a chimpanzee, as was one of the two best-known species of *Proconsul*, *P. nyanzae*. Females appear to have been smaller, but these smaller specimens may represent a second species (Leakey and Walker, 1997; Leakey et al., 1988). The smaller forms were slightly larger than the best-known *Proconsul* species, *P. heseloni,* which was about the size of a siamang. Here, the two sizes are considered to represent males and females of *A. turkanensis* (Leakey et al., 1988).

Afropithecus is currently the second-best known postcranially of early Miocene hominoids. Postcrania are known from both sexes, and from the sites of Buluk and Kalodirr (Leakey and Walker, 1985; Leakey et al., 1988). The Buluk postcranial specimen is a phalanx. The Kalodirr postcranial remains include specimens associated with cranial elements (parts of an ulna, fibula, and first, third and fourth metatarsals), as well as specimens that are unassociated, preserving parts of numerous joint complexes of the upper and lower limbs, including bones of the elbow, wrist, hand, fingers, ankle, foot and toes (Leakey et al., 1988). A detailed functional analysis of these bones is in preparation by M. Rose, C. Ward, A. Walker and M. Leakey. The *Afropithecus* sample provides enough information to reconstruct a fairly accurate picture of locomotor behavior in this genus, and the even more complete sample of *Proconsul* fossils provides an even better picture.

Another well-known primitive ape (Begun et al., 1997b) is *Kenyapithecus,* from the middle Miocene of Kenya. The sample of postcrania attributed to *Kenyapithecus* has been expanding, and now includes most skeletal elements (McCrossin, 1994; McCrossin and Benefit, 1997; McCrossin et al., this volume), but many of these important fossils have only recently been discovered and remain unpublished (Ward and Brown, 1996; Rose, pers. comm.).

Afropithecus and *Proconsul,* along with *Kenyapithecus* and sometimes other genera, have been interpreted to be basal members of the hominoid lineage, but were probably not sister taxa (Figure 1) (reviews and references in Begun et al., 1997b). This hypothesis is supported by a recent phylogenetic analysis of 240 characters from the cranial and postcranial skeletons of nine fossil and four extant hominoid taxa (Begun et al., 1997b). Although the phyletic relationships of these taxa relative to one another are equivocal, their position relative to extant hominoids is more stable; placing any of these taxa within the extant hominoid clade increases tree length by 4% or more. It is therefore reasonable to hypothesize that these three taxa represent an early radiation of stem hominoids that existed throughout the early and middle Miocene.

Recently, based on newly discovered and yet unpublished fossils, McCrossin (1997) has claimed that *Kenyapithecus* may have shared more postcranial apomorphies with ex-

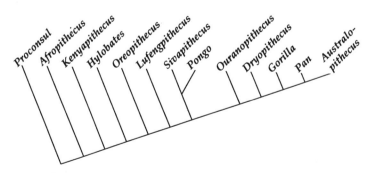

Figure 1. Hypothesis of hominoid relationships based on a cladistic analysis of 240 cranial and postcranial characters (from Begun et al., 1997b). The tree length is 446 steps and the consistency index is 63. The relationships among taxa illustrated here differs from some other published hypotheses, but most phylogenies propose that *Proconsul* and *Afropithecus* were not sister taxa (reviews and discussions in Begun et al., 1997a).

tant apes than do *Proconsul* or *Afropithecus*. McCrossin reports that new *Kenyapithecus* fossils from Maboko Island exhibit derived postcranial features, such as a straighter humeral shaft, than previously reported for this taxon. McCrossin interprets these features to mean a more modern great ape-like locomotor adaptation for *Kenyapithecus* than is found in *Afropithecus* or *Proconsul*, including an emphasis on forelimb-dominated arboreality. This new interpretation contrasts with previous arguments for terrestrial adaptations in *Kenyapithecus*, which is suggested by many postcranial features, including a retroflexed humeral shaft (Benefit and McCrossin, 1995; McCrossin, 1994, McCrossin and Benefit, 1994, 1997). These apparently contrasting morphologies confuse functional and phylogenetic interpretation of *Kenyapithecus*, and so a more precise interpretation of its affinities will have to wait until the new fossils are published. Nevertheless, *Kenyapithecus* appears to be the closest sister taxon to extant hominoids, and is considered a middle Miocene stem hominoid here (reviews and references in Begun et al., 1997).

Afropithecus and *Proconsul* are strikingly similar in preserved parts of their postcranial skeletons, exhibiting no evidence for substantial differences in locomotor adaptation. Their faces, teeth and reconstructed dietary adaptations, however, are quite different. They illustrate the pattern of mosaic evolution characterizing not only the early hominoid lineage, but much of hominoid evolution. It appears that the generalized locomotor adaptations of early hominoids permitted ecological and dietary specialization without necessitating substantial postcranial change. It may be that the generalized nature of early hominoid locomotor adaptations even facilitated the initial radiation of apes, with selection producing considerable postcranial modification later in hominoid evolution.

3. COMPARATIVE POSTCRANIAL ANATOMY OF *AFROPITHECUS* AND *PROCONSUL*

3.1. Elbow

Afropithecus is known from a small piece of distal humeral articular surface, including part of the capitulum and the entire width of the zona conoidea (Figure 2; Leakey et al., 1988). Although small, this fossil preserves one of the most functionally diagnostic regions of the distal humerus (Morbeck, 1983; Rose, 1983, 1988, 1993a). *Proconsul* is known from more complete distal humeri (Napier and Davis, 1959; Rafferty et al., 1995; Senut, 1980; Walker and Pickford, 1983; Walker and Teaford, 1988; Walker et al., 1985). Both *Afropithecus* and *Proconsul* resemble extant hominoids in having a well-defined, deep zona conoidea medial to the capitular surface for articulation with the margins of the radial head (Figure 2), although not as pronounced as in African apes. In contrast, most monkeys tend to have a much flatter, broader zona conoidea. This difference between apes and most monkeys in humeral morphology reflects two fundamentally different patterns of habitual loading and mobility at the elbow (Napier and Davis, 1959; Rose, 1987, 1993a), and the presence of a well-defined zona conoidea in *Afropithecus* and *Proconsul* suggests that it arose early in the hominoid lineage.

Hominoids have fairly symmetrical radial heads that maintain similar amounts of humeral contact during all phases of pronation-supination, reflecting the habitually varied forelimb postures of these animals during climbing, hanging, and bridging arboreal locomotion. In addition, the rounded capitulum and deep zona conoidea of hominoids provide joint surfaces capable of effectively resisting loads from directions other than relatively simple axial loading. In contrast, cercopithecoids have an asymmetrical configuration of

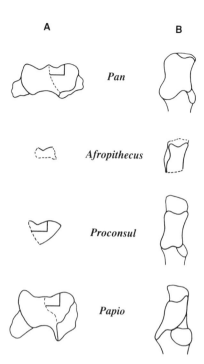

Figure 2. The distal humerus and proximal ulna in extant and fossil taxa. A) Distal view of left humeri, *Afropithecus* is known from the fragment (KNM-WK 17008) shown here, comparable areas on other taxa are illustrated. Only the area for articulation with the radius is illustrated for *Proconsul.* B) Anterior view of proximal left ulnae, *Afropithecus* (KNM-WK 18395) is missing the proximal end of the olecranon and the distal end of the trochlear notch. Modified from Rose (1987).

the humeroradial joint that maximizes chondral contact and joint congruence between the humerus and radius during pronation, the position in which the elbow is typically loaded during palmigrade quadrupedalism, at the expense of articular contact in supinated postures (Rose, 1987).

The zona conoidea of *Proconsul* and *Afropithecus* resembles that of other hominoids, suggesting that they used their elbows in varied positions of pronation and supination during locomotion. Further support for this functional hypothesis comes from the rest of the humeroradial joint of *Proconsul,* which more closely resembles that of extant apes than most monkeys (Morbeck, 1983; Rose, 1983, 1993b, 1997).

A similar pattern of weight-bearing during all phases of forelimb pronation and supination may also be implied by the proximal ulna. Trochlear notches of *Afropithecus* and *Proconsul* ulnae were more symmetrical than those of Old World monkeys with weak, but more proximodistally-oriented, keels (Rose, 1987, 1993a, 1997).

The derived elbow morphology suggested by the fragmentary elbow joint remains of *Afropithecus,* and seen more extensively in *Proconsul* and other fossil hominoids, resembles that of extant apes in important ways, implying a similar pattern of varied elbow use during locomotion. *Afropithecus* and *Proconsul,* as with extant hominoids, appeared to have used their elbows in a variety of postures, from pronation to supination, during locomotion. Although chimps and gorillas maintain a pronated hand posture during terrestrial quadrupedalism, they employ varied elbow postures during arboreal climbing and bridging activities. The antiquity of this morphology suggests that adaptation to variable pronation-supination postures of the elbow in hominoids evolved before advanced suspensory adaptations (see below), and is sufficiently generalized to accommodate a range of locomotor specializations.

Proconsul, however, differed from extant hominoids in other ways, including having a longer olecranon process. Although the olecranon process is broken in *Afropithecus*, the preserved contour of its posterior ulna suggests that the olecranon process extended somewhat further proximally than is seen in extant apes, probably to the extent seen in *Proconsul* (Figure 2; Rose, 1993b, 1997). Among extant anthropoids, longer olecranons are associated with habitually flexed elbows (Harrison, 1982; Rose, 1993). As the elbow moves into extension, a proximally elongate olecranon results in more rapid shortening of the triceps muscle, moving the triceps towards the end of its length-tension curve where it would have less contractile power (Pitman and Peterson, 1989). A shorter olecranon allows powerful contraction of the triceps muscle near full elbow extension. *Proconsul*, and probably *Afropithecus*, appears to have been adapted for at least a moderate amount of habitual elbow flexion during locomotion, rather than the extended elbow postures of great apes when arm-hanging or knuckle-walking. This suggests that although *Proconsul*, and perhaps *Afropithecus*, was a habitual climber, it did not rely on arm-hanging as extensively as extant apes.

3.2. Hand

The emphasis on a range of pronation-supination in apes and pronated hand postures in monkeys is also seen in bones associated with the midcarpal joint. Jenkins (1981) demonstrated structural adaptations of the extant hominoid wrist that allow significant midcarpal supination, an important adaptation for brachiation. These morphologies are expressed most strongly in hylobatids and orangutans, but are also seen in chimpanzees, and to a lesser extent gorillas. Midcarpal supination is permitted by an embrasure between the capitate and trapezoid for the scaphoid/centrale that widens palmarly in apes. This partly is a result of the laterally-facing facet for the scaphoid/centrale on the capitate in suspensory apes. In monkeys, the dorsally-facing joint surface for the scaphoid/centrale results in a narrowing of the embrasure for the scaphoid/centrale towards the palm, limiting movement of these bones palmarly, and restricting midcarpal supination.

The scaphoid/centrale facet on capitates of *Afropithecus* and *Proconsul* resembles that of most monkeys, and faces dorsolaterally, with evidence for only a small gap against the trapezoid (Figure 3). Of course, without knowing what the other bones involved would have looked like, any conclusions about joint function are tentative. The narrow, laterally-facing capitate head, however, implies limited midcarpal supination in *Afropithecus* and *Proconsul*, but probably free pronation at this joint. Because this wrist morphology is also found in most monkeys, this pattern of restricted midcarpal supination is probably primitive for catarrhines.

Because midcarpal supination is an important component of forelimb suspension in extant apes, it appears that *Afropithecus* and *Proconsul* did not rely as heavily on forelimb-dominated suspension during locomotion as do many extant apes or even *Sivapithecus*. Instead, they were probably habitually palmigrade. This monkey-like midcarpal anatomy, however, would not affect their ability to climb. The only limitation it would impose, compared with that seen in extant apes, is on midcarpal pronation, which tends to occur when moving from branch to branch when hanging below a support by the forelimbs, and is only extensively developed in the most arboreal extant apes, gibbons and orangutans.

The pattern of weight-bearing through the carpus of *Afropithecus* appears to have differed from that of most extant catarrhines. The lunate exhibits an atypical pattern of relationships among some intercarpal joints. The lunate of *Afropithecus*, like that of *Proconsul*, is compressed proximodistally (Figure 3). The capitate and triquetral facets are almost

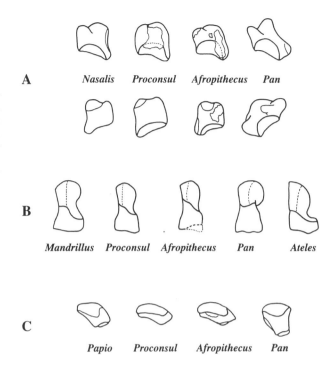

Figure 3. Some carpal bones of extant and fossil taxa. A) Right trapezia in palmar (above) and dorsal (below) views, distal end is down. *Afropithecus* and *Proconsul* share a sellar metacarpal facet that extends onto the dorsal aspect of the bone with extant hominoids; this is a feature not found on most monkey trapezia. Modified from Rose (1992). B) Right capitates in dorsal view, distal end is down, lateral side is right. Suspensory anthropoids are represented by *Ateles*. Contour of lateral sides is generally related to morphology of embrasure between capitate and trapezoid. Figure modified from Rose (1984). C) Right lunates in palmar view, radial facet is down, triquetral facet upper left, capitate facet upper right.

parallel to the radial facet. This shape is only seen in certain New World monkeys today (Rose, 1993b), and may be primitive for hominoids, if not all catarrhines. Because a joint surface must remain normal to transarticular loads, the flattened lunate shape implies that weight could be effectively transferred through the lunate from both the capitate and triquetral to the radius, when the wrist is neutral with respect to radial and ulnar deviation. The angle between the triquetral and radial facets is slightly more oblique in cercopithecids, and much more so in apes.

The triquetral facet on great ape lunates is set roughly normal to the radial facet, and almost parallel to the scaphoid surface (Figure 3C). It seems that in extant hominoids weight may be transferred to the radius primarily through the capitate and then lunate when the wrist is neutral with respect to pronation and supination, and through the triquetral to the lunate to the scaphoid when the wrist is adducted. The carpal joint configuration seen in Miocene apes and New World monkeys indicates a common pattern of load transmission through the carpus that differs from that of other anthropoids, and probably represents the primitive condition for catarrhines.

Extant apes have altered their wrist even further by proximodistal elongation of the hamate allowing increased mobility between the triquetral and hamate (Sarmiento, 1988; Lewis, 1989). They also have reduced the length of the ulnar styloid process, further increasing the capacity for ulnar deviation by providing more space between the ulna and carpus (Lewis, 1969, 1972, 1974, 1989). In most monkeys, weight is transferred via bony contact directly from the triquetral to the ulna during palmigrade postures, and it may be that relatively little passes directly through the triquetrolunate joint to the radius, explaining the oblique orientation of this joint surface with respect to the radial lunate surface in cercopithecoids (Beard et al., 1986; Cartmill and Milton, 1977; Lewis, 1989; Sarmiento, 1988). Articular facets for the ulnar styloid are visible on the pisiform and triquetral of *Proconsul*, suggesting that they also had direct bony contact here (Beard et al., 1986). Although this may indicate the capacity for weight transfer through the medial portion of the wrist, the long, proximally-facing facet for the ulnar styloid on the pisiform may indicate a greater range of ulnar deviation in *Proconsul* than is typical for most monkeys, except atelines and *Alouatta*. *Proconsul* shows signs of enhanced ulnar deviation when compared with cercopithecoids, probably reflecting a greater reliance on climbing and varied limb postures during arboreal locomotion than is typical for cercopithecoids, but did not have the capacity for ulnar deviation found in extant hominoids. Although no triquetral or pisiform is known for *Afropithecus*, because the *Afropithecus* lunate so closely resembles that of *Proconsul*, the other carpal bones may have been similar morphologically and functionally.

The *Afropithecus* hand is also known from the second and fourth metacarpals, as well as an associated first metacarpal and trapezium; these elements are also known for *Proconsul*. *Afropithecus* and *Proconsul* have morphologically functionally similar trapezia and first metacarpals. The joint between the trapezium and first metacarpal of these fossil apes is sellar (Figure 3A) and functioned most like that of extant hominoids (Rafferty, 1990; Rose, 1992). Greater ranges of abduction-adduction and rotation were possible at this joint in all living and fossil apes than in monkeys, as a result of greater curvature and incongruity of the opposing joint surfaces in the hominoids (Rafferty, 1990). This similarity among hominoids probably represents a synapomorphy, and reflects an habitual reliance on enhanced grasping capabilities in even the earliest apes.

Although the grasping capabilities of *Afropithecus* and *Proconsul* resembled those of extant hominoids, they appear to have had slightly more gracile pollical phalanges (Begun, 1994) and metacarpals. Still, the pollical phalanx and metacarpals were more robust than those of Old World monkeys, suggesting strength of manual grasping intermediate between that of apes and cercopithecoids. As with the foot (see below), this suggests that grasping was more important than for most monkeys, but was not as well-developed as in extant apes, who rely more heavily on it when engaged in forelimb suspensory and bridging activities. Again, this is consistent with the hypothesis that early hominoids were frequent climbers, but were not adapted for extensive forelimb suspension.

The second and fourth metacarpals are also informative. Hominoids have the metacarpal bases facing directly proximally, most likely correlated with the neutral or flexed position of the carpometacarpal joints during climbing and, in the case of African apes, knuckle-walking. Most monkeys, on the other hand, have the metacarpal bases facing proximodorsally. Because joint surfaces must be oriented normal to transarticular loads, a proximodorsal orientation probably reflects habitual extension at the carpometacarpal joints, as occurs during palmigrade or mildly digitigrade progression. *Afropithecus* and *Proconsul* resemble monkeys in the proximodorsal orientation of their metacarpal bases, suggesting habitual palmigrady on large supports or when terrestrial. Combined with information from the thumb, this evidence suggests that fairly powerful manual grasping was an important adaptation in these apes, and that *Afropithecus* and *Proconsul* were adapted for palmigrade quadrupedality as well as climbing.

3.3. Foot

The ankle of *Afropithecus* is not only known from a nearly complete fibula, but from a nearly complete talus. These elements resemble those known for *Proconsul nyanzae*. Great apes have an asymmetrical talar trochlea with a shallower groove than do most monkeys. The tali of *Proconsul, Afropithecus,* and most other Miocene apes are more symmetrical with deeper trochleas than those of extant apes (Figure 4B), which would have

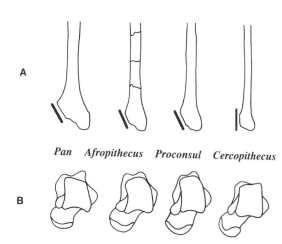

Figure 4. Talocrural bones of extant and fossil taxa. A) Anterior view of left fibulae. B) Dorsal view of right tali. See text for discussion.

provided less conjunct rotation of the foot during flexion and extension, and suggests more habitual plantigrady during locomotion.

Evidence of hallucal size and morphology of *Afropithecus* comes from the fibula, talus and the first metatarsal itself. The *Afropithecus* fibular shaft was more gracile than those of extant apes relative to the size of its talar facet, but was more robust than those of most monkeys (Figure 4A). A large fibula implies the presence of well-developed hallucal flexor muscles. A strong flexor hallucis longus is also indicated by the pronounced groove on the posterior margin of the talar trochlea (Figure 4B).

In addition, the talar facet on the fibula is set at an angle relative to the fibular shaft (Figure 4A). This condition is found in extant apes and the most arboreally adapted monkeys, including atelines and the fossil colobine *Rhinocolobus turkanensis*. An angled talofibular joint provides a joint surface contact area that is oriented obliquely with respect to the talar trochlear surface, perhaps indicating weight-bearing when the foot is in everted postures, such as when it is grasping a limb, and/or the knee joint is abducted when the foot is in a plantigrade position. All of these postures occur when climbing and require the presence of a grasping hallux. At least as importantly, talar facet orientation probably also reflects the relative size of the hallucal flexors, and their angle of pull relative to the fibular shaft.

The first metatarsal of *Afropithecus* and *Proconsul* was not as large as that of apes, but more robust than that of monkeys (Figure 5A). The insertion site of peroneus longus is pronounced, implying a strong muscle and correspondingly strong hallucal adduction. *Proconsul* halluces are long relative to those of comparably-sized monkeys, and are more strongly built, and the hallucal terminal phalanges are broad relative to those of other anthropoids (Begun et al., 1994; Strasser, 1993). Robust hallucal metatarsals and phalanges imply fairly powerful grasping abilities in *Afropithecus,* especially when combined with evidence of powerful hallucal flexors. The cuneiform facet has a definite concave medial portion, suggesting habitual abduction, though not as strong as in most extant apes. While hallucal size and strength in *Afropithecus* and *Proconsul* appear to have exceeded those of extant monkeys, they were not as well-developed as in extant apes. The sesamoid grooves on the head of the first metatarsal are pronounced, as in monkeys. This morphology suggests that hallucal grasping was probably not as strong as in extant apes, and that the feet were frequently used in plantigrade postures. Most likely, feet were used

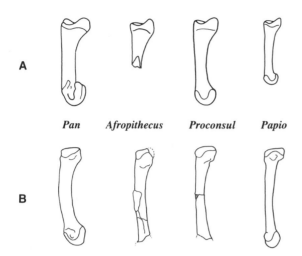

Figure 5. Metatarsals of extant and fossil taxa. A) Medial view of left first metatarsals, distal is down. B) Medial view of left fourth metatarsals, distal is down. See text for discussion.

in grasping during climbing, but probably not extensively in bridging or suspensory activities.

3.4. Phalanges

Many phalanges are known for *Afropithecus*, and even more for *Proconsul*. The phalanges of *Proconsul* have been described and analyzed in detail (Begun, 1994). The *Afropithecus* phalanges do not differ appreciably from those of *Proconsul*, so Begun's conclusions about the locomotor repertoire of *Proconsul*, based on phalangeal morphology, are equally valid for *Afropithecus*.

Proconsul manual and pedal phalanges exhibit only mild curvature and relatively weak secondary shaft characteristics, suggesting that these early Miocene apes did not rely on suspensory postures as extensively as do extant apes (Begun, 1994). Instead, they were probably above-branch arboreal quadrupeds. Again, like other parts of the postcranium, the *Afropithecus* and *Proconsul* phalanges suggest that these apes were well-adapted climbers, but probably not as suspensory as extant apes. In addition, metacarpal and phalangeal morphology suggests slightly hyperextended postures of the carpometacarpal and metacarpophalangeal joints at least some of the time, probably on larger arboreal supports and when terrestrial.

Extant hominoids exhibit clearly differentiated manual phalanges compared with pedal phalanges, reflecting the divergent use of the hands and feet in climbing and suspension. Monkeys, on the other hand, tend to have morphologically similar manual and pedal phalanges, reflecting similar use of the hand and foot in locomotion. There appears to have been only minor differentiation between manual and pedal phalanges in *Proconsul* (Begun, 1994), and in *Afropithecus*. The hint of a difference in *Proconsul* and *Afropithecus* may indicate the beginnings of specialization for more forelimb-dominated climbing activities in these early apes, but the derived condition found in extant apes had not yet taken place. Again, *Proconsul* and *Afropithecus* phalangeal morphology reiterates the functional scenario revealed in other parts of the skeletons.

4. LOCOMOTOR BEHAVIOR OF THE EARLIEST HOMINOIDS AND EVOLUTION OF THE APE SKELETON

The known parts of the *Afropithecus turkanensis* skeleton are morphologically, and presumably functionally, equivalent to those of *Proconsul nyanzae* and *P. heseloni*. In fact, the postcrania known for the presumptive males of *Afropithecus turkanensis* are difficult to distinguish morphologically from those of *Proconsul nyanzae*. Because parts of the elbow, hands, and feet are known for both genera, and show no appreciable differences, it is reasonable to assume that the rest of their postcranial skeletons were similar as well. These animals appear to have had similar locomotor adaptations. In a direct comparison of similar body parts, no obvious functional distinctions can be made.

All of the information from these genera can be used to provide a picture of their locomotor adaptations. It is clear that *Afropithecus* and *Proconsul* were primarily pronograde quadrupeds. They both exhibit signs of habitual extension at the carpometacarpal, tarsometatarsal, and metapodial-phalangeal joints. There also are pronounced sesamoid grooves on the metapodial heads, as in most monkeys, and only a modest amount of differentiation between hand and foot phalanges, suggesting roughly similar use of the hands and feet during locomotion.

Proconsul, being better known, provides additional evidence that it was primarily a pronograde animal. *Proconsul* had a narrow thoracic cage, as seen in the cercopithecoid-like curvature of the ribs, the narrow iliac blades, and the ventrally-situated lumbar vertebral transverse processes (Ward, 1993; Ward et al., 1993). These morphologies suggest that the scapulae were oriented in a roughly parasagittal position on the sides of the rib cage, and that the glenohumeral joints were oriented ventrally. In this position, the thoraco- and scapulohumeral muscles are aligned to produce primarily flexion-extension movements of the limbs. Extant hominoids, along with atelines to a lesser extent, have broader thoracic cages with laterally-facing shoulder joints, enabling their upper limb muscles to produce more effective abduction-adduction movements, critical for many forelimb-dominated climbing and suspensory behaviors. *Proconsul* was built more like monkeys, and emphasized primarily flexion-extension of its limbs during propulsive movements in locomotion.

Proconsul also had a long, flexible torso like most monkeys, in contrast to the short, stiff backs of extant apes. Short backs decrease bending moments about the lower vertebrae, and provide for effective attachment and function of latissimus dorsi to adduct the humerus, an important movement during suspensory locomotion, without risking excessive stresses on the lower back (Ward, 1993; Ward et al., 1993). Because *Afropithecus* limbs were so similar to those of *Proconsul*, it probably resembled *Proconsul* in axial morphology and function as well.

Despite adaptations for habitual pronogrady, morphological adaptations to climbing are clearly present in these earliest hominoids. Both of these genera exhibit numerous adaptations to climbing, and to a more diverse array of limb postures during locomotion than typifies Old World monkeys. The elbows of *Afropithecus* and *Proconsul* were adapted to weight-bearing in a variety of pronation-supination postures, albeit habitually slightly flexed elbow postures. The *Proconsul* hip and knee were similar, and lack adaptations to loading in stereotypical postures seen in Old World monkeys, even when arboreal. Although early hominoids exhibit better developed grasping capabilities of the hand and foot and more mobile wrists and ankles than do most monkeys, they were not as well-adapted for suspensory locomotion as are extant hominoids. Thus, early Miocene apes probably did not engage in as much suspensory activities as their modern counterparts.

In fact, the mobile elbow, hip and knee, and moderately well-developed grasping abilities of *Proconsul* and *Afropithecus* (where known), suggest that climbing was an important component of their locomotor repertoire. An arboreal milieu necessarily has randomly oriented supports, requiring diverse limb postures to negotiate effectively. Cercopithecoids are able to negotiate among arboreal supports by above-branch quadrupedalism, combined with some leaping and bridging. It appears that the earliest apes increased their emphasis on climbing and bridging behaviors, perhaps because of their relatively large body sizes. Male *Afropithecus* individuals and *Proconsul nyanzae* were comparable to modern chimpanzees in size. Because of this, they would not have been able to locomote as monkeys do in the trees. Large individuals would have had to distribute their weight over a greater number of supports to reach terminal branches where nuts or seeds might be located. They would also have to bridge gaps and/or be terrestrial more frequently than many monkeys. Even the smaller species, *Proconsul heseloni,* seems to have been adapted to increased climbing and bridging, and decreased running and leaping, than is typical for most monkeys. Kelley (1995) has linked tail loss with the need for alternate forms of balance in trees, and has argued that postcranial adaptations of early apes represents a solution to this problem. This may be why *Proconsul*, and perhaps other early apes, lacked external tails (Ward et al., 1991). Certainly, early hominoids practiced a form of above-branch arboreality that involved substantial limb mobility for weight distribution and balance. Many morphologies shared by early and later hominoids may represent adaptations to this more generalized locomotor repertoire than the derived, forelimb-dominated locomotor modes seen among extant species (Kelley, 1995). Only later in evolution did the shoulder joint and torso change form to become adapted for forelimb suspension and forelimb-dominated climbing.

The generalized form of hominoid postcranial anatomy and locomotion exhibited by *Afropithecus* and *Proconsul* seems to have been typical for most early Miocene apes. *Turkanapithecus*, which is known from several postcranial bones, may have been slightly better adapted for arboreal climbing than were *Afropithecus* and *Proconsul*, but still, it shares their general body form in their preserved parts. Even *Kenyapithecus* shares most early hominoid morphologies, and morphological characters suggestive of increased terrestriality (McCrossin, 1994, 1997) represent relatively minor variations on the early hominoid theme. Among early to middle Miocene hominoids, only *Morotopithecus* appears to differ much from the early hominoid pattern, and more closely resembles extant apes (Gebo et al., 1997; Sanders and Bodenbender, 1994; Walker and Rose, 1968; Ward, 1993).

Such a generalized set of adaptations for arboreal locomotion appears to have been sufficiently flexible to allow early Miocene apes to exploit a variety of dietary niches. It would not have made them less effective on the ground than in the trees, and it would have permitted them to reach the terminal branches of trees to exploit resources there as well. The fact that early apes did exploit an array of dietary niches is also illustrated clearly by *Afropithecus* and *Proconsul*.

5. CRANIAL ANATOMY AND DIETARY ADAPTATIONS OF *AFROPITHECUS* AND *PROCONSUL*

The overwhelming similarities between the postcranial anatomy of *Proconsul* and *Afropithecus* contrast markedly with the disparity in their apparent dietary adaptation and associated craniofacial structure.

Afropithecus appears to have been a sclerocarp forager, based on its craniofacial and dental morphologies (Leakey and Walker, 1997; McCrossin and Benefit, 1997) . *Afropi-*

thecus mimics the suite of craniofacial and dental characters summarized by Kinzey (1992) for pitheciins, and noted for *Kenyapithecus* (Benefit and McCrossin, 1995; McCrossin, 1994; McCrossin and Benefit, 1994, 1997). *Afropithecus* canines are large, laterally splayed, and have strong roots with short crowns, an arrangement that would allow incisor and canine functions to be uncoupled, so that the incisors can be used for cropping foods and the canines for forceful puncturing (Kinzey, 1992). This evidence, combined with the relatively low crowns and the apparent minimal sexual dimorphism of *Afropithecus* canines, supports the argument that *Afropithecus* canines were used more for food preparation than agonistic intraspecific display. These morphologies parallel those in pitheciins, such as *Chiropotes,* which use their canines to puncture hard fruits or seeds (Kinzey, 1992). *Afropithecus*, however, has large incisors such as those generally found in frugivores (Kay, 1984). The central upper incisors are much larger than the lateral. The incisors are procumbent, and the upper ones are set in a prognathic premaxilla and occlude with the robust, styliform lower incisors.

The molar morphology of *Afropithecus* is consistent with its craniofacial form. *Afropithecus* has short molar shearing crests, consistent with a diet of frugivory or seed predation (Leakey and Walker, 1997; McCrossin and Benefit, 1997). It also has thick molar enamel with extremely heavy tooth wear. The pronounced frontal trigon diminishes in size with age, a feature correlated with a powerful temporalis muscle (Leakey and Walker, 1997; Simons, 1987). In addition, the mandible is strongly buttressed, also indicating that the masticatory apparatus of *Afropithecus* was powerful (Leakey and Walker, 1997). Taken together, its craniofacial and dental morphology indicate that *Afropithecus* was adapted to exploiting a diet of hard foods, much like pitheciins.

This reconstruction of diet in *Afropithecus* differs from that inferred for *Proconsul*, which was probably a more generalized frugivore. Dental and craniofacial anatomy of *Proconsul* is only well-known for two of the four species, *P. heseloni* and *P. nyanzae.* The functional discussion here is based on these species.

Proconsul had moderate shearing crests on its molars (Kay, 1977; Kay and Ungar, 1997) implying a frugivorous diet. Molar microwear studies (Walker et al., 1994) further support this interpretation. The interpretation that *Proconsul* was a frugivore is supported by its inferred brain-body size ratio (Walker et al., 1993), which is larger than expected for a folivore. *Proconsul* species are all characterized by a moderate degree of body size dimorphism, about 1.3:1, similar to that found in *Pan troglodytes* (Walker et al., 1993; Rafferty et al., 1995). Canine dimorphism, however, is strong, with males having large, blade-like canines and females having small, conical ones (Walker, 1997). The canines are more vertically implanted than in *Afropithecus*, and are in line with the postcanine tooth row. Thus, the canines of *Proconsul* were probably more important for intraspecific displays than diet, unlike the condition found in *Afropithecus*.

Proconsul incisors are vertically implanted, and appear small relative to molar size (Harrison, 1982). This size relationship, however, could be a result of the proposed relative molar megadontia (Rafferty et al., 1995; Ruff et al., 1989; Teaford et al., 1993) of *Proconsul*, rather than diminished incisor size. The *Proconsul* face and mandible are more gracile than that of *Afropithecus*, and there is no frontal trigon. It appears that the masticatory apparatus was not as well-developed in *Proconsul*, and the tooth wear was not as severe, supporting the interpretation of a diet of softer fruits in *Proconsul* than found in *Afropithecus*.

This divergent craniofacial morphology of *Afropithecus* and *Proconsul* suggests quite different ecological and dietary niches for these two hominoids. These very different skulls, however, are associated with strikingly similar postcranial skeletons. This disparity illustrates the mosaic nature of hominoid evolution. Clearly, the generalized form of arboreal locomotion practiced by these two apes allowed them to exploit widely different food sources without necessitating any evolutionary change in the postcranial skeleton.

6. SUMMARY AND CONCLUSIONS

Afropithecus are *Proconsul* are the two best-known genera of early Miocene apes. They are broadly similar in size, and strikingly so in postcranial morphology. Still, while they appear to have had similar locomotor adaptations, they were divergent in craniofacial form and inferred dietary specialization. In no two hominoid genera is the disparity greater between postcranial similarity and craniofacial and dental dissimilarity.

Because these genera appear to share no particular phyletic relationship, and because their skeletons are broadly similar to those of many other early to middle Miocene apes, it is reasonable to assume that the pattern of postcranial anatomy they share was close to the primitive hominoid condition. Many extant hominoid postcranial synapomorphies were already present in these earliest apes, but those that reflect the most specialized adaptations to forelimb-dominated climbing and suspensory locomotion seen in extant apes do not appear to have evolved until much later in hominoid evolution. The disparity in craniofacial and dental form between *Afropithecus* and *Proconsul* suggests that early hominoid locomotor adaptations did not limit or constrain ecological diversification, and may even have facilitated the initial hominoid radiation.

ACKNOWLEDGMENTS

I would like to thank the editors for inviting me to be a part of the Primate Locomotion 1995 symposium and volume. I would also like to express my admiration and appreciation of the late Warren Kinzey for his pioneering contributions to the field of primate studies, which formed the basis for much of the work presented here. I would like to express my gratitude to Mike Rose for inviting me to work on the *Afropithecus* fossils, to Alan Walker and Mark Teaford for inviting me to work on the *Proconsul* fossils, and to Meave Leakey, the Government of Kenya and the directors and staff of the National Museums of Kenya for the opportunity to work on these and other fossils, and to the editors and reviewer for their helpful suggestions. This research was supported by the National Science Foundation and by the L. S. B. Leakey Foundation.

REFERENCES

Beard KC, Teaford MF, and Walker A (1986) New wrist bones of *Proconsul africanus* and *P. nyanzae* from Rusinga Island, Kenya. Folia Primatol. *47:*97–118.

Begun DR (1994) Comparative and functional anatomy of *Proconsul* phalanges from the Kaswanga Primate Site, Rusinga Island, Kenya. J. Hum. Evol. *26:*89–165.

Begun DR, Ward CV, and Rose MD, eds. (1997a) Function, Phylogeny and Fossils: Miocene Hominoid Evolution and Adaptations. New York: Plenum Press.

Begun DR, Ward CV, and Rose MD (1997b) Events in hominoid evolution. In DR Begun, CV Ward and MD Rose (eds.): Function, Phylogeny and Fossils: Miocene Hominoid Evolution and Adaptations. New York: Plenum Press, pp. 389–415.

Benefit BR, and McCrossin ML (1995) Miocene hominoids and hominid origins. Ann. Rev. Anthropol. *24:*237–256.

Cartmill M, and Milton K (1977) The lorisiform wrist joint and the evolution of "brachiating" adaptations in the Hominoidea. Am. J. Phys. Anthropol. *47:*249–272.

Clark WE Le Gros, and Leakey LSB (1951) The Miocene Hominoidea of East Africa. Br. Mus. Nat. Hist. Fossil Mamm. Afr. *1*:1–117.

Fleagle J, Stern J Jr., Jungers W, Susman R, Vangor A, and Wells J (1981) Climbing: A biomechanical link with brachiation and with bipedalism. Symp. Zool. Soc. Lond. *48*:359–375.

Gebo DL, MacLatchy LM, Kityo R, Deino A, Kingston J, and Pilbeam D (1997) A hominoid genus from the early Miocene of Uganda. Science *276*:401–404.

Harrison T (1982) Small-Bodied Apes from the Miocene of East Africa. Ph.D. Dissertation, University of London.

Hopwood AT (1933) Miocene primates from British East Africa. Ann. Mag. Nat. Hist. *11*:96–98.

Hunt KD (1992) Social rank and body size as determinants of positional behavior in *Pan troglodytes.* Primates *33*:347–357.

Jenkins FA (1981) Wrist rotation in primates: A critical adaptation for brachiators. Symp. Zool. Soc. Lond. *48*:429–451.

Kay RF (1977) Diets of early Miocene African hominoids. Nature *268*:628–630.

Kay RF, and Ungar PS (1997) Dental evidence for diet in some Miocene catarrhines with comments on the effects of phylogeny on the interpretation of adaptation. In DR Begun, CV Ward and MD Rose (eds.): Function, Phylogeny and Fossils: Miocene Hominoid Evolution and Adaptations. New York: Plenum Press, pp. 131–151.

Keith A (1923) Man's posture, its evolution and disorders. Brit. Med. J. *1*:451–672.

Kelley J (1995) A functional interpretive framework for the early hominoid postcranium. Am. J. Phys. Anthropol. *Supplement. 20*:125.

Kinzey WG (1992) Dietary and dental adaptations in the Pitheciinae. Am. J. Phys. Anthropol. *88*:499–514.

Leakey M, and Walker A (1997) *Afropithecus* - function and phylogeny. In DR Begun, CV Ward and MD Rose (eds.): Function, Phylogeny and Fossils: Miocene Hominoid Evolution and Adaptations. New York: Plenum Press, pp. 225–239.

Leakey MG (1985) Early Miocene cercopithecids from Buluk, Northern Kenya. Folia Primatol. *44*:1–14.

Leakey RE, and Leakey MG (1986) A new Miocene hominoid from Kenya. Nature *324*:143–145.

Leakey RE, Leakey MG, and Walker AC (1988) Morphology of *Afropithecus turkanensis* from Kenya. Am. J. Phys. Anthropol. *76*:289–307.

Lewis OJ (1969) The hominoid wrist joint. Am. J. Phys. Anthropol. *30*:251–268.

Lewis OJ (1972) Osteological features characterizing the wrists of monkeys and apes, with a reconsideration of this region in *Dryopithecus (Proconsul) africanus.* Am. J. Phys. Anthropol. *36*:45–58.

Lewis OJ (1974) The wrist articulations of the Anthropoidea. In FA Jenkins (ed.): Primate Locomotion. New York: Academic Press, pp. 143–169.

Lewis OJ (1989) Functional Morphology of the Evolving Hand and Foot. Oxford: Oxford University Press.

McCrossin ML (1994) The phylogenetic relationships, adaptations and ecology of *Kenyapithecus.* Ph.D. Dissertation, University of California, Berkeley.

McCrossin ML (1997) New postcranial remains of *Kenyapithecus* and their implications for understanding the origins of hominoid terrestriality. Am. J. Phys. Anthropol. *Supplement. 24*:164.

McCrossin ML, and Benefit BR (1994) Maboko Island and the evolutionary history of Old World monkeys and apes. In RS Corruccini and RL Ciochon (eds.): Integrative Paths to the Past. Englewood Cliffs: Prentice Hall, pp. 95–122.

McCrossin ML, and Benefit BR (1997) On the relationships and adaptations of *Kenyapithecus,* a large-bodied hominoid from the middle Miocene of eastern Africa. In DR Begun, CV Ward and MD Rose (eds.): Function, Phylogeny and Fossils: Miocene Hominoid Evolution and Adaptations. New York: Plenum Press, pp. 241–267.

Morbeck ME (1983) Miocene hominoid discoveries from Rudabánya: Implications from the postcranial skeleton. In RL Ciochon and RS Corruccini (eds.): New Interpretations of Ape and Human Ancestry. New York: Plenum Press, pp. 369–404.

Napier J, and Davis P (1959) The forelimb skeleton and associated remains of *Proconsul africanus.* Foss. Mamm. Afr. *16*:1–70.

Nengo IO, and Rae TC (1992) New hominoid fossils from the early Miocene site of Songhor, Kenya. J. Hum. Evol. *23*:423–429.

Pitman MI, and Peterson L (1989) Biomechanics of skeletal muscle. In M Nordin and VH Frankel (eds.): Basic Biomechanics of the Musculoskeletal System. Philadelphia: Lea and Febiger, pp. 89–111.

Rafferty KL (1990) The functional and phylogenetic significance of the carpometacarpal joint of the thumb in anthropoid primates. M.A. Thesis, New York University.

Rafferty KL, Walker A, Ruff CB, Rose MD, and Andrews PJ (1995) Postcranial estimates of body weight in *Proconsul,* with a note on distal tibia of *P. major* from Napak, Uganda. Am. J. Phys. Anthropol. *97*:391–402.

Rose MD (1983) Miocene hominoid postcranial morphology: Monkey-like, ape-like, neither, or both? In RL Ciochon and RS Corruccini (eds.): New Interpretations of Ape and Human Ancestry. New York: Plenum Press,
pp. 405–417.

Rose, MD (1984) Hominoid postcranial specimens from the Middle Miocene Chinji Formation, Pakistan. J. Hum.
Evol. *13:*503–516.

Rose MD (1987) Another look at the anthropoid elbow. J. Hum. Evol. *17:*193–224.

Rose MD (1992) Kinematics of the trapezium-1st metacarpal joint in extant anthropoids and Miocene hominoids.
J. Hum. Evol. *22:*255–266.

Rose MD (1993a) Functional anatomy of the primate elbow and forearm. In DL Gebo (ed.): Postcranial Adaptation in Nonhuman Primates. De Kalb: Northern Illinois Press, pp. 70–95.

Rose MD (1993b) Locomotor anatomy of Miocene hominoids. In DL Gebo (ed.): Postcranial Adaptation in Nonhuman Primates. De Kalb: Northern Illinois Press, pp. 252–272.

Rose MD (1994) Quadrupedalism in Miocene hominoids. J. Hum. Evol. *26:*387–411.

Rose MD (1997) Functional and phylogenetic features of the forelimb in Miocene hominoids. In DR Begun, CV
Ward and MD Rose (eds.): Function, Phylogeny and Fossils: Miocene Hominoid Evolution and Adaptations. New York: Plenum Press. pp. 79–100.

Ruff CB, Walker A, and Teaford MF (1989) Body mass, sexual dimorphism and femoral proportions of *Proconsul*
from Rusinga and Mfangano Islands, Kenya. J. Hum. Evol. *18:*515–536.

Sanders WJ and Bodenbender BE (1994) Morphological analysis of lumbar vertebra UMP 67–28. J. Hum. Evol.
*26:*203–237.

Sarmiento EE (1988) Anatomy of the hominoid wrist joint: its evolutionary and functional implications. Int. J. Primatol. *9:*281–345.

Senut B (1980) New data on the humerus and its joints in Plio-Pleistocene hominids. Coll. Anthrop. *1:*87–93.

Simons EL (1987) New faces of *Aegyptopithecus*, early human forebear from the Oligocene of Egypt. J. Hum.
Evol. *16:*273–290.

Stern JT (1971) Functional myology of the hip and thigh of cebid monkeys and its implications for the evolution
of erect posture. Bibl. Primatol. Basel: Karger *14:*1–318.

Stern JT, Wells JP, Vangor AK, and Fleagle JG (1977) Electromyography of some muscles of the upper limb in
Ateles and *Lagothrix*. Yrbk. Phys. Anthropol. *20:*498–507.

Strasser E (1993) Kaswanga *Proconsul* foot proportions. Am. J. Phys. Anthropol. *Supplement. 16:*191.

Teaford MF, Walker A, and Mugaisi GS (1993) Species discrimination in *Proconsul* from Rusinga and Mfangano
Islands. In WH Kimbel and LB Martin (eds.): Species, Species Concepts and Primate Evolution. New
York: Plenum Press, pp. 373–392.

Walker A (1997) *Proconsul* - function and phylogeny. In DR Begun, CV Ward and MD Rose (eds.): Function, Phylogeny and Fossils: Miocene Hominoid Evolution and Adaptations. New York: Plenum Press, pp. 209–224.

Walker A, Falk D, Smith R, and Pickford M (1983) The skull of *Proconsul africanus*: Reconstruction and cranial
capacity. Nature *305:*525–527.

Walker A, and Pickford M (1983) New postcranial fossils of *Proconsul africanus* and *Proconsul nyanzae*. In R
Ciochon and R Corruccini (eds.): New Interpretations of Ape and Human Ancestry. New York: Plenum
Press, pp. 325–351.

Walker A, and Rose MD (1968) Fossil hominoid vertebra from the Miocene of Uganda. Nature *217:*980–981.

Walker A, and Teaford MF (1988) The Kaswanga Primate Site: An Early Miocene hominoid site on Rusinga Island, Kenya. J. Hum. Evol. *17:*539–544.

Walker A, Teaford MF, and Leakey RE (1985) New information concerning the R114 *Proconsul* site, Rusinga Island, Kenya. In J Else and P Lee (eds.): Proc. Xth Congress of the Int. Primatol. Soc., Vol. 1: Primate Evolution. Cambridge: Cambridge University Press, pp. 143–149.

Walker A, Teaford MF, Martin L, and Andrews P (1993) A new species of *Proconsul* from the early Miocene of
Rusinga/Mfangano Islands, Kenya. J. Hum. Evol. *25:*43–56.

Walker A, Teaford MF, and Ungar PS (1994) Enamel microwear differences between species of *Proconsul* from
the early Miocene of Kenya. Am. J. Phys. Anthropol. *Supplement. 18:*202–203.

Ward CV (1993) Torso morphology and locomotion in *Proconsul nyanzae*. Am. J. Phys. Anthropol. *92:*291–328.

Ward CV (1997) Functional and phyletic implications of the hominoid trunk and hindlimb. In DR Begun, CV
Ward and MD Rose (eds.): Function, Phylogeny and Fossils: Miocene Hominoid Evolution and Adaptations. New York: Plenum Press, pp. 101–130.

Ward CV, Walker A, and Teaford MF (1991) *Proconsul* did not have a tail. J. Hum. Evol. *21:*215–220.

Ward CV, Walker A, Teaford MF, and Odhiambo I (1993) A partial skeleton of *Proconsul nyanzae* from Mfangano
Island, Kenya. Am. J. Phys. Anthropol. *90:*77–111.

Ward S, and Brown B (1996) Forelimb of *Kenyapithecus africanus* from the Tugen Hills, Baringo District, Kenya.
Am. J. Phys. Anthropol. *Supplement. 22:*240.

19

FOSSIL EVIDENCE FOR THE ORIGINS OF TERRESTRIALITY AMONG OLD WORLD HIGHER PRIMATES

Monte L. McCrossin,[1] Brenda R. Benefit,[1] Stephen N. Gitau,[1] Angela K. Palmer,[1] and Kathleen T. Blue[2]

[1]Department of Anthropology
Southern Illinois University
Carbondale, Illinois 62901
[2]Department of Anthropology
University of Chicago
Chicago, Illinois 60637

1. INTRODUCTION

Preference for terrestrial substrates is one of the most significant adaptive differences between some members of the radiation of Old World higher primates and the anthropoids of the Neotropics (Le Gros Clark, 1959; Napier and Napier, 1967, 1985; Fleagle, 1988; Martin, 1990). Adaptations for terrestriality are most conspicuous among savanna baboons (*Papio* - Rose, 1977), geladas (*Theropithecus* – Jolly, 1967; Dunbar and Dunbar, 1974), and humans (Napier, 1967). Varying degrees of semi-terrestriality and terrestriality are also present among the African great apes (*Gorilla* – Remis, 1995 and *Pan* - Hunt, 1992; Doran, 1993) and some of the Asian colobines (*Presbytis entellus* – Ripley, 1967 and *Rhinopithecus roxellana* – Davison, 1982), guenons (*Cercopithecus aethiops* and *Erythrocebus patas* – Hall, 1965), mandrills and drills (*Mandrillus* – Jouventin, 1975), mangabeys (*Cercocebus* – Waser, 1984), and macaques (e.g., *Macaca nemestrina* – Caldicott, 1986). In contrast, terrestrial adaptations are notably absent from the otherwise diverse adaptive array of New World anthropoids.

Substrate preference exhibits clear linkages with mode of locomotion among Old World higher primates (Ashton and Oxnard, 1964; Rose, 1973; Fleagle, 1988). The positional behavior of tree-dwelling Old World monkeys ranges from fairly deliberate pronograde quadrupedalism to more active and agile styles of arboreal quadrupedalism supplemented by leaping and arm-swinging (Ripley, 1967; Fleagle, 1977; Rollinson and Martin, 1981). Ground-dwelling cercopithecoids, in contrast, are primarily cursorial quadrupeds when on the ground and pronograde quadrupeds when in the trees (Rose, 1977). Modes of locomotion differ more dramatically between the arboreal hominoids of Asia and the terrestrially adapted African hominoids. Gibbons progress from branch to branch through brachiation (Fleagle, 1974) and orang-utans use all four limbs to clamber through the trees (MacKinnon, 1974; Cant, 1987) while gorillas and chimpanzees knuckle-walk over the ground and use a diverse repertoire of positional behaviors, including orthograde climbing and arm-swinging while in the trees (Hunt, 1992; Doran, 1993; Remis, 1995).

Beyond these obvious correlates between substrate preference and mode of locomotion, terrestrially adapted Old World monkeys and apes have been observed to differ from their arboreal relatives in several aspects of their ecology (tolerance and occasionally even preference for more open environments, larger day ranges and home-ranges, greater geographic distribution, and greater niche separation among sympatric primate species), behavior (more allomothering, more aggressive anti-predator behaviors, and foraging strategies directed toward harvesting and consuming a more diverse spectrum of foods), social organization (more precocial infants, greater likelihood of having multi-male groups, and larger group size), and anatomy (increased body size and higher levels of sexual dimorphism in canine height and body weight) (DeVore, 1963; Crook and Gartlan, 1966; Eisenberg et al., 1972; Clutton-Brock and Harvey, 1977). A few of these changes in ecology and behavior, especially a shift toward preference for open environments and an eclectic (omnivorous) foraging strategy, have long been implicated in the origin of bipedal Hominidae (Jolly, 1970; Lovejoy, 1981; Wolpoff, 1982; White, 1995).

Many of the distinctions drawn between ground-dwelling and tree-dwelling catarrhines have also been observed in comparisons between sympatric semi-terrestrial and arboreal lemurs at Antserananomby, Madagascar (Sussman, 1974). According to Sussman (1974), the semi-terrestrial ring-tailed lemur (*Lemur catta*) differs from the arboreal brown lemur (*L. fulvus rufus*) ecologically in its use of more arid habitats, behaviorally in its more varied diet, and socially in its larger group size. The existence of these same differences in both arboreal and terrestrial strepsirhines and catarrhines tends to support viewing these contrasts as either direct causes or consequences of terrestriality, rather than the result of common inheritance due to close phylogenetic relationships. Among some African cercopithecines, however, the influence of phylogenetic heritage seems to over-ride the effects wrought by differences of substrate preference. Rowell (1966) and Struhsaker (1969) have demonstrated that comparisons between arboreal and terrestrial cercopithecines tend not to support the correlations between terrestriality and larger group size seen in some other primate groups.

Although there is an extensive literature seeking to explain the origins of terrestrial bipedalism in early hominids (Jolly, 1970; Lovejoy, 1981; Wolpoff, 1982; White, 1995), the change from life in the trees to life on the ground among non-human catarrhines has received much less attention. While the evolutionary history of some other fundamental aspects of primate adaptation are fairly well understood, such as diurnal and nocturnal activity patterns among living primates (Martin, 1979), it is not known how many times terrestrial adaptations have been independently acquired in the evolutionary history of cercopithecoids (Napier, 1970; Rollinson and Martin, 1981; Andrews and Aiello, 1984; Benefit, 1987; Pickford and Senut, 1988). Similarly, the antiquity of terrestrial adaptations among large-bodied Miocene hominoids and its relevance for understanding the emergence of knuckle-walking and bipedalism among African apes and humans are debated (Washburn, 1968; Conroy and Fleagle, 1973; Tuttle, 1974; Gebo, 1992; McCrossin, 1997).

Most research concerning catarrhine adaptations for life on the ground has focussed upon the functional anatomy of living species (e.g., Jolly, 1967; Tuttle, 1967). The comparative anatomy of modern primates provides information concerning osteological corre-lates of particular positional behaviors. Osteological correlates of locomotor pattern are crucial for reconstructing the substrate preferences of fossil primates (e.g., Oxnard, 1963; Jolly, 1967; Knussman, 1967; Tuttle, 1967). Neontological comparisons, however, cannot address the contextual circumstances surrounding the origins of catarrhine terrestriality.

It is most fortunate, therefore, that recent advances in our understanding of the fossil record of early Old World monkeys and apes provide much new data concerning the ori-gins of terrestriality among catarrhines. Consideration of these new data, some presented here for the first time, provide new ideas concerning the timing, ecological context, and adaptive correlates of the origins of terrestriality among Old World higher primates. In this paper, we attempt to clarify the origins of terrestriality among catarrhines by reviewing postcranial indicators of semi-terrestriality, paleoenvironmental setting, considerations of body size, and reconstruction of the dietary adaptations of *Victoriapithecus* and *Kenyapi-thecus*.

1.1. Suggested Causes of the Transition from Arboreality to Terrestriality

Three major factors have been advanced as primary causes of the shift from life in the trees to life on the ground among Old World higher primates. First, it is widely thought that the transition from arboreality to terrestriality was spurred on by a widespread change in environments (Brain, 1981). The arboreality of the archaic catarrhines of the Oligocene (Conroy, 1976; Fleagle, 1983) and the formative hominoids of the early Miocene (Napier and Davis, 1959; Rose, 1993) is inextricably linked with the fact that they inhabited densely forested environments (Bowen and Vondra, 1974; Andrews and Van Couvering, 1975; Van Couvering 1980; Bown et al. 1982). The terrestrial adaptations of many of the cercopithecoids (Jolly, 1967; Birchette, 1981; Ciochon, 1993) and the early hominids (Lovejoy, 1981) of the Plio-Pleistocene of eastern Africa are widely regarded as resulting from colonization of the patchily vegetated savannas. Second, the shift from life in the trees to life on the ground is correlated with an increase in body size, with terrestrial pri-mates tending to be larger than their arboreal relatives (Andrews, 1983; Fleagle, 1985; Doran, 1993). Third, a fundamental change in diet and foraging strategy is commonly thought to have been a primary motive force in the advent of terrestriality, especially among hominoids (Jolly, 1970; Coursey, 1973; Szalay, 1975; White, 1995). The first ground-dwelling catarrhines are thought to have exploited terrestrial food sources such as grass seeds (Jolly, 1970), tuberous roots (Coursey, 1973), and animal carcasses (Szalay, 1975), in contrast to the concentration of arboreal primates upon foods available in the trees, primarily fruit and leaves.

It is unclear, however, to what degree occupation of open environments, increased body size, and consumption of terrestrial foods might be consequences rather than causes of life on the ground. The fossil record provides evidence that allows separation of cause from consequence in factors associated with the advent of terrestriality among Old World higher primates.

The earliest clear evidence of terrestrial adaptations among Old World higher primates is seen in two genera from the middle Miocene site of Maboko Island in Kenya (Benefit and McCrossin, 1989; Feibel and Brown, 1991; McCrossin and Benefit, 1994), the cercopithe-coid *Victoriapithecus* (Von Koenigswald, 1969; Benefit, 1987, 1993, 1994; Benefit and McCrossin, 1991, 1993, 1997; McCrossin and Benefit, 1992) and the large-bodied homi-noid *Kenyapithecus* (Le Gros Clark and Leakey, 1951; Leakey, 1962, 1967; McCrossin and

Benefit, 1993, 1997; McCrossin, 1994a). The fossil primate community from Maboko Island also includes two bushbabies -*Komba* and a new genus (McCrossin, 1990, 1992a), two small-bodied primitive catarrhines — *Simiolus* and cf. *Limnopithecus* (Benefit, 1991), and an oreopithecid — *Mabokopithecus* (Von Koenigswald, 1969; McCrossin, 1992b; Benefit et al., 1998). *Victoriapithecus* has also been identified at the early Miocene site of Napak in Uganda (Pilbeam and Walker, 1968; Pickford et al. 1986) as well as from two other sites from the middle Miocene of Kenya: Baragoi (Ishida, 1986) and Nyakach (Pickford and Senut, 1988). *Kenyapithecus* also occurs at the Kenyan middle Miocene sites of Fort Ternan (Leakey, 1962), Majiwa and Kaloma (Pickford, 1982a), Nyakach (Pickford, 1985), Baragoi (Ishida et al. 1984) and Kipsaramon (Brown et al., 1991).

2. TAXONOMY AND PHYLOGENETIC RELATIONSHIPS OF *VICTORIAPITHECUS* AND *KENYAPITHECUS*

Von Koenigswald (1969) named two species of *Victoriapithecus*, *V. macinnesi* and *V. leakeyi*, based on a small sample of cercopithecoid dentognathic and postcranial remains from Maboko Island. Because of the retention of a crista obliqua on its upper molars, von Koenigswald (1969) suggested that *Victoriapithecus* was a basal cercopithecid and assigned the genus to a distinct subfamily (Victoriapithecinae), ancestral to both Cercopithecinae and Colobinae.

Delson (1973, 1975) reassessed the Maboko Island cercopithecoids and suggested that *Victoriapithecus macinnesi* was an early colobine while *V. leakeyi* was an early cercopithecine. Delson (1975) perceived colobine-like arboreal adaptations in a distal humerus (KNM-MB 3) and an intermediate phalanx (KNM-MB 93) from Maboko and attributed them to *V. macinnesi*. Delson (1975) referred all other cercopithecoid postcranial remains then known from Maboko to *V. leakeyi*, describing cercopithecine-like terrestrial adaptations for another distal humerus (KNM-MB 19), two proximal ulnae (KNM-MB 2 and 32), a calcaneum (KNM-MB 16), a proximal phalanx (KNM-MB 12) and three intermediate phalanges (KNM-MB 13, 21, and 22). Analysis of the greatly expanded sample of dentognathic remains of *Victoriapithecus* (Pickford, 1984; Benefit and McCrossin, 1989), however, has indicated that only one species of cercopithecoid, *V. macinnesi*, is present at Maboko Island (Benefit, 1987, 1993, 1994) and Delson (Strasser and Delson, 1987) has subsequently abandoned his hypothesis that Colobinae and Cercopithecinae are represented in the craniodental assemblage from Maboko. Reassessment of the range of variation in *Victoriapithecus* postcranial remains (Harrison, 1989), including several new specimens collected on Maboko from 1982 to 1984 (Pickford, 1986), also support the existence of a single species (but see Senut, 1986a and Pickford and Senut, 1988 for arguments in support of two species). The absence of numerous craniodental synapomorphies of Colobinae and Cercopithecinae demonstrates that *Victoriapithecus* belongs to the sister-taxon (Victoriapithecidae) of Cercopithecidae (Benefit, 1987, 1993, 1994; Benefit and McCrossin, 1991, 1993, 1997).

The taxonomy of *Kenyapithecus* is a matter of long dispute. The genus was proposed by Leakey (1962) based on material from Fort Ternan, with *K. wickeri* as the type species. Subsequently, Leakey (1967) transferred other specimens from Maboko Island, including the type specimen of *Sivapithecus africanus* (Le Gros Clark and Leakey, 1951), to the genus as *K. africanus*.

Although this taxonomic arrangement has been confirmed by most workers who have studied the original fossil samples (Ishida et al., 1984; Ishida, 1986; Brown et al.,

1991; McCrossin and Benefit, 1993, 1994, 1997; McCrossin, 1994a), there have been repeated suggestions of generic distinctions between *Kenyapithecus wickeri* and *K. africanus* (Simons and Pilbeam, 1978; Pickford, 1985; Andrews, 1992; Harrison, 1992). According to Pickford (1985), *K. wickeri* differs from *K. africanus* in having a palatal process that is more highly arched, a maxillary sinus that is positioned at a higher level, an anterior root of the zygomatic arch that originates at a higher level above the molar alveolar plane and flares more markedly, a palatal process that is more highly arched, a maxillary sinus that is positioned at a higher level, an I^1 that lacks a distinct lingual pillar, a more vertically oriented canine socket, a P^3 with one (rather than two labial roots), a P^4 with weaker transverse ridges, and M^{1-2} without a lingual cingulum. Detailed metric and morphological examination of these features, however, demonstrates that the suggested differences between *K. wickeri* and *K. africanus* do not exceed levels of variation observed within extant hominoid genera (McCrossin, 1994a; Benefit and McCrossin, 1994; McCrossin and Benefit, 1997). In addition, *K. wickeri* and *K. africanus* share several derived features compared with other Miocene hominoids (McCrossin, 1994a; McCrossin and Benefit, 1997), including a distinctively proclined mandibular symphysis (McCrossin and Benefit, 1993) and lower molar roots that extend inferiorly almost to the basal margin of the mandibular corpus (Brown, 1997).

The phylogenetic relationships of *Kenyapithecus* have also elicited much interest. Influential suggestions of the phylogenetic relationships of *Kenyapithecus* include affinities to the great ape and human clade in general (Greenfield, 1979; Ward and Pilbeam, 1983; Andrews, 1985; McCrossin and Benefit, 1993) and to *Homo* specifically (Leakey, 1962, 1967; Andrews, 1971; Walker and Andrews, 1973; Simons and Pilbeam, 1978). Another recent suggestion is that *Kenyapithecus*, *Sivapithecus*, and *Dryopithecus* are members of an archaic radiation of apes antecedent to the last common ancestor of living hominoids (McCrossin, 1994a; McCrossin and Benefit, 1994, 1997; Benefit and McCrossin, 1995; Pilbeam, 1997). Finally, it has recently been suggested that *Kenyapithecus* exhibits craniodental and postcranial features indicating that it was the earliest known member of the African great ape and human clade (Ishida, 1986; Brown and Ward, 1988; McCrossin, 1997). A phylogenetic tree showing the inferred relationships of *Victoriapithecus* and *Kenyapithecus* is shown in Figure 1.

3. SKELETAL INDICATORS OF SEMI-TERRESTRIALITY

Despite numerous pervasive differences in postcranial morphology indicative of their respective cercopithecoid and hominoid affinities (von Koenigswald, 1969; Harrison, 1989; McCrossin and Benefit, 1992; McCrossin, 1994a), many skeletal indicators of semi-terrestriality are shared by both *Victoriapithecus* and *Kenyapithecus*. Functional implications of these features have been considered in greater detail elsewhere (McCrossin, 1992b, 1994a; McCrossin and Benefit, 1994, 1997) and will only be briefly reviewed here.

3.1. *Victoriapithecus*

The proximal humerus of *Victoriapithecus* (Figure 2) was until recently known only from an isolated articular end, KNM-MB 12044 (Senut, 1986a; Harrison, 1989). In isolation, this specimen was interpreted as indicating that the articular surface of the humeral head extended farther proximally than the greater tubercle (Senut, 1986a; Harrison, 1989).

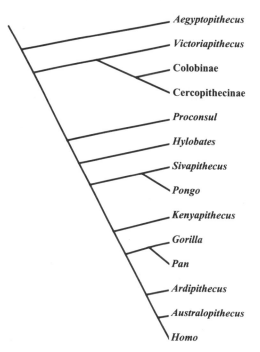

Figure 1. Branching diagram of the phylogenetic relationships of *Victoriapithecus* and *Kenyapithecus* to other Old World higher primates. The sister-group relationship of *Victoriapithecus* to Cercopithecidae is supported by synapomorphic cercopithecoid features of the cranium, facial skeleton, dentition, and postcranium and by the fact that living colobines and cercopithecines share several additional features, especially in the deciduous dentition and permanent molars, not seen in *Victoriapithecus* (von Koenigswald, 1969; Benefit, 1987, 1993, 1994; Benefit and McCrossin, 1991, 1993, 1997; Harrison, 1989; McCrossin and Benefit, 1992). Affinity of *Kenyapithecus* to the African great ape and human clade is supported by the subnasal pattern, the morphology of dp_3, the humerofemoral index, the volar slant of the inferior radioulnar joint, the presence of a metacarpal torus, and the articular morphology of the calcaneum, cuboid, and navicular (Ishida, 1986; Brown and Ward, 1988; McCrossin and Benefit, 1993; McCrossin, 1994a, 1997; McCrossin et al., 1998).

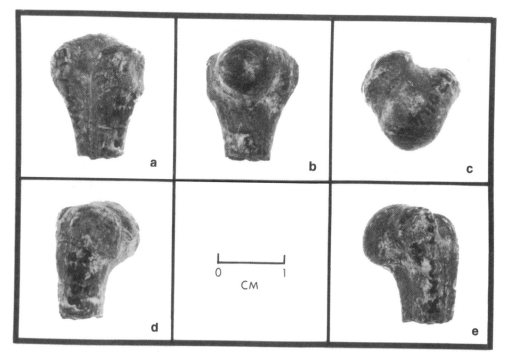

Figure 2. Isolated left proximal humerus of *Victoriapithecus* (KNM-MB 12044) in anterior (a), posterior (b), proximal (c), lateral (d), and medial (e) views.

Senut (1986a) and Harrison (1989) suggested that this perceived configuration related to a degree of shoulder mobility and arboreality for *Victoriapithecus* comparable to that seen in living arboreal guenons and colobines. Because the proximal disposition of the humeral head and greater tubercle are determined in part by the curvature of the humeral shaft (McCrossin, 1994a), however, it is not possible to assess this feature with proximal humerus specimens isolated from their shafts, including KNM-RU 17376 (*Proconsul africanus* or *Dendropithecus macinnesi*; Gebo et al., 1988), KNM-MB 21206 (*Mabokopithecus*; McCrossin, 1992b), and GSP 28062 (?Catarrhini indet.; Rose, 1989) as well as KNM-MB 12044 (*V. macinnesi*; Senut, 1986a; Harrison, 1989).

An almost complete humerus of *Victoriapithecus* recently collected from Maboko Island reveals that the greater tubercle actually extends above the surface of the humeral head (Figure 3). This proximal extension of the greater tubercle is due in large part to the curvature of the humeral shaft, especially in the region of the deltopectoral crest. In extant hominoids, colobines (except some individuals of *Presbytis entellus* as well as the extinct species *Mesopithecus pentelici* and *Cercopithecoides williamsi*; Birchette, 1981, 1982), guenons (except *Cercopithecus aethiops* and *Erythrocebus patas*) and New World monkeys, the humeral head extends above the greater tubercle (McCrossin, 1994a: Figure 28). In terrestrial cercopithecoids, such as *Erythrocebus*, *Papio*, and *Theropithecus*, the greater tubercle extends high above the humeral head. Possible allometric influences were examined by Birchette (1982:151) who concluded that proximal extension of the greater tubercle above the humeral head appears not to be related to "mechanical demands placed on supraspinatus by the increase in limb weight associated with a general increase in body

Figure 3. Right humerus (missing capitulum and part of trochlea) of *Victoriapithecus* (KNM-MB 21809) in lateral view. Curvature of the shaft in the deltopectoral region results in proximal extension of greater tubercle slightly above the articular surface. (The scale bar equals one centimeter.)

weight" and thus is not a function of large body size in *Erythrocebus*, *Papio*, and *Theropithecus*.

Jolly (1967) suggested that elevation of the greater tubercle above the humeral head is an adaptation to terrestrial walking and running. This arrangement serves to resist passive retraction of the arm that results from weight-bearing at the glenohumeral joint and enhances forceful protraction of the arm by providing for elongation of the moment arm of *m. supraspinatus* (Jolly, 1967). Electromyographic studies, however, indicate that the most important function of *m. supraspinatus* may be to stabilize the shoulder during the stance phase of quadrupedalism (Larson and Stern, 1992). The lower lesser tubercle and shorter moment arm of *m. supraspinatus* in platyrrhines, hominoids, and arboreal cercopithecoids is related to greater mobility at the glenohumeral joint because it results in faster (albeit weaker) protraction of the humerus (Jolly, 1967; Birchette, 1982). The greatest degree of humeral head projection, seen in atelines, *Hylobates*, and *Pongo*, allows for extreme mobility and rotatory capabilities at the glenohumeral joint.

The distal humerus of *Victoriapithecus* (Figure 4) exhibits a combination of features, all of which seem to be shared derived similarities with living cercopithecoids (Pickford and Senut, 1988; Harrison, 1989). The distal humerus articular surface of *Victoriapithecus* (Figure 4) resembles modern cercopithecoids in that, overall, it is narrow, the trochlea is narrower than the capitulum, and a strong trochlear keel is present medially (Pickford and Senut, 1988; Harrison, 1989). The broad and somewhat flattened capitulum of *Victoriapithecus* (Pickford and Senut, 1988; Harrison, 1989) indicates that primary transference of weight is through the humeroradial joint while the narrow trochlea and strong medial trochlear keel reflect emphasis on restriction of the humeroulnar joint to hinge-like flexion and extension motions (Napier and Davis, 1959; Jenkins, 1973).

One of the most distinctive features of the *Victoriapithecus* distal humerus is the strong posterior inclination of the medial epicondyle (Pickford and Senut, 1988). Delson (1973, 1975) claimed that the variation in medial epicondyle inclination seen in the humeral remains of *Victoriapithecus* supported assigning one specimen to an arboreal colobine and another specimen to a terrestrial cercopithecine. Harrison (1989:18), however,

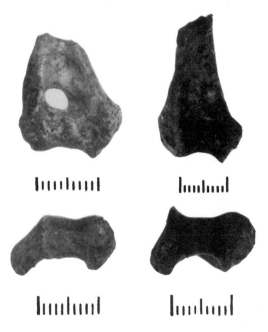

Figure 4. Left distal humeri of *Victoriapithecus* (left column: KNM-MB 3; right column: KNM-MB 19) in posterior (top row) and distal (bottom row) views. The strong medial trochlear keel and posterior orientation of the medial epicondyle in KNM-MB 19 are some of the most compelling indicators of semi-terrestriality for *Victoriapithecus*. The more medially oriented medial epicondyle of KNM-MB 3 was once regarded as indicating that it represented a more arboreal taxon, but the difference in medial epicondyle orientation seen in KNM-MB 3 and 19 does not exceed intraspecific variation in living species. (The scale bars equal one centimeter.)

has shown that the variation in *Victoriapithecus* medial epicondyle inclination "does not exceed the variation seen in extant catarrhine species."

In hominoids and platyrrhines, the medial epicondyle is large and medially directed, reflecting the premium placed on digital grasping (Fleagle and Simons, 1982). In contrast, in cercopithecoids, especially the semi-terrestrial and terrestrial cercopithecines, the medial epicondyle is abbreviated and posteromedially oriented (Jolly, 1967). This shortening and posterior reflection of the medial epicondyle in ground-dwelling cercopithecines has been related to a reduction in the mass of the carpal and digital flexors, which take their origin from the entepicondyle, and to an increase in the moment arm of *m. pronator teres* around the axis of pronation (Jolly, 1967; Birchette, 1982).

The proximal ulna of *Victoriapithecus* (Figure 5) is dominated by a moderately long and dorsally deflected ulnar olecranon (von Koenigswald, 1969; Harrison, 1989). This combination of a moderately long and retroflexed olecranon is seen among semi-terrestrial and terrestrial cercopithecoids (Jolly, 1967). Arboreal cercopithecoids usually have longer and straighter olecranons (Oxnard, 1963; Jolly, 1967). Angles of olecranon retroflexion differ greatly among extant cercopithecoids, with means between 25 (*Procolobus*) and 37 degrees (*Rhinopithecus*) for modern colobine genera (Birchette, 1982: Table 25) while individuals of *Papio* range from 40 to 60 degrees (Jolly, 1972).

The olecranon is the site of insertion of *m. triceps brachii*, the major extensor of the forearm, and "the triceps is at its greatest mechanical advantage when its line of action lies farthest from the axis of the elbow joint, i.e., when it is perpendicular to the axis of the olecranon" (Birchette, 1982:241). Thus, retroflection of the olecranon may help to enhance the thrust produced by *m. triceps brachii* when the elbow is in postures approaching maximum extension (Jolly, 1967).

Figure 5. Right proximal ulna of *Victoriapithecus* (KNM-MB 32) in lateral (left) and medial (right) views. The slight dorsal angulation of the olecranon process with respect to the longitudinal axis of the ulnar shaft is a resemblance to living semi-terrestrial Old World monkeys. (The scale bars equal one centimeter.)

The radial neck is relatively very short, indicating that "*Victoriapithecus* was probably capable of rapid locomotion on level surface substrates" (Harrison, 1989:25). The length of the radial neck corresponds to the moment arm of *m. biceps brachii* and consequently relates to the speed and forcefulness of elbow flexion (Napier and Davis, 1959). Among arboreal anthropoids, a premium is placed on the force produced by *m. biceps brachii* in elbow flexion during hoisting and climbing and the radial neck is relatively long (Conroy, 1976). Terrestrial cercopithecoids, in contrast, rely upon rapid flexion of the elbow during the recovery phase of cursorial quadrupedalism and the radial neck is relatively short (Jolly, 1967). Recently, Reno et al. (1997:197) have questioned the proposed correlation between radial neck length and substrate preference because total length of the radius is correlated to body mass and "lengthening of the radius attendant to increasing body mass is differentially expressed by distal (rather than proximal) growth of the radius". According to Reno et al. (1997:197), "any association of relative radial neck length with substrate preference such as that reported in monkeys" by Harrison (1989) "merely reflects selection on increased body mass in a terrestrial environment". Radial neck length among anthropoids, however, does not appear to be wholly dictated by allometric effects. When indexed against the minimum diameter of the radial head, the radial necks of arboreal anthropoids are relatively long while those of semi-terrestrial and terrestrial cercopithecoids are relatively short (Harrison, 1989: Fig. 11).

The ischial body of *Victoriapithecus* is relatively long, as in semi-terrestrial and terrestrial cercopithecoids (McCrossin and Benefit, 1992:280, Table 1, Fig. 1). The length of the ischium corresponds to the lever arm of the ischiocrural musculature, upon which the

Figure 6. Right femur (lacking distal epiphysis) of *Victoriapithecus* (KNM-MB 20230) in anterior view. The low femoral neck-shaft angle seen in this specimen indicates mainly parasagittal movements of the thigh in adducted postures. (The scale bar equals one centimeter.)

torque produced by the hamstrings depends (Waterman, 1929; Leutenegger, 1970). In addition, a raised line from the origin of *m. gemellus* extends continuously along the ischial body of *Victoriapithecus*, from the ischial spine to the tuberosity (McCrossin and Benefit, 1992). This configuration is also seen in semi-terrestrial cercopithecoids such as *Cercopithecus aethiops* and *Erythrocebus patas* while markings from the origin *of m. gemellus* are more proximally restricted in arboreal Old World monkeys such as *Lophocebus albigena* and *Presbytis rubicundus*.

Until recently, the femur of *Victoriapithecus* was known only from four very poorly preserved fragments of the femoral head, one preserving a small portion of the femoral neck (Harrison, 1989). An almost complete femur of *Victoriapithecus*, lacking only the distal epiphysis, was recently collected from Maboko Island (Figure 6). The previously collected fragments and the new femur show that the femoral head of *Victoriapithecus* is relatively small, its articular surface continues onto the posterior and proximal sides of the femoral neck, the femoral neck is quite short, and the neck angle is low. Although shared by both tree-dwelling and ground-dwelling cercopithecoids, these features are associated with greater efficiency of parasagittal movements of the thigh (Jenkins and Camazine, 1977) and may be related to adaptations for walking and running over level substrates.

The entocuneiform of *Victoriapithecus* exhibits a first metatarsal facet that is mainly confined to the distal end and extends for only a short distance medially (Harrison, 1989). This morphology is also seen in living terrestrial cercopithecoids, such as *Erythrocebus*, *Papio*, and *Theropithecus*, and is associated with a limited range of hallucial abduction (Jolly, 1967). Arboreal cercopithecoids, in contrast, have a much more medially extensive first metatarsal facet on the entocuneiform, allowing a great range of hallucial abduction (Napier and Davis, 1959).

Preliminary indications from study of the greatly expanded sample of cercopithecoid pedal specimens from Maboko Island (Strasser, 1997:222) re-affirm that the functional attributes linking *Victoriapithecus* to living Old World monkeys are "related to stabilizing the foot at the upper and lower ankle joints, increased pronation of the forefoot and reduced hallucial abduction". The morphocline polarity of five cercopithecine-like characters in the foot skeleton of *Victoriapithecus* are unresolved and "*may* reflect a shared habitus rather than a phyletic link" (Strasser, 1997:222 – her emphasis).

Delson (1973, 1975) divided *Victoriapithecus* intermediate phalanges into two categories. One specimen, KNM-MB 93, perceived as being long and slender, was referred to *V. macinnesi* and regarded as representing an arboreally adapted colobine (Delson, 1973, 1975). Three other specimens (KNM-MB 13, 21 and 22), described as being relatively short and robust, were attributed to *V. leakeyi*, a species that Delson (1973, 1975) claimed was a terrestrially adapted cercopithecine. Reassessment of these and additional specimens, however, reveals that the range of variation seen in *Victoriapithecus* does not exceed that seen in living Old World monkey species (Harrison, 1989). In fact, the "relatively short and stout" proportions of the phalanges "indicate that *Victoriapithecus* was suited to moving effectively on a terrestrial substrate, and implies that terrestrial digitigrady may have been an important component of its locomotor repertoire" (Harrison, 1989:38).

In summary, the postcranial skeleton of *Victoriapithecus* exhibits numerous indicators of semi-terrestriality and few, if any, adaptations for the degree of arboreality seen among arboreal guenons and colobines. This interpretation differs from that proposed by Delson (1973, 1975), who presented one set of cercopithecoid postcrania from Maboko Island as belonging to an arboreally adapted colobine and another group of specimens as belonging to a terrestrially adapted cercopithecine. Although Senut's (1986a; Pickford and

Senut, 1988) and Harrison's (1989) studies did not support Delson's (1973, 1975) division of the Maboko Island cercopithecoids into colobines and cercopithecines, they also perceived the presence of both arboreal and terrestrial adaptations in the sample of *Victoriapithecus* postcrania. Senut (1986a) interpreted these differences as indicating the presence of two species. Harrison's (1989) results, in contrast, were seen as supporting Benefit's (1987) contention that only one cercopithecoid species, *V. macinnesi*, is present at Maboko Island. Harrison's (1989:50) perception of arboreal adaptations in the shoulder anatomy of *Victoriapithecus* led him to speculate that *Victoriapithecus* "may have been capable of agile climbing and clambering in the multi-dimensional small-branch milieu of the upper canopy". As discussed previously, however, the isolation from the humeral shaft of the KNM-MB 12044 proximal humerus does not allow interpretation of the disposition of the greater tubercle and the kinematics of *m. supraspinatus* of *Victoriapithecus*. While we have long recognized that the semi-terrestriality of *Victoriapithecus* involved utilization of both arboreal and terrestrial substrates and resources (Benefit, 1987; McCrossin and Benefit, 1992, 1994), we see no evidence from the postcranial skeleton for agile arboreal climbing similar to that of arboreal guenons and colobines (*contra* Harrison, 1989).

3.2. *Kenyapithecus*

Semi-terrestrial adaptations of the limb skeleton of *Kenyapithecus* from Maboko Island and Fort Ternan have only recently been recognized (McCrossin, 1994a,b; McCrossin and Benefit, 1992, 1994, 1997; Benefit and McCrossin, 1995). Previously, postcranial remains of *Kenyapithecus* from Maboko Island and Fort Ternan, as well as from Baragoi,

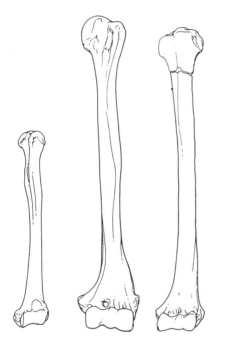

Colobus   Pan   Kenyapithecus

Figure 7. Left humerus of *Colobus guereza*, *Pan paniscus*, and *Kenyapithecus* composite reconstruction (including the following specimens: KNM-MB 24729, left proximal humerus; BMNH M. 16334, left humerus shaft; KNM-FT 2751, right distal humerus – reversed) in anterior view. The humeral shaft and proximal end of *Kenyapithecus* from Maboko Island were collected in 1933 and 1992, respectively. A jagged reciprocal pattern of breakage allows them to be conjoined. (The scale bar equals five centimeters.)

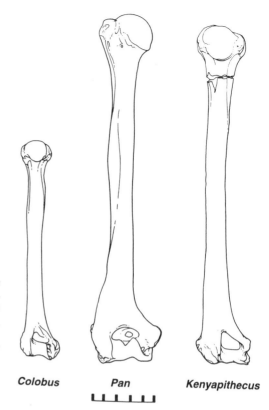

Figure 8. Left humerus of *Colobus guereza, Pan paniscus,* and *Kenyapithecus* composite reconstruction (including the following specimens: KNM-MB 24729, left proximal humerus; BMNH M. 16334, left humerus shaft; KNM-FT 2751, right distal humerus – reversed) in posterior view. The greater tubercle extends proximally farther than the articular surface of the humeral head in *Kenyapithecus.* (The scale bar equals five centimeters.)

have been interpreted as indicating *Proconsul*-like arboreal quadrupedal adaptations (Andrews and Walker, 1976; Rose, 1983, 1993; Harrison, 1992; Rose et al., 1996).

The most compelling features indicative of semi-terrestriality for *Kenyapithecus* concern the humerus (Figures 7-8). The head of the humerus is directed posteriorly as in most anthropoids, including terrestrially adapted cercopithecoids. This pattern of glenohumeral articulation is most effective in the predominantly rectilinear protraction and retraction of the humerus employed by all quadrupedal primates, but most efficiently by the cursorial ground-dwelling cercopithecines. The posterior orientation of the humeral head of *Kenyapithecus* differs from the medial orientation of the humeral head seen in extant hominoids and atelines. Medial orientation of the humeral head is most extreme in *Gorilla* and *Pan,* and enables maintenance of semi-pronated forearm postures during pronograde quadrupedalism (Larson, 1988).

Like cercopithecoids and most ceboids, the humeral shaft of *Kenyapithecus* collected from Maboko Island in 1933 (Le Gros Clark and Leakey, 1951) is flexed approximately one-third of the way down its length, due to strong development of the deltopectoral crest. Napier and Davis (1959) suggested that flexion of the humeral shaft is associated with quadrupedal modes of locomotion among primates. As preserved, however, the shaft of another *Kenyapithecus* humerus collected from Maboko Island in 1996 is quite straight in medial view (Figure 9), like that of living hominoids (McCrossin, 1997).

The greater tubercle of *Kenyapithecus* (Figure 10) extends slightly farther superiorly than the level of the articular surface of the humeral head (McCrossin, 1994a; McCrossin and Benefit, 1997). The articular surface of the humeral head of *Kenyapithecus* is also

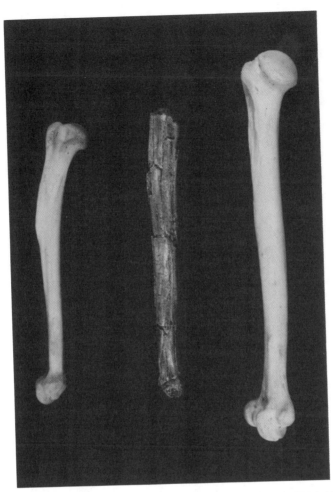

Figure 9. Right humeral shaft of *Kenyapithecus* collected from Maboko Island in 1996 (middle), compared with humeri of *Papio* (left) and *Homo* (right) in medial view. As preserved, this *Kenyapithecus* specimen shares a straight humeral shaft with modern hominoids, including humans, rather than the flexed humeral shaft of cercopithecoids.

quite flattened proximally. Among agile arboreal anthropoids, especially suspensory forms such as *Ateles*, *Lagothrix*, and *Hylobates*, the proximal surface of the humeral head is strongly domed, forming a ball-like articulation for the glenoid cavity of the scapula.

In order to gauge the overall phenetic resemblance of the *Kenyapithecus* proximal humerus to other anthropoids, a principal components analysis was performed using five natural logarithm transformed variables (McCrossin, 1994a). The first component accounts for 88% of the variance and is derived from the first four variables (proximodistal height of the head, mediolateral breadth of the head, lesser tubercle diameter, and depth of the bicipital groove) while the breadth of the bicipital groove is the primary determinant of the second component, which accounts for 8% of the variance (McCrossin, 1994a: Fig. 32). A plot of the first and second principal components (Figure 11) shows *Kenyapithecus* positioned closest to terrestrial and semi-terrestrial cercopithecoids such as *Mandrillus* (including both the mandrill and drill), *Papio, Presbytis entellus, Theropithecus gelada*, and *Macaca nemestrina*.

Distally, the humeral medial epicondyle of *Kenyapithecus* also resembles that of *Victoriapithecus* and terrestrially adapted cercopithecoids in being very short and posteriorly oriented (Figure 12). The angle of posterior reflection of the medial epicondyle seen in *K. wickeri* (54 degrees) is most comparable to values for *Erythrocebus patas* (mean = 51 degrees, mean ± 1 s.d. = 46–57 degrees; Fleagle and Simons, 1982: Table 2) and *Macaca mulatta* (mean = 56 degrees, range = 40–69 degrees; Harrison, 1989: Fig. 6). The medial epicondyle is more medially oriented in other Oligo-Miocene non-cercopithecoid catarrhines, including *Aegyptopithecus zeuxis* (23–27 degrees; DPC 1026, DPC 1275, CGM 40123, CGM 40855), *Pliopithecus vindobonensis* (13 degrees; Individual I; Zapfe, 1960), *Simiolus enjiessi* (26 degrees; KNM-WK 17009), and *Dendropithecus macinnesi* (28 degrees; KNM-RU 1675) (Fleagle and Simons, 1982: Table 1; Rose et al., 1992: Table 3). Comparable values are seen in arboreal quadrupeds such as *Cebus apella* (mean = 35 degrees; mean ± 1 s.d. = 30–39 degrees) and *Alouatta seniculus* (mean = 14 degrees, range = 12–44 degrees) (Fleagle and Simons, 1982: Table 2; Harrison, 1989: Fig. 6). It is reasonable, therefore, to reconstruct the ancestral catarrhine condition for medial epicondyle orientation as spanning the range from approximately 10 to 40 degrees, based on commonality of distribution among platyrrhines (as an outgroup), propliopithecids, pliopithecids, small-bodied primitive catarrhines, and arboreally adapted cercopithecoids. Divergently derived conditions are seen in the more medially directed medial epicondyles reconstructed for the last common ancestor of living hominoids (with angles of approximately 0 to 10 degrees) and the higher medial epicondyle angles that were independently evolved by terrestrially adapted cercopithecoids and *Kenyapithecus*. Consequently, the moderately medioposteriorly angled medial epicondyle of *Dryopithecus brancoi* (35 degrees for Rud 53; Rose et al., 1992: Table 3) reflects conservative retention of arboreal quadrupedalism like that seen in *Proconsul* while the retroflexed medial epicondyle of *Kenyapithecus* is a derived condition, related to terrestrial adaptations.

The olecranon process of a proximal ulna of *Kenyapithecus* (KNM-MB 24737) is moderately long, most like that of cercopithecines (Figure 13), and unlike the extremely reduced olecranon of modern hominoids (McCrossin, 1994a:147–153). The olecranon appears to be somewhat retroflexed judging from the posteriorly directed sloping of the anterior border from the anconeal process to the insertion scar of *m. triceps brachii*. In anterior view, the superior margin of the articular surface for the humeral trochlea is asymmetrical, with a greater proximal extension present on the lateral side of the anconeal process.

The presence of a greater amount of articular surface on the lateral (than medial) side of the anconeal process in *Kenyapithecus* is shared with terrestrial cercopithecoids while arboreal cercopithecoids exhibit mediolateral symmetry of the superior margins of the trochlear surface (Birchette, 1982:252–253; Fig. 13). In terrestrial cercopithecoids, the lateral side of the anconeal process articulates with a posterior extension of articular surface on the lateral margin of the olecranon fossa of the distal humerus. This lateral articular buttressing serves to resist lateral displacement at the elbow joint.

A distal radius of *Kenyapithecus* collected from Maboko Island in 1997 provides the first knowledge of the inferior radioulnar joint for a large-bodied hominoid from the middle to late Miocene (McCrossin et al., 1998). Although critical to reconstruction of forelimb kinematics and phylogenetic affinities (Jenkins and Fleagle, 1975; O'Connor, 1975; McHenry and Corruccini, 1983), the inferior radioulnar joint of Miocene catarrhines has previously been known only for *Proconsul* from Site R114 on Rusinga Island (Napier and Davis, 1959) and *Pliopithecus* from Neudorf an der March (Zapfe, 1960). The distal ra-

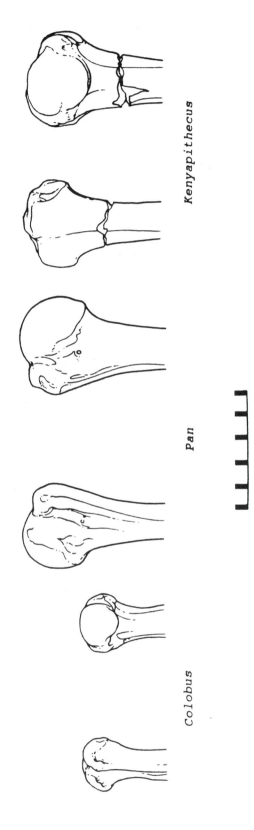

Figure 10. Humeral head projection above greater tubercle x 100/greater tubercle diameter (McCrossin, 1994a). Some details of sample composition are as follows: *Presbytis* combines *P. entellus* (N = 2), *P. cristatus* (N = 10), and *P. rubicundus* (N = 1); *Cercopithecus* includes *C. aethiops* (N = 9), *C. mitis* (N = 6), *C. ascanius* (N = 2), *C. cephus* (N = 1), *C. diana* (N = 1), *C. lhoesti* (N = 1), *C. neglectus* (N = 1), and *C. nictitans* (N=1). The vertical line indicates the median value, the boxed area encloses the inter-quartile range, the horizontal line extends to 1.5 times the inter-quartile range, and the small boxes are values which lie beyond 1.5 times the inter-quartile range. The greater tubercle of *Kenyapithecus* projects slightly above the level of the humeral head, most like cercopithecoids such as *Lophocebus, Presbytis entellus, Macaca,* and *Cercopithecus aethiops.*

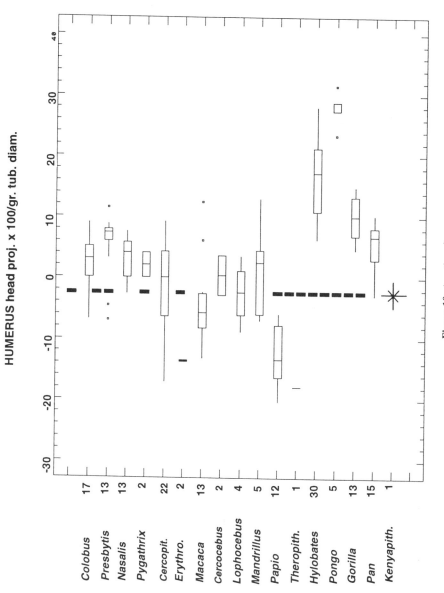

Figure 10. *(continued)*

dius of *Kenyapithecus* strongly resembles modern hominoids, especially *Gorilla* and *Pan*, in several features (McCrossin et al., 1998). The medial margin of the *Kenyapithecus* distal radius forms a dorsoventrally crescentic and concave articular embrasure for the distal ulna (McCrossin et al., 1998). The similarly crescentic inferior radioulnar joint of living hominoids allows an extensive range of forearm pronation and supination (O'Connor, 1975). Despite erosion, the dorsal ridges of the distal radius appear to have been strongly developed (McCrossin et al., 1998). According to Tuttle (1974: Fig. 8), the dorsal ridges of the distal radius "are implicated in the close-packed position" of the radioscaphoid joint of *Pan troglodytes*. The radius of *Kenyapithecus* appears to exhibit the marked volar slant of the distal end that serves to limit dorsiflexion of the wrist during knuckle-walking and distinguishes *Gorilla* and *Pan* from ceboids, *Pliopithecus*, cercopithecoids, *Proconsul*, *Hylobates*, and *Pongo* (Napier and Davis, 1959; Zapfe, 1960; Jenkins and Fleagle, 1975; O'Connor, 1975; McHenry and Corruccini, 1983). In addition, the articular surfaces for the scaphoid and lunate are deeply concave (McCrossin et al., 1998), unlike the relatively flat surfaces of platyrrhines, *Pliopithecus*, cercopithecoids, *Proconsul*, *Hylobates*, and *Pongo* (Napier and Davis, 1959; Zapfe, 1960; Tuttle, 1974).

A complete third metacarpal of *Kenyapithecus* (Figure 14) has a strong transverse dorsal ridge, or metacarpal torus, adjacent to the distal articulation for the proximal phalanx. A similar transverse dorsal ridge is seen on the second through fifth metacarpals of *Gorilla* and *Pan*, but is absent from homologues of *Hylobates* and *Pongo* (Tuttle, 1974: Fig. 7). Because of its presence on the second through fifth metacarpals of *Gorilla* and *Pan*, the metacarpal torus is thought to be a derived feature that is a potential indicator of a knuckle-walking pattern of locomotion in fossil taxa (Tuttle, 1967). During knuckle-walking, the metacarpal torus prevents hyperextension at the metacarpophalangeal joint (Tuttle, 1967, 1974). The presence of a transverse dorsal ridge on the third metacarpal of *Kenyapithecus* appears to be indicative, therefore, of postures of the metacarpophalangeal joint similar to those employed by *Gorilla* and *Pan*.

In plantar view, the medial portion of the entocuneiform facet of a left first metatarsal of *Kenyapithecus* (KNM-MB 24728) is quite flat (Figure 15). A similarly flat configuration is also seen in *Papio hamadryas*, but the medial portion of the entocuneiform facet is proximally recurved in *Colobus guereza* and *Pan troglodytes* (Figure 15). The absence

---

Figure 11. Plot of a principal components analysis of the proximal humerus, based on five dimensions transformed to their natural logarithms: 1) depth of the bicipital groove, 2) superoinferior height of the head, 3) mediolateral breadth of the head, 4) lesser tubercle diameter, and 5) breadth of the bicipital groove (McCrossin, 1994a). Abbreviations for plotted means of taxa (and their sample sizes) are as follows: Ag = *Aegyptopithecus zeuxis* (DPC 1275; Fleagle and Simons, 1982), Al = *Alouatta* (N = 8), At = *Ateles* (N = 3), Ca = *Cercopithecus aethiops* (N = 6), Cb = *Cebuella* (N = 1), Cc = *Cercocebus* (N = 2), Ch = *Chiropotes* (N = 1), Cj = *Cacajao* (N = 2), Cl = *Colobus* (N = 17), Cs = *Cebus* (N = 3), Cx = *Cercopithecus* spp. (*C. ascanius*, N = 2; *C. cephus*, N = 1; *C. diana*, N = 1; *C. lhoesti*, N = 1; *C. mitis*, N = 9; *C. neglectus*, N = 1; *C. nictitans*, N = 1), Er = *Erythrocebus* (N = 2), Gr = *Gorilla* (N = 13), Hl = *Hylobates* (N = 30), Kn = *Kenyapithecus* (KNM-MB 24729 and BMNH M. 16334), Lp = *Lophocebus* (N = 4), Md = *Mandrillus* (N = 5), Mf = *Macaca fascicularis* (N = 9), Mn = *Macaca nemestrina* (N = 5), Mp = *Miopithecus* (N = 2), Ns = *Nasalis* (N = 13), Ny = cf. *Mabokopithecus* (KNM-MB 21206; McCrossin, 1992a), Pe = *Presbytis entellus* (N = 2), Pg = *Pygathrix* (N = 2), Pl = *Pliopithecus vindobonensis* (OE 304; Ginsburg and Mein, 1980), Pn = *Pan* (N = 15), Po = *Pongo* (N = 5), Pp = *Papio* (N = 12), Px = *Presbytis* spp. (*P. cristatus*, N = 10; *P. rubicundus*, N = 1), Ru = *Dendropithecus macinnesi* or *Proconsul africanus* (KNM-RU 17376; Gebo et al., 1988), Sg = *Saguinus* (N = 1), Sm = *Saimiri* (N = 2), Th = *Theropithecus* (N = 1). *Kenyapithecus* lies closest to semi-terrestrial and terrestrial forms such as *Mandrillus*, *Papio*, and *Presbytis entellus*.

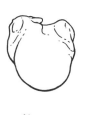

Nasalis   *Papio*   *Pan*   *Kenyapithecus*

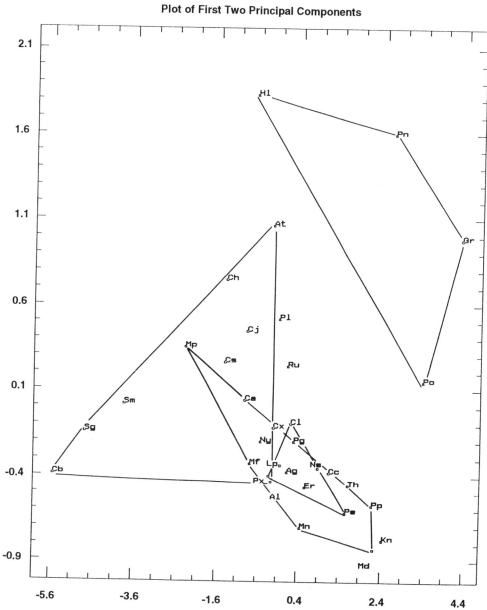

Plot of First Two Principal Components

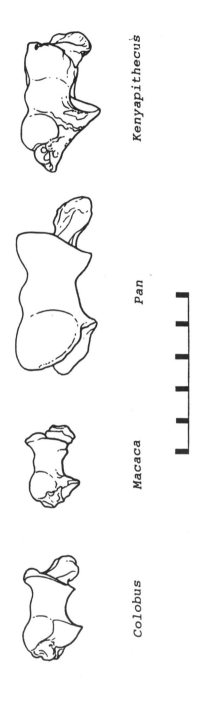

Figure 12. Retroflection of the medial epicondyle of the humerus in degrees (McCrossin, 1994a). The vertical line is the mean and the box encloses the mean ± 1 standard deviation for *Erythrocebus patas*, *Macaca fascicularis*, and *M. nemestrina* (data are from Fleagle and Simons, 1982). The vertical line is the median and the box encloses the entire range for *Colobus polykomos*, *Presbytis rubicundus*, *Miopithecus talapoin*, *Cercopithecus aethiops*, *C. mitis*, *C. nictitans*, *Macaca mulatta*, *Papio anubis*, *Theropithecus gelada*, *Hylobates lar*, *Pongo pygmaeus*, *Gorilla gorilla*, and *Pan troglodytes* (data for all but *C. aethiops* and *C. nictitans* from Harrison, 1989). Data for *Aegyptopithecus zeuxis*, *Pliopithecus vindobonensis*, *Simiolus enjiessi*, *Dendropithecus macinnesi*, *Proconsul africanus*, and *Dryopithecus brancoi* are from Rose et al. (1992). The posteriorly directed medial epicondyle of *Kenyapithecus* finds its closest match with semi-terrestrial and terrestrial forms such as *Erythrocebus*, *Macaca*, and *Theropithecus*.

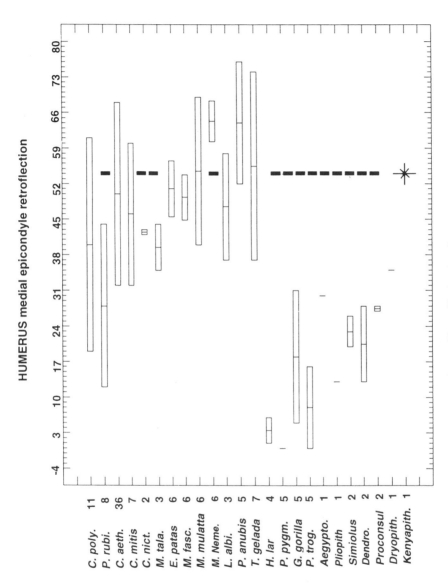

Figure 12. *(continued)*

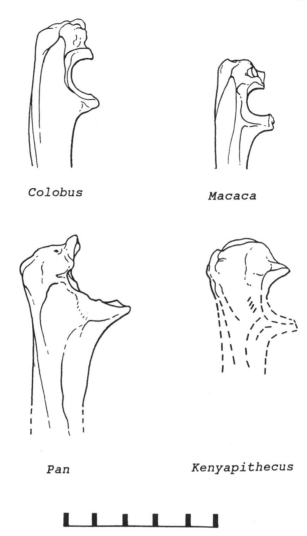

Colobus   Macaca

Pan   Kenyapithecus

Figure 13. Left proximal ulna of *Colobus guereza*, *Macaca nigra*, *Pan paniscus*, and *Kenyapithecus africanus* (KNM-MB 24737 – dashed lines indicate reconstructed coronoid process and distal portion of trochlear notch) in medial view. (The scale bar equals five centimeters.)

of articular surface wrapping medially around the entocuneiform may indicate that the hallux of *Kenyapithecus* was habitually adducted as in baboons during terrestrial quadrupedalism.

A left cuboid of *Kenyapithecus* (Figure 16) shows derived resemblances to African apes and humans in several features (McCrossin, 1997). The cuboid of *Kenyapithecus* is relatively short and strongly wedged and the proximal end possesses a moderately well developed peg-like process for articulation with a reciprocal concavity on the distal surface of the calcaneum (McCrossin, 1997). In many respects, the cuboid of *Kenyapithecus* differs from primitive character states seen in the cuboids of propliopithecids, *Proconsul*, and *Sivapithecus* (Langdon, 1986).

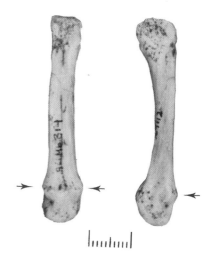

Figure 14. Right third metacarpal of *Kenyapithecus* collected from Maboko Island in dorsal (left) and medial (right) views. The strongly developed metacarpal torus of *Kenyapithecus* (arrows indicate position) is a point of resemblance to *Gorilla* and *Pan* and suggests resistance of hyperextension at the metacarpophalangeal joint. (The scale bar equals one centimeter.)

An intermediate phalanx of *Kenyapithecus* (KNM-MB 28393) is relatively short and stout with little dorsoventral curvature of the shaft and well marked flanges on the ventral surface for insertion of the digital flexors (Figure 17). A bivariate plot of proximodistal length and proximal mediolateral breadth (Figure 18) shows that the intermediate phalanx of *Kenyapithecus* is relatively short and robust, like that of semi-terrestrial catarrhines, es-

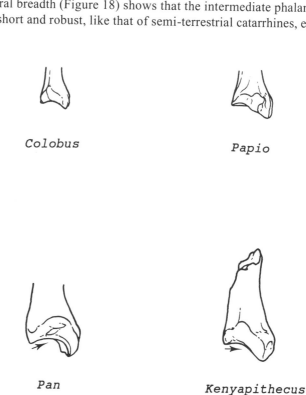

Figure 15. Proximal end of left first metatarsal of *Colobus guereza*, *Papio hamadryas*, *Pan troglodytes*, and *Kenyapithecus africanus* (KNM-MB 24728) in plantar view. Arrows indicate proximal recurvature of entocuneiform facet in *Pan* and comparatively flat surface in *Kenyapithecus*. (The scale bar equals five centimeters.)

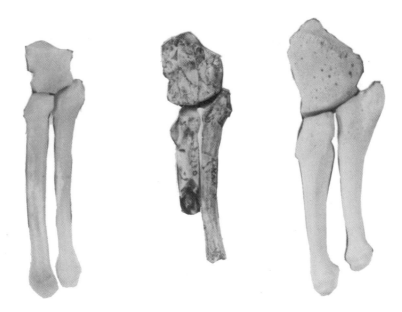

Figure 16. Left cuboid, fourth metatarsal, and fifth metatarsal of *Papio* (left), *Kenyapithecus* (middle), and *Homo* (right) in dorsal view. The cuboid of *Kenyapithecus* resembles *Gorilla*, *Pan*, and *Homo* in that, overall, it is relatively short (proximodistally), strongly wedged and the proximal peg (for articulation with the calcaneum) is moderately large.

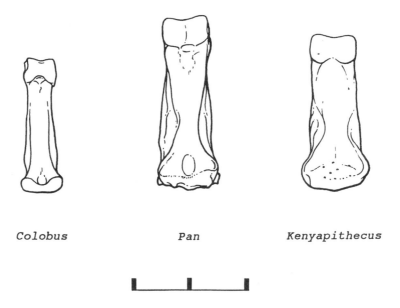

Colobus   Pan   Kenyapithecus

Figure 17. Intermediate phalanx of *Colobus guereza*, *Pan troglodytes*, and *Kenyapithecus africanus* (KNM-MB 28393) in ventral view. The intermediate phalanx of *Kenyapithecus* closely resembles that of *Pan* in terms of its robusticity, curvature, and the development of secondary shaft features, especially the ridges for the phalangeal flexors. (The scale bar equals two centimeters.)

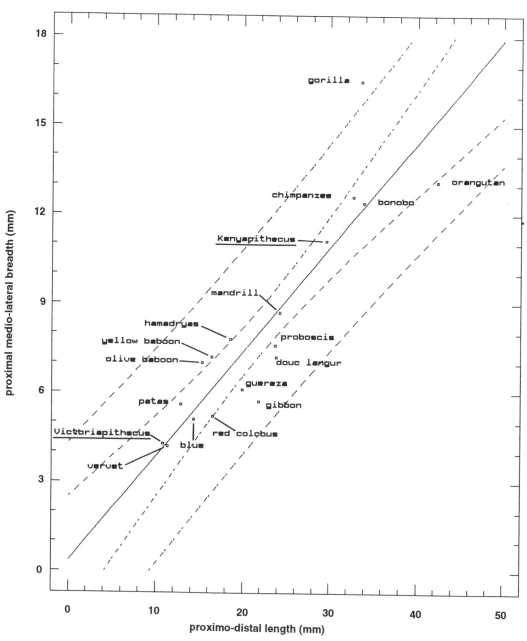

Figure 18. Plot of mean proximodistal length (mm) and mean proximal mediolateral breadth (mm) of intermediate phalanges (McCrossin, 1994a). The relatively short and robust intermediate phalanges of semi-terrestrial and terrestrial primates, including the vervet monkey (*Cercopithecus aethiops*, N = 46), the patas monkey (*Erythrocebus patas*, N = 17), savanna baboons (*Papio anubis*, N = 22; *P. cynocephalus*, N = 23; *P. hamadryas*, N = 2), the mandrill (*Mandrillus sphinx*, N = 2), the bonobo (*Pan paniscus*, N = 2), the chimpanzee (*Pan troglodytes*, N = 18), and the gorilla (*Gorilla gorilla*, N =13) fall close to or above the regression line. In contrast, the relatively long and gracile intermediate phalanges of the blue monkey (*Cercopithecus mitis*, N = 30), the red colobus (*Colobus badius*, N = 16), the guereza (*Colobus guereza*, N = 30), the douc langur (*Pygathrix nemaeus*, N = 2), the proboscis monkey (*Nasalis larvatus*, N = 5), the white-handed gibbon (*Hylobates lar*, N = 14) and the orangutan (*Pongo pygmaeus*, N = 6) fall below the regression line. The intermediate phalanges of *Victoriapithecus* are most similar to those of the vervet monkey while that of *Kenyapithecus* is intermediate between those of bonobos and chimpanzees.

pecially *Pan*. The intermediate phalanges of arboreal catarrhines, such as red colobus, douc langurs, gibbons, and orang-utans, are relatively longer and more gracile.

3.3. Summary

In summary, several indicators of semi-terrestriality are shared by *Victoriapithecus* and *Kenyapithecus*. The most conspicuous of these are the proximal extension of the greater tubercle of the humerus, the posterior orientation of the humeral medial epicondyle, the retroflection of the ulnar olecranon process, the distally restricted nature of the articulation between the entocuneiform and the hallucal metatarsal, and the robusticity of the intermediate phalanges. Overall, these similarities reflect shared emphases on stability and forceful protraction of the shoulder, a premium on rapid flexion and extension at the elbow, and a reduction of cheiridial grasping.

In addition to these similarities, *Victoriapithecus* and *Kenyapithecus* exhibit aspects of morphology that reveal fundamental differences in their styles of locomotion. In virtually every aspect of its postcranial anatomy, *Victoriapithecus* resembles living semi-terrestrial cercopithecoids, especially vervet monkeys. *Kenyapithecus*, in contrast, exhibits numerous features indicative of greater joint mobility than is seen in cercopithecoids and many of these features are shared with living hominoids. Hominoid-like features of *Kenyapithecus* include the relative size of the humeral head, the gracility of the humeral shaft, the breadth of the humeral trochlea, the development of the humeral lateral trochlear keel, the depth of the *zona conoidea*, the crescentic inferior radioulnar joint, and the angle of the femoral neck (McCrossin, 1994a; McCrossin et al., 1998). These features may relate to scansorial activities, including forelimb-dominated movements such as hoisting, stable pronation and supination of the forearm throughout the entire range of elbow flexion and extension, as well as hindlimb participation in vertical climbing. Intriguingly, *Kenyapithecus* also shares several distinctive postcranial features with *Gorilla* and *Pan*. The most important of these features concern aspects of the design of the inferior radioulnar joint and the metacarpophalangeal joint that appear to correspond to anatomical complexes related to knuckle-walking.

4. PALEOENVIRONMENT

There have been numerous attempts to reconstruct the environment of the middle Miocene of eastern Africa based on faunal and floral evidence from Maboko Island and Fort Ternan (Andrews and Walker, 1976; Andrews and Nesbit Evans, 1979; Van Couvering, 1980; Nesbit Evans et al., 1981; Shipman et al., 1981; Pickford, 1983, 1987; Bonnefille, 1984, 1985; Shipman, 1986; Pickford and Senut, 1988; Retallack et al., 1990; Cerling et al., 1991; Retallack, 1992; Solounias and Moelleken, 1993). The most important contributions to understanding the paleoenvironments at these sites are the original descriptions of fauna and flora at Maboko Island (e.g., MacInnes, 1936, 1942; Whitworth, 1958; Cifelli et al., 1986; Thomas, 1985; Winkler, 1994) and Fort Ternan (e.g., Lavocat, 1964, 1988, 1989; Hooijer, 1968; Churcher, 1970; Gentry, 1970; Hillenius, 1978; Pickford, 1982b; Thomas, 1984; Denys and Jaeger, 1992; Retallack, 1992; Tong and Jaeger, 1993, Van der Made, 1996).

Unfortunately, there has been an oversimplifying tendency in the secondary literature to equate the paleoenvironment (abiotic and biotic contexts) of Maboko Island and Fort Ternan with the paleoecology (e.g., diet, locomotion, and habitat preference) of *Ken-*

yapithecus (Andrews and Nesbit Evans, 1979; Nesbit Evans et al., 1981). One paleoenvironmental reconstruction even claims to impose restrictions on the possible phylogenetic affinities and substrate preferences of *Kenyapithecus* (Cerling et al., 1991). In addition, identification of fauna by non-specialists has led to confusion concerning the paleoenvironment of Maboko Island and Fort Ternan. For example, Andrews et al. (1981) list the presence of the anomalurid (scaly-tail "flying squirrel") *Zenkerella* at Maboko, ostensibly as an indicator of forest. Re-examination of the Maboko rodent assemblage, however, has shown that no anomularids are present at the site (Winkler, 1994). In fact, the specimen from Maboko identified as *Zenkerella* by Andrews et al. (1981) is actually a bathyergid (Cifelli et al., 1986). Thus, Andrews et al. (1981) used a subterranean rodent to indicate the presence of a tree-top milieu at Maboko (Cifelli et al., 1986).

Faunal lists of Maboko Island and Fort Ternan (Harrison, 1992: Table 5) also contain multiple errors and omissions concerning the primates, carnivores, rodents, lagomorphs, and artiodactyls actually known from these sites (Lavocat, 1964, 1988; Thomas, 1984; Cifelli et al., 1986; Gentry, 1990; Benefit, 1991; McCrossin, 1992a, 1994a; Denys and Jaeger, 1992; Tong and Jaeger, 1993, Van der Made, 1996). Some recent speculations concerning the paleoenvironment, biostratigraphy, and paleobiogeography of Maboko Island and Fort Ternan (Harrison, 1992) are based, therefore, on an unacceptably inaccurate foundation.

The paleoenvironments of Maboko Island and Fort Ternan are interesting because both sites register the effects of sweeping changes from the primarily forested environments that prevailed during the early Miocene in western Kenya. Many changes in the fauna and flora probably reflect the emergence of more open country environments that were effected by greater seasonality of climate than existed in the early Miocene (Van Couvering and Van Couvering, 1976; Shipman et al., 1981; Pickford, 1983; Benefit, 1987; Pickford and Senut, 1988; McCrossin and Benefit, 1994). Global [G] and sub-Saharan African [SSA] first appearances include hippopotamids [G], bovids [SSA], climacocerid giraffoids [G], and the amebelodont gomphothere *Choerolophodon* [SSA] at Maboko Island and C4 grasses [G], ostriches [G], advanced giraffoids (i.e., *Giraffokeryx*) [SSA], and hyaenids [SSA] at Fort Ternan (MacInnes, 1936, 1942; Arambourg, 1945; Crusafont and Aguirre, 1971; Andrews and Walker, 1976; Hamilton, 1978; Tassy, 1979; Thomas, 1979, 1985; Pickford, 1982b; Gentry, 1990; Retallack et al., 1990).

Because antelopes make their first appearance in sub-Saharan Africa at Maboko Island (Whitworth, 1958; Thomas, 1979) and are numerically dominant at Fort Ternan (Gentry, 1970), their adaptations are of particular relevance to understanding the paleoenvironment of the middle Miocene of eastern Africa. At least four different antelopes are known from Maboko Island, including a boselaphine resembling *Eotragus*, an antilopine (cf. *Gazella*), an indeterminate caprine, and *Nyanzameryx* (Whitworth, 1958; Gentry, 1970; Thomas, 1979, 1985). Thomas (1985) referred *Nyanzameryx* to the Family Climacoceridae, but examination of the type specimen indicates that it is a bovid rather than a giraffoid. The frontal appendages of *Nyanzameryx*, in particular, are horn cores rather than ossicones as they are separated from the frontal squama by a pedicle and the surface superior to the pedicle is vesiculated (unlike the continuously smooth cortical surface of giraffoid ossicones). Five bovids are known from Fort Ternan (Gentry, 1970, 1990; Thomas, 1979, 1984): two boselaphines – cf. *Eotragus* and *Kipsigicerus labidotus*, an antilopine or aepycerotine – cf. *Gazella* or *Aepyceros*, and two caprines – *Hypsodontus tanyceras* and *Caprotragoides potwaricus*.

Gentry (1970) demonstrated a series of morphological distinctions of the postcranial skeleton between highly cursorial antelopes inhabiting open country, such as savanna grasslands, and slower moving antelopes living in more closed environments, such as woodland and forest. In particular, Gentry focussed on clear distinctions in the structure of the proximal and distal femur between open country and closed country antelopes. In his analysis, Gentry demonstrated a series of cursorial adaptations for life in open country habitats for the caprine *Hypsodontus tanyceras* and retention of adaptations for life in woodland and forest habitats for the boselaphine *Kipsigicerus labidotus*. Specifically, *H. tanyceras* exhibits cursorial adaptations of the femur, most notably an articular surface of the head with "an anteroposteriorly long lateral part" and a distal end in which "the medial side of the patellar fossa projects strongly anteriorly" (Gentry, 1970:279–280, Fig. 12b). Cursorial adaptations similar to those of *H. tanyceras* are also seen in the distal femur anatomy of *Caprotragoides potwaricus*. Using a specimen from Nyakach, Thomas (1985:82, Pl.2–3) has shown that strong projection of the medial keel of the patellar groove is evident in *C. potwaricus*. Dental microwear analysis demonstrates that *H. tanyceras* was a grazer (Shipman et al., 1981) and *K. labidotus* was a grazer or mixed feeder (Solounias and Moelleken, 1993). Assessment of correlations between muzzle shape and dietary preference (Janis and Ehrhardt, 1988) also indicates grazing or mixed feeding for *K. labidotus* (Solounias and Moelleken, 1993).

Andrews and Walker (1976:299, 302) suggested that a wide range of environments, including tracts of "open and lightly wooded country" as well as "evergreen forest" might have been present near Fort Ternan. The existence of "moderately dense bushland" at Fort Ternan was based on the presence of an ostrich and a springhare, *Megapedetes* (Andrews and Walker, 1976:301). The presence of woodland and forest was supported by the identification of a leaf impression of Sterculiaceae, "a family of mostly woodland trees and shrubs", gastropods suggestive of "woodland" (*Burtoa*) and "evergreen forest" (*Maizania, Homorus*), an elephant shrew (*Rhynchocyon*), a lorisine, and a scaly-tail "squirrel" (*Paranomalurus*) (Andrews and Walker, 1976:299–302).

Shipman and colleagues (1981) suggested that a perceived difference in quality of preservation between remains attributed to "*Ramapithecus*" (= *Kenyapithecus*) *wickeri* and "*Dryopithecus*" (= *Proconsul*) *nyanzae* by Andrews and Walker (1976) indicated that the former lived in savanna and woodland environments near the site while the latter lived in forest environments farther away, on the slopes of the Tinderet volcano. The validity of this idea has been cast into doubt by the demonstration that the specimens attributed to "*Dryopithecus*" *nyanzae* by Andrews and Walker (1976) probably represent *Kenyapithecus wickeri* (Pickford, 1985).

Terrestrial gastropod assemblages have been interpreted as indicating the presence of "semi-arid woodland with gallery forest" at Maboko Island and "upland humid woodland" at Fort Ternan (Pickford and Senut, 1988:43). Further support for Gentry's (1970) proposal of a grassy woodland environment at Fort Ternan comes from the identification of grass pollen and phytoliths representing at least three species (Bonnefille, 1984, 1985; Retallack et al., 1990; Retallack, 1992).

5. BODY SIZE

Body size is an integral aspect of primate adaptation, with general correlations to diet and locomotion (Fleagle and Mittermeier, 1980; Kay, 1984; Fleagle, 1985; Smith and Jungers, 1997). In addition, knowledge of body mass is critical to understanding important trends such as encephalization and sexual dimorphism (Radinsky, 1974, 1982; Fleagle et al., 1980). Furthermore, the distribution of body size is a fundamental attribute of primate faunas and allows appreciation of differences in the structure of primate guilds and communities through time (Fleagle, 1978, 1985; Mihlbachler et al., 1996). Finally, dimensions that are isometrically correlated with body weight are the most appropriate independent variables for assessment of allometric trends. Consequently, a variety of methods for estimating body weights from tooth and limb bone measurements have been developed and applied to many fossil primates (Gingerich, 1977; Gingerich et al., 1982; Conroy, 1987; Dagosto and Terranova, 1992).

As is the case with many fossil primates, including *Aegyptopithecus zeuxis* and *Proconsul africanus* (Conroy, 1987), body weight estimates derived from the dentitions of *Victoriapithecus* and *Kenyapithecus* yield greater values than those derived from postcranial remains. Regressions of first molar area reconstruct a body weight of approximately 5.0 to 5.5 kg for *Victoriapithecus* (Gingerich et al., 1982; Conroy, 1987). Postcranial dimensions indicate that *Victoriapithecus* individuals ranged from 2.0 to 5.5 kg, with estimates of average female body weight at 3.0 kg and average male body weight at 4.5 kg (Harrison, 1989).

A series of 28 logarithm-transformed molar and postcranial dimensions have been used to estimate the body weight of *Kenyapithecus* (McCrossin, 1994a: Table 44). Estimates based on postcranial measurements vary greatly, from a minimum of 17.3 kg based on the mean of seven astragalar dimensions to a maximum of 36.2 kg based on the dimensions of the intermediate phalanx (McCrossin, 1994a: Table 44). Nevertheless, there is close agreement between mean estimates based on molar dimensions, 31.8, and mean estimates based on seven postcranial elements, 28.4 kg (Figure 19). In light of the high levels of intraspecific variation (mainly due to sexual dimorphism) of body weight seen among modern catarrhine species, these estimates are not very different from those of 23.0 kg and 18.4 kg derived from the midshaft and head diameters, respectively, of the femur collected from Maboko Island in 1933 (Ruff et al., 1989). Apparently, therefore, *Kenyapithecus* was slightly larger than the mandrill (*Mandrillus sphinx*), the largest living Old World monkey, but somewhat smaller than the bonobo (*Pan paniscus*), the smallest living great ape.

Body weights in the size range of *Victoriapithecus* and *Kenyapithecus* are not encountered among insectivorous primates, which are substantially smaller, but are more commonly associated with fruit- and leaf-eating species (Kay, 1984). Although overlapping the size ranges of semi-terrestrial and terrestrial species, the body weights of *Victoriapithecus* and *Kenyapithecus* are also well within the range of arboreal primates. In terms of locomotion, large arboreal primates tend to engage less frequently in bouts of leaping and instead rely upon bridging gaps between branches by suspension or clambering (Fleagle and Mittermeier, 1980; Fleagle, 1985). However, most primates in the size range of *Kenyapithecus* spend at least part of their lives on the ground, sometimes simply to cross patches of open ground between clumps of trees, but more commonly in foraging for foods, such as grass corms and fallen fruit (Napier, 1967).

The wide range of body weights estimated from various postcranial elements may corroborate the high degree of sexual dimorphism inferred for *Victoriapithecus* and *Kenyapithecus* on the basis of two size categories of the upper canine (Benefit, 1987; Pickford, 1985; McCrossin, 1994a). Sexual dimorphism of catarrhine canine height and body size has been linked to social systems involving male-male competition (Leutenegger and Kelly, 1977; Leutenegger, 1978; Plavcan and van Schaik, 1997). Among some African cercopithecins, however, the correlation between sexual dimorphism and social organization is not straightforward (Gautier-Hion and Gautier, 1985) and it is possible that the greater body size and possession of large canines seen among male cercopithecins may be more

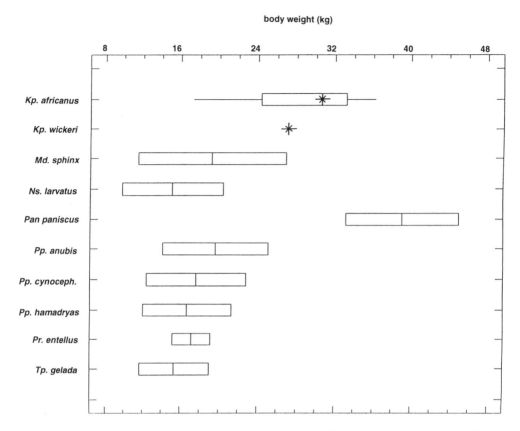

Figure 19. Estimated body weights of *Kenyapithecus wickeri* and *K. africanus* compared with body weights of some extant catarrhines. Estimate for *K. wickeri* is from regression of M^1 area to body weight in a sample of 49 extant primates (McCrossin, 1994a: Table 44). Estimates for *K. africanus* come from regressions to body weight of M^1 area, M_1 area, the mean of five humerus dimensions, the mean of two ulna dimensions, the mean of three femur dimensions, the mean of two patella dimensions, the mean of seven astragalus dimensions, the mean of six pollical proximal phalanx dimensions, and one intermediate phalanx dimension (McCrossin, 1994a: Table 44). All regressions have r-squared values greater than 0.90. Body weights of male and female *Mandrillus sphinx*, *Nasalis larvatus*, *Pan paniscus*, *Papio anubis*, *P. cynocephalus*, *P. hamadryas*, *Presbytis entellus*, and *Theropithecus gelada* are from Fleagle (1988). For *K. africanus* estimates, the plotting conventions are as in Figure 10. The vertical line indicates the mean and the box encloses the weights of females and males for extant catarrhines. Body weight estimates of *Kenyapithecus* exhibit a range similar to that seen in living catarrhines of similar body weight. Overall, *Kenyapithecus* body weight appears to have overlapped the range from the largest of living Old World monkeys (*Mandrillus sphinx*) to the smallest of living great apes (*Pan paniscus*).

closely related to solitary defense against predators in social systems characterized by female philopatry (Rowell and Chism, 1986). According to DiFiore and Rendall (1994), the female philopatry of most living cercopithecoids is a derived pattern of social dispersal for catarrhines. Thus, the body size and canine sexual dimorphism inferred for *Victoriapithecus*, together with results from paleodemography (Benefit, 1994), may indicate that the distinctive social organization of modern cercopithecoids, involving male transfer out of the natal group, had evolved by the middle Miocene.

The social organization of living hominoids are more diverse than those of extant cercopithecoids. The body size and canines of *Hylobates* are monomorphic and gibbons defend their territories against interlopers of the same sex. Living great apes exhibit strong sexual dimorphism in canine length and body size. Orang-utans have a distinct social organization, involving few interactions between adult males and females, while gorillas and chimpanzees are more gregarious and dispersal patterns always involve female transfer from the natal group. Within this context, the high degree of sexual dimorphism seen in the canine and body size of *Kenyapithecus* may reflect male-male competition but this is by no means certain.

6. DIET

Although most models of cercopithecoid origins predicted facultative folivory for Old World monkeys (Jolly, 1966; Napier, 1970; Delson, 1973, 1975; Andrews, 1981; Temerin and Cant, 1983), analysis of molar shear crests indicates that *Victoriapithecus* mainly consumed hard fruits (Kay, 1977; Benefit, 1987, 1990). A frugivorous diet for *Victoriapithecus* is also indicated by the relatively broad proportions of the upper central incisors (Benefit, 1987, 1990, 1993) and by the anatomy of the skull, particularly the strength and position of markings for *m. temporalis* (Benefit and McCrossin, 1991, 1993, 1997) (Figure 20).

Previous reconstructions of the dietary adaptations of *Kenyapithecus* focussed on its thick-enameled molars, between which hard food items were presumed to have been ground (Andrews, 1971; Walker and Andrews, 1973; Andrews and Walker, 1976; Simons and Pilbeam, 1978; Kay, 1981). Models for the diet of *Kenyapithecus* that placed an emphasis on the role of molar grinding often involved an expectation of reduction in upper canine and lower incisor size. Although *Kenyapithecus* has been characterized as having small upper canines (Yulish, 1970; Conroy, 1972; Andrews and Walker, 1976), it is now recognized that the upper canines are sexually dimorphic, with smaller specimens (such as KNM-FT 46; Leakey, 1962) representing females and larger specimens (such as KNM-FT 39, attributed to "*Dryopithecus*" (= *Proconsul*) *nyanzae* by Andrews and Walker, 1976) belonging to males (Pickford, 1985; McCrossin, 1994a; McCrossin and Benefit, 1997). Si-

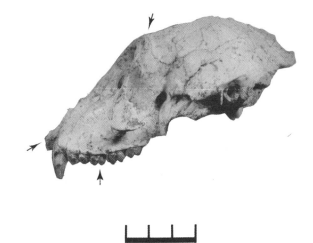

Figure 20. Skull of *Victoriapithecus macinnesi* (KNM-MB 29100; Benefit and McCrossin, 1997) in left lateral view. (Scale bar equals three centimeters.) Indicators of frugivory are marked by arrows (in clockwise order, from the top): anterior convergence of strongly marked temporal lines, relatively short shearing crests on molars, and relatively long premaxilla with broad incisors (known from isolated specimens) (Benefit, 1987, 1990, 1993; Benefit and McCrossin, 1991, 1993, 1997).

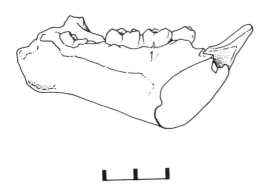

Figure 21. Juvenile mandible of *Kenyapithecus africanus* (KNM-MB 20573; McCrossin and Benefit, 1993) in medial view, showing cross-section of mandibular symphysis. (The scale bar equals two centimeters.) The strong procumbency of the lateral incisor is a resemblance to pitheciines and indicates that *Kenyapithecus* used a specialized anterior dentition to consume hard fruits and nuts (McCrossin and Benefit, 1993, 1994, 1997; McCrossin, 1994a).

mons and Pilbeam (1978:149, 152) confidently concluded that "*Ramapithecus*" (= *Kenyapithecus*) *wickeri* "had remarkably small lower incisors" that "were relatively unimportant in food preparation", but this inference was based on observation of empty alveoli rather than knowledge of the actual size of the lower incisors.

More recent attempts to reconstruct the diet of *Kenyapithecus* include new insights concerning the functional morphology of the maxilla, upper incisors, canines and premolars, mandible (Figure 21), and lower incisors (McCrossin and Benefit, 1993, 1994, 1997; McCrossin, 1994a). The maxilla of *Kenyapithecus* is distinctive in that it exhibits a prominent canine jugum and a deep post-canine fossa together with an anteriorly positioned root of the zygomatic process (Leakey, 1962, 1967; McCrossin, 1994a; McCrossin and Benefit, 1997). Based on comparisons of isolated specimens, the upper lateral incisors of *Kenyapithecus* are substantially narrower (mesiodistally), thinner (labiolingually), and lower-crowned than the upper central incisors (McCrossin, 1994a). The upper canines of *Kenyapithecus* are quite robust and are inferred to have been implanted in a pattern involving external rotation of the root (Ward and Pilbeam, 1983; McCrossin, 1994a). Compared to *Proconsul* and the small-bodied primitive catarrhines of the African early and middle Miocene, the upper premolars of *Kenyapithecus* are large relative to the size of the first upper molar (McCrossin, 1994a; McCrossin and Benefit, 1997). The mandibular corpus of *Kenyapithecus* is robust and marked by strong lateral buttressing while the mandibular symphysis is markedly proclined and dominated by the presence of a massive inferior transverse torus (McCrossin and Benefit, 1993). Strain gauge tests of primate mandibles indicate that the inferior transverse torus resists anteroinferiorly directed bending moments during incisal biting (Hylander, 1984). The lower incisors of *Kenyapithecus* are strongly procumbent and narrower (mesiodistally), thicker (labiolingually), and higher-crowned than those of extant apes (McCrossin and Benefit, 1993; McCrossin, 1994a).

This suite of features, especially the presence of markedly heteromorphic upper incisors, tusk-like upper canines, enlarged upper premolars, and procumbent lower incisors (McCrossin and Benefit, 1993, 1997; McCrossin, 1994a), is quite similar to adaptations for seed-predation seen in the pitheciines—the saki (*Pithecia*), the bearded saki (*Chiropotes*) and the uakari (*Cacajao*) (Mittermeier and van Roosmalen, 1981; van Roosmalen et al., 1988; Kinzey, 1992; Kinzey and Norconk, 1993). The presence of these features in *Kenyapithecus* appears to be related to a diet involving consumption of hard fruits and nuts (McCrossin and Benefit, 1993, 1994, 1997; McCrossin, 1994a; Benefit and McCrossin, 1995). An earlier stage in the trend toward a pitheciine-like dentognathic apparatus is recorded for *Afropithecus* (McCrossin, 1994a; McCrossin and Benefit, 1993, 1994, 1997). *Afropithecus* possesses enlarged upper premolars and a deep post-canine fossa in combina-

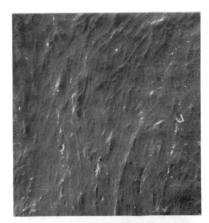

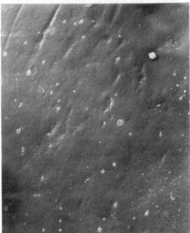

Figure 22. Scanning electron micrographs of microwear on Facet 9 (hypoconid) of the second lower molar in *Victoriapithecus* (KNM-MB 24530 – top; magnified 220 times) and *Kenyapithecus* (KNM-MB 20200 – bottom; magnified 150 times). Both *Victoriapithecus* and *Kenyapithecus* show the extensively pitted microwear that is characteristic of a diet of hard fruits, in contrast to the heavily scratched enamel, and inferred folivory, of sympatric small-bodied primitive catarrhines (*Simiolus*) and oreopithecids (*Mabokopithecus*) (Palmer et al., 1998).

tion with externally rotated and tusk-like canines while catastrophic exposure of dentine on the molars of *Afropithecus* finds its match only in pitheciines among primates, fossil and modern (McCrossin, 1994a; McCrossin and Benefit, 1993, 1994, 1997).

Dental microwear analyses corroborate reconstructions of preference for hard fruits, seeds, and nuts by *Victoriapithecus* and *Kenyapithecus* based on functional assessments of craniodental morphology (Palmer et al., 1998). Both *Victoriapithecus* and *Kenyapithecus* resemble hard object feeders in terms of the high percentage of microwear features that are pits (Palmer et al., 1998; Figure 22). A more specialized mode of hard object feeding for *Kenyapithecus* may be indicated by the fact that its microwear pits are much wider than those of *Victoriapithecus* (Palmer et al., 1998). In contrast, ostensibly folivorous primates from Maboko, such as *Simiolus* and *Mabokopithecus*, exhibit a great number of parallel scratches (Palmer et al., 1998).

7. CONCLUSIONS

Due to their integral phylogenetic relationships (Figure 1), the presence of semi-terrestrial adaptations in *Victoriapithecus* and *Kenyapithecus* are especially significant for understanding the origins of terrestriality among cercopithecoids and hominoids.

Because parsimony indicates that morphology and adaptations shared by victoriapi-
thecids and either, or both, cercopithecid subfamilies are most likely to be primitive for
Cercopithecoidea (Benefit, 1987, 1993, Harrison, 1989; Benefit and McCrossin, 1991,
1993, 1997; McCrossin and Benefit, 1992), a predominantly semi-terrestrial substrate
preference probably characterized ancestral Old World monkeys. The possibility that
semi-terrestriality was the primitive substrate preference of cercopithecids (Benefit, 1987;
Pickford and Senut, 1988; Strasser, 1988; Harrison, 1989) is also corroborated by the fact
that many fossil colobines, including *Mesopithecus* (von Beyrich, 1861) and *Cercopithecoides* (Birchette, 1981), were also adapted for life on the ground. Consequently, the arboreality of most living colobines may be a comparatively recent adaptive novelty
(Benefit, 1987; Pickford and Senut, 1988; Strasser, 1988; Harrison, 1989) rather than the
conservative retention of an ancestral arboreality once reconstructed for cercopithecids
(Jolly, 1970; Napier, 1970).

Consideration of the paleoenvironment, body size, and diet of *Victoriapithecus* allows us to separate cause from consequence in the origin of terrestrial adaptations among
Old World monkeys. Unlike the open grassland habitats currently occupied by the most
terrestrially adapted cercopithecoids, such as savanna baboons and geladas, the middle
Miocene environment reconstructed for Maboko Island shows that the semi-terrestriality
of ancestral Old World monkeys probably originated in a woodland milieu (Benefit, 1987;
Pickford and Senut, 1988). Although "increased body mass in a terrestrial environment" is
expected among ground-dwelling cercopithecoids (Reno et al., 1997:197), body size estimates of *Victoriapithecus* show that an increase in body size did not influence the earliest
phases of the transition from life in the trees to life on the ground among Old World monkeys. Thus, the adoption of a diet of hard fruits and seeds in seasonally dry woodlands by
Victoriapithecus may have been the driving force behind the origin of semi-terrestriality in
cercopithecoids (Benefit, 1987, 1990).

According to Rose (1993:269), "obvious morphological correlates that are indicators
of terrestrial locomotor modes" have not been "demonstrated convincingly for any Miocene hominoids" but "such a demonstration—for known or yet-to-be found Miocene
hominoids—will be essential for an understanding of their paleoecology, and of hominid
and African-ape evolution". Indeed, it has recently been claimed that postcranial remains
"from Maboko Island...indicate little change from the generalized arboreal quadrupedalism present in the early Miocene hominoids like *Proconsul*" (Andrews, 1992:643).

Although substantially augmenting knowledge of torso and limb structure, additional postcranial material of *Proconsul* (Walker and Pickford, 1983; Beard et al., 1986;
Langdon, 1986; Ruff et al., 1989; Ward et al., 1993, 1995) has done little to modify the
original conclusion that it was a generalized arboreal quadruped (Napier and Davis, 1959).
The postcranial anatomy, locomotor pattern and substrate preference of *Dryopithecus* are
less clearly understood. Suspensory locomotion, especially orangutan-like positional behaviors, have recently been suggested for *Dryopithecus* (Moya-Sola and Kohler, 1996),
but these claims rest partly on reconstructed, rather than known, intermembral proportions. Moreover, the posteromedial orientation of the humeral medial epicondyle, the retention of a bony joint between the ulna and wrist, and the weak development of the
hamulus on the hamate indicate that *Dryopithecus* lacked the specialized digital grasping
and wrist mobility of suspensory hominoids and instead was primarily an arboreal quadruped, perhaps with some enhanced adaptations for climbing compared with *Proconsul*. A
broad array of "generalized, varied, mostly arboreal" locomotor patterns have also been
reconstructed for *Sivapithecus* (Rose, 1984:503), but recent interpretations of the elbow
and carpometacarpal joints may also indicate some terrestrial behaviors (Senut, 1986b;

Spoor et al., 1992). The pervasive semi-terrestrial adaptations inferred here for *Kenyapithecus* are in marked contrast, therefore, to the apparently "generalized morphology and underlying generalized capabilities" usually reconstructed for Miocene catarrhines such as *Proconsul*, *Dryopithecus*, and *Sivapithecus* (Rose, 1983:415).

Contrary to expectations of a linkage to the spread of African savanna caused by the Messinian Event (Hsu et al., 1973) approximately 5 million years ago (Brain, 1981), the shift from life in the trees to life on the ground among apes and humans was initiated about 15 million years ago (McCrossin, 1994a; Benefit and McCrossin, 1995). With regard to environmental context, this change in substrate preference first occurred in an African ape inhabiting mainly woodland, not savanna, environments. Although an·increase in body size is widely thought to be directly correlated with terrestrial adaptations among apes and humans (Napier, 1967; Andrews, 1983; Fleagle, 1985), body size estimates for *Kenyapithecus* fall within the range of both tree-dwelling and ground-dwelling catarrhines. Colonization of savanna environments and attainment of body size in excess of that seen in arboreal primates may have played a role, therefore, in the evolution of more specialized terrestrial adaptations among humans and gorillas, but seem not to be involved with the origin of semi-terrestriality in Hominoidea.

Evidence from *Kenyapithecus* (McCrossin and Benefit, 1993, 1997) indicates that a change to a pitheciine-like diet and sclerocarp foraging strategy may have been the most profound impetus for the origin of terrestrial adaptations among apes and humans. Unlike other platyrrhines that almost always move about in trees, *Pithecia*, *Chiropotes*, and *Cacajao* often descend to the ground (Ayres, 1986, 1989; Walker, 1996). Ayres (1986, 1989) and Walker (1996) have suggested that utilization of terrestrial substrates by *Pithecia* and *Cacajao* may be related to their sclerocarpivory and attempts to locate their preferred foods. As noted by Walker (1996:355), "in the habitats of *Chiropotes* and *Cacajao*, the various vegetation types diverge greatly in height", "emergent trees are often used for feeding", and "this necessitates frequent ascent and descent for traveling between trees". Although *Cacajao* is primarily a canopy primate (Walker, 1996), during the driest part of the year Ayres (1986) observed uakaris feeding on the ground approximately 30 to 40% of the time. The terrestriality of *Cacajao* may be permitted by the scarcity of predators in the seasonally flooded forests (Ayres, 1986), but uakaris "occur in large groups, typically in the middle to upper canopy, with numerous individuals to watch for predators" (Walker, 1996:358).

The possibility of a causal relationship between sclerocarp foraging and the advent of terrestriality is compelling because it exists in such distantly related anthropoids as the uakaris of the flooded forests of Amazonia and *Kenyapithecus*, the large-bodied hominoid best known from the middle Miocene of Maboko Island. The terrestrial behaviors of uakaris are nascent and occur primarily in association with feeding activities and during locomotion between trees. Indeed, the postcranial skeleton of *Cacajao* lacks any of the indicators of terrestrial locomotion seen among Old World higher primates, including *Kenyapithecus* (McCrossin, 1994a). It is tempting to view the postcranial adaptations for life on the ground seen in *Kenyapithecus* as the product of a dietary and foraging strategy involving seed-predation, perhaps originally evolving in an *Afropithecus*-like ancestor (McCrossin, 1994a; McCrossin and Benefit, 1993, 1997) but without some of the autapomorphies of that taxon. Postcranial remains of *Afropithecus* have been interpreted as being *Proconsul*-like and indicative of pronograde arboreal quadrupedalism (Rose, 1993), but their fragmentary preservation does not enable confident determination of substrate preference. In any event, the shift from the trees to the ground among African apes and humans seems not to have been initiated by the consumption of grass, tuberous roots, or animal carcasses, but by the harvesting of hard fruit and nuts, presumably fallen from trees, by *Kenyapithecus*.

Some interesting insights emerge from a preliminary comparison of the environment, diet, and body size of *Kenyapithecus* and the earliest known hominids, *Ardipithecus ramidus* from 4.4 million year old deposits at Aramis, Ethiopia (White et al., 1994; Wolde-Gabriel et al., 1994) and *Australopithecus anamensis* from 3.5 to 4.2 million year old sediments at Kanapoi, Allia Bay, and Sibilot in Kenya (Leakey et al., 1995). The dominant mammals from Aramis are colobine monkeys and a kudu, *Tragelaphus* sp. (WoldeGabriel et al., 1994). The Aramis site is reconstructed as dense woodland based in large part on the assumption that the colobines were arboreally adapted. One of the Aramis colobines, however, is tentatively allied to Colobinae sp. A (WoldeGabriel et al., 1994). A skeleton of Colobinae sp. A from Leadu has been analyzed by Ciochon (1993). His results show detailed and pervasive resemblances to semi-terrestrially adapted forms, especially *Presbytis entellus*. Thus, the reconstruction of a dense woodland environment at Aramis is partly based on direct analogy to the substrate and habitat preferences only of living African colobines rather than the morphology of the fossil taxa present. According to Wolpoff (1996), *Ardipithecus* can be estimated to have weighed approximately 42 kg.

The most numerous mammal species at Kanapoi are *Parapapio*, a macaque-like cercopithecine, and *Tragelaphus kyaloae*, a kudu (Leakey et al., 1995). These faunal elements are suggested to indicate dry and open woodland or bushland conditions with gallery forest. *Australopithecus anamensis* may have weighed approximately 47–55 kg (Leakey et al., 1995).

Surprisingly, therefore, there are preliminary indications that the earliest hominids remained denizens of woodland environments and that more open environments were not colonized until substantially later on, perhaps 3.5 million years ago. Moreover, substantial increases in body weight that accompany the first appearance of bipedalism seem to be consequences rather than causes of life on the ground (McCrossin, 1997).

Explicit dietary reconstructions have not yet been made for *Ardipithecus ramidus* and *Australopithecus anamensis*, but it is interesting to see the degree to which detailed elements of a *Kenyapithecus*-like dentognathic morphology are retained by the earliest known hominids. Similarities to *Kenyapithecus* are especially evident in the dp_3 morphology of *Ardipithecus* and the proclined long axis of the mandibular symphysis of the type-specimen of *Australopithecus anamensis* (McCrossin and Benefit, 1993, 1994, 1997; White et al., 1994; Leakey et al., 1995).

Whatever the phylogenetic relationships of *Kenyapithecus* ultimately prove to be, the fossil record has been very kind to us by providing a large-bodied hominoid caught in the act of undergoing the transition from life in the trees to life on the ground. If *Kenyapithecus* is an archaic avatar of modern great apes, then the ground-dwelling capabilities of the living African apes and humans are an adaptive palimpsest, with ancient adaptations for semi-terrestriality being obscured by the more recent and obvious specializations of knuckle-walking and bipedalism. On the other hand, if *Kenyapithecus* turns out to be specially related to African apes and humans, then it would appear that the first steps on a long evolutionary path toward the committed terrestrial bipedalism of humans were taken 15 million years ago, by a creature possessing few, if any, of the features we expected of a pioneer of life on the ground.

It is not currently known whether the middle Miocene advent of semi-terrestriality in cercopithecoids and hominoids was synchronously accompanied by the pervasive constellation of changes in ecology and behavior associated with life on the ground among living catarrhines. Nevertheless, this transition in substrate preference among early representatives of the two major catarrhine clades represents an adaptive departure of tremendous magnitude from New World anthropoids and almost all prosimians, with profound repercussions for the subsequent evolutionary career of Old World monkeys, apes, and humans.

ACKNOWLEDGMENTS

We gratefully acknowledge the Office of the President of the Republic of Kenya and the National Museums of Kenya for permission to conduct excavations on Maboko Island. We thank M.G. Leakey and the curatorial staff of Paleontology Division of the National Museums of Kenya, especially M. Muungu, E. Mbua, F. Kyalo, A. Ibui, and N. Malit. We extend our thanks to our field crew on Maboko Island, especially B. Onyango, C. Omondi Agak, J. Onyango Miumi, and C. Obote Odhiambo. We thank the following agencies for financial support: National Science Foundation, L.S.B. Leakey Foundation, Fulbright Collaborative Research Program, Wenner-Gren Foundation for Anthropological Research, Rotary International Foundation, Office of Research Development and Administration of Southern Illinois University, the Boise Fund of Oxford University, and the Lowie Fund of the University of California at Berkeley. We thank T. Gatlin for illustrations and three anonymous reviewers for their helpful comments. Finally, we thank E. Strasser, J. Fleagle, A. Rosenberger, and H. McHenry for giving us the opportunity to participate in the conference at the University of California at Davis and contribute to this volume.

REFERENCES

Andrews P (1971) *Ramapithecus wickeri* mandible from Fort Ternan, Kenya. Nature *228*:537–540.
Andrews P (1981) Species diversity and diet in monkeys and apes during the Miocene. In CB Stringer (ed.): Aspects of Human Evolution. London: Taylor and Francis, pp. 25–61.
Andrews P (1983) The natural history of *Sivapithecus*. In RL Ciochon and RS Corruccini (eds.) New Interpretations of Ape and Human Ancestry. New York: Plenum Press, pp. 441–464.
Andrews P (1985) Family group systematics and evolution among catarrhine primates. In E Delson (ed.): Ancestors: The Hard Evidence. New York: AR Liss, pp. 14–22.
Andrews P (1992) Evolution and environment in the Hominoidea. Nature *360*:641–646.
Andrews P, and Aiello L (1984) An evolutionary model for feeding and positional behavior. In DJ Chivers, BA Wood, and A Bilsborough (eds.): Food Acquisition and Processing in Primates. New York: Plenum Press, pp. 429–466.
Andrews P, and Nesbit Evans E (1979) The environment of *Ramapithecus* in Africa. Paleobiology 5:22–30.
Andrews P, and Van Couvering JAH (1975) Paleoenvironments in the East African Miocene. Contrib. Primatol. 5:62–103.
Andrews P, and Walker A (1976) The primate and other fauna from Fort Ternan, Kenya. In GLl Isaac and ER McCown (eds.): Human Origins: Louis Leakey and the East African Evidence. Menlo Park, CA: WA Benjamin, pp. 279–304.
Andrews P, Meyer G, Pilbeam DR, Van Couvering JA, and Van Couvering JAH (1981) The Miocene fossil beds of Maboko Island, Kenya: Geology, age, taphonomy, and paleontology. J. Hum. Evol. *10*:35–48.
Arambourg C (1945) *Anancus osiris*, un nouveau Mastodonte du Pliocène inferieur d'Egypt. Bull. Soc. Geol. Fr. *15*:479–495.
Ashton EH, and Oxnard CE (1964) Locomotor patterns in primates. Proc. Zool. Soc. Lond. *142*:1–28.
Ayres JM (1986) Uakaris and Amazonian Flooded Forest. Ph.D. Dissertation, University of Cambridge.
Ayres JM (1989) Comparative feeding ecology of the uakari and bearded saki, *Cacajao* and *Chiropotes*. J. Hum. Evol. *18*:697–716.
Beard KC, Teaford MF, and Walker A (1986) New wrist bones of *Proconsul africanus* and *P. nyanzae* from Rusinga Island, Kenya. Folia Primatol. *47*:97–118.
Benefit BR (1987) The Molar Morphology, Natural History, and Phylogenetic Position of the Middle Miocene Monkey *Victoriapithecus*. Ph.D. Dissertation, New York University.

Benefit BR (1990) Fossil evidence for the dietary evolution of Old World monkeys. Am. J. Phys. Anthropol. *81*:191.

Benefit BR (1991) The taxonomic status of Maboko small apes. Am. J. Phys. Anthropol. *12*:50–51.

Benefit BR (1993) The permanent dentition and phylogenetic position of *Victoriapithecus* from Maboko Island, Kenya. J. Hum. Evol. *25*:83–172.

Benefit BR (1994) Phylogenetic, paleodemographic, and taphonomic implications of *Victoriapithecus* deciduous teeth. Am. J. Phys. Anthropol. *95*:277–331.

Benefit BR, and McCrossin ML (1989) New primate fossils from the middle Miocene of Maboko Island, Kenya. J. Hum. Evol. *18*:493–497.

Benefit BR, and McCrossin ML (1991) Ancestral facial morphology of Old World higher primates. Proc. Natl. Acad. Sci. USA *88*:5267–5271.

Benefit BR, and McCrossin ML (1993) The facial anatomy of *Victoriapithecus* and its relevance to the ancestral cranial morphology of Old World monkeys and apes. Am. J. Phys. Anthropol. *92*:329–370.

Benefit BR, and McCrossin ML (1994) Comparative study of the dentition of *Kenyapithecus africanus* and *K. wickeri*. Am. J. Phys. Anthropol. Suppl. *18*:55.

Benefit BR, and McCrossin ML (1995) Miocene hominoids and hominid origins. Annu. Rev. Anthropol. *24*:237–256.

Benefit BR, and McCrossin ML (1997) Earliest known Old World monkey skull. Nature *388*:368–371.

Benefit BR, Gitau SN, McCrossin ML, and Palmer AK (1998) A mandible of *Mabokopithecus clarki* sheds new light on oreopithecid evolution. Am. J. Phys. Anthropol. Suppl. *26:*109.

Birchette MG (1981) Postcranial remains of *Cercopithecoides*. Am. J. Phys. Anthropol. *54*:201.

Birchette MG (1982) The Postcranial Skeleton of *Paracolobus chemeroni*. Ph.D. Dissertation, Harvard University.

Bonnefille R (1984) Cenozoic vegetation and environments of early hominids in East Africa. In RO Whyte (ed.): The Evolution of the East Asian Environment, Volume 2: Palaeobotany, Palaeozoology, and Palaeoanthropology. Hong Kong: University of Hong Kong, pp. 579–612.

Bonnefille R (1985) Evolution of the continental vegetation: The paleobotanical record from East Africa. S. Afr. J. Sci. *81*:267–270.

Bowen BE, and Vondra CF (1974) Paleoenvironmental interpretations of the Oligocene Gabal el Qatrani Formation, Fayum Depression, Egypt. Annals of the Geological Survey of Egypt *4*:115–138.

Bown TM, Kraus MJ, Wing SL, Fleagle JG, Tiffany B, Simons EL, and Vondra CF (1982) The Fayum forest revisited. J. Hum. Evol. *11*:603–632.

Brain CK (1981) Hominid evolution and climatic changes. S. Afr. J. Sci. *77*:104–105.

Brown B (1997) Miocene hominoid mandibles: Functional and phylogenetic perspectives. In DR Begun, CV Ward, and MD Rose (eds.): Function, Phylogeny, and Fossils: Miocene Hominoid Evolution and Adaptations. New York: Plenum, pp. 153–171.

Brown B, and Ward S (1988) Basicranial and facial topography in *Pongo* and *Sivapithecus*. In JH Schwartz (ed.) Orang-utan Biology. London: Oxford University Press, pp. 247–260.

Brown B, Hill A, and Ward S (1991) New Miocene large hominoids from the Tugen Hills, Baringo District, Kenya. Am. J. Phys. Anthropol. Suppl. *12*:55.

Butler PM (1984) Macroscelidea, Insectivora and Chiroptera from the Miocene of East Africa. Palaeovertebrata *14*:117–200.

Caldicott JO (1986) An ecological and behavioral study of the pig-tailed macaque. Contrib. Primatol. *21*:1–259.

Cant JG (1987) Effects of sexual dimorphism in body size on feeding postural behavior of Sumatran orangutans (*Pongo pygmaeus*). Am. J. Phys. Anthropol. *74*:143–148.

Cerling TE, Quade J, Ambrose SH, and Sikes NE (1991) Fossil soils, grasses, and carbon isotopes from Fort Ternan, Kenya: Grassland or woodland? J. Hum. Evol. *21*:295–306.

Churcher CS (1970) Two new upper Miocene giraffids from Fort Ternan, Kenya, East Africa: *Palaeotragus primaevus* n. sp. and *Samotherium africanum* n. sp. In LSB Leakey and RJG Savage (eds.): Fossil Vertebrates of Africa, Vol. 2. London: Academic Press, pp. 1–106.

Cifelli RL, Ibui AK, Jacobs LL, and Thorington RW (1986) A giant tree squirrel from the late Miocene of Kenya. J. Mamm. *67*:274–283.

Ciochon RL (1993) Evolution of the cercopithecoid forelimb: Phylogenetic and functional implications from morphometric analyses. Univ. Calif. Publ. Geol. Sci. *138*:1–251.

Clutton-Brock TH, and Harvey PH (1977) Primate social organization and ecology. Nature *250*:539–542.

Conroy GC (1972) Problems in the interpretation of *Ramapithecus*: With special reference to anterior tooth reduction. Am. J. Phys. Anthropol. *37*:41–47.

Conroy GC (1976) Primate postcranial remains from the Oligocene of Egypt. Contrib. Primatol. *8*:1–134.

Conroy GC (1987) Problems of body-weight estimation in fossil primates. Int. J. Primatol. *8*:115–137.

Conroy GC, and Fleagle JG (1972) Locomotor behavior in living and fossil pongids. Nature *237*:103–104.

Coursey DG (1973) Hominid evolution and hypogeous plant foods. Man *8*:634–635.

Crook JH, and Gartlan JS (1966) On the evolution of primate societies. Nature *210*:1200–1203.

Crusafont M, and Aguirre E (1971) A new species of *Percrocuta* from the middle Miocene of Kenya. Abh. Hess. L.-Amt. Bodenforsch. *60*:51–58.

Dagosto M, and Terranova CJ (1992) Estimating the body size of Eocene primates: A comparison of results from dental and postcranial variables. Int. J. Primatol. *13*:307–344.

Davison GWH (1982) Convergence with terrestrial cercopithecines by the monkey *Rhinopithecus roxellanae*. Folia Primatol. *37*:209–215.

Delson E (1973) Fossil Colobine Monkeys of the Circum-Mediterranean Region and the Evolutionary History of the Cercopithecidae (Primates, Mammalia). Ph.D. Dissertation, Columbia University.

Delson E (1975) Evolutionary history of the Cercopithecidae. Contributions to Primatology *5*:167–217.

Denys C, and Jaeger J-J (1992) Rodents of the Miocene site of Fort Ternan (Kenya). First part: Phiomyids, bathyergids, sciurids and anomalurids. N. Jb. Geol. Paleont. Abh. *185*:63–84.

DeVore I (1963) A comparison of the ecology and behavior of monkeys and apes. In SL Washburn (ed.): Classification and Human Evolution. Chicago: Aldine, pp. 301–319.

DiFiore A, and Rendall D (1994) Evolution of social organization: A reappraisal for primates by using phylogenetic methods. Proc. Natl. Acad. Sci. USA *91*:9941–9945.

Doran DM (1993) Sex differences in adult chimpanzee positional behavior: The influence of body size on locomotion and posture. Am. J. Phys. Anthropol. *91*:99–116.

Dunbar RIM, and Dunbar EP (1974) Ecological relations and niche separation between sympatric terrestrial primates in Ethiopia. Folia Primatol. *21*:36–60.

Eisenberg JF, Muckenhirn NA, and Rudran R (1972) The relation between ecology and social structure in primates. Science *176*:863–874.

Feibel CS, and Brown FH (1991) Age of the primate-bearing deposits on Maboko Island, Kenya. J. Hum. Evol. *21*:221–225.

Fleagle JG (1974) Dynamics of a brachiating siamang (*Hylobates* (*Symphalangus*) *syndactylus*). Nature *248*:259–260.

Fleagle JG (1977) Locomotor behavior and skeletal anatomy of sympatric Malaysian leaf-monkeys (*Presbytis obscura* and *Presbytis melalophos*). Yrbk. Phys. Anthropol. *20*:440–453.

Fleagle JG (1978) Size distributions of living and fossil primate faunas. Paleobiology *4*:67–76.

Fleagle JG (1983) Locomotor adaptations of Oligocene and Miocene hominoids and their phyletic implications. In RL Ciochon and RS Corruccini (eds.): New Interpretations of Ape and Human Ancestry. New York: Plenum Press, pp. 301–323.

Fleagle JG (1985) Size and adaptation in primates. In WL Jungers (ed.): Size and Scaling in Primate Biology. New York: Plenum Press, pp. 1–19.

Fleagle JG (1988) Primate Adaptation and Evolution. New York: Academic Press.

Fleagle JG, and Mittermeier RA (1980) Locomotor behavior, body size, and comparative ecology of seven Surinam monkeys. Am. J. Phys. Anthropol. *52*:301–314.

Fleagle JG, and Simons EL (1982) The humerus of *Aegyptopithecus*: A primitive anthropoid. Am. J. Phys. Anthropol. *59*:175–193.

Fleagle JG, Kay RF, and Simons EL (1980) Sexual dimorphism in early anthropoids. Nature *287*:328–330.

Gautier-Hion A, and Gautier J-P (1985) Sexual dimorphism, social units and ecology among sympatric forest guenons. In J Ghesquiere, RD Martin, and F Newcombe (eds.): Human Sexual Dimorphism. London: Taylor and Francis, pp. 61–77.

Gebo DL (1992) Plantigrady and foot adaptation in African apes: Implications for hominid origins. Am. J. Phys. Anthropol. *89*:29–58.

Gebo DL, Beard KC, Teaford MF, Walker A, Larson SG, Jungers WL, and Fleagle JG (1988) A hominoid proximal humerus from the early Miocene of Rusinga Island, Kenya. J. Hum. Evol. *17*:393–401.

Gentry A (1970) The Bovidae of the Fort Ternan fossil fauna. In LSB Leakey and RJG Savage (eds.) Fossil Vertebrates of Africa, Vol 2. London: Academic Press, pp. 243–324.

Gentry AW (1990) Ruminant artiodactyls of Pasalar, Turkey. J. Hum. Evol. *19*:529–550.

Gingerich PD (1977) Correlation of tooth size and body size in living hominoid primates, with a note on relative brain size in *Aegyptopithecus* and *Proconsul*. Am. J. Phys. Anthropol. *47*:395–398.

Gingerich PD, Smith BH, and Rosenberg K (1982) Allometric scaling in the dentition of primates and prediction of body weight from tooth size in fossils. Am. J. Phys. Anthropol. *58*:81–100.

Ginsburg L, and Mein P (1980) *Crouzelia rhodanica*, nouvelle espece de primate Catarrhinien et essai sur la position systematique des Pliopithecidae. Bull. Mus. Nat., Paris 2:57–85.

Greenfield LO (1979) On the adaptive pattern of "*Ramapithecus*". Am. J. Phys. Anthropol. *50*:527–548.

Hall KRL (1965) Behavior and ecology of the wild patas monkey, *Erythrocebus patas*, in Uganda. J. Zool. *148*:15–87.

Hamilton WR (1978) Fossil giraffes from the Miocene of Africa and a revision of the phylogeny of the Giraffoidea. Phil. Trans. Roy. Soc. Lond. *283*:165–229.

Harrison T (1989) New postcranial remains of *Victoriapithecus* from the middle Miocene of Kenya. J. Hum. Evol. *18*:3–54.

Harrison T (1992) A reassessment of the taxonomic and phylogenetic affinities of the fossil catarrhines from Fort Ternan, Kenya. Primates *33*:501–522.

Hillenius D (1978) Notes on chameleons IV. A new chameleon from the Miocene of Fort Ternan, Kenya (Chamaeleonidae, Reptilia). Beaufortia *28*:9–15.

Hooijer DA (1968) A rhinoceros from the late Miocene of Fort Ternan, Kenya. Zoologische Mededelingen *43*:77–92.

Hsu KJ, Ryan WBF, and Cita MB (1973) Late Miocene desiccation of the Mediterranean. Nature *242*:240–244.

Hunt K (1992) Positional behavior of *Pan troglodytes* in the Mahale Mountains and Gombe Stream National Parks, Tanzania. Am. J. Phys. Anthropol. *87*:83–105.

Hylander W (1984) Stress and strain in the mandibular symphysis of primates: A test of competing hypotheses. Am. J. Phys. Anthropol. *64*:1–46.

Ishida H (1986) Investigation in northern Kenya and new hominoid fossils. Kagaku *56*:220–226.

Ishida H, Pickford M, Nakaya H, and Nakano Y (1984) Fossil anthropoids from Nachola and Samburu Hills, Samburu District, Northern Kenya. Afr. Stud. Monogr. (Kyoto Univ.) Suppl. *2*:73–86.

Janis CM, and Ehrhardt D (1988) Correlation of relative muzzle width and relative incisor width with dietary preference in ungulates. Zool. J. Linnean Soc. *92*:267–284.

Jenkins FA (1973) The functional anatomy and evolution of the mammalian humero-ulnar articulation. Am. J. Anat. *137*:281–298.

Jenkins FA, and Camazine SM (1977) Hip structure and locomotion in ambulatory and cursorial carnivores. J. Zool., Lond. *181*:351–370.

Jenkins FA, and Fleagle JG (1975) Knuckle-walking and the functional anatomy of the wrists in living apes. In RH Tuttle (ed.): Primate Functional Morphology and Evolution. The Hague: Mouton, pp. 213–227.

Jolly CJ (1966) Introduction to the Cercopithecoidea with notes on their use as laboratory animals. Symp. Zool. Soc. Lond. *17*:427–457.

Jolly CJ (1967) The evolution of baboons. In H Vagtborg (ed.): The Baboon in Medical Research, Volume 2. Austin: University of Texas Press, pp. 23–50.

Jolly CJ (1970) The seed eaters: A new model of hominid differentiation based on a baboon analogy. Man *5*:5–26.

Jolly CJ (1972) The classification and natural history of *Theropithecus* (*Simopithecus*) (Andrews, 1916), baboons of the African Plio-Pleistocene. Bull. Brit. Mus. (Nat. Hist.), Geol. *22*:1–122.

Jouventin P (1975) Observations sur la socio-ecologie du mandrill. La Terre et La Vie *29*:493–532.

Kay RF (1977) Post-Oligocene evolution of catarrhine diets. Am. J. Phys. Anthropol. *47*:141–142.

Kay RF (1981) The nut-crackers: A new theory of the adaptation of the Ramapithecinae. Am. J. Phys. Anthropol. *55*:141–151.

Kay RF (1984) On the use of anatomical features to infer foraging behavior in extinct primates, In PS Rodman and JGH Cant (eds.): Adaptations for Foraging in Nonhuman Primates. New York: Columbia University Press, pp. 21–53.

Kinzey WG (1992) Dietary and dental adaptations in the Pitheciinae. Am. J. Phys. Anthropol. *88*:499–514.

Kinzey WG, and Norconk MA (1993) Physical and chemical properties of fruit and seeds eaten by *Pithecia* and *Chiropotes* in Surinam and Venezuela. Int. J. Primatol. *14*:207–227.

Knussman R (1967) Humerus, Ulna und Radius der Simiae. Bibl. Primatol. *5*:1–399.

Langdon JH (1986) Functional morphology of the Miocene hominoid foot. Contrib. Primatol. *22*:1–225.

Larson SG (1988) Subscapularis function in gibbons and chimpanzees: Implications for interpretation of humeral torsion in hominoids. Am. J. Phys. Anthropol. *76*:449–462.

Larson SG, and Stern JT (1992) Further evidence for the role of supraspinatus in quadrupedal monkeys. Am. J. Phys. Anthropol. *87*:359–363.

Lavocat R (1964) Fossil rodents from Fort Ternan, Kenya. Nature *202*:1131.

Lavocat R (1988) Un rongeur Bathyergide nouveau remarquable du Miocene de Fort Ternan (Kenya). C. R. Acad. Sci. Paris *306*:1301–1304.

Lavocat R (1989) Osteologie de la tete de *Richardus excavans* Lavocat, 1988. Palaeovertebrata *19*:73–80.

Le Gros Clark WE (1959) The Antecedents of Man: An Introduction to the Evolution of the Primates. Edinburgh: Edinburgh University Press.

Le Gros Clark WE, and Leakey LSB (1951) The Miocene Hominoidea of East Africa. Fossil Mammals of Africa *1*:1–117.

Leakey LSB (1962) A new lower Pliocene fossil primate from Kenya. Ann. Mag. Nat. Hist. 4:689–696.

Leakey LSB (1967) An early Miocene member of Hominidae. Nature 213:155–163.

Leakey MG, Feibel CS, McDougall I, and Walker A (1995) New four-million-year-old hominid species from Kanapoi and Allia Bay, Kenya. Nature 376:565–571.

Leutenegger W (1970) Das Becken der rezenten Primaten. Morph. Jahrb 115:1–101.

Leutenegger W (1978) Scaling of sexual dimorphism in body size and breeding system in primates. Nature 272:610–611.

Leutenegger W, and Kelly JT (1977) Relationship of sexual dimorphism in canine size and body size to social, behavioral and ecological correlates in anthropoid primates. Primates 18:117–136.

Lovejoy CO (1981) The origin of man. Science 211:341–350.

MacInnes DG (1936) A new genus of fossil deer from the Miocene of Africa. J. Linn. Soc. Lond. Zool. 39:521–530.

MacInnes DG (1942) Miocene and post-Miocene Proboscidea from East Africa. Trans. Zool. Soc., Lond. 25:33–103.

MacKinnon J (1974) The behavior and ecology of wild orang-utans (Pongo pygmaeus). Anim. Behav. 22:3–74.

Martin RD (1979) Phylogenetic aspects of prosimian behavior. In GA Doyle and RD Martin (eds.): The Study of Prosimian Behavior. New York: Academic Press, pp. 45–77.

Martin RD (1990) Primate Origins and Evolution: A Phylogenetic Reconstruction. Princeton NJ: Princeton University Press.

McCrossin ML (1990) Fossil galagos from the middle Miocene of Kenya. Am. J. Phys. Anthropol. 81:265–266.

McCrossin ML (1992a) A new species of bushbaby from the middle Miocene of Maboko Island, Kenya. Am. J. Phys. Anthropol. 89:215–233.

McCrossin ML (1992b) An oreopithecid proximal humerus from the middle Miocene of Maboko Island, Kenya. Int. J. Primatol. 13:659–677.

McCrossin ML (1994a) The Phylogenetic Relationships, Adaptations, and Ecology of Kenyapithecus. Ph.D Dissertation, University of California at Berkeley.

McCrossin ML (1994b) Semi-terrestrial adaptations of Kenyapithecus. Am. J. Phys. Anthropol. Suppl. 18:142–143.

McCrossin ML (1997) New postcranial remains of Kenyapithecus and their implications for understanding the origins of hominoid terrestriality. Am. J. Phys. Anthropol. Suppl. 24:164.

McCrossin ML, and Benefit BR (1992) Comparative assessment of the ischial morphology of Victoriapithecus macinnesi. Am. J. Phys. Anthropol. 87:277–290.

McCrossin ML, and Benefit BR (1993) Recently discovered Kenyapithecus mandible and its implications for great ape and human origins. Proc. Natl. Acad. Sci. USA 90:1962–1966.

McCrossin ML, and Benefit BR (1994) Maboko Island and the evolutionary history of Old World monkeys and apes. In RS Corruccini and RL Ciochon (eds.): Integrative Paths to the Past: Paleoanthropological Advances in Honor of F. Clark Howell. Englewood Cliffs, NJ: Prentice-Hall, pp. 95–124.

McCrossin ML, and Benefit BR (1997) On the relationships and adaptations of Kenyapithecus, a large-bodied hominoid from the middle Miocene of eastern Africa. In DR Begun, CV Ward, and MD Rose (eds.): Function, Phylogeny and Fossils: Miocene Hominoid Evolution and Adaptations. New York: Plenum, pp. 241–267.

McCrossin ML, Benefit BR, and Gitau SN (1998) Functional and phylogenetic analysis of the distal radius of Kenyapithecus, with comments on the origin of the African great ape and human clade. Am. J. Phys. Anthropol. Suppl. 26:158–159.

McHenry HM, and Corruccini RS (1983) The wrist of Proconsul africanus and the origin of hominoid postcranial adaptations. In RL Ciochon and RS Corruccini (eds.): New Interpretations of Ape and Human Ancestry. New York: Plenum, pp. 353–367.

Mihlbachler MC, McCrossin ML, and Benefit BR (1996) Body size distribution and the evolution of African primate community structure. Am. J. Phys. Anthropol. Suppl. 22:168.

Mittermeier RA, and Van Roosmalen MGM (1981) Preliminary observations on habitat utilization and diet in eight Surinam monkeys. Folia Primatol. 36:1–39.

Moya-Sola S, and Kohler M (1996) A Dryopithecus skeleton and the origins of great ape locomotion. Nature 379:156–159.

Napier JR (1967) Evolutionary aspects of primate locomotion. Am. J. Phys. Anthropol. 27:333–342.

Napier JR (1970) Paleoecology and catarrhine evolution. In JR Napier and PH Napier (eds.): Old World Monkeys. London: Academic Press, pp. 55–95.

Napier JR, and Davis PR (1959) The forelimb skeleton and associated remains of Proconsul africanus. Fossil Mammals of Africa 16:1–69.

Napier JR, and Napier PH (1967) A Handbook of Living Primates. New York: Academic Press.

Napier JR, and Napier PH (1985) The Natural History of the Primates. Cambridge, MA: The MIT Press.

Nesbit Evans EM, Van Couvering JAH, and Andrews P (1981) Palaeoecology of Miocene sites in Western Kenya. J. Hum. Evol. *10*:99–116.

Norconk MA (1996) Seasonal variation in the diets of white-faced and bearded sakis (*Pithecia pithecia and Chiropotes satanas*) in Guri Lake, Venezuela. In MA Norconk, AL Rosenberger, and PA Garber (eds.): Adaptive Radiations of Neotropical Primates. New York: Plenum Press, pp. 403–423.

O'Connor BL (1975) The functional morphology of the cercopithecoid wrist and inferior radioulnar joints, and their bearing on some problems in the evolution of the Hominoidea. Am. J. Phys. Anthropol. *43*:113–122.

Oxnard CE (1963) Locomotor adaptations in the primate forelimb. Symp. Zool. Soc. Lond. *10*:165–182.

Palmer AK, Benefit BR, McCrossin ML, and Gitau, SN (1998) Paleoecological implications of dental microwear analysis for the middle Miocene primate fauna from Maboko Island, Kenya. Am. J. Phys. Anthropol. Suppl. *26*:175.

Pickford M (1982a) New higher primate fossils from the middle Miocene deposits at Majiwa and Kaloma, western Kenya. Am. J. Phys. Anthropol. *58*:1–19.

Pickford M (1982b) On the origins of Hippopotamidae together with descriptions of two new species, a new genus and a new subfamily from the Miocene of Kenya. Geobios *16*:193–217.

Pickford M (1983) Sequence and environments of the lower and middle Miocene hominoids of western Kenya. In RL Ciochon and RS Corruccini (eds.): New Interpretations of Ape and Human Ancestry. New York: Plenum Press, pp. 421–439.

Pickford M (1984) Kenya Palaeontology Gazetteer, Volume 1 – Western Kenya. Nairobi: National Museums of Kenya, Department of Sites and Monuments Documentation.

Pickford M (1985) A new look at *Kenyapithecus* based on recent discoveries in western Kenya. J. Hum. Evol. *14*:113–143.

Pickford M (1987) Fort Ternan (Kenya) paleoecology. J. Hum. Evol. *16*:305–309.

Pickford M, and Senut B (1988) Habitat and locomotion in Miocene cercopithecoids. In F Bourliere, A Gautier-Hion, and J Kingdon (eds.): A Primate Radiation: Evolutionary History of the African Guenons. Cambridge: Cambridge University Press, pp. 35–53.

Pickford M, Senut B, Hadoto D, Musisi J, and Kariira C (1986) Nouvelle decouvertes dans le Miocene inferieur de Napak, Ouganda Oriental. C. R. Acad. Sci. Paris *302*:47–52.

Pilbeam DR (1997) Research on Miocene hominoids and hominid origins: The last three decades. In DR Begun, CV Ward, and MD Rose (eds.): Function, Phylogeny, and Fossils: Miocene Hominoid Evolution and Adaptations. New York: Plenum Press, pp. 13–28.

Pilbeam DR, and Walker A (1968) Fossil monkeys from the Miocene of Napak, northeast Uganda. Nature 220:657–660.

Plavcan JM, and van Schaik C (1997) Interpreting hominid behavior on the basis of sexual dimorphism. J. Hum. Evol. *32*:345–374.

Radinsky L (1974) The fossil evidence of anthropoid brain evolution. Am. J. Phys. Anthropol. *41*:15–27.

Radinsky L (1982) Some cautionary notes on making inferences about relative brain size. In E Armstrong and D Falk (eds.): Primate Brain Evolution: Methods and Concepts. New York: Plenum Press, pp. 29–37.

Remis MJ (1995) Effects of body size and social context on the arboreal activities of lowland gorillas in the Central African Republic. Am. J. Phys. Anthropol. *97*:413–434.

Reno PL, McCollum MA, and Lovejoy CO (1997) Anthropoid radial neck length and its implications for hominid locomotor behavior. Am. J. Phys. Anthropol. Suppl. *24*:197.

Retallack GJ, Dugas DP, and Bestland EA (1990) Fossil soils and grasses of a middle Miocene East African grassland. Science *247*:1325–1328.

Retallack GJ (1992) Middle Miocene fossil plants from Fort Ternan (Kenya) and evolution of African grasslands. Paleobiology *18*:382–400.

Ripley S (1967) The leaping of langurs: A problem in the study of locomotor behaviors. Am. J. Phys. Anthropol. *26*:149–170.

Rollinson JMM, and Martin RD (1981) Comparative aspects of primate locomotion, with special reference to arboreal cercopithecines. Symp. Zool. Soc. Lond. *48*:377–427.

Rose MD (1973) Quadrupedalism in primates. Primates *14*:337–358.

Rose MD (1977) Positional behavior of olive baboons (*Papio anubis*) and its relationship to maintenance and social activities. Primates *18*:59–116.

Rose MD (1983) Miocene hominoid postcranial morphology: Monkey-like, ape-like, neither, or both? In RL Ciochon and RS Corruccini (eds.): New Interpretations of Ape and Human Ancestry. New York: Plenum Press, pp. 405–417.

Rose MD (1984) Hominoid specimens from the middle Miocene Chinji Formation, Pakistan. J. Hum. Evol. *13*:503–516.

Rose MD (1989) New postcranial specimens of catarrhines from the middle Miocene Chinji Formation, Pakistan: Descriptions and a discussion of proximal humeral functional morphology in anthropoids. J. Hum. Evol. *8*:131–162.

Rose MD (1993) Locomotor anatomy of Miocene hominoids. In D Gebo (ed.): Postcranial Adaptation in Nonhuman Primates. DeKalb: Northern Illinois University Press, pp. 252–272.

Rose MD, Leakey MG, Leakey RE, and Walker AC (1992) Postcranial specimens of *Simiolus enjiessi* and other primitive catarrhines from the early Miocene of Lake Turkana, Kenya. J. Hum. Evol. *22*:171–237.

Rose MD, Nakano Y, and Ishida H (1996) *Kenyapithecus* postcranial specimens from Nachola, Kenya. African Study Monographs (Kyoto University) Supplementary Issue *24*:1–56.

Rowell TE (1966) Forest living baboons in Uganda. J. Zool., Lond. *149*:344–364.

Rowell TE, and Chism J (1986) Sexual dimorphism and mating systems: Jumping to conclusions. In M Pickford and B Chiarelli (eds.): Sexual Dimorphism in Living and Fossil Primates. Florence: Il Sedicesimo, pp. 107–111.

Ruff CB, Walker A, and Teaford MF (1989) Body mass, sexual dimorphism and femoral proportions of *Proconsul* from Rusinga and Mfangano Islands, Kenya. J. Hum. Evol. *18*:515–536.

Senut B (1986a) Nouvelle decouvertes de restes post-craniens de primates Miocenes (Hominoidea and Cercopithecoidea) sur le site Maboko au Kenya occidental. C. R. Acad. Sci. Paris *303*:1359–1362.

Senut B (1986b) New data on Miocene hominoid humeri from Pakistan and Kenya. In JG Else and PC Lee (eds.): Primate Evolution. Cambridge: Cambridge University Press, pp. 151–161.

Shipman P (1986) Paleoecology of Fort Ternan reconsidered. J. Hum. Evol. *15*:193–204.

Shipman P, Walker A, Van Couvering JAH, and Hooker PJ (1981) The Fort Ternan hominoid site, Kenya: Geology, taphonomy, and paleoecology. J. Hum. Evol. *10*:49–72.

Smith RJ, and Jungers WL (1997) Body mass in comparative primatology. J. Hum. Evol. *32*:523–559.

Simons EL, and Pilbeam DR (1978) *Ramapithecus* (Hominidae, Hominoidea). In VJ Maglio and HBS Cooke (eds.): Evolution of African Mammals. Cambridge, MA: Harvard University Press, pp. 147–153.

Solounias N, and Moelleken SMC (1993) Tooth microwear and premaxillary shape of an archaic antelope. Lethaia *26*:261–268.

Spoor CF, Sondaar PY, and Hussein ST (1992) A new hominoid hamate and first metacarpal from the late Miocene Nagri Formation of Pakistan. J. Hum. Evol. *21*:413–424.

Strasser E (1988) Pedal evidence for the origin and diversification of cercopithecid clades. J. Hum. Evol. *17*:225–245.

Strasser E (1997) A cladistic analysis of the cercopithecid foot. Am. J. Phys. Anthropol. Suppl. *24*:222.

Strasser E, and Delson E (1987) Cladistic analysis of cercopithecid relationships. J. Hum. Evol. *16*:81–99.

Struhsaker TT (1969) Correlates of ecology and social organization among African cercopithecines. Folia Primatol. *11*:80–118.

Sussman RW (1974) Ecological distinction in sympatric species of *Lemur*. In RD Martin, GA Doyle, and AC Walker (eds.): Prosimian Biology. London: Duckworth, pp. 75–108.

Szalay FS (1975) Hunting-scavenging protohominids: A model for hominid origins. Man *10*:420–429.

Tassy P (1979) Les proboscideans (Mammalia) du Miocene d'Afrique orientale: Resultats preliminaires. Bull Soc. Geol. France *21*:265–269.

Temerin LA, and Cant JGH (1983) The evolutionary divergence of Old World monkeys and apes. Am. Nat. *122*:335–351.

Thomas H (1979) Les bovides Miocenes des rifts est-Africains: Implications paleobiogeographiques. Bull. Soc. Geol. France *21*:295–299.

Thomas H (1984) Les bovides ante-Hipparions des Siwaliks Inferieurs (Plateau du Potwar, Pakistan). Mem. Soc. Geol. France *145*:1–65.

Thomas H (1985) Les Giraffoidea et les Bovidae miocenes de la Formation Nyakach (Rift Nyanza, Kenya). Palaeontographica (A) *183*:64–89.

Tong H, and Jaeger J-J (1993) Muroid rodents from the middle Miocene Fort Ternan locality (Kenya) and their contribution to the phylogeny of muroids. Palaeontographica Abt. A *229*:51–73.

Tuttle RH (1967) Knuckle-walking and the evolution of hominoid hands. Am. J. Phys. Anthropol. *26*:171–206.

Tuttle RH (1974) Darwin's apes, dental apes and the descent of man: Normal science in evolutionary anthropology. Curr. Anthropol. *15*:389–398.

Van Couvering JAH (1980) Community evolution in East Africa during the Late Cenozoic. In AK Behrensmeyer and AP Hill (eds.): Fossils in the Making: Vertebrate Taphonomy and Paleoecology. Chicago: University of Chicago Press, pp. 272–298.

Van Couvering JAH, and Van Couvering JA (1976) Early Miocene mammal fossils from East Africa: Aspects of geology, faunistics and paleoecology. In GLl Isaac and E McCown (eds.): Human Origins: Louis Leakey and the East African Evidence. Menlo Park, CA: WA Benjamin, pp. 155–207.

Van der Made J (1996) *Albanohyus*, a small Miocene pig. Acta Zoologica Cracoviensia *39*:293–303.

Van Roosmalen MGM, Mittermeier RA, and Fleagle JG (1988) Diet of the bearded saki (*Chiropotes satanas chiropotes*): A neotropical seed predator. Am. J. Primatol. *14*:11–35.

von Beyrich H (1861) Uber *Semnopithecus pentelicus*. Phys. Abh. K. Akad. Wiss., Berl. *1860*:1–26.

von Koenigswald GHR (1969) Miocene Cercopithecoidea and Oreopithecoidea from the Miocene of East Africa. Fossil Vertebrates of Africa *1*:39–51.

Walker A, and Andrews P (1973) Reconstruction of the dental arcade of *Ramapithecus wickeri*. Nature *224*:213–214.

Walker A, and Pickford M (1983) New postcranial fossils of *Proconsul africanus* and *Proconsul nyanzae*. In RL Ciochon and RS Corruccini (eds.): New Interpretations of Ape and Human Ancestry. New York: Plenum Press, pp. 325–351.

Walker SE (1996) The evolution of positional behavior in the saki-uakaris (*Pithecia*, *Chiropotes*, and *Cacajao*). In MA Norconk, AL Rosenberger, and PA Garber (eds.): Adaptive Radiations of Neotropical Primates. New York: Plenum Press, pp. 335–367.

Ward CV, Walker A, Teaford MF, and Odhiambo I (1993) Partial skeleton of *Proconsul nyanzae* from Mfangano Island, Kenya. Am. J. Phys. Anthropol. *90*:77–111.

Ward CV, Ruff CB, Walker A, Teaford MF, Rose MD, and Nengo IO (1995) Functional morphology of *Proconsul* patellas from Rusinga Island, Kenya, with implications for other Miocene-Pliocene catarrhines. J. Hum. Evol. *29*:1–19.

Ward SC, and Pilbeam DR (1983) Maxillofacial morphology of Miocene hominoids from Africa and Indo-Pakistan. In RL Ciochon and RS Corruccini (eds.): New Interpretations of Ape and Human Ancestry. New York: Plenum Press, pp. 211–238.

Waser P (1984) Ecological differences and behavioral contrasts between two mangabey species. In PS Rodman and JGH Cant (eds.): Adaptations for Foraging in Non-human Primates. New York: Columbia University Press, pp. 211–238.

Washburn SL (1968) Speculations on the problems of man's coming to the ground. In B Rothblatt (ed.): Changing Perspectives on Man. Chicago: University of Chicago Press, pp. 191–206.

Waterman HC (1929) Studies on the evolution of the pelvis of man and other primates. Bull. Am. Mus. Nat. Hist. *58*:585–642.

White TD (1995) African omnivores: Global climatic change and Plio-Pleistocene hominids and suids. In E Vrba (ed.) Paleoclimate and Evolution, With Emphasis on Human Origins. New Haven: Yale University Press, pp. 369–384

White TD, Suwa G, and Asfaw B (1994) *Australopithecus ramidus*, a new species of early hominid from Aramis, Ethiopia. Nature *371*:306–312.

Whitworth T (1958) Miocene ruminants of East Africa. Fossil Mammals of Africa *15*: 1–50.

Winkler AJ (1994) Middle Miocene rodents from Maboko Island, western Kenya: Contributions to understanding small mammal evolution during the Neogene. J. Vert. Paleo. *14*:53A.

Woldegabriel G, White TD, Suwa G, Renne P, de Heinzelin J, Hart WK, and Heiken G (1994) Ecological and temporal placement of early Pliocene hominids at Aramis, Ethiopia. Nature *371*:330–333.

Wolpoff MH (1982) *Ramapithecus* and hominid origins. Curr. Anthropol. *23*:501–510.

Wolpoff MH (1996) Human Evolution. New York: McGraw-Hill.

Yulish SM (1970) Anterior tooth reduction in *Ramapithecus*. Primates *11*:255–263.

Zapfe H (1960) Die Primatenfunde aus der miozanen Spaltenfullung von Neudorf an der March (Devinska Nova Ves), Tschechoslowakei. Mit anhang: Der Primatenfund aus dem Miozan von Klein Hadersdorf in Niederosterreich. Schweiz. Palaeontol. Abh. *78*:1–293.

20

ECOLOGICAL MORPHOLOGY OF *AUSTRALOPITHECUS AFARENSIS*

Traveling Terrestrially, Eating Arboreally

Kevin D. Hunt

Department of Anthropology
Indiana University
Bloomington, Indiana 47405

1. INTRODUCTION

Kinzey (1976, 1977, 1978) was quick to appreciate the utility of integrating system-atic ecological research with study of positional behavior and morphology, a synthetic area of scholarly pursuit now distinguished by its own appellation, ecological morphology (Wainwright and Riley, 1994). Primatologists so often focus on food and food-gathering behaviors as keys to understanding primate anatomy because primate activity budgets are dominated by feeding. Across habitats ranging from thicket woodland to closed canopy forest, chimpanzees consistently dedicate half of their activity budget to feeding (Table 1). The next most common activity, "resting," might as well be called "digesting." As a sim-plifying first-assumption, an ecological perspective takes the view that the hominoid body is a food-getting machine, and ignores the presumably lesser selective roles played by in-traspecific aggression and predator avoidance (though predation may be more important for smaller primates: van Schaik, 1983; van Schaik and van Hooff, 1983; van Schaik et al., 1983a, b; van Schaik and van Noordwijk, 1989). This view holds as significant the rar-ity of predation on hominoids (Cheney and Wrangham, 1986), and that intraspecific agonism is less a threat to survival than is starvation.

For those who find this perspective comfortable, feeding hypotheses (Du Brul, 1962; Jolly, 1970; Tuttle, 1975, 1981; Rose, 1976, 1984, 1991; Stern, 1976; Wrangham, 1980; Tuttle et al., 1991; Hunt, 1994a) for the origin of bipedalism (and/or the divergence of apes and humans) are particularly appealing. Jolly's seed eating hypothesis (1970) main-tained that the prevalence of grass seeds in dry habitats selected for a suite of human char-acteristics. The small diameter and even distribution of grass seeds, he argued, demanded sustained bipedalism during postural collection. Shuffling bipedalism was seen as a high-

Table 1. Activity budgets for Gombe, Mahale, and Kibale chimpanzees

Activity	Gombe (Hunt, 1989)	Gombe* (Wrangham, 1977)	Mahale (Hunt, 1989)	Kibale (Ngogo) (Ghiglieri, 1984)
Food	49.0	55.7	46.9	57.3
Rest, socialize	36.2	30.3	40.3	31.5
Travel	14.8	13.9	12.8	11.1

*Males only

profile variation on gelada scooting, advantageous for moving between food resources while sustaining a higher reach. The manipulation of small-diameter seeds selected for considerable manual dexterity, preadapting hominids for tool use.

Savanna baboons were observed to be bipedal when feeding on small food items, though grass seeds themselves were not common (Rose, 1976, 1984, 1991). Wrangham (1980) noted that gathering fruit from bushes, in particular, elicited bipedalism among chimpanzees. He offered a locomotor corollary to Rose's small-object feeding hypothesis: bipedal locomotion saves energy by eliminating the action of raising the upper body to feed bipedally after walking between resources quadrupedally. Jolly agreed, suggesting that specialization on small-diameter fruits, rather than seeds, might have been the critical selective pressure that resulted in the evolution of bipedalism (Jolly and Plog, 1987). The small-object postural feeding hypothesis (Rose, 1976, et seq.) derived from the research of Jolly, Rose and Wrangham postulates that bipedalism evolved as a terrestrial feeding posture advantageous for reaching into trees, and that bipedal locomotion evolved to reduce energy costs when traveling between densely packed feeding sites.

Recently, Hunt (1994a, 1996) added an arboreal component to the small-object postural feeding hypothesis, noting that small-object feeding elicits bipedalism in trees as well as on the ground. The small diameter of supports in the small trees in which small-diameter fruits tend to be found appears to encourage bipedalism arboreally as much as fruit diameter. The data from which these conclusions were made are examined below in more detail than is available in Hunt (1992a, 1996).

2. METHODS

I observed chimpanzees for 571 hours at the Mahale Mountains National Park and for 130 hours at the Gombe Stream National Park (Hunt, 1992a). Sixteen thousand three hundred and three instantaneous, 2-minute focal observations were made on 26 well-habituated prime adults spanning all social ranks. Twenty-five positional behavior variables were monitored, including positional mode, behavioral context of the mode, and a number of feeding parameters. Two thousand eighty seven observations were made on Gombe baboons over 83 hours using identical methods.

I identified one of 65 locomotor or postural modes in a target animal at each 2-minute point. I recorded positional behavior, location in canopy, height, size of supports, proximity to others, and food type. Because the chimpanzees were well-habituated, I was able to make observations during all hours of the day, in all contexts, and with no decrease in quality of observation when individuals were on the ground.

Bipedalism was defined as posture or locomotion in which it was judged that more than half of the body weight was borne by the hind limbs in compression. If neither forelimb nor any other part of the body other than the hind feet touched a support, I labeled the mode "unassisted bipedalism." When a forelimb or other body part supported some but less than half of the body weight, I called the mode "assisted bipedalism." I put behaviors for which I judged that more than half of the weight was borne by an abducted forelimb in an "arm-hanging" mode; arm-hanging is not part of this analysis.

3. RESULTS

3.1. Contexts of Chimpanzee Bipedalism

Bipedalism was not a common chimpanzee behavior. Ninety-seven instances of bipedalism among 21 individuals were sampled in 700 hours of observation. No two observations were made in consecutive time-points. *Ad libitum* observations (i.e., observations on non-target individuals, observations between time-points) were used as supplemental evidence, but tables and figures include only systematic observations (i.e., focal individual, time-point samples). By far, the most common context of bipedalism was feeding (Figure 1).

"Move in patch" typically occurred when an individual moved from one harvesting perch in a fruit tree to another within the same tree. Not uncommonly, a target individual did not even cease chewing during the move. If this behavior is pooled with "feed," which seems reasonable, a full 80% of chimpanzee bipedalism was in the context of feeding. The feeding function of bipedalism is dramatically illustrated by comparing bipedal contexts to the daily activity budget. Whereas half of the typical chimpanzee day is occupied with feeding (Table 1), *over 80%* of bipedalism was in a feeding context. Bipedalism was clearly a feeding mode both on the ground and in the trees (Figure 2). Rose found that bipedalism among baboons had the same feeding function (Rose, 1976).

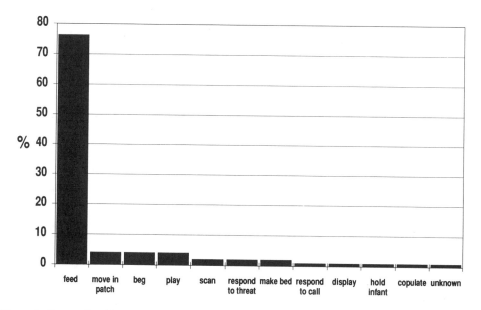

Figure 1. Contexts in which bipedalism was observed in Tanzanian (Gombe and Mahale) chimpanzees. Of 97 observations of bipedalism, feeding and moving within a feeding patch constituted 80.4% of all bipedal behavior.

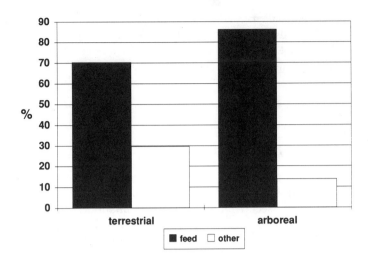

Figure 2. Percentage of bipedalism engaged in during feeding (dark columns) versus other activities (light columns) on the ground (left) and in trees (right). Note that bipedalism functioned as a feeding posture whether chimpanzees were on the ground or in the trees (N=97).

Perhaps the greatest surprise is that after feeding, the next two most common contexts of bipedalism are "beg" and "play" (Figure 1), two behaviors that have not been implicated as selective pressures for the evolution of bipedality (Rose, 1991). Chimpanzees stood bipedally occasionally to get a better view of their environment ("scan"); such scanning was sampled twice, totaling 4.1% of bipedal behavior. "Respond to threat" (4.1%) describes behavior in which an individual stood bipedally, often with one or both forelimbs lightly contacting a tree, to be ready to climb the trunk quickly if a displaying male or an altercation drifted in their direction. Chimpanzees were not uncommonly bipedal to free their hands to manipulate foliage while making a nest, or night bed (4.1%). Chimpanzees sometimes stood bipedally to listen to distant vocalizations (not sampled here, but observed *ad libitum*), and while responding to calls with their own vocalizations (2%). Display, though impressive when it occurs, is actually a rare behavior among chimpanzees (2%). Observations reported here do not include those made in feeding camp, where dominance displays in contests for provisioned food, or in response to unusually large social groupings, have allowed the frequent filming of bipedalism. Chimpanzees often support clinging newborns with a forelimb when walking, but utilizing both forelimbs for support and thereby requiring bipedalism ("hold infant") was rare (2%). Bipedal infant-carrying is probably so rare because infants become capable of gripping mother's fur within days of birth, after which little support is needed. I suspect, based on *ad hoc* observations, that mothers walk little the first few days after giving birth, reducing the need for helping a newborn to cling during its helpless phase. "Copulate" and "unknown" are rare contexts for chimpanzee bipedalism as well (2% each).

3.2. Postural versus Locomotor Bipedalism

Although postural behavior is more common than locomotor behavior in chimpanzees—85% of chimpanzee positional behavior is postural (Table 2)—this balance does not hold for bipedalism, where the postural component predominates. Postural bipedalism

Table 2. Proportion of locomotion vs. posture among chimpanzees

Population	N	Posture	Locomotion
Mahale	11,471	84.8	15.2
Gombe	2,910	83.3	16.7

made up 95% of bipedal observations (Figure 3). This proportion is identical to that reported by Rose (1976) for baboons.

3.3. Is Bipedalism a Terrestrial Behavior?

Bipedalism is sometimes assumed to be a behavior naturally elicited by terrestriality. Chimpanzee behavior refutes this assumption. Chimpanzees were actually more likely to be bipedal when arboreal than when terrestrial; 61% of all bipedal behavior was arboreal (Figure 4). Nor is bipedalism limited to large branches in the tree core. Nearly 30% of bipedalism was observed among the terminal branches (Figure 4).

Although terrestriality did not elicit bipedalism, some might expect *locomotor* bipedalism to have been more common terrestrially than arboreally. That was not the case. Over 80% of locomotor bipedalism was arboreal. Of the 6 bipedal locomotor bouts in the sample, 5 were arboreal. Chimpanzees were moving arboreally within a feeding patch in 4 of 6 observations (Figure 1). The fifth observation was making a night nest. The sole terrestrial observation of bipedal locomotion was in the context of "playing" (Figure 1).

3.4. Forelimb-Suspension and Bipedalism Are Linked

Bipedalism was stabilized by an arm-hanging-like (=unimanual forelimb-suspension) support from a forelimb nearly 60% of the time (Figure 5). It appears that the inherent instability of bipedalism (Kummer, 1991) is amplified arboreally by the small, flexible nature of the support. While bipedal posture has two advantages—it increases the height

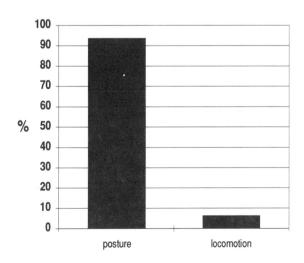

Figure 3. Percent of postural bipedalism compared to locomotor bipedalism. Although locomotion typically makes up 15% of all positional behavior, of 97 observations of bipedalism only 6 (6%) were locomotor.

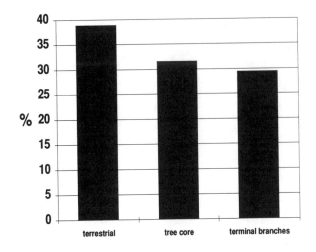

Figure 4. Percent of total bipedal bouts observed on the ground versus in the tree core (central part of tree), versus in the terminal branches (within 1 m of tree edge). Note that only 38% of bipedality was terrestrial. It was arboreality, not terrestriality, that most commonly elicited bipedalism among chimpanzees. Notably, bipedalism was nearly as common among tiny twigs in the terminal branches as it was in the tree core or on the ground. The common assumption that a stable substrate elicits bipedalism is clearly wrong.

of the reach and does not require a "gear-change" expense[*] (*sensu* Wrangham, 1980) when moving—it requires additional stabilization arboreally to be an effective harvesting posture. Assisted bipedalism involved stabilizing the body with the forelimb in a fully abducted forelimb-suspension gestalt.

The link between forelimb-suspension and bipedalism was most powerful among the terminal branches of trees. A forelimb oriented as in arm-hanging (though bearing less weight) stabilized bipedal posture in 93% of observations among terminal branches (N=27). Bipedalism occurred on significantly smaller branches than did other postures (12.2 cm vs. 15.0 cm, Mann-Whitney U test, U=123,620, P=0.0001, $N_{1,2}$=64,5375), probably because small trees offer few large branches stable enough for sitting or unassisted bipedal standing. In the tree core, or central part of the tree, 52% (N=23) of bipedalism was assisted by a forelimb. Arboreal bipedal locomotion was relatively rare (4.1% of all bipedal episodes), and consisted exclusively of short-stride-length shuffling.

Terrestrial bipedalism was less often assisted, presumably because the firm substrate did not require an extra stabilizing contact point. Terrestrial postural bipedalism was *unassisted* nearly 2/3 of the time, whereas bipedalism was assisted 3/4 of the time arboreally (Figure 6). During terrestrial gathering both hands were often used to harvest fruits. Not infrequently, one hand was used to pull down and hold an otherwise inaccessible fruit-bearing limb, so that the posture became a terrestrial arm-hanging-bipedalism. A few terrestrial bipedal bouts were locomotor bipedalism in the context of moving between feeding sites at the same tree (4.1%). That is, short-distance within-site shuffling, rather than long distance travel, was the most common context for locomotor bipedalism.

Fruit was the most common food resource harvested by bipedal chimpanzees (Figure 7). Among other foods harvested bipedally, only ants might be called common. Manipulation of an ant-dipping tool often required both hands, and therefore sites from which chimpanzees harvested ants tended to be those in which they could work while bipedal.

---

[*] That is, changing postural or locomotor mode entails some expense. In this case, the cost of raising or lowering the torso is an energetic expense that must be balanced against the increased energy expenditure required to locomote using an inefficient mode (bipedalism). If the difference between the energy expended to locomote bipedally versus quadrupedally is less than the energy expended to lower the torso to allow quadrupedal locomotion, and then raise the torso to a bipedal posture at a new feeding site, there is selection for the individual to locomote bipedally.

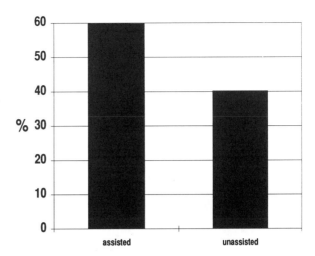

Figure 5. Frequency of assisted versus unassisted bipedalism. Nearly 60% of bipedalism in chimpanzees was assisted by a forelimb, most often in an forelimb-suspension-like manner. These figures do not include observations in which the forelimb was judged to be bearing more than half the body weight, a posture labeled arm-hanging (=forelimb-suspension). Supported forelimb-suspension made up another 3.6% of all positional behavior (N=11,393).

3.5. Characteristics of Foods Harvested Bipedally

Bipedalism occurred both terrestrially and arboreally when chimpanzees fed from *Garcinia huillensis, Harungana madagascarensis, Monanthotaxis poggei* and *Grewia* sp. Together these four species of trees constituted 27% of all bipedal feeding episodes and 48% of the bipedal episodes in which the plant material being eaten could be identified. At Mahale, *Garcinia huillensis* rarely reaches 15 m in height. It lives in forest edge habitats

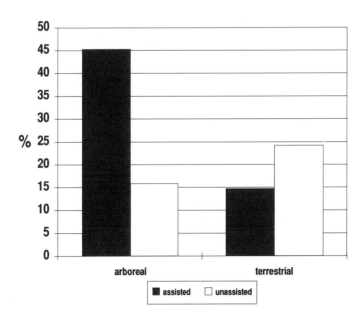

Figure 6. Frequency of assisted (=stabilized by a forelimb) and unassisted (support by hind limbs only) bipedalism in trees versus on the ground. Bipedalism is mostly unassisted on the ground, and mostly assisted in the trees. Unassisted bipedalism should be preferred because it allows a higher feeding rate, since both hands can be used for gathering (Rose, 1976). The higher frequency of assisted bipedalism among terminal branches is presumed to be due to the instability of bipedalism on small-diameter branches.

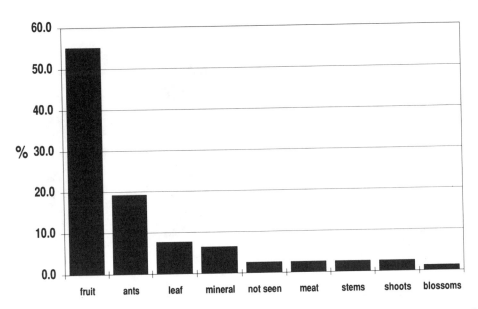

Figure 7. Food items harvested bipedally by chimpanzees. Note that aside from fruits, only ants are commonly harvested bipedally.

and lake-side environments. Its fruits are approximately 2 cm in diameter. *Harungana madagascarensis* is a many-branched understory tree ranging from several meters to 12 m in height. It is widely distributed in East Africa, typically in forest edge environments (Hamilton, 1982). Its fruits are 3–5 mm in diameter. It is typically found in monospecific stands (pers. obs.). *Monanthotaxis poggei* is a 1–2 m tall shrubby plant found in forest edge environments. It often occurs in dense stands. The fruits are approximately 1 cm in diameter. *Grewia* is a 5 m open-forest tree with 1 cm fruits.

All of these trees are short, forest-edge or open-forest trees. The fruits of each species are small (~2 cm, 0.4 cm, 1 cm and 1 cm respectively). The trees occur more often in monospecific stands than trees chimpanzees harvested fruits from in the more closed-forest part of their range. Three of the four (not *Garcinia*) are dry, fibrous, difficult to masticate fruits.

Although I did not sample bipedal gathering *both* arboreally and terrestrially in any other tree, other small trees with small fruits elicited bipedalism either arboreally or terrestrially much more commonly than did large trees. Bipedal food collecting was significantly more common among small (mature height of ≤ 15 m) trees (Figure 8) with small fruits (44 vs. 8, Fisher's Exact test, P<0.001, χ^2=27.8, df=1), suggesting that fruit diameter and tree height are the critical factors eliciting bipedalism. It is difficult to distinguish between the effects of small trees and small fruits, since all but one small tree also had small (≤ 2 cm) fruit. When plant-foods gathered during bipedal bouts were identified, 28 of 33 fruits (85%) were ≤ 2 cm in diameter (Figure 9).

3.6. Why Do Chimpanzees Climb Trees?

Chimpanzee bipedalism and other behaviors might be argued to be poor predictors of behavior among the more terrestrial hominids. Chimpanzees may forage in trees be-

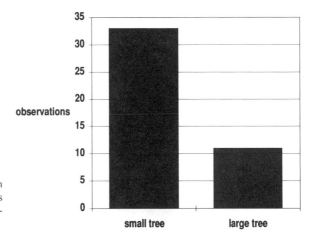

Figure 8. Although chimpanzees fed in small trees (N=1439) and large trees (N=1536) almost equally, most observations of bipedalism were in small trees.

cause they prefer arboreality, whereas australopithecines might have preferred terrestriality. Chimpanzee foraging data suggest that chimpanzees go into trees not because they prefer them, but because their most preferred foods are found there. Early hominids may have been similarly constrained.

Chimpanzees spent 70% of the time they were in trees feeding. Furthermore, chimpanzees entered terminal branches almost exclusively to feed; feeding constituted nearly 90% of terminal branch activity (Figure 10).

Because different chimpanzees spend differing amounts of time in trees, we may draw some generalizations about arboreality that can provide a model for canopy use in protohominids. Social rank, body size and sex appear to determine arboreality in chimpanzees (Hunt, 1992b, 1994b). Larger chimpanzees spent less time in trees than did smaller chimpanzees. In a multiple regression that included social rank, body size, and canopy height, with social rank factored out, large males positioned themselves lower in the canopy (R=0.30, P<0.0002, N=6,600).

When like-rank males of different body sizes are divided into 2 classes, large males fed from smaller tree species than did small males (Figure 11; large and small individuals

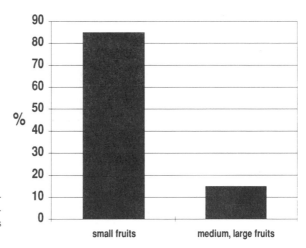

Figure 9. Frequency of bipedalism compared when eating small versus medium-sized and large fruits (N=33). Small fruits are associated with bipedalism.

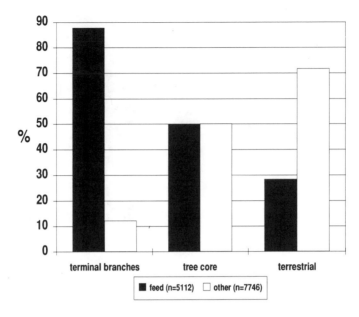

Figure 10. Frequency of feeding in different canopy locations. Chimpanzees do most of their feeding in trees, and they enter terminal branch sites almost exclusively to feed. Apparently there is little food on the ground that chimpanzees prefer.

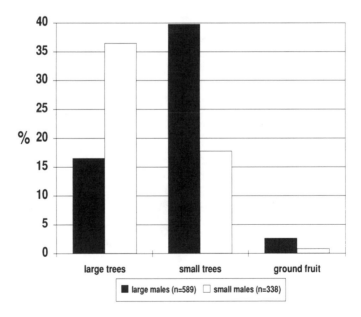

Figure 11. Comparison of feeding site preferences for large and small males, matched for social rank. Large males preferentially fed in small trees and on the ground. Large males minimized climbing, hypothetically because feeding sites lower in the canopy were more valuable to them.

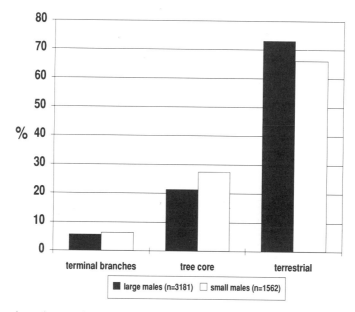

Figure 12. Comparison of canopy location between large and small males, matched for social rank. Large males spent more time on the ground.

were *matched for social rank*). *Matched for social rank*, large males spent more time on the ground than did small males (Figure 12). *Matched for social rank*, large males climbed less often, 0.8% of all behavior versus 3.2% for small males (χ^2=6.65, df=1, P=0.01, N=494, 476). The likeliest explanation for this pattern is that arboreal positional behavior is less demanding for smaller males.

Because these observations are limited to males, sex and infant care did not contribute to the differences. Individual males were matched for social rank, which means that body size alone caused these differences in canopy use. Comparisons of male and female chimpanzees show the same results, though it is not possible to adjust for infant-care and social rank differences. Females spend more time in the trees (Figure 13). It is likely that body size contributes to female arboreality by making arboreality less energetically demanding. Arboreality may also be predator-avoidance strategy for females, though the evidence from males suggests not.

4. AUSTRALOPITHECINE ECOLOGY: CRANIODENTAL EVIDENCE

Craniofacial shape, robusticity of the masticatory apparatus, cusp morphology, tooth size, enamel thickness, and incisor size are among the evidence that has been brought to bear on reconstructing australopithecine diets, most notably in a synthesis by Kay (1985). Although the correlation between each of these variables and diet is low, if considered together they provide a detailed model of australopithecine diet.

Molar dental microwear has not been examined in *Australopithecus afarensis*, leaving *A. africanus* microwear as an admittedly unsatisfactory stand-in (Walker, 1981; Teaford and Walker, 1984; Grine and Kay, 1988; Kay and Grine, 1988; Teaford, 1994). The

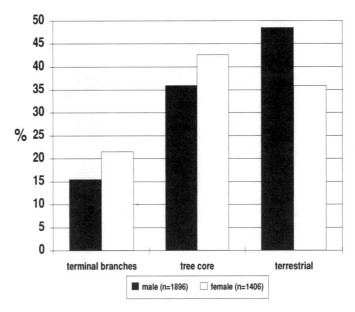

Figure 13. Comparison of canopy location in males and females. Females are more arboreal and spend more time among the terminal branches.

similarity of dentition of the two (Tobias, 1980), however, makes this unsatisfactory proxy more acceptable. Walker (1981) interpreted the *Pan-* or *Mandrillus*-like microwear of both *A. africanus* and *A. robustus* as meaning they were frugivores. Kay and Grine (1988) found that in both microwear feature width and relative frequency of pits versus scratches suggested the same thing. Molar microwear feature width falls between *Alouatta palliata* and *Cebus nigrivittatus* (Kay and Grine, 1988). Pit:scratch frequency comparisons place them between orangutans and chimpanzees (Kay and Grine, 1988). With the exception of howlers, these primates eat fruit at least 57% of the time (Table 3). Howlers concentrate on leaves. All four species include at least 10% leaves in their diet. With the exception of pith/herbs for chimpanzees, other dietary items are uncommon.

Pit frequencies for *Australopithecus* fall intermediate between orangutans and chimpanzees suggesting a low frequency of seed-eating and oral nut-cracking (Kay and Grine, 1988; Table 3). Orangutans open nuts orally, though it is not clear how common the be-

Table 3. Primate diets

Species	Insects	Leaf	Meat	Fruit	Piths/herbs	Flowers	Bark	Other
Pan troglodytes[1]	5.6	10.3	1.0	57.0	22.6	0.7	0.0	–
Alouatta palliata[2]	0.0	64.0	0.0	12.0	0.0	18.0	0.0	0.0
Pongo pygmaeus[3]	1.0	26.0	0.0	58.0	–	–	13.0	2.0
Cebus spp.[4]	20.0	15.0	0.0	65.0	0.0	0.0	0.0	–
Mean	7.0	29.0	0.3	48.0	6.0	5.0	3.0	1.0

[1]Hunt (1989); feeding time, based on 3,891 feeding records of *Pan troglodytes schweinfurthii* at Mahale.
[2]Glander (1978); feeding time.
[3]Rodman (1984); feeding time, Kutai, Kalimantan 40,022 min. observations.
[4]Hladik and Hladik (1969); dry weight of stomach contents; records for *Cebus nigrivittatus* are 100% fruit (Fooden, 1964).

havior is; chimpanzees rarely crack nuts orally. Other evidence suggests that nut-cracking was not common in australopithecines. Peters (1987) argued that a small modelled gape in australopithecines and a relative lack of crenulations suggest low levels of nut-cracking. Although some nut-cracking cannot be ruled out, a small gape and unpitted teeth suggest that *A. africanus* and *A. robustus* were not specialized for this dietary item. A high frequency of pitting is also found in animals that eat foods that have adhering grit (Teaford and Walker, 1984; Teaford, 1994). The low frequency of pitting in australopithecine molars makes the utilization of roots and tubers and other below-ground resources with adhering grit unlikely.

Macrowear adds an exclamation point to indications of some leaf-eating in australopithecines. In *A. afarensis* the incisors are unevenly worn so that the occlusal surface has an undulating appearance in frontal view (Puech et al., 1984). Such wear results from stripping, such as pulling a twig through the mouth to remove leaves.

An objective but somewhat crude estimate of the australopithecine diet might be gained by simply averaging the frequency of dietary items in primates that have microwear most similar to australopithecines (Table 3). Frugivory is suggested, with leaves as an important secondary component of the diet.

Dental microwear of *A. africanus* narrows the possible australopithecine diet somewhat from what is indicated by dental morphology.[†] Among primates, summed (i.e., I1 and I2) incisor size is correlated with food object size and are larger in frugivores than graminivores or folivores (Hylander, 1975; Kay and Hylander, 1978). Primates that eat small fruits have small incisors. Fruits require incisal penetration of an often fibrous and/or abrasive husk; leaves do not. The extra wear on incisors cause by fruit processing requires larger incisors to maintain a functioning incisal edge in older animals. A regression of the natural logarithm of maxillary incisor width and log body weight for 57 primates yields a regression line with good separation between principally frugivorous cercopithecines and principally folivorous colobines (Hylander, 1975; Kay and Hylander, 1978). Australopithecines fall below both the regression line for apes alone (Kay 1985) and for pooled primates (Kay and Hylander, 1978; using McHenry, 1992 body weights).

The two earliest hominid species are not represented by lateral incisors, and cannot be directly compared to Kay and Hylander's (1978) results. Both, however, have I^1s quite near the mean of *A. afarensis* (10.63 mm, White et al., 1981). The I^1 of *Ardipithecus ramidus* is 10.0 mm (White et al., 1994), and the I^1 of *A. anamensis* is 10.5 mm (Leakey et al., 1995). The small early australopithecine incisor dimensions as expressed in their relation to the primate and ape regression line suggest that they consumed smaller food items than extant primate frugivores, including orangutans and chimpanzees.

Australopithecus afarensis molar areas fall above the regression line of cheek tooth area compared to body mass (log-log plot) for hominoids (Kay, 1985). The residual is nearly equidistant between orangutans and *A. africanus* (McHenry, 1984; Kay, 1985), and well above the values for chimpanzees. Among extant primates, species that open nuts and seeds and/or eat hard-to-masticate fibrous fruits have large molars (orangutans and capuchins, Kay, 1981, 1985).

Australopithecines have thick enamel (Kay, 1983), even thicker than the thickest-enamelled of extant primates (i.e., mangabeys, capuchins, and orangutans). Thick enamel is correlated with consumption of hard and brittle foods (Teaford, 1985) such as seeds,

---

[†] Newly discovered fossils (*Ardipithecus ramidus*, White et al., 1995; *A. anamensis*, Leakey et al., 1995) are yet to be thoroughly evaluated, but where possible they are discussed. Otherwise, discussion centers on the more abundant and better studied fossils of *A. afarensis* and *A. africanus*.

nuts or hard-husked fruits. Thick enamel also occurs among mammals that consume foods that are abrasive, or foods that have adhering grit, as subterranean items do. Chimpanzees and gorillas have thin enamel with well-developed shearing blades (Kay, 1981), a morphology correlated with folivory and believed to function to finely comminute leaves into a fine, more digestible gruel. Because australopithecines have thicker enamel and shorter shearing blades than do chimpanzees and gorillas (Kay 1983), their diet must have contained considerably fewer leaves and piths than does the diet of African apes.

Compared to apes, *A. afarensis* has taller, more squared-off, more robust zygomae (White et al., 1981). The diameter of the mandibular corpus is also greater. A forward shift of the zygomae and thickening of the mandibular corpus mean that australopithecines could produce greater occlusal forces than those of extant apes, or could sustain high masticatory pressures for longer (Jolly, 1970). Feeding on hard-to-masticate items is suggested. Reduced prognathism suggests a reduced gape for australopithecines compared to apes.

If we consider all of this evidence at once, a rather precise modelling of the australopithecine diet is possible. Small incisors and the small gape, compared to living apes, suggest that australopithecines consumed smaller food items than do African apes or orangutans. Molar microwear suggests that australopithecines rarely engaged in nut-cracking, and that they did not consume below-ground foods. Their microwear suggests a frugivorous diet supplemented with piths or leaves. Their low-cusped molars with short shearing blades, however, offer contrary evidences, suggesting a lower proportion of piths and leaves. Large, thickly enameled molars suggests a specialization on fibrous fruits. A robust face suggests hard to masticate food items, again implying a specialization on fibrous fruits. Australopithecine craniodental fossils unambiguously suggest a diet high in small-diameter, fibrous fruits. The proportion of leaves in the diet is less clearly indicated, but it seems likely they were an important dietary item, though one clearly less important than fruits.

5. AUSTRALOPITHECINE POSITIONAL BEHAVIOR

5.1. Postcranial Evidence for Arm-Hanging.

The australopithecine torso is broad, shallow and cone-shaped (Schmid, 1983, 1991), the glenoid fossa cranially oriented (Robinson, 1972; Stern and Susman, 1983), the cross sectional area of the vertebral column quite small (Robinson, 1972; Jungers, 1988), and the brachial index chimpanzee-like (Kimbel et al., 1994). These features are adaptations to arm-hanging (Hunt, 1991a).

The australopithecine wrist is mobile (McHenry, 1991a), an adaptation that reduces stress on the wrist when suspending from vertical supports (Hunt, 1991a). Thin twigs bend to vertical when chimpanzees hang from them, necessitating a mobile wrist to maintain grip.

5.2. Postcranial Evidence for Vertical Climbing

The bicipital groove in which the biceps tendon rests is large (Robinson, 1972; Lovejoy et al., 1982), implying a large biceps muscle. The supracondylar ridge (proximal attachment of the extensor carpi radialis and brachioradialis muscles) is huge (White et al., 1993). The large muscles implied by these skeletal features suggest an ability to perform a powerful pull-up action, seen in extant hominoids most often during vertical climbing (Hunt, 1991b).

The convex joint surface of the *A. afarensis* medial cuneiform indicates a rudimentary ablity to abduct the first toe (Stern and Susman, 1983; Deloison, 1991; pers. obs.) *contra* Latimer and colleagues (Latimer et al., 1982). *Australopithecus afarensis* also has long, curved toes (Tuttle, 1981; Stern and Susman, 1983) and an antero-posteriorly short, rounded lateral femoral condyle (in the smaller specimens; Tardieu, 1983; McHenry, 1986, 1991a). A strongly developed fibular groove for the tendon of the peroneus longus muscle suggests ape-like great-toe flexion (Tuttle, 1981; Deloison, 1991), as might be used to grip branches when standing arboreally or climbing. Alternatively, a robust peroneus longus muscle may stabilize a more mobile foot, or support a ligamentously poorly supported arch; if so, bipedalism would be that much less energetically efficient, since muscular support would be necessary for toe-off, rather than a non-energy-consuming ligamentous support. A plantar set, or at least greater mobility (Latimer and Lovejoy, 1990), of the ankle allows full plantarflexion of the foot. Gombe and Mahale chimpanzees plantarflexed their feet when they used their toes grip a branch to support body weight with the hind limb in tension. Curved pedal phalanges (Tuttle, 1981) and a *third* pedal digit longer than the first or second (Stern and Susman, 1983) are gripping adaptations. Such pedal gripping, especially with the lateral 4 toes only, is used by Tanzanian chimpanzees during arm-hanging to increase stability among slender terminal branches. Climbing adaptations in the hindlimb are not limitied to the foot. A long moment arm for the hamstrings (Stern and Susman, 1983) increases the power of hip extension, implying a better climbing adaptation than in modern humans.

5.3. Postcranial Evidence for Both Arm-Hanging and Vertical Climbing

The deltoid tuberosity is large and laterally flaring, suggesting a large deltoid (White et al., 1993; Kimbel et al., 1994), and the coracoid process (proximal attachment of biceps brachii and distal attachment of pectoralis minor muscle) is large (Robinson, 1972). Chimpanzees use the deltoid to raise the arm during vertical climbing and when reaching out to pluck fruits when arm-hanging.

Australopithecine fingers are curved and have large flexor sheath ridges (Stern and Susman, 1983). The fingers are more human-like than ape-like in length; but the thumb is short (=chimpanzee-like) with a chimpanzee-like articulation (Tuttle, 1981; McHenry, 1991a). Arm and leg length proportions are intermediate between those of modern humans and chimpanzees, even when the diminutive stature of the fossils is considered (Jungers, 1982, 1991). These features are adaptations to powerful gripping of cylindrical surfaces, such as occurs in chimpanzees only during arm-hanging and vertical climbing (Hunt, 1992a).

5.4. Postcranial Evidence for Bipedalism

In both general morphology and detail, the pelvis and the hind limb morphology of *A. afarensis* and later hominids indicate bipedalism (Johanson and Edey, 1981; Lovejoy, 1988; Latimer and Lovejoy, 1989, 1990; McHenry, 1991a). The lumbar vertebrae are lordotic (Abitbol, 1987), the sacral alae are expanded, and the pelvis has a very human gestalt (Lovejoy, 1988; McHenry, 1991a). The femur has a deep patellar groove, and at least some specimens have an elliptical lateral condyle. The calcaneus is essentially modern

(Latimer and Lovejoy, 1989). The great toe is robust and the foot has well developed transverse and longitudinal arches (Latimer and Lovejoy, 1990; Langdon et al., 1991).

5.5. Morphology at Odds with Refined Locomotor Bipedalism

Other features in *A. afarensis* suggest a bipedalism that is not as refined as that of modern humans. Although the os coxae are human-like in appearance, *A. afarensis* has smaller sacro-iliac ligaments than modern humans (Stern and Susman, 1983). The width of the AL 288-1 pelvis is proportionally greater than the femoral neck length, suggesting that when compared to modern humans, *A. afarensis* had a greater joint reaction force at the hip and a lower mechanical advantage for muscles that prevent the hip from collapsing when one foot is off the ground (Jungers, 1991). A relatively small acetabulum/femoral head in australopithecines compounds the stresses caused by wide hips, creating even more stress in the hip joint.

Wide hips also cause the moment arm of the body weight of *A. afarensis* to be increased over that of modern humans, increasing the stress on the diaphyseal/femoral neck junction when bearing weight (Hunt, 1994a) and decreasing energetic efficiency during walking by requiring greater muscular activity (Jungers, 1991). The extraordinarily wide hips of AL 288 (Berge and Kazmierczak, 1986; Rak, 1991) and STS 14 (Robinson, 1972) are not obstetric adaptations (Rosenberg and Trevathan, 1995; Stoller, 1995). The broad hips of these hominids are due mostly to an unusually broad pelvic inlet. The pelvic index averaged 77.6 for modern human females (Tague and Lovejoy, 1986), but is 57.6 in AL 288-1. Although a large biacetabular breadth in modern humans is a necessary adaptation for giving birth to large-headed offspring, cephalopelvic reconstruction of AL 288-1 shows a considerable gap between the fetal head and pelvic inlet walls opposite the acetabula (Figure 14). In other words, the pelvis of australopithecines is much broader than could possibly be necessary for parturition.

Other features suggest reduced locomotor competence as well. *Australopithecus afarensis* has quite short hind limbs for its weight and height, suggesting greater energy expenditure per unit distance traveled (Jungers, 1991). The lumbar vertebrae and lumbosacral articular surface of other australopithecines are small, whether in proportion to

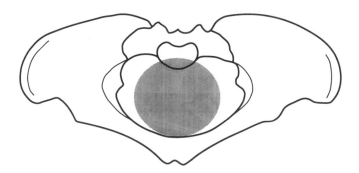

Figure 14. Reconstructed pelvis of *A. afarensis* (redrawn from Lovejoy and Tague, 1986 and from Lovejoy, 1988) with a hypothetical fetal head superimposed. The pelvis is much broader than necessary for parturition. Broad hips and short hind limbs are hypothesized to function to lower the center of gravity. A low center of gravity confers greater balance, which is particularly valuable when standing or moving on small and therefore unstable arboreal supports.

body weight, hip width or nearly any other measure that has been attempted (Robinson, 1972; Jungers, 1988, 1991; McHenry, 1991a; Rak, 1991).

6. DISCUSSION

We can bring four lines of evidence to bear on the question of australopithecine bipedalism: (1) australopithecine craniodental data; (2) australopithecine postcranial anatomy; (3) the contexts that elicit bipedalism in living animals (chimpanzees in this case); and, (4) differences between large and small chimpanzees.

Molar feature width and low pitting argue strongly against nut-cracking or specialization on below-ground resources for australopithecines. Fruit and leaves are most consistent with their microwear, but these data conflict with dental morphology. Thick enamel, low molar shearing quotients, large molar areas and robust craniofacial morphology indicate a diet high in hard items, but not particularly high in leaves or piths. Because below-ground resources are precluded by the microwear data, and because nut-cracking is not strongly indicated, hard-husked and/or fibrous fruits that require extensive mastication fit best with the microwear data. Small incisor breadths indicate an australopithecine specialization on small diameter food items (Rose, 1976).

If australopithecines were frugivores, as these craniodental data suggest, they could forsake arboreality only if they lived in a habitat where most fruits were found in bushes shorter than about 2 m. Is there evidence that australopithecines had forsaken arboreality? In a word, no.

Above the waist, australopithecines are intermediate in morphology between chimpanzees and humans. Chimpanzees are adapted to unimanual suspension (arm-hanging) and vertical climbing. The positional repertoire of *A. afarensis* might then be reconstructed as containing about half as much arboreal activity as chimpanzees. Curved, powerful fingers accord well with an arm-hanging adaptation. Short fingers suggest that the supports from which *A. afarensis* suspended themselves were of smaller diameter than those commonly used by extant chimpanzees. Small trees with small branches are indicated as the most common feeding sites for australopithecines.

Below the waist, australopithecines had short limbs, long, curved toes, a more gripping great toe than that of modern humans, and extraordinarily wide hips. Short lower limbs bring the most massive part of the body, the torso, closer to the substrate, thereby lowering the center of gravity. Wide hips, especially in combination with the cone-shaped torso, lowers the center of gravity by increasing the mass of the lower portion of the torso, essentially allowing internal organs to rest lower in the torso. Both these features increase arboreal competence by making balancing on branches easier. Among apes, gripping toes are frequently used to stabilize the body. The strong toes of *A. afarensis* may have been used to grip small twigs during arboreal bipedal gathering.

Smaller joint surface areas and a less robustly ligamented pelvis than in modern humans indicate that *A. afarensis* was more prone to fatigue or injury during powerful and sustained bipedal locomotion. Bipedal carrying, with its imposition of greater stress on the musculoskeleton, must have been less common than it is in modern humans (Jungers, 1988, 1991; Hunt, 1994a). *Australopithecus afarensis* probably engaged in less sustained bipedal walking and less bipedal carrying than modern humans do.

Bipedalism is neither a locomotor behavior nor strictly a terrestrial behavior among chimpanzees; over 90% of Mahale chimpanzee bipedalism was postural, and more than half of all their bipedal episodes were arboreal. Among Mahale chimpanzees, bipedalism was overwhelmingly a feeding posture; 80% of all bipedalism occurred during feeding. Other behaviors that have been hypothesized to have exerted selective pressure on pre-bipedal protohominids to become bipedal—searching for predators, threatening, hunting—each constituted <5% of bipedal episodes among Mahale chimpanzees.

Chimpanzees fed bipedally from short trees with small fruits. Bipedal standing allowed individuals to reach higher in trees when feeding terrestrially, to bring more fruits within their reach. It also allows faster gathering (Jolly, 1970; Jolly and Plog, 1987). By using both hands, a flow of fruits can be maintained that takes full advantage of chimpanzee chewing capacities. This is particularly important for small fruits, since it is picking, not chewing, that constrains feeding rate. Terrestrial gathering is practical only when feeding from short trees or bushes. Even the lowest branches of trees >15 m were too high for the fruit to be reached. In some trees (e.g., *Harungana*) chimpanzees held the tree lower with one forelimb while they fed with the other. The effect was a terrestrial arm-hanging/bipedalism.

Small trees elicited bipedalism when they were fed from arboreally as well, but for different reasons. The small diameter of the branches discouraged sitting as a collecting posture. The small twigs in these trees bent to vertical under the weight of chimpanzees, making the substrate impossible to sit on. Among the terminal branches of these trees chimpanzees used as many contact points as possible. By extending the hip and fully abducting the arm they maximized the reach so that a larger number of branches were within grasp.

Short-stride-length arboreal movement and terrestrial bipedal shuffling are advantageous for collecting fine-grained resources such as small evenly distributed fruits in small trees. Frequent short distance (i.e., ~1 m) travel is necessary when harvesting small fruits both in the trees and on the ground. Feeding sites are depleted quickly so that postures that allow a switch to locomotion with little energy cost are preferred (Wrangham, 1980). Sitting postures during feeding are often engaged in on small-diameter supports, yet a squatting shuffling (gelada-style) to an adjacent branch is impossible. When bearing weight, terminal branch supports are often bent to such a degree that they are a meter or more lower than otherwise. Movement to adjacent small branches can be accomplished more effectively with bipedal locomotion, since it allows a hindlimb to be raised to the as yet unweighted level of the adjacent branch. Bipedal locomotion on small substrates is much like walking on an extremely soft cushion. Legs must be raised very high with each step. It is not practical for a harvester to avoid the costs of small branch collecting by breaking off a branch and retreating to a more stable perch, since any one branch contains too little food to constitute a meal. The combination of greater stability and easier transition to locomotion makes bipedalism a favored collecting strategy among small branches common among smaller trees.

Among chimpanzees, large individuals spent more time on the ground, less time among the terminal branches, and fed from smaller trees. Because vertical climbing is disproportionately more expensive for large animals (Taylor et al., 1972), the most straightforward interpretation of these data is that large individuals must balance the benefits of climbing (obtaining high-quality foods) with the costs of vertical ascent (a larger energy expenditure per unit weight, compared to smaller individuals). Because climbing is more expensive for big chimpanzees, arboreal foods are less valuable. As large-bodied primates to begin with, we might expect that compared to other primates chimpanzees prefer to minimize climbing, and large chimpanzees make more compromises to minimize climbing than do small chimpanzees. This leads to the expectation that adult male australopithecines should forage on the ground most often, since their body weight was nearly double that of females (McHenry, 1991b). Females might be expected to be more arboreal.

7. CONCLUSION

The synthesis of chimpanzee ecology and australopithecine functional morphology yields a postural feeding hypothesis that suggests that australopithecines were semi-arboreal, postural bipeds that specialized on gathering small, hard-husked fruit in short-statured trees.

Large males hypothetically found it more advantageous to gather terrestrially, whereas females were about half as arboreal as chimpanzees. Short hind limbs and wide hips are splendid adaptations for tree movement because they lower the center of gravity, increase stability, and improve climbing mechanics.

Bipedalism is hypothesized to have evolved initially in conjunction with arm-hanging, as a feeding posture effective for collecting small diameter fruits from small trees. Refinements to allow efficient bipedalism in the pelvis and hind limb of *A. afarensis* suggest that their terrestrial locomotion was fully bipedal, albeit less efficient than the bipedalism of modern humans. This suggests that their mode of travel on the ground was fully upright bipedalism.

Chimpanzees tend to travel terrestrially, and to feed arboreally. The woodland and open-forest habitat suggested for australopithecines would have required even more terrestrial movement than is found among extant chimpanzees. The ecological morphology of australopithecines suggests that they traveled bipedally terrestrially, but fed arboreally using a number of chimpanzee-like suspensory behaviors in addition to sitting and bipedal standing.

ACKNOWLEDGMENTS

I am grateful to the National Science Foundation, the Wenner-Gren Foundation for Anthropological Reserarch, the California State University, Sacramento Foundation and the University of California, Davis for sponsoring the *Primate Locomotion - 1995* conference, out of which this paper grew. I thank the editors of this volume, E. Strasser, J. Fleagle, A. Rosenberger, and H. McHenry for inviting me to speak at the conference and to write this manuscript. I also thank the anonymous reviewers and the editors for their comments on the manuscripts, which have improved it substantially. Research was aided by the L.S.B. Leakey Foundation, NSF BNS-86-09869, Richard W. Wrangham, and Harvard University.

REFERENCES

Abitbol M (1987) Evolution of the lumbosacral angle. Am. J. Phys. Anthropol. *72*:361–372.

Berge C, and Kazmierczak J-B (1986) Effects of size and locomotor adaptations on the hominid pelvis: Evaluation of australopithecine bipedality with a new multivariate method. Folia Primatol. *46*:185–204.

Cheney DL, and Wrangham RW (1986) Predation. In BB Smuts, DL Cheney, RM Seyfarth, RW Wrangham, and TT Struhsaker (eds.): Primate Societies. Chicago: University of Chicago Press, pp. 227–239.

Deloison Y (1991) Les Australopithèques marchaient-ils comme nous? In Y Coppens and B Senut, (eds.): Origine(s) de la Bipédie chez les Hominidés, Cah. Paléoanthrop. Paris: Editions du CNRS, pp. 177–186.

Du Brul EL (1962) The general phenomenon of bipedalism. Am. Zool. *2*:205–208.

Fooden J (1964) Stomach contents and gastro-intestinal proportions in wild-shot Guinan monkeys. Am. J. Phys. Anthropol. *22*:227–232.

Gantt DG (1983) The enamel of Neogene hominoids: Structural and phyletic implications. In RL Ciochon and RS Corruccini (eds.): New Interpretations of Ape and Human Ancestry. New York: Plenum Press, pp. 249–298.

Ghiglieri MP (1984) The Chimpanzees of Kibale Forest. New York: Columbia University Press.

Glander KE (1978) Howling monkey's feeding behaviour and plant secondary compounds: A study of strategies. In GG Montgomery (ed.): The Ecology of Arboreal Folivores. Washington: Smithsonian Inst. Press, pp. 561–573.

Grine FE, and Kay RF (1988) Early hominid diets from quantitative image analysis of dental microwear. Nature. *333*:65–68.

Hamilton AC (1982) Environmental History of East Africa: A Study of the Quaternary. London: Academic Press.

Hladik A, and Hladik CM (1969) Rapports trophiques entre vegetation et primates dans la foret de Barro Colorado (Panama). Terre et Vie *23*:25–117.

Hunt KD (1989) Positional behavior in *Pan troglodytes* at the Mahale Mountains and Gombe Stream National Parks, Tanzania. Ph.D. Dissertation., University of Michigan.

Hunt KD (1991a) Positional behavior in the Hominoidea. Int. J. Primatol. *12*:95–118.

Hunt KD (1991b) Mechanical implications of chimpanzee positional behaviour. Am. J. Phys. Anthropol. *86*:521–536.

Hunt KD (1992a) Positional behaviour of *Pan troglodytes* in the Mahale Mountains and Gombe Stream National Parks, Tanzania. Am. J. Phys. Anthropol. *87*:83–107.

Hunt KD (1992b) Social rank and body weight as determinants of positional behavior in *Pan troglodytes*. Primates *33*:347–357.

Hunt KD (1994a) The evolution of human bipedality: Ecology and functional morphology. J. Hum. Evol. *26*:183–202.

Hunt KD (1994 b) Body size effects on vertical climbing among chimpanzees. Int. J. Primatol. *15*:855–865.

Hunt KD (1996) The postural feeding hypothesis: An ecological model for the evolution of bipedalism. S. Afr. J. Sci. *92*:77–90.

Hylander WL (1975) Incisor size and diet in anthropoids with special reference to Cercopithecidae. Science *189*:1095–1098.

Johanson DC, and Edey M (1981) Lucy: The Beginnings of Humankind. New York: Simon and Schuster.

Jolly CJ (1970) The seed-eaters: A new model of hominid differentiation based on a baboon analogy. Man. *5*:1–26.

Jolly CJ and Plog F (1987) Physical Anthropology and Archeology. New York: Knopf.

Jungers WL (1982) Lucy's limbs: Skeletal allometry and locomotion in *Australopithecus afarensis*. Nature. *297*:676–678.

Jungers WL (1988) Relative joint size and hominoid locomotor adaptations with implications for the evolution of hominid bipedalism. J. Hum. Evol. *17*:247–265.

Jungers WL (1991) A pygmy perspective on body size and shape in *Australopithecus afarensis* (AL 288–1, "Lucy"). In Y Coppens and B Senut (eds): Origine(s) de la Bipédie chez les Hominidés, Cah. Paléoanthrop. Paris: Editions du CNRS, pp. 215–224.

Kay RF (1981) The nut-crackers – A new theory of the adaptations of the Ramapithecinae. Am. J. Phys. Anthropol. *55*:141–151.

Kay RF (1985) Dental evidence for the diet of *Australopithecus*. Ann. Rev. Anthropol. *14*:15–41.

Kay RF, and Grine FE (1988) Tooth morphology, wear and diet in *Australopithecus* and *Paranthropus* from southern Africa. In F Grine (ed.): Evolutionary History of the Robust Australopithecines. Chicago: Aldine, pp. 427–447.

Kay RF, and Hylander WL (1978) The dental structure of mammalian folivores with special reference to Primates and Phalangeroidea (Marsupialia). In GG Montgomery (ed.): The Biology of Arboreal Folivores. Washington, D.C.: Smithsonian Inst. Press, pp. 173–191.

Kimbel WH, Johanson DC, and Rak Y (1994) The first skull and other new discoveries of *Australopithecus afarensis* at Hadar, Ethiopia. Nature *368*:449–451.

Kinzey WG (1976) Positional behavior and ecology in *Callicebus torquatus*. Yrbk. Phys. Anthropol. *20*:468–480.

Kinzey WG (1977) Diet and feeding behaviour of *Callicebus torquatus*. In TH Clutton-Brock (ed.): Primate Ecology. London: Academic Press, pp. 127–152.

Kinzey WG (1978) Feeding behaviour and molar features in two species of titi monkey. In DJ Chivers and J Herbert (eds.): Recent Advances in Primatology, Vol. 1. London: Academic Press, pp. 373–385.

Kummer B (1991) Biomechanical foundations of the development of human bipedalism. In Y Coppens and B Senut (eds): Origine(s) de la Bipédie chez les Hominidés, Cah. Paléoanthrop. Paris: Editions du CNRS, pp. 1–8.

Langdon JH, Bruckner J, and Baker HH (1991) Pedal mechanics and bipedalism in early hominids. In Y Coppens and B Senut (eds): Origine(s) de la Bipédie chez les Hominidés, Cah. Paléoanthrop. Paris: Editions du CNRS, pp. 159–167.

Latimer B (1991) Locomotor adaptations in *Australopithecus afarensis*: The issue of arboreality. In Y Coppens and B Senut (eds.): Origine(s) de la Bipédie chez les Hominidés, Cah. Paléoanthrop. Paris: Editions du CNRS, pp. 169–176.

Latimer B, and Lovejoy CO (1989) The calcaneus of *Australopithecus afarensis* and its implications for the evolution of bipedality. Am. J. Phys. Anthropol. *78*:369–386.

Latimer B, and Lovejoy CO (1990) Hallucal tarsometatarsal joint in *Australopithecus afarensis*. Am. J. Phys. Anthropol. *82*:125–133.

Latimer B, Lovejoy CO, Johanson DC, and Coppens, Y (1982) Hominid tarsal, metatarsal and phalangeal bones recovered from the Hadar formation: 1974–1977 collections. Am. J. Phys. Anthropol. *53*:701–719.

Leakey MG, Feibel CS, McDougall I, and Walker AC (1995) New four-million-year-old hominid species from Kanapoi and Allia Bay, Kenya. Nature *376*:565–571.

Lovejoy CO (1981) The origin of man. Science. *211*:341–350.

Lovejoy CO (1988) The evolution of human walking. Sci. Am. *259*:118–125.

Lovejoy CO, Johanson DC, and Coppens Y (1982) Hominid upper limb bones recovered from the Hadar Formation: 1974–1977 collections. Am. J. Phys. Anthropol. *57*:637–649.

McHenry HM (1986) The first bipeds: A comparison of the *A. afarensis* and *A. africanus* postcranium and implications for the evolution of bipedalism. J. Hum. Evol. *15*:177–191.

McHenry HM (1991a) First steps? Analyses of the postcranium of early hominids. In Y Coppens and B Senut (eds): Origine(s) de la Bipédie chez les Hominidés, Cah. Paléoanthrop. Paris: Editions du CNRS, pp. 133–141.

McHenry HM (1991b) Sexual dimorphism in *Australopithecus afarensis*. J. Hum. Evol. *20*:21–32.

McHenry HM (1992) Body size and proportions of early hominids. Am. J. Phys. Anthropol. *87*:407–431.

Peters CR (1987) Nut-like oil seeds: Food for monkeys, chimpanzees, humans and probably ape-men. Am. J. Phys. Anthropol. *73*:333–363.

Puech P-F, and Albertini H (1984) Dental microwear and mechanisms in early hominids from Laetoli and Hadar. Am. J. Phys. Anthropol. *65*:87–92.

Rak Y (1991) Lucy's pelvic anatomy: Its role in bipedal gait. J. Hum. Evol. *20*:283–290.

Robinson JT (1972) Early Hominid Posture and Locomotion. Chicago: University of Chicago Press.

Rodman PS (1984) Foraging and social systems of orangutans and chimpanzees. In PS Rodman and JGH Cant (eds.): Adaptations for Foraging in Nonhuman Primates: Contributions to an Organismal Biology of Prosimians, Monkeys, and Apes. New York: Columbia University Press, pp. 134–160.

Rose MD (1976) Bipedal behaviour of olive baboons (*Papio anubis*) and its relevance to an understanding of the evolution of human bipedalism. Am. J. Phys. Anthropol. *44*:247–261.

Rose MD (1984) Food acquisition and the evolution of positional behaviour: The case of bipedalism. In DJ Chivers , BA Wood, and A Bilsborough (eds.): Food Acquisition and Processing in Primates. New York: Plenum Press, pp. 509–524.

Rose MD (1991) The process of bipedalization in hominids. In Y Coppens and B Senut, (eds): Origine(s) de la Bipédie chez les Hominidés, Cah. Paléoanthrop. Paris: Editions du CNRS, pp. 37–48.

Rosenberg KR, and Trevathan W (1995) Bipedalism and human birth: The obstetrical dilemma revisited. Evol. Anthropol. *4*:161–168.

Sarmiento EE (1988) Anatomy of the hominoid wrist joint: Its evolutionary and functional implications. Int. J. Primatol. *9*:281–345.

Schmid P (1983) Eine Reconstrucktion des skelettes von A.L. 288–1 (Hadar) und deren konsequenzen. Folia Primatol. *40*:283–306.

Schmid P (1991) The trunk of the Australopithecines. In Y Coppens and B Senut, (eds): Origine(s) de la Bipédie chez les Hominidés, Cah. Paléoanthrop. Paris: Editions du CNRS, pp. 225–234.

Stern JT, Jr (1976) Before bipedality. Ybk. Phys. Anthropol. *19*:59–68.

Stern JT, Jr, and Susman RL (1983) The locomotor anatomy of *Australopithecus afarensis*. Am J. Phys. Anthropol. *60*:279–317.

Stoller M (1995) The Obstetric Pelvis and Mechanism of Labor in Nonhuman Primates. Ph.D. Dissertation, The University of Chicago.

Tague RG, and Lovejoy CO (1986) The obstetric pelvis of A.L. 288–1 (Lucy) J. Hum. Evol. *15*:237–255.

Tardieu C (1983) L'articulation du genou. Analyse morpho-fonctionelle chez les primates. Application aux hominides fossiles. Paris: Centre National de la Recherche Scientifique.

Taylor CR, Caldwell SL, and Rowntree VJ (1972) Running up and down hills: Some consequences of size. Science *178*:1096–1097.

Teaford MF (1985) Molar microwear and diet in the genus *Cebus*. Am. J. Phys. Anthropol. *66*:363–370.

Teaford MF (1994) Dental microwear and dental function. Evol. Anthropol. *3*:17–30.

Teaford MF, and Walker AC (1984) Quantitative differences in dental microwear between primate species with different diets and a comment on the presumed diet of *Sivapithecus*. Am. J. Phys. Anthropol. *64*:191–200.

Tobias PV (1980) A survey and synthesis of the African hominids of the late Tertiary and early Quaternary periods. In L Konigsson (ed.): Current Argument on Early Man. Oxford: Pergamon, pp. 86–113.

Tuttle RH (1975) Parallelism, brachiation and hominoid phylogeny. In WP Luckett and FS Szalay (eds.): The Phylogeny of the Primates: A Multidisciplinary Approach. New York: Plenum, pp. 447–480.

Tuttle RH (1981) Evolution of hominid bipedalism and prehensile capabilities. Phil. Trans. Roy. Soc. *292*:89–94.

Tuttle RH, Webb DM, and Tuttle NI (1991) Laetoli footprint trails and the evolution of hominid bipedalism. In Y Coppens and B Senut (eds.): Origine(s) de la Bipédie chez les Hominidés, Cah. Paléoanthrop. Paris: Editions du CNRS, pp.187–198.

van Schaik CP (1983) Why are diurnal primates living in groups? Behaviour *87*:120–143.

van Schaik CP and van Hooff J (1983) On the ultimate causes of primate social systems. Behaviour *85*:91–117.

van Schaik CP, van Noordwijk MA, Boer RJ de, Tonkelaar I den. (1983) The effect of group size on time budgets and social behaviour in wild long-tailed macaques. Behav. Ecol. Sociobiol. *13*:173–181.

van Schaik CP, van Noordwijk MA, Warsono B, and Sutriono E (1983) Party size and early detection of predators in Sumatran forest primates. Primates. *24*:211–221.

van Schaik CP, and van Noordwijk MA (1989) The special role of male *Cebus* monkeys in predation avoidance and its effect on group composition. Behav. Ecol. Sociobiol. *24*:265–276.

Wainwright PC, and Reilly SM (1994) Ecological Morphology: Integrative Organismal Biology. Chicago: University of Chicago Press.

Walker AC (1981) Diet and teeth—dietary hypotheses and human evolution. Phil. Trans. Roy. Soc. London B. *292*:57–64.

White TD, Johanson DC, and Kimbel WH (1981) *Australopithecus africanus*: Its phyletic position reconsidered. S. Afr. J. Sci. *77*:445–470.

White TD, Suwa G, and Asfaw B (1994) *Australopithecus ramidus*, a new species of early hominid from Aramis, Ethiopia. Nature *371*:306–312.

White TD, Suwa G, Hart WK, Walter RC, WoldeGabriel G, de Heinselin J, Clark JD, Asfaw B, and Vrba E (1993) New discoveries of *Australopithecus* at Maka in Ethiopia. Nature *366*:261–265.

Wrangham RW (1977) Feeding behaviors of chimpanzees in Gombe National Park, Tanzania. In TH Clutton-Brock (ed.): Primate Ecology. London: Academic Press, pp. 503–538.

Wrangham RW (1980) Bipedal locomotion as a feeding adaptation in gelada baboons, and its implications for hominid evolution. J. Hum. Evol. *9*:329–331.

21

TIME AND ENERGY: THE ECOLOGICAL CONTEXT FOR THE EVOLUTION OF BIPEDALISM

Robert A. Foley and Sarah Elton

Human Evolutionary Biology Research Group
Department of Biological Anthropology
University of Cambridge
Downing Street
Cambridge, CB2 3DZ England

1. INTRODUCTION

This paper is concerned with explanations for the evolution of bipedalism. Its general point is a very simple one—that the occurrence of bipedalism is context specific. The pattern of hominid evolution, as much as that of any other lineage, reflects the costs and benefits of the way an animal is structured and behaves, and this ratio is entirely dependent upon when and where it is occurring. Historically the context for bipedalism has been the general characteristics of the environment—savanna grasslands, open environments, patchy woodland versus forest. This remains important, but here we shall add a new consideration, that of time budgets, which provides a more specific ecological context for considering the energetics of bipedalism.

The paper will first discuss the various contexts that do need to be taken into account when considering the origins of bipedalism. It extends a model developed earlier (Foley, 1992), based on a cost-benefit analysis, which suggested that the advantages of bipedalism should be placed into the specific ecological context of the Pliocene hominids. That context was specifically the effect of more arid, seasonal and open environments with more patchy and dispersed resources. The major change predicted in the model was an increase in day range length, and that the energetics of bipedalism are directly linked to increased ranging area (Foley, 1992). In quantitative terms, it was suggested (Foley, 1992: Figure 5.3) that bipedalism allowed a 50 kg hominid to exploit a day range of 16 km for the same energy as a male chimpanzee (45 kg) uses for the maximally observed day range length of 10 km (Rodman, 1984), a result that Leonard and Robertson (1997) have recently replicated. The earlier model will be developed here to incorporate a new element,

Primate Locomotion, edited by Strasser *et al.*
Plenum Press, New York, 1998

time. Although the energetic advantages of bipedalism allow longer day ranges, the extension of a day range length has to occur in the context of a finite resource, that of time, and in particular hours of daylight (Foley, 1995). Time, as Dunbar (1992) has argued, is the hidden constraint in behavioral ecology. To model correctly the energetics of bipedalism in an ecological context it is therefore necessary to construct a time budget of daily activities. This is done here, with particular emphasis on how such activities are distributed terrestrially and arboreally. This distribution is important as it is not just the energetic advantages that accrue from being bipedal as a terrestrial primate, but also the costs of climbing. The model of daily energetics indicates that a very substantial part of the day has to be spent terrestrially before the benefits of bipedalism exceed the costs. In the final part of the paper, the implications for reconstructions of scenarios for the origins of bipedalism are discussed.

2. ADAPTATION AS PROBLEM-SOLVING: THE COSTS AND BENEFITS OF BIPEDALISM

Ecological context is highly specific, not just in terms of the habitat, but also the activities of the animal concerned. Two species of primate can live in the same environment but be morphologically different because their diets, activities and time budgets vary. The same trait could thus be both adaptive or maladaptive, depending upon circumstances, and no adaptation is for all times and places. The lack of context is perhaps the greatest disparity between what might be called the pre-modern synthesis and more recent approaches to hominid evolution. The essence of a neo-Darwinian approach is that traits are advantageous in particular circumstances, rather than in absolute terms. This implies that any feature of an organism, be it behavioral or morphological, will be an evolutionary response to precise settings of time and place. Or, to put it another way, that such traits are not only advantageous in particular environments, they are also disadvantageous in others. To explain the evolution of features such as bipedalism during the course of hominid evolution thus requires a detailed consideration of the contexts in which it is likely to occur.

Within modern evolutionary biology there are two broad conceptual frameworks that provide the basis for attempting to place adaptive evolutionary events into context. The first of these is what might be called the optimal problem-solving approach (Maynard Smith, 1978; Foley, 1987). Features evolve because they solve the problems of survival faced by organisms in particular environments, and selection is the evolutionary mechanism that leads to what might be called problem-solving optima. This approach to a wide ranging series of adaptive problems in hominid evolution has been developed extensively (Foley, 1987). Complementary to this idea is that of cost-benefit algorithms. Adaptations may be treated as benefits to an organism. All such adaptive traits, however, impose costs on an organism. These costs may be associated with development or maintenance; they may be energetic or structural; they may be behavioral. They may be direct costs or opportunity costs—that is, costs that come into account because the evolution of one feature will reduce the opportunities for other evolutionary changes. Cost-benefit analysis has proved to be very productive as a means of analyzing evolutionary strategies, particularly in terms of foraging behavior, where direct observation and measurement of energetic and time inputs and outputs is possible. Both problem-solving and cost-benefit analysis are, in effect, ways of making more practicable the basic Darwinian premise that features that develop over the course of evolution, features that are selected for, provide specific advantages—adaptations—in particular contexts.

Bipedalism is a classic example. Among mammals bipedalism is very rare, and as a specialized adaptation in primates occurs uniquely in hominids. The changes in the musculoskeletal system that result from bipedalism are very extensive, and it has been argued that bipedalism is the basis for many other non-locomotory changes in hominid evolution. The benefits that arise from bipedalism range from increased manual dexterity, enhanced neural and general thermoregulation, decreased locomotor energetic expenditure, greater anti-predatory vigilance, intra-specific display, etc. Such an array of benefits can at times seem at odds with the rarity of bipedalism among primates. Only when the costs of bipedalism are considered—that is, the disadvantages—does it become clearer why bipedalism may both have evolved among hominids and, which is equally important, not evolved among other species. The costs, in turn, are as context dependent as the benefits.

Discussions of bipedalism have generally focused on two types of advantage that may accrue. The first of these is the direct locomotor advantage, that is, the energetic benefits that arise from being able to walk upright habitually. There has been considerable debate as to whether these advantages exist (Taylor and Rowntree, 1973; Rodman and McHenry, 1980; Steudel, 1994; Isbell and Young, 1996) and, as a result, a second stream of explanations relate to the non-locomotory advantages. In particular, it has been argued that bipedalism leads to greater manual dexterity (Darwin, 1871), vigilance, and thermoregulatory efficiency (Newman, 1970; Wheeler 1984, 1985, 1991). These indirect advantages will not be incorporated here. Instead, in modeling the direct energetic costs and benefits two novel themes in particular will be developed. The first is that at least some of the costs imposed by bipedalism arise from a reduced ability to climb. The second is that bipedalism, as with any locomotor behavior, occurs in the context of a daily schedule of activities, and models should therefore be built around such a time-budgeting set of parameters.

Application of cost benefit analysis to problems in long term evolution is bound to be more problematic than analyses of living populations. Direct observation in the field or in the laboratory of actual energetic costs is obviously an important part of any such analysis, and with animals long extinct this is not possible. The alternative that will be used here will be to build a model using data and observations drawn from living species—in particular, chimpanzees and humans. The limitations of such an approach should be kept in mind.

3. THE CONTEXTS OF BIPEDALISM

Several contexts for assessing the costs and benefits of bipedalism can be described.

3.1. Taxonomic Context

There is now a relatively rich fossil record for hominids within Africa covering the Plio-Pleistocene, representing a number of distinct taxa. The evidence they provide illustrates the diversity of hominid locomotor strategies.

3.1.1. Ardipithecus ramidus. This is the earliest known hominid (> 4.4 Myr), for which only limited postcranial material has yet been published (White et al., 1994). On the basis of described material it cannot be ascertained whether *A. ramidus* was more bipedal than extant apes, and it may be that this taxon either does not belong within the hominid lineage or that it represents a form prior to any major changes in locomotor anatomy.

3.1.2. Australopithecus anamensis. This hominid is also over 4.0 Myr, and has a relatively primitive cranial morphology (Leakey et al., 1995). The associated tibia, however, has elements that link it with other early hominids, which may indicate some level of bipedalism.

3.1.3. Australopithecus afarensis. This is the most extensively known of the early hominids; two lines of evidence indicate that it is a relatively bipedal hominid. The postcranium of AL-288, especially the pelvis, indicates more habitual bipedalism (Lovejoy, 1979), while the footprint trail at Laetoli also suggests a bipedal hominid. These observations lead to the conclusion that at least some level of bipedalism was present in the early hominids by shortly after 4.0 Myr, but there is considerable debate as to whether this is universal among hominids, is fully established, precludes other locomotor behavior, and is different from that found in modern humans (Lovejoy, 1980; Jungers, 1982; Senut and Tardieu, 1985). The most likely consensus is that some hominids were active bipedally on the ground for some of the time, but that their overall body size and shape still shared a number of features with African apes (see papers by Hunt and Tuttle et al., this volume).

3.1.4. Australopithecus africanus. In overall morphology this taxon is likely to be relatively similar to or less like that of modern humans than *A. afarensis*. Interpretation has been strongly influenced by the complete innominate Sts14, which shows clear similarities with later bipedal hominids. A recent discovery of a partial foot showing a divergent hallux has led to some questioning of the extent of bipedalism in *A. africanus,* although this position is complicated by uncertainty concerning the taxonomic integrity of specimens assigned to this taxon (Clarke, 1985).

3.1.5. Australopithecus robustus and Allies. There is little doubt that these more robust and later (< 2.7 Myr) taxa are at least as bipedal as the earlier australopithecines, and many elements of their dietary adaptations have indicated a predominantly terrestrial way of life (Susman and Brain, 1988; Grine, 1989).

3.1.6. Homo habilis. The evidence of the Olduvai specimens has generally led people to suppose that early *Homo* was fully bipedal, but this position has been called into question to some extent by reinterpretations of the OH36 ulna and the postcranial anatomy of OH62, both of which are relatively ape-like in morphology (Aiello and Dean, 1990).

3.1.7. Homo ergaster. The partial skeleton WT15000 from West Turkana provides the most complete evidence for hominid postcrania for the Plio-Pleistocene, and although there are a number of small differences, nonetheless it is clear that by 1.6 Myr at least one lineage, as represented by *H. ergaster*, is fully adapted to bipedalism in ways that are not dissimilar to modern humans (Walker and Leakey, 1993).

In overall terms, therefore, the paleontological evidence shows that the context for considering the evolution of bipedalism lies during the Plio-Pleistocene in sub-Saharan Africa. At the beginning of the Pliocene, the remains from Aramis might be taken to indicate essentially non-bipedal or partially bipedal hominids. The early australopithecines are best thought of as being ape-like in morphology, but with a considerably greater set of adaptations for bipedalism than the living African apes. The later australopithecines are probably fully terrestrial and bipedal, but with some adaptive differences from later *Homo*, while early *Homo* itself shows a mixed suite of adaptations. Full human bipedalism was present by 1.6 Myr. It should be stressed that the recognition of multiple taxa changes the way we should look at bipedalism. It is likely that during the Pliocene, when the hominids were diversifying, so too was their locomotor behavior. The presence of evidence for bipedalism in one taxon does not necessarily imply that it was universal among all.

3.2. Paleoenvironmental Context

The paleoenvironmental context of these hominids has itself been subject to considerable discussion. An earlier consensus that all the hominids existed in relatively open environments has been somewhat eroded in recent years (see Reed, 1997, for a recent summary). Earliest hominid sites from the Middle Awash and Hadar have all been interpreted as showing at least some tree cover, and recently the same interpretation has been applied to the southern African sites. The earliest hominids are now considered to have occupied habitats that were not fully forest, but would certainly have contained at least some tree cover. The amount of tree cover is likely to have varied considerably from location to location and through time. There is little doubt that some hominids after 2.0 Myr were living in fully open, semi-arid environments. Overall, the paleoenvironmental context for the evolution of bipedalism is thus likely to have been a mixed tropical African environment: savanna in the sense that grass would have been extensive, but also with significant levels of tree and bush cover, and in all likelihood in relatively close proximity of water (see Table 1). These environments are likely to have resources that were highly seasonal in distribution, patchily and unevenly distributed, and occurring both in trees and on the ground. In addition, the level of thermal stress is likely to have been considerable.

3.3. Phylogenetic Context

All evolutionary change is a function of the interaction between the existing components of a species and its current environment. Initial phylogenetic conditions thus play an important role in determining the nature of evolutionary change, for the existing phenotype provides the framework for the level of costs and benefits imposed by any particular strategy. Phylogenetic context is almost certainly one of the reasons why baboons and hominids have such divergent locomotor strategies despite very similar habitats, for the costs and benefits of changing the existing phenotype would have been very different (Foley, 1987). With regard to bipedalism, there is a great deal of uncertainty about what might be the phylogenetic starting conditions. There is virtually a complete absence of ap-

Table 1. Reconstructions of the environments of early hominid taxa in Africa. Reconstructions based primarily on Reed (1997)

Taxon	Habitats
A. ramidus	Woodland
A. anamensis	?
A. afarensis	Closed to open woodland, with edaphic grassland/shrubland/deltaic floodplain/water and trees present
A. africanus	Woodland and bushland with riverine forest
A. aethiopicus	Bushland to open woodland/edaphic grassland
A. robustus	Open or wooded-bushed grassland/edaphic grassland
A. boisei	Woodland to scrubland/edaphic grassland
H. habilis + H. rudolfensis	Open grasslands/dry shrublands/edaphic grasslands
H. ergaster	Shrubland/riparian woodland/arid open landscapes

propriately situated fossil material, so what is known is largely estimated from modern humans, later fossil hominids and extant apes. It has been proposed that the last common ancestor of the hominids and African apes was variously a knuckle-walker, brachiator or generalized clamberer/climber (see Hunt, this volume, for a discussion of this problem). While it is difficult to test for these various models, it would seem more probable that the ancestors of the first hominids either approximated the more generalized locomotor behavior of *Pan* or were even less specialized.

3.4. Behavioral Context

Bipedalism is usually treated as an anatomical problem by paleoanthropologists, but at heart the problem is behavioral. Bipedalism is a means to an end, that end being the avoidance of predators, the acquisition of mates and food, and the balancing of an overall energy budget. Bipedalism will evolve when the costs of moving in this way are less than the benefits. The key context is, therefore, the way in which an animal spends its day; time is the 'hidden constraint in primate behavioral ecology' (Dunbar, 1992). It is essential to realize that the key context in which bipedalism must be assessed is how a hominid would have budgeted its day under different ecological, phylogenetic and energetic contexts. Dunbar (1992) has developed a model in which a primate day is partitioned up into four activities—feeding, traveling, resting and socializing. Each of these is essential for the satisfactory functioning of a population. Feeding and traveling time are the highest priority, but there are also limits on how much resting and socializing time can be varied in a typical activity budget. For example, among savanna baboons the range of time budgets is: feeding = 23–56%, traveling = 17–36%, resting = 5–60%, and socializing = 22–57%. Teleki (1989, quoted in Williamson, 1997) gives for chimpanzees from Gombe a feeding time of 42.8%, traveling time of 13.4%, resting time of 18.9%, and 24.9% socializing time. Williamson (1997) gives for both species of *Pan* and for different environments ranges of 29.7–67% for feeding, 13–27.5% for traveling, and 30–43% for resting/socializing. When considering the energetics of bipedalism, it is for this scale of activity budget that energetic efficiency has to be considered.

The background for modeling the costs and benefits of the evolution of bipedalism can thus be approximated as that of a generalized ape living in a relatively well-treed environment in Africa. This ape would be capable of both arboreal and terrestrial foraging and movement, and would have to acquire food efficiently across a patchy and hot environment. The problem is to model the energetics of such an animal under different ecological conditions, to determine when it may pay to become a more 'committed biped'. The novel element is placing this setting into the framework of a daily time budget such as would be expected for a typical social primate.

4. THE MODEL

There are two components to the model, one related to time budgeting and one to energetics. The first component is a partitioning of a 12 hour day into percentage time spent feeding and traveling. For baboons and chimpanzees these two activities may account for up to half of a day (Dunbar, 1992; Isbell and Young, 1996; Williamson, 1997). As feeding and traveling are the most energetically expensive activities, and the most affected by locomotor abilities, the time spent on these activities is varied in the model. Both feeding time and traveling time are varied between a minimum of 20% and a maxi-

Table 2. Parameters used in the model, showing range of values and coefficients applied in the simulations

Parameter	Organism	Value/range
Length of day (tropical environments)		720 min.
Body weight		30 kg
Feeding time range[1]		20–40%
Traveling time range[1]		20–40%
Resting/socializing time range[1]		40–60%
Energetic costs of feeding on ground (kJ/min)[2]	modern human	5.15
	early biped	5.67
	quadrupedal ape	6.44
Energetic costs of feeding in trees (kJ/min)[3]	modern human	9.41
	early biped	9.41
	quadrupedal ape	7.06
Energetic costs of traveling on ground (kJ/min)	modern human[4]	10.29
	early biped[5]	11.32
	quadrupedal ape[6]	12.86
Energetic costs of traveling in the trees (kJ/min)	modern human[7]	18.81
	early biped[7]	18.81
	quadrupedal ape[8]	14.11
Energetic costs of resting and socializing activities (all habitats and species) (kJ/min)[9]		2.91

[1]Time ranges taken from Dunbar (1992).
[2]Energetic cost of feeding on ground estimated at half the terrestrial traveling rate.
[3]Energetic cost of feeding in trees estimated at half the arboreal traveling rate.
[4]Energetic cost of traveling on the ground calculated using data obtained Elton et al. (in press; see Table 3).
[5]Energetic cost of traveling on the ground calculated as 10% more costly than for the modern human.
[6]Energetic cost of traveling on the ground calculated as 25% more costly than for the modern human.
[7]Energetic cost of traveling in the trees calculated using data from Elton et al. (in press; see Table 3).
[8]Energetic cost of traveling in the trees estimated as 75% of the modern human cost.
[9]Resting and socializing costs estimated using the modern human standing cost, using data from Elton et al. (in press; see Table 3).

mum of 40% (Dunbar, 1992; Williamson, 1997; Table 2). A control value of 25% for feeding time and 20% for traveling time is used when the other is being varied. A case can be made that social behavior may also be sensitive to pattern of locomotion (see, for example, Jablonski and Chaplin, 1993), but this is not pursued here. A further constraint on the model is that it only explores strategies that reduce the costs of locomotion in particular contexts; ways in which energetic benefits through access to additional resources may accrue have not been considered, although a case may be made that this was a significant factor in the evolution of bipedalism (Hunt, 1996).

The second component of the model is the energetics of the hominids and chimpanzees as a comparative animal for a quadrupedal ape. There are considerable data available on human energy expenditure (see Ulijaszek, 1995, for a recent summary). There are virtually none available on the energy costs of climbing, however, and therefore an experimental study was carried out on the energetics of standing (which is used here as a surrogate for resting/socializing), walking and climbing in a mixed sex sample (see Elton et al., in press, for details; Table 3). Data on chimpanzee energy expenditure are scarce, most of which returns to an early study by Taylor and Rowntree (1973) on a juvenile chimpanzee. There are no data on climbing costs in chimpanzees, although Caldwell et al. (1972) found that moving vertically was around double the energetic costs of traveling horizontally.

Table 3. Energetics of the activities used in the model. These estimates
are derived from a mixed sex study involving controlled
exercise (Elton et al., in press)

Model	Standing kJ/min/kg	Walking kJ/min/kg	Climbing kJ/min/kg
Energy per kg	0.097	0.343	0.627
Small biped (30 kg)	2.91	10.29	18.81

In the absence of unequivocal data the following estimates were used. Standing, walking and climbing costs were calculated in kJ per minute per kg using the results of the study by Elton et al. (in press; Table 3). Energy expenditure during feeding time was calculated at half the walking rate for terrestrial feeding, and half the climbing rate for arboreal feeding. This rate was selected to reflect the probability that feeding would involve some movement on whatever substrate was being used. Energy expenditure during traveling time was estimated using the walking rate for terrestrial travel, and the climbing rate for arboreal travel. Energy expenditure during resting and socializing time was estimated for modern humans using the rates for energy expenditure whilst standing. Overall daily (12 hour) energy expenditure is the sum of time spent in each activity. In the model, the percentage of time spent on the ground was varied from 20% to 100%. This affected overall energy expenditure as traveling and feeding time spent in the trees were more energetically expensive than that on the ground.

Three types of 'creature' were modeled: one that basically has the same rate of expenditure as a modern human, using the rates described above; a less efficient, early biped; and, a chimpanzee-like quadruped. The climbing costs of the less-efficient biped were set at the "modern human" rate, and the cost of bipedalism in this creature was 10% more than that of the modern human (Table 2). The chimpanzee-like quadruped's terrestrial efficiency was less than that of a biped, following Rodman and McHenry's (1980) results showing that chimpanzees are around a third less efficient on the ground than are humans. A more conservative estimate of a 25% increase in energy expenditure was used here. Conversely, it was assumed that this creature would have been more efficient in the trees than a modern human, so the climbing rate for the chimpanzee was estimated at 75% of the modern human rate. In the model described here a body size of 30 kg was used. This is a relatively small body size, and is appropriate for some of the very earliest hominids and for chimpanzee females. Increased body size would, in these models, not affect the outcome other than by increasing overall energy expenditure across all model creatures (but see Steudel, 1994, for a discussion of the allometric factors relating to locomotion in primates). The main parameters of the model are shown in Table 2.

4.1. Results

Applying the model to the creatures described above shows how differences in activity patterns affect energy expenditure, and these can be used to explore the effects of time budgets on the costs and benefits of bipedalism.

4.1.1. Feeding Time. As discussed above, in practical terms, using baboon and chimpanzee analogues, a large social primate can expect to feed for about 20% of its time when conditions are relatively good and to increase feeding time to as much as 40% when food is either of poor quality, takes time to process, or is hard to find. Figure 1 shows the effect

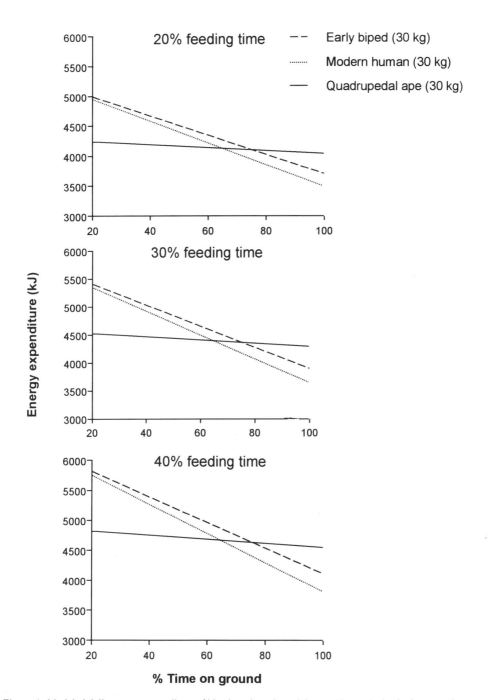

Figure 1. Modeled daily energy expenditure of bipeds and quadrupedal apes. The vertical axis shows total energy expenditure across all activities (feeding, traveling, resting, socializing). The horizontal axis shows increasing percentage of time spent on the ground. As time spent terrestrially increases, overall energy expenditure decreases, but that of bipeds at a faster rate. All graphs use 20% travel time. The top graph shows the situation when 20% of the time is spent feeding (relatively good conditions); the middle graph when 30% is spent feeding, and the bottom when 40% is spent feeding (very harsh conditions).

of increasing feeding time for the three model creatures, with traveling time held constant at 20%. As can be seen, for all three (modern humans, early bipeds, and chimpanzee-like quadrupeds), the costs of foraging decrease with increasing time spent on the ground. This reflects the greater costs of arboreal feeding in the model. When most of the feeding (and proportionately, travel) is done in the trees, then the chimpanzee has a considerable energetic advantage. As the percentage of time spent on the ground increases the gap between the bipeds and the quadrupeds closes, and eventually the bipeds have an energetic advantage. From the point of view of the context for the evolution of bipedalism, the question is, at what stage does the cross-over occur. The answer is when more than 60% of activities are spent on the ground for a modern biped, and more than 70% for a less efficient biped. This is the case for a relatively inactive biped (20% feeding time), and increasing the amount of feeding time raises the overall levels of energy expenditure, but has little effect on the point at which the cross over occurs relative to an equally active quadruped. Even a less efficient biped will have an energetic advantage when the time it spends on the ground exceeds 75%.

4.2. Traveling Time

The effects in relation to increasing traveling time are similar (Figure 2). The transition from an advantage to quadrupeds to an advantage to bipeds again occurs when over 60% time is spent on the ground for modern human levels of efficiency, and at 75% for a less efficient biped. Similarly, as travel time increases (now holding feeding time constant at 20%), overall levels of energy expenditure rise for all model creatures.

4.3. Integrated Patterns

What these results indicate is that, other things being equal, the critical zone, in terms of time budgets, for the advantage shifting from a quadruped to a biped occurs when terrestrial activity lies between 60 and 80% (Figure 3). Bringing in other factors may bring this value down, but the transition point does appear to be relatively stable for the range of daily activities a typical social primate is likely to employ.

5. DISCUSSION

It should be remembered that these are models based on limited energetic data and a firm application of the principle of uniformitarianism; nonetheless, some interesting implications arise from considering bipedalism in the context of time budgets.

The first of these is that the results hinge on there being an advantage to bipedalism as a mode of locomotion on the ground, and an increased cost when arboreal activity is involved. The first of these has been relatively well documented (Taylor and Rowntree, 1973; Rodman and McHenry, 1980; Foley, 1992; Steudel, 1994; Leonard and Robertson, 1997). Climbing for a biped is energetically expensive, but at this stage we do not know whether this is very much greater than for a chimpanzee-like animal. Biomechanical principles would seem to indicate that chimpanzees would be able to climb more efficiently, and they certainly have major advantages in terms of speed, opportunity costs, and risk of accidents. Given as an assumption the additional climbing costs associated with bipedalism, it is interesting, if not unexpected, to note that higher levels of terrestrial activity promote bipedalism.

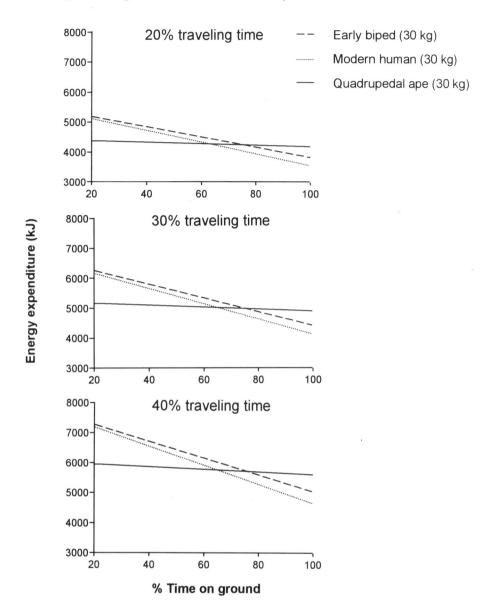

Figure 2. Modeled daily energy expenditure of bipeds and quadrupedal apes. The vertical axis shows total energy expenditure across all activities (feeding, traveling, resting, socializing). The horizontal axis shows increasing percentage of time spent on the ground. As time spent terrestrially increases, overall energy expenditure decreases, but that of bipeds at a faster rate. All graphs use 25% feeding time. The top graph shows the situation when 20% of the time is spent traveling (relatively good conditions); the middle graph when 30% is spent traveling, and the bottom when 40% is spent traveling (very harsh conditions).

The significance of the model lies in the fact that when time budgets are considered, some level of quantification is possible. A time budget sets constraints on the level of energetic advantage that is required. Although other things may have varied in the past in ways that we cannot comprehend today, ultimately all diurnal primates are limited by what they can achieve in a twelve hour period of daylight. A social primate must spend time maintain-

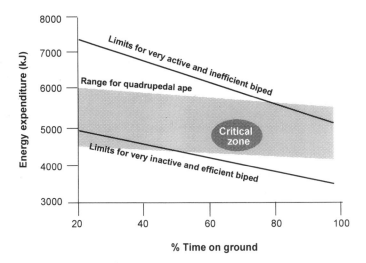

Figure 3. Summary graph. The upper limit for the quadrupedal ape is set by a large male chimpanzee under conditions of high energy expenditure, the lower limit by a small female chimpanzee under low energy expenditure conditions. The two lines for the biped set the outer limits to energy expenditure. The critical zone, derived from the model discussed in text, is where the switch to bipedalism is most likely to be beneficial.

ing social relationships, and an overly active animal is likely both to run into energy deficit and to increase the probability of accidents and predation, and hence there are limits to the amount of time that can be spent feeding and traveling. Those limits set the real ecological constraints within which the costs and benefits of bipedalism should be considered.

This model specifies a link between the habitat and the behavior of early hominids that may go some way towards resolving the discussions about bipedalism and habitat. The clearest result shown here is that bipedalism does have a significant advantage over quadrupedalism, but only when a very considerable amount of time is spent on the ground. As long as the ancestral populations were feeding and traveling in the trees for as much as 40% of the time, then bipedalism will be more of a cost than a benefit. In other words, it would take a predominantly terrestrial life to lead to bipedalism, unless other factors came into play. Furthermore, traveling time has a greater affect than feeding time on the switch to bipedalism with increased terrestriality, and thus if traveling time is high and feeding time low, then bipedalism might develop if most of the traveling is on the ground, as might be the case in scattered woodland.

What implications does this model have for our thinking on the evolution of bipedalism among Plio-Pleistocene hominids? The first is that the major change is a behavioral one. The strategy of feeding on the ground is likely to be independent of bipedalism for a considerable period of time. As long as the ancestral populations were still feeding at least half the time in the trees, then bipedalism will be more of a cost than a benefit. By the time bipedalism does evolve then it is likely that the hominids would have been well established as very competent terrestrial primates. Although evolutionary change can occur very rapidly, it is likely that a long history of terrestrial activity may precede the first evidence of bipedalism in the fossil record.

The second implication is that it is traveling time that is critical here. The model did not explore in detail minor variations in time spent differentially on the ground when feeding or traveling, but nonetheless it is the traveling time that is likely to be energetically critical. What might be significant is not the relative amounts of time, but the overall increase in the percentage of time spent both feeding and traveling at the expense of resting and socializing time. Dunbar (1992) has shown that in hotter and drier environments baboons will spend more time traveling, and that this is time taken from that available for resting or socializing. Thus, a hominid living in a more arid environment would spend more time traveling in search of food and traveling between dispersed patches. Overall, these populations are likely to be more active and to have larger day ranges than forest dwelling apes. The hypothesis that the energetic advantages of bipedalism are brought into play primarily by increased day range lengths was proposed by one of us (Foley, 1991), and has been supported by some recent reanalyses (Leonard and Robertson, 1997). These time budget models support the hypothesis that the energetic advantages of bipedalism are especially important in the context of increased day range lengths (Foley 1992), when time stress can become a critical problem. It is interesting to note that a maximum day range length of around 16 km arises from the day length model (Foley, 1992), the time budget model presented here, and Leonard and Robertson's (1997) analyses.

A third implication of some importance is that while bipedalism may be the primary observable change at the divergence of hominids from the other African apes, it may be a consequence of another and equally significant change. Terrestriality is less energetically costly than arboreality, but it still takes time to forage and, furthermore, food is likely to be more widely dispersed. The key change might therefore be the ability to spend more time in feeding and traveling behavior, in order to be generally more active across the course of a day (Foley, 1995). This higher level of activity is the sort of pressure that might also integrate well with the idea that early hominids have very high thermoregulatory costs and that thermoregulation has played a major role in shifting hominid anatomy and physiology (Wheeler, 1984, 1985, 1991a,b).

Finally, it may be worthwhile returning to the larger time scale of hominid evolution to discuss the locomotor and paleoenvironmental evidence for Plio-Pleistocene hominids in the light of these time budget models. As discussed earlier, the very early Pliocene hominid, *Ardipithecus ramidus*, has not been published in sufficient detail to determine its locomotor repertoire, but there is convincing evidence that the other three taxa of early hominids—*A. anamensis, A. afarensis,* and *A. africanus*—were neither fully bipedal like modern humans, nor exhibited chimpanzee-like quadrupedalism. They perhaps approximated the inefficient small-bodied biped modeled here. If so, it can be inferred that they were likely to have been spending at least 65% of their time on the ground. It has been argued that they may have been living in substantially wooded environments, but while this may have been the case, it is unlikely that this reflects specialized arboreality. Put the other way, however, the level of bipedalism found in the early australopithecines is consistent with as much as 35% of daily activity involving time spent in the trees. Later australopithecines, with greater robusticity and megadontic specializations, in addition to relatively good evidence associating them with more open habitats, are, on the basis of these models, expected to be specialist bipeds, a conclusion consistent with the known anatomical evidence. *Homo ergaster* is anatomically fully bipedal, and it would be inferred to have been a complete terrestrial specialist with a time budget that would reflect this. Perhaps the greatest remaining area of uncertainty is whether early representatives of *Homo*, living around 2.5 Myr, would have still retained the pattern found in the earlier australopithecines, or whether they already possessed the derived condition found in both later *Homo* or the robust australopithecines. Certainly models of land use at Olduvai, which have been linked to *Homo habilis*, imply a terrestrial way of life (Peters and

Blumenschine, 1995), but, as this paper has shown, terrestriality may be more widespread than bipedalism.

6. CONCLUSIONS

Simulation models should always be treated with great caution, for they simulate our hypotheses, not the past. Nonetheless, the model developed and discussed here does perhaps provide some new insights into the evolution of bipedalism. Stress was laid on the importance of context, for it is only in relation to very specific contexts that the costs and benefits of any adaptation can be assessed. It was argued that the context for bipedalism was ecological and, in particular, the way in which extinct hominoid populations deployed their activities across a day—in other words, their time budget. By specifying the time budget for a twelve hour set of activities it was possible to take into account not just the benefits of bipedalism, but also the costs of losing climbing ability.

The time and energy model examined how the percent of time spent on the ground influenced the adaptive value of bipedalism in relation to the relative and absolute amount of time spent feeding and traveling. The principal conclusion drawn was that at least 60% of daily activities would have to be spent terrestrially before the energetic advantages of bipedalism outweighed the loss of climbing ability. This result implies that, other things being equal, hominids may well have had a long ancestry of terrestrial activity prior to the evolution of bipedalism. Extensive foraging on the ground, and traveling larger distances with greater day ranges, even in relatively closed habitats, may have been the ecological heritage of the first hominids, and the essential pre-requisite for successful and ultimately bipedal adaptation to drier and more open environments.

ACKNOWLEDGMENTS

We thank P.C. Lee for comments, Charles Fitzgerald for help with the computer models, and RAF is grateful to Henry McHenry and Elizabeth Strasser for the invitation to contribute to the conference in Davis. Elizabeth Strasser, Kevin Hunt and a number of anonymous reviewers provided helpful comments on an earlier draft.

REFERENCES

Aiello LC, and MC Dean (1990) An Introduction to Human Evolutionary Anatomy. London: Academic Press.
Clarke RJ (1985) *Australopithecus* and early *Homo* in southern Africa. In E Delson (ed.): Ancestors: the Hard Evidence. New York: Alan Liss, pp. 171–177.
Darwin C (1871) Descent of Man and Selection in Relation to Sex. London: Murray.
Dunbar RIM (1992) Time: A hidden constraint on the behavioral ecology of baboons. Behav. Ecol. and Sociobiol. *31*:35–49.
Elton S, Foley RA, and Ulijaszek SJ (in press) How much does it cost a human biped to climb and clamber? Ann. Hum. Biol.
Foley RA (1987) Another Unique Species: Patterns of Human Evolutionary Ecology. Harlow: Longman.
Foley RA (1992) Evolutionary ecology of fossil hominids. In EA Smith and B Winterhalder (eds.): Evolutionary Ecology and Human Behavior. Chicago: Aldine de Gruyter, pp. 131–164.
Foley RA (1995) Humans Before Humanity: An Evolutionary Perspective. Oxford: Blackwells Publishers.
Grine FE, Editor (1989) The Evolutionary History of the "Robust" Australopithecines. Chicago: Aldine de Gruyter.

Hunt KD (1996) The postural feeding hypothesis - an ecological model for the evolution of bipedalism. S. Afr. J. Sci. 92:77–90.

Hunt KD (this volume) Ecological morphology of Australopithecus afarensis: Traveling terrestrially, eating arboreally. In E Strasser, JG Fleagle, AL Rosenberger, and HM McHenry (eds.): Primate Locomotion: Recent Advances. New York: Plenum Press, pp. 397–418.

Isbell LA, and Young TP (1996) The evolution of bipedalism in hominids and reduced group-size in chimpanzees - alternative responses to decreasing resource availability. J. Hum. Evol. 30:389–397.

Jablonski NG, and Chaplin G (1993) Origin of habitual terrestrial bipedalism in the ancestor of the Hominidae. J. Hum. Evol. 24:259–280.

Jungers WL (1982) Lucy's limbs: Skeletal allometry and locomotion in Australopithecus afarensis. Nature 297:676–678.

Leakey MG, Feibel CS, McDougall I, and Walker A (1995) New four million year old hominid species from Kanapoi and Allia Bay, Kenya. Nature 376:565–571.

Leonard WR, and Robertson ML (1997) Rethinking the energetics of bipedalism. Curr. Anthropol. 38:304–309.

Lovejoy CO (1979) A reconstruction of the pelvis of AL-288 (Hadar Formation). Am. J. Phys. Anthropol. 50:460.

Lovejoy CO (1980) Hominid origins: The role of bipedalism. Am. J. Phys. Anthropol. 52:250.

Maynard Smith J (1978) Optimization theory in evolution. Ann. Rev. Ecol. Syst. 9:31–56.

Newman RW (1970) Why man is such a sweaty and thirsty naked animal: A speculative review. Hum. Biol. 42:12–27.

Peters CR, and Blumenschine R (1995) Landscape perspectives on possible landuse patterns for early hominids in the Olduvai Basin. J. Hum. Evol. 29:321–362.

Reader J (1988) Missing Links. London: Pelican.

Reed KE (1997) Early hominid evolution and ecological change through the African Plio-Pleistocene. J. Hum. Evol. 32:289–322.

Rodman PS (1984) Foraging and social systems of orangutans and chimpanzees. In PS Rodman and JGH Cant (eds.): Adaptations for Foraging in Non-Human Primates. New York: Columbia University Press, pp. 134–160.

Rodman PS, and McHenry HM (1980) Bioenergetics and origins of bipedalism. Am. J. Phys. Anthropol 52:103–106.

Senut B, and Tardieu C (1985) Functional aspects of Plio-Pleistocene hominid limb bones: Implications for phylogeny and taxonomy. In E Delson (ed.): Ancestors: The Hard Evidence. New York: Alan Liss, pp. 193–201.

Steudel K (1994) Locomotor energetics and hominid evolution. Evol. Anthropol. 3:42–48.

Susman RL, and Brain TM (1988) New first metatarsal (SKX5017) from Swartkrans and the gait of Paranthropus robustus. Am. J. Phys. Anthropol 77:7–16.

Taylor CR, and Rowntree VJ (1973) Running on two legs or four: Which consumes more energy. Science 179:186–187.

Tuttle RH, Hallgrímsson B, and Stein T (this volume) Heel, squat, stand, stride: Function and evolution of hominoid feet. In E Strasser, JG Fleagle, AL Rosenberger, and HM McHenry (eds.): Primate Locomotion: Recent Advances. New York: Plenum Press, pp. 435–448.

Ulijaszek SJ (1995) Human Energetics in Biological Anthropology. Cambridge: Cambridge University Press.

Walker AC, and Leakey RE, Editors (1993) The Nariokotome Skeleton. Cambridge: Harvard University Press.

Wheeler P (1984) The evolution of bipedality and loss of functional body hair in hominids. J. Hum. Evol. 13:91–98.

Wheeler P (1985) The loss of functional body hair in man: The influence of thermal environment, body form and bipedality. J. Hum. Evol. 14:23–28.

Wheeler PE (1991a) The influence of bipedalism on the energy and water budgets of early hominids. J. Hum. Evol. 21:117–136.

Wheeler PE (1991b) The thermoregulatory advantages of hominid bipedalism in open equatorial environments: The contribution of increased convective heat loss and cutaneous evaporative cooling. J. Hum. Evol. 21:107–116.

White TD, Suwa G, and Asfaw B (1994) Australopithecus ramidus, a new species of early hominid from Aramis, Ethiopia. Nature 371:306–312.

Williamson D (1997) Hominid Socioecology. Ph.D. Dissertation, University College London.

22

HEEL, SQUAT, STAND, STRIDE

Function and Evolution of Hominoid Feet

Russell H. Tuttle,[1] Benedikt Hallgrímsson,[2] and Tamara Stein[1]

[1]Department of Anthropology
The University of Chicago
1126 East 59th Street
Chicago, Illinois 60627
[2]Department of Anatomy
University of Puerto Rico
GPO Box 5067
San Juan, Puerto Rico 00936

1. INTRODUCTION

Primate feet are remarkably diverse due to natural selection for a notable variety of positional behavior in a wide spectrum of arboreal and terrestrial niches (Schultz, 1963). Although positional behavior embraces both posture and locomotion, special features of primate feet are customarily related almost exclusively to locomotor adaptations and behavior, with posture treated secondarily or not considered at all (Tuttle et al., in press). Based on comparative functional morphological studies of extant apes and Pliocene-Recent hominids, we reason that squatting and bipedal standing were important components of the selective complex that produced the human foot, which has been associated more commonly with bipedal locomotion (Latimer and Lovejoy, 1989) instead of posture per se.

2. HEELS

Asian apes—orangutans and gibbons—have relatively modest heels, whereas humans and the African apes—gorillas, chimpanzees and bonobos—sport robust heels (Schultz, 1963; Sarmiento, 1983, 1994; Figure 1). This difference is commonly ascribed to the arboreality of Asian apes versus the terrestriality of African apes and people, with particular emphasis on locomotor differences among them (Tuttle, 1970, 1972).

Primate Locomotion, edited by Strasser *et al.*
Plenum Press, New York, 1998

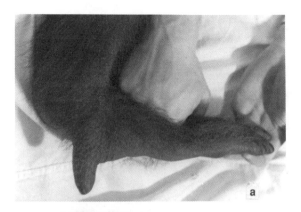

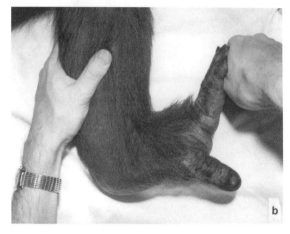

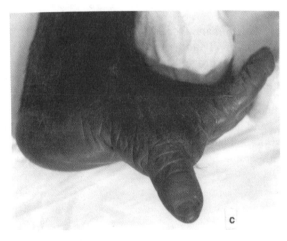

Figure 1. Medial views of left feet of an orangutan (*a*), a chimpanzee (*b*), and a gorilla (*c*). The African apes (*b* and *c*) have extensive passive excursions of metatarsophalangeal joints II-V (Tuttle, 1970), which would facilitate human-like bipedal locomotion if the toes were reduced in length.

Our feet must bear the entire bodily mass; they are on the front line against ground reaction forces when we stand, squat, crouch and move. Their peculiar morphology reflects the special demands that are placed upon them (Carrier et al., 1994).

As we stand, our weight is evenly balanced between the feet (Nicole and Paul, 1988). Within each foot, approximately half of each pedal load falls on the heel and the other half

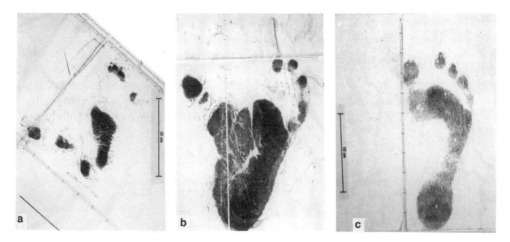

Figure 2. Bipedal right footprints of a gibbon (*a*), a chimpanzee (*b*), and a human who had never worn footgear (*c*). Note the absence of a heel impression in the gibbon and the flat-footed impression of the chimpanzee versus the arched plantigrade foot of the human.

is distributed among six contact points under the metatarsal heads. Two sesamoid bones beneath the first metatarsal head and the heads of the lateral four metatarsals share the load on the distal foot during stance. The axis of balance of the foot bisects the heel and sole longitudinally before passing between the second and third toes (Morton, 1935). Squatting can shift weight from the pedal ball to the heel, thereby eccentrically loading it.

When we walk, the pedal load shifts medially so that the leverage axis falls between the great and second toes (Morton, 1935). After heel strike, one's weight shifts to the ball of the foot during the first third of the gait cycle (Nicol and Paul, 1988). Accordingly, human footprints are characterized by prominent heel and ball impressions (Figure 2). Because weight shifts medially during stance phase, the great toe is the final contact of the foot before it swings clear of the ground. The hallucal pad is usually prominent in human footprints, but commonly the lateral four toes leave fainter impressions. Because of the medial longitudinal arch, much of the central sole is unimpressive in human footprints. The lateral border of the sole may be evident, but lightly, due to the quick shift from the heel to the medial ball as the foot acts as a compact lever.

Running is the supreme mechanical challenge to the ball and toes, especially when we sprint quasi-digitigrade. Long, down-curved toes would be an impediment to running hominids who might attempt to orient them strategically in the direction of travel.

The large human heel serves as a powerful lever for the triceps surae muscle, which plantarflexes the foot at the ankle joint during bipedal walking and running (Basmajian and De Luca, 1985). But, the robust heel probably also evolved to serve postural functions: squatting and standing.

3. SQUATTING AND BIPEDALISM

African apes, particularly gorillas, also have robust heels (Schultz, 1963), yet they rarely engage in bipedal locomotion. They squat for long periods of time, however, to forage and to rest (Tuttle and Watts, 1985; Remis, 1995).

Squatting and bipedalism, versus sitting, increase the height that foragers can reach overhead and keep the rump, thighs and legs off wet and otherwise uncomfortable substrates. Moreover, squatters are better prepared for locomotion than were they sitting or reclining. Accordingly, we should consider the possibility that robustness of the heel evolved in our lineage in response to squatting, bipedal foraging, and short-distance bipedal travel before early hominids were fully adapted to obligate bipedalism and long-distance travel. In this scenario, the toes may have retained arboreal features while the heel and ankle joint were more truly humanoid (Tuttle et al., in press; Tuttle, submitted).

4. HOMINOID FOOT POSTURES

Most primates are semiplantigrade, a posture that is facilitated by mobile transverse tarsal joints (Gebo, 1993a). They stand and walk with the proximal heel raised above the substrate (Morton, 1935; Gebo, 1992). Accordingly, the African apes and people are exceptional in habitually exhibiting full contact of the plantar surface of the heel during quadrupedal and bipedal walking and squatting.

Like catarrhine monkeys and prosimians, gibbons generally elevate the heel when moving rapidly bipedally on the ground (Morton, 1935; Tuttle et al., 1992; Gebo, 1993b), and probably also on branches. In arboreal contexts, gibbons must grasp the substrate with their toes, which is accompanied by flexed knees and hips and an elevated proximal heel (Tuttle, 1972). Schmitt and Larson (1995) reported that 3 captive lar gibbons exhibited mid-foot/heel contact—the heel touched the substrate after or simultaneously with the midfoot—as they walked bipedally at unspecified speeds on a variety of branches of unspecified diameter. Moreover, 2 adult lar gibbons evidenced complete heel contact, following mid-foot strike, in >85% of cases as they walked bipedally on the ground in an experimental setting (Schmitt and Larson, 1995). Contrarily, Tuttle et al. (1992) recorded no heel contact among 12 grounded bipedal captive lesser apes (11 lar gibbons and 1 siabon, 2.5–9 years-old), which moved rapidly (1.175 ± 0.589 m/sec) on pressure-sensitive paper runners (™Shutrak).

As usual, orangutans are unique and controversial (Tuttle and Cortright, 1988). Gebo (1992, 1993b, 1996) claimed that although grounded orangutans exhibit heel-strike plantigrady, they are not truly plantigrade *sensu* African apes and humans, while Meldrum (1993) countered that they probably are. Schmitt and Larson (1995) reported that 5 ambulating orangutans exhibited heel-strike at the end of swing phase.

The variability of facultative terrestrial foot postures among orangutans confounds those who would classify them neatly. The combination of grossly elongate, ventrally curved second-to-fifth toes and a puny hallux permits a spectrum of postures, ranging from one in which only the lateral aspect of the sole and fifth digit are on the ground to full plantigrade placement of the heel and extension of digits II-V, with the hallux serving as a strut. When standing bipedally, the heels of captives may be respectably plantigrade, or the subject may rise onto toe-tips so that the heel is free of the floor (Tuttle, 1970; Tuttle and Cortright, 1988; Figure 3).

Orangutans seem not to emulate the characteristic semiplantigrade posture of cercopithecoid monkeys and gibbons. Moreover, although, like orangutans, Visoke mountain gorillas sometimes tightly flex their toes and invert their feet to walk on the lateral soles, instead of being fully plantigrade (Tuttle and Watts, 1985), we do not exclude them from the plantigrades.

Figure 3. Orangutan foot postures. Walking on the lateral side of the foot and fifth toe (*a, b*) (note plantigrade heel in *a*), standing bipedally with heels plantigrade, toes extended, and halluces widely abducted (*c*).

5. HEELS, SQUATTING, AND BIPEDALISM

Habitual terrestrial bipedalism is unique to humans among hominoid primates. Accordingly, the large heels and plantigrade postures of African apes cannot be explained as adaptations to bipedalism.

There is no apparent feature of African pongid terrestrial quadrupedism that would require a large heel and plantigrade posture. A prominent proximal calcaneus provides purchase for the triceps surae muscle, which plantarflexes the ankle. Our EMG studies revealed marked activity in the triceps surae muscle when chimpanzee and gorilla subjects jumped to reach incentives overhead and while the chimpanzee stood bipedally on toetips. This mechanism is probably important also during arboreal vertical climbing of both species (Tuttle et al., in press).

The triceps surae muscle, however, showed only low or nil EMG potentials as the gorilla and chimpanzee stood and walked quadrupedally. Accordingly, we infer that the prominent calcanean tubers of African apes are not primarily an adaptation for terrestrial quadrupedism (Tuttle et al., in press). Instead, selection for a secure squatting platform on the ground and on large boughs may have been an important factor in the evolution of their feet.

6. BEHAVIOR

Watts found that, on average, Visoke mountain gorillas squat terrestrially 31.6% of feeding time and arboreally 3.8% of feeding time. They sit nearly twice as much (60%) as they squat while feeding on the ground, but they squat predominantly while feeding in trees. Indeed, a silverback never sat as he fed in trees, and squatting accounted for 86% of arboreal feeding postures overall among the Visoke group. Since Visoke gorillas devote approximately 60% of the average daily activity period to feeding, we may assume that their heels bear considerable loads over the course of the day (Tuttle and Watts, 1985).

Watts noted that Visoke gorillas rarely run. Running is virtually restricted to play and agonistic behavior. Bipedalism is also rare. Bipedally, they ran more than they walked over short distances. Bipedal feeding and foraging are uncommon in trees and on the ground. Silverbacks, however, stand bipedally on the ground to collect important food sources overhead, while smaller gorillas climb for them (Tuttle and Watts, 1985).

At the Bai Hokôu Study Site in the Central African Republic, western gorillas squatted arboreally more in the wet season than in the dry season (Remis, 1995). During arboreal positional behavior, squatting accounted for 29% in females, 26% in lone males, and 18% in group males. In the dry season, arboreal squatting was rare: 3% of female positional behavior and 7% of group male positional behavior. Unlike Visoke gorillas, those at Bai Hokôu sat more than they squatted arboreally: in the wet season, females, 36%, lone males, 40%, and group males, 57%, and in the dry season, females, 63%, and group males, 82%. Visibility was too poor for Remis (1995) to obtain reliable data on terrestrial squatting, sitting and bipedalism.

Bai Hokôu gorillas exibited low levels of arboreal bipedalism: 0% of arboreal positional behavior in the dry season; and in the wet season, females, 6%, lone males, 5%, and group males, 3% of arboreal positional behavior.

Stein's (1995) 8-week (30-hr) focal-animal sampling (30-min intervals) and scan sampling study (≤2 min each 10 min; Table 1) of positional behavior in a group of 9 western gorillas at the Brookfield Zoo revealed frequencies of squatting that are similar to those from the Bai Hokôu Study Site. The Brookfield group included 1 silverback (38 yr), 1 black-backed male (9 yr), 3 adult females (12–33 yrs), and 4 immature subjects (2.8–6 yrs). Most of their postural behavior occurred on the horizontal cement floor and spacious platforms of their holding area (21 hrs) or on the hilly floor and low, broad boughs in the exhibit (9 hrs).

We treated Stein's data on the Brookfield gorillas together regardless of sex or age of the individual. To facilitate comparison of the Brookfield and Bai Hokôu gorillas, we combined Remis's (1995) data for each positional activity regardless of sex, social grouping or season, i.e., we recalculated frequencies of positional activity based on frequencies and sample sizes in Remis's (1995) tables 8 and 10.

The Brookfield gorillas squatted 12% of the observation period versus 21% by Bai Hokôu gorillas. While feeding, the Brookfield gorillas squatted 35% of the time, which is quite similar to the frequency of squatting while feeding (39%) by Bai Hokôu gorillas. Further, Chi-square tests indicate that squatting among the Brookfield gorillas is significantly associated with feeding behavior (χ^2=46, N=229, P<0.001; Table 1).

Table 1. Frequencies of positional behaviors during all observations compared to during only feeding behavior[1]

Observed group	All behavior				Feeding behavior			
	Squat	Sit	Bipedal	Other	Squat	Sit	Bipedal	Other
Brookfield Zoo Gorillas[2]	12 N=229	53	6	29	34 N=34	59	0[3]	6
Bai Hoköu Gorillas[4]	21 N=1464	48	4	27	39 N=432	53	3	5
Tanzanian Chimpanzees[5]	29 N=16303	34[6]	0.4	37	44 N=4666	43	NA	13
Visoke Gorillas[7]	NA	NA	NA	NA	35 N=NA	60	0	5

[1]Frequencies are given in percentages (%).
[2]Data from Stein (1995). Chi-square tests indicate that squatting is significantly associated with feeding behavior (χ^2=46, P<0.0001).
[3]Because this is a zoo population, we would not expect bipedal feeding behavior as per Hunt (1991).
[4]Data from Remis (1995, Tables 8 and 10). Observations were of arboreal positional behavior only.
[5]Data from Hunt (1991, Table 2; 1992, Table 1). Hunt's "sit-(in)" category was added to his "squat" category to calculate our "squat" category.
[6]Because we used "sit-(in)" as squatting behavior, sitting behavior may be underestimated while squatting behavior may be overestimated.
[7]Data from Tuttle and Watts (1985, Table 3). Terrestrial and arboreal behaviors were combined.

Hunt's (1991) subjects squatted much less than they sat: 0.7% versus 62% of all positional behavior, respectively. Moreover, forelimb-assisted squatting occurred most often on vertical and nearly-vertical supports. Nonetheless, his category "sit (in)," which entailed flexion of the hind limbs and accounted for 28% of positional behavior, probably often loaded their heels, in addition to their ischial tuberosities. Accordingly, we disagree with Hunt (1992) that sitting produces little stress and that squatting is too infrequent to have affected chimpanzee adaptive morphology. The plantigrade heels of chimpanzees may relate to their sitting with acutely flexed hind limbs and squatting on the ground and on boughs (Tuttle et al., in press).

Hunt (1994) concluded that among wild adult Tanzanian chimpanzees bipedalism is a feeding adaptation, and that it is overwhelmingly postural: 86% of their arboreal bipedal activity occurred while feeding; 70% of their terrestrial bipedal activity occurred while feeding; 95% of their bipedalism was postural, and only 5% was locomotor. They most commonly employed bipedalism to feed on small fruits of low, open-forest trees by reaching either from the ground or from a lower branch in the tree. Hunt (1994) noted that by standing bipedally on the ground, chimpanzees increase their foraging rate because both hands can harvest to keep the mouth full.

Most of the 5% locomotor bipedalism entailed shuffling between fresh feeding sites under fruiting trees. The Tanzanian chimpanzees rarely employed bipedalism during social display (1%) or to scan the environment (2%) (Hunt, 1994). Doran (1993) noted that Taï Forest chimpanzees also employ forelimb-unassisted bipedalism chiefly during terrestrial foraging, e.g., to gather and to carry nuts short distances to terrestrial cracking sites.

7. FOSSIL HOMINID HEELS

The earliest evidence of robust heels in a fully plantigrade hominid foot is the 3.5-Ma Laetoli Site G bipedal trackways (Clarke, 1979; Leakey and Hay, 1979; Day and

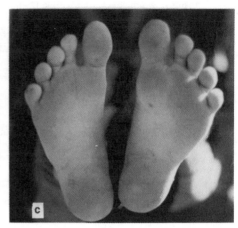

Figure 4. Cast of Laetoli G-1/37 (*a*), left print of a Peruvian Indian in mud (*b*), and soles of a 14-year-old male Peruvian Indian who had never worn footgear (*c*).

Wickens, 1980; White, 1980; Capecchi, 1984; Drake and Curtis, 1987; Harris, 1987; Leakey, 1987; Robbins, 1987; Tuttle, 1987).

In all discernible features, the pedal morphology of the Laetoli track-makers is like that of *Homo sapiens* (Tuttle, 1996; Figure 4). Their foot indices, which indicate foot length versus foot breadth, fit comfortably within a global sample of human foot indices (Tuttle, 1987). The hallux is aligned with the lateral four toes, and the gap between it and the second toe is quite human, particularly when compared to undeformed feet of persons who have never worn footgear (Tuttle et al., 1990; Feibel et al., 1996).

The lateral four toes are arrayed relative to the hallux and to each other as in a modern human foot, and none extends notably beyond the tip of the hallux. The toes of Laetoli G hominids are approximately 30% of total foot length, which is not significantly different from mean relative toe length of never-shod Peruvian Indians and Tanzanian Hadzabe (Tuttle et al., 1991; Musiba et al., 1997).

The Laetoli G hominid prints evidence a medial longitudinal arch. Apparently, the transfer of body weight during bipedal walking was quite human—from robust heel strike, more lightly along the lateral sole, then more heavily medially across the ball of the foot so that the brunt of toe-off was borne by the hallux, which, unlike the lateral toes, regularly left prominent impressions in the substrate. Contrarily, African pongid prints are flat-footed with the toes arrayed quite differently from those of human prints (Tuttle et al., 1992; Figure 2).

Mean widths of heel impressions for two (G-1 and G-3) of the three individuals that walked in the moist volcanic ash at Laetoli Site G are 65.8 ± 5.4 mm and 75.9 ± 5.1 mm, respectively (Tuttle, 1987), which is notably larger than mean heel breadths of Hadzabe adults (58.8 ± 6.2 mm) and juveniles (50.9 ± 6.1 mm) (Musiba et al., 1997).

It is likely that heel size of the Laetoli hominids is somewhat exaggerated by the pedal impressions. Apparently, G-1 placed its heels quite deliberately in the substrate; its heel impressions are generally quite deep (Tuttle, 1996; Deloison, 1991, 1992).

Three 3-Ma calcanei (AL 333–8; AL 333–37; AL 333–55) from Hadar, Ethiopia, allow morphological studies and mechanical modelling of heels in one species of Pliocene hominid: *Australopithecus afarensis* (Latimer et al., 1982; Latimer and Lovejoy, 1989).

Latimer and Lovejoy (1989) found that 2 calcanean tubers (AL 333–8, AL 333–55) sport sufficient robusticity and minimum volume to attest to notable bipedality in Hadar *Australopithecus afarensis*. Their minimum cross-sectional areas fall above those of chimpanzees and gorillas, the latter of which were certainly much heavier than were the Hadar hominids, and at the lower extremity of a human range, based on calcanei of persons who had worn western footgear.

Minimum volumes of the calcanean tubers, expressed as a product of minimum cross-sectional area and length, sharply separate the Hadar hominids from chimpanzees and female gorillas, on the one hand, and male gorillas and humans, on the other. Indices of minimum calcanean volume to body weight place the Hadar calcanean tubers closer to those of humans than to those of African apes (Latimer and Lovejoy, 1989).

Accordingly, one may fairly conclude that the bony heels of Hadar *Australopithecus afarensis* are relatively robust and tend toward the human condition in this feature. Moreover, statements to the contrary (Susman et al., 1984; Deloison, 1985) notwithstanding, the 3 Hadar calcanean specimens evidence not only a medial plantar process but also a lateral plantar process, which is characteristic of humans and is absent in apes (Latimer and Lovejoy, 1989).

Were Hadar *Australopithecus afarensis* obligately bipedal on the ground? Most probably, but this inference is more compellingly supported by other features of the pelvic limb (Tuttle, 1981; Johanson et al., 1982; Lovejoy et al., 1982; Latimer et al., 1987; Latimer and Lovejoy, 1990a; Coppens, 1991; Jungers, 1991; Langdon et al., 1991; Latimer, 1991; McHenry, 1991; Vancata, 1991) than by the calcanean tuber per se. Hadar hominid calcanean morphology is consistent with this model, but it does not unequivocally command commitment to it, since the very features that would facilitate bipedalism also enable terrestrial and arboreal squatting. Nevertheless, we readily acknowledge bipedal standing and short distance walking by Hadar *Australopithecus afarensis* because of the total morphological pattern of their pelvic limbs.

Whether Hadar *Australopithecus afarensis* walked like we do and whether they could crouch, sprint and course quasidigitigrade like an athletic modern person are moot puzzles that cannot be resolved by calcanean studies alone. Within the foot, there is highly suggestive evidence that no matter how posturally bipedal the Hadar *Australopithecus afarensis* may have been, the bipedal component of their locomotion was not as advanced as those of *Homo sapiens* and probably also those of Pleistocene populations of *Homo*, or perhaps even those of penecontemporaneous *Australopithecus sensu lato*, who lived in more open habitats.

Although Lamy (1986) concluded from talonavicular morphology that, like other eastern African species of Plio-Pleistocene Hominidae, Hadar *Australopithecus afarensis* had longitudinally-arched feet, Gomberg (1985; Gomberg and Latimer, 1984) concluded from calcaneonavicular morphology that they did not have humanoid pedal arches. This disagreement recalls earlier controversy regarding whether *Homo habilis* (OH-8) and hominids from Koobi Fora had relatively compact humanoid tarsal arches (Day and Napier, 1964; Day and Wood, 1968; Day, 1976; Lamy, 1983; Wood, 1974a,b, 1976) or more mobile pongoid transverse tarsal joints (Oxnard, 1972, 1973, 1984; Lisowski, 1976; Oxnard and Lisowski, 1980). On balance, we express reasonable doubt that Hadar *Australopithecus afarensis* had fully humanoid transverse tarsal joints.

Further distally, the feet of Hadar *Australopithecus afarensis* depart more tellingly from those of *Homo sapiens*. Latimer and Lovejoy (1990b: 23) reported that the orientation of the basal articular surfaces of the proximal phalanges, the potential dorsoplantar excursions of the metatarsophalangeal joints, and the shape and orientation of the metatarsal heads of digital rays II-V not only confirm "a dramatic commitment to terrestrial bipedality" in Hadar *Australopithecus afarensis* but also "contravene any significant pedal grasping."

Contrarily, via more refined, quantitative studies, Duncan et al. (1994) demonstrated that the metatarsophalangeal joints of Hadar *Australopithecus afarensis* are not as humanoid as had been claimed by Latimer and Lovejoy (1990b). Indeed, Hadar *Australopithecus afarensis* fall midway between African apes and humans in orientation of the articular surfaces of their pedal proximal phalanges; the potential dorsoplantar excursions of their metatarsophalangeal joints are greater than estimates by Latimer and Lovejoy (1990b); and, actual orientation of the metatarsal heads are unmeasurable due to taphonomic damage and incompleteness of the Hadar fossils (Duncan et al., 1994). Moreover, because *Pan troglodytes* and *Homo sapiens* share dorsally oriented metatarsal heads and both *Pan gorilla* and *Pongo pygmaeus* sport plantar orientations of the metatarsal heads Duncan and coworkers (1994) rightly question the diagnostic value of this feature to indicate bipedalism versus prehensility in the Hominoidea.

We all agree that the degree of dorsiflexion suggested by the metatarsophalangeal joints of Hadar *Australopithecus afarensis* would facilitate bipedal locomotion (Tuttle, 1981; Latimer and Lovejoy, 1990b; Duncan et al., 1994). But, the degrees of metatarsophalangeal plantarflexion evidenced by Hadar *Australopithecus afarensis*—whether they be the stenotic guesstimates of Latimer and Lovejoy (1990b) or the wider excursions predicted by Duncan et al. (1994)—would facilitate pedal prehension of arboreal substrates, particularly trunks, boughs and larger branches. Further, the permissive metatarsophalangeal dorsiflexion of Hadar *Australopithecus afarensis* may have served them well during climbing, reaching overhead while bipedal, and walking short distances bipedally (Tuttle, 1981).

In Hadar *Australopithecus afarensis*, the proximal phalanges and, to a lesser extent the middle phalanges, of pedal digits II-V are curved downward (Tuttle, 1981); some sport prominent ridges for attachment of fibrous flexor sheaths (Latimer et al., 1982); and, their heads face plantarly, which accentuates overall curvature of the articulated digits. Further, the toes were probably relatively long in comparison with those of modern human feet (McHenry, 1986, 1991; Tuttle, 1988; Latimer and Lovejoy, 1990b).

All in all, the second-to-fifth toes of Hadar *Australopithecus afarensis* possessed prehensile capacities that would serve them well in climbing, squatting and standing bipedally on arboreal substrates, particularly in the absence of a markedly prehensile hallux. The full range of diameters of the substrates that Hadar *Australopithecus afarensis* could have comfortably grasped cannot be estimated reliably because we do not know whether their metatarsals were elongate, relatively short, or intermediate in length. Although pongoid grasps of twigs and small branches may have been problematic for Hadar *Australopithecus afarensis*, their toes, soles and heels attest to versatility vis-à-vis larger arboreal supports, probably in a positional repertoire that is not emulated closely by any extant hominoid species.

8. SCENARIO

Our Miocene arboreal hylobatian ancestors probably had small or modest heels, though other features of the hind limbs and torso predisposed their terrestrial descendents to bipedalism instead of quadrupedism (Tuttle, 1994). Hunt (1994) argued persuasively that certain features of Pliocene hominids—represented by *Australopithecus afarensis*—are adaptations to foraging bipedally on small fruits in low open-forest trees from terrestrial and low-branch vantage points. We would add that squatting to forage and to feed on the ground, to dig, to process hard-shelled foods and during rest probably also selected for prominent heels and plantigrade postures in our Pliocene ancestors, notably before the evolution of long-distance bipedal walking and sustained running in open habitats. Accordingly, in the early phases of hominid calcanean enlargement, ground reaction forces acting on the heel during squatting and bipedal foraging activities should be viewed as a major factor that supplemented traction of the triceps surae muscle, particularly during arboreal climbing.

During squatting, the gastrocnemius muscle was probably silent due to acute flexion of the knee and the soleus was probably not needed to maintain balance even though the ankle may have been dorsiflexed. Speculation on this point is moot since proportions of the lower limb segments would determine, to some extent, the distribution of weight on the heel versus distal areas of the foot.

Reduction of pedal digits II-V, which is essential for athletic human running, is not evidenced by Hadar *Australopithecus afarensis*, but toe length and absence of down-curved digits is consistent with fully human bipedal locomotion by hominids that made the 3.5-Ma footprint trails at Laetoli Site G, Tanzania (Tuttle, 1985, 1996; Tuttle et al., 1991; Musiba et al., 1997). These observations challenge to the conspecific status of Hadar *Australopithecus afarensis* and the Laetoli printmakers.

Although the Laetoli footprints proffer no evidence to the contrary, we resist the temptation to send them racing across the savanna. Instead, we await the discovery of pelvic limb skeletons and additional trackways before deciding whether and how to set them on a faster course than that evidenced by the site G trackways (Tuttle, 1987, 1994, submitted; Tuttle et al., 1990, 1992). We somewhat more confidently state that their feet appear to be better adapted for terrestrial walking, perhaps over notable distances, than those of Hadar *Australopithecus afarensis*.

Unfortunately, studies of squatting facets on tali and distal tibiae will not resolve questions on the role of squatting in early hominid evolution. Their absence is insufficient to deny frequent squatting, and their presence is consistent with other positional behaviors, particularly in active creatures like the Plio-Pleistocene Hominidae (Trinkaus, 1995).

Precisely when a completely human foot that functioned like ours evolved is unanswerable because of the paucity of pedal specimens of early *Homo* spp. that preceded the Neandertalians, whose robust, virtually human feet are well-documented and attest to a locomotor repertoire basically like ours (Trinkaus, 1983). Like modern African apes, Holocene humans living in many different cultures, and probably also their Plio-Pleistocene predecessors, the Neandertalians, commonly engaged their large heels in prolonged bouts of squatting (Trinkaus, 1975).

ACKNOWLEDGMENTS

This investigation was supported by NSF grants GS-3209, SOC75–02478 and BNS 8540290, a Public Health Service research career development award (1-KO4-GM16347–01) from the National Institutes of Health, the Guggenheim Foundation, Brookfield Zoo, and NIH grant RR-00165 from the Division of Research Resources to the Yerkes Regional Primate Research Center, which is fully accredited by the American Association of Laboratory Animal Care. We are especially grateful for the assistance of J. Malone, E. Regenos, J. Perry, Dr. G.H. Bourne, Dr. F.A. King, R. Pollard, S. Lee, R. Mathis, J. Roberts, Dr. M. Keeling, Dr. M. Vitti, and J. Hudson.

R. Tuttle thanks D.C. Johanson and M.D. Leakey for opportunities to study the Hadar postcranial specimens and the Laetoli footprints, respectively, and colleagues and staffers of the British Museum (Natural History); Institut royal des Sciences naturelles de Belgique, Brussels; Musée de l'Homme, Paris; Museum of the Department of Antiquities, Jerusalem; National Museum of Geology and Palaeontology, Zhagreb; National Museums of Kenya, Nairobi; Peabody Museum, Harvard University, Cambridge, MA; Rheinisches Landesmuseum, Bonn; Rockefeller Museum, Jerusalem; Tel Aviv University, Tel Aviv; Transvaal Museum, Pretoria; University of New Mexico, Albuquerque; and University of the Witwatersrand Medical School, Johannesburg, who facilitated his studies on Plio-Pleistocene hominid specimens in their keeping.

REFERENCES

Basmajian JV, and De Luca CJ (1985) Muscles Alive, 5th ed., Baltimore: Williams and Wilkins.

Capecchi V (1984) Reflections on the footprints of the hominids found at Laetoli. Anthropologischer Anzeiger *42*:81–86.

Carrier DR, Heglund NC, and Earls KD (1994) Variable gearing during locomotion in the human musculoskeletal system. Science *265*:651–653.

Clarke RJ (1979) Early hominid footprints from Tanzania. S. Af. J. Sci. *75*:148–149.

Coppens Y (1991) L'évolution des hominidés, de leur locomotion et de leurs environnnements. In Y Coppens and B Senut (eds.): Origine(s) de la Bipédie chez les Hominidés, Cah. Paléoanthrop. Paris: Editions du CNRS, pp. 295–301.

Day MH (1976) Hominid postcranial remains from the East Rudolf succession: A review. In Y Coppens, FC Howell, GL Isaac, and REF Leakey (eds.): Earliest Man and Environments in the Lake Rudolf Basin. Chicago: University of Chicago Press, pp. 507–521.

Day MH, and Napier JR (1964) Hominid fossils from Bed I Olduvai Gorge, Tanganyika: Fossil foot bones. Nature *201*:967–970.

Day MH, and Wickens EH (1980) Laetoli Pliocene hominid footprints and bipedalism. Nature *286*:385–387.

Day MH, and Wood BA (1968) Functional affinities of the Olduvai hominid 8 talus. Man *3*:440–455.

Deloison Y (1985) Comparative study of calcanei of primates and *Pan-Australopithecus-Homo* relationship. In PV Tobias (ed.): Hominid Evolution: Past, Present and Future. New York: Liss, pp. 143–147.

Deloison Y (1991) Les Australopitheques marchaient-ils comme nous? In Y Coppens and B Senut (eds.): Origine(s) de la Bipédie chez les Hominidés, Cah. Paléoanthrop. Paris: Editions du CNRS, pp. 177–186.

Deloison Y (1992) Empreintes de pas à Laetoli (Tanzanie). Leur apport à une meilleure connaissance de la locomotion des Hominidés fossiles. C. R. Acad. Sci., Sér. II, *315*:103–109.

Drake R, and Curtis GH (1987) K-Ar geochronology of the Laetoli fossil localities. In MD Leakey and JM Harris (eds.): Laetoli: A Pliocene Site in Northern Tanzania. Oxford: Clarenden Press, pp. 48–52.

Doran DM (1993) Sex differences in adult chimpanzee positional behavior: The influence of body size on locomotion and posture. Am. J. Phys. Anthropol. *91*:99–115.

Duncan AS, Kappelman J, and Shapiro LJ (1994) Metatarsophalangeal joint function and positional behavior in *Australopithecus afarensis*. Am. J. Phys. Anthropol. *93*:67–81.

Feibel CS, Agnew N, Latimer B, Demas M, Marshall F, Waane AC, and Schmid P (1996) The Laetoli hominid footprints - a preliminary report on the conservation and scientific restudy. Evol. Anthropol. *4*:149–154.

Gebo DL (1992) Plantigrady and foot adaptation in African apes: Implications for hominid origins. Am. J. Phys. Anthropol. *89*:29–58.

Gebo DL (1993a) Functional morphology of the foot in primates. In DL Gebo (ed.): Postcranial Adaptation in Nonhuman Primates. DeKalb: Northern Illinois University Press, pp. 175–196.

Gebo DL (1993b) Reply to Meldrum. Am. J. Phys. Anthropol. *91*:382–385.

Gebo DL (1996) Climbing, brachiation, and terrestrial quadrupedalism: Historical precursors of hominid bipedalism. Am. J. Phys. Anthropol. *101*:55–92.

Gomberg DN (1985) Functional differences of three ligaments of the transverse tarsal joint in hominoids. J. Hum. Evol. *14*:553–562.

Gomberg DN, and Latimer B (1984) Observations on the transverse tarsal joint of *A. afarensis*. Am. J. Phys. Anthropol. *63*:164.

Harris JM (1987) Summary. In MD Leakey and JM Harris (eds.): Laetoli: A Pliocene Site in Northern Tanzania. Oxford: Clarenden Press, pp. 524–531.

Hunt KD (1991) Mechanical implications of chimpanzee positional behavior. Am. J. Phys. Anthropol. 86:521–536.

Hunt KD (1992) Positional behavior of *Pan troglodytes* in the Mahale Mountains and Gombe Stream National Parks, Tanzania. Am. J. Phys. Anthropol. 87:83–105.

Hunt KD (1994) The evolution of human bipedality: Ecology and functional morphology. J. Hum. Evol. 26:183–202.

Johanson DC, Taieb M, and Coppens Y (1982) Pliocene hominids from the Hadar Formation, Ethiopia (1973–1977): Stratigraphic, chronologic, and paleoenvironmental contexts, with notes on hominid morphology and systematics. Am. J. Phys. Anthropol. 57:373–402.

Jungers WL (1991) A pygmy perspective on body size and shape in *Australopithecus afarensis* (AL 288–1, "Lucy"). In Y Coppens and B Senut (eds.): Origine(s) de la Bipédie chez les Hominidés, Cah. Paléoanthrop. Paris: Editions du CNRS, pp. 215–224.

Lamy P (1983) Le système podal de certains hominidés fossiles du Plio-Pleistocène d'Afrique de l'est: étude morpho-dynamique. L'Anthropologie (Paris) 87:435–464.

Lamy P (1986) The settlement of the longitudinal plantar arch of some African Plio-Pleistocene hominids: A morphological study. J. Hum. Evol. 15:31–46.

Langdon JH, Bruckner J, and Baker HH (1991) Pedal mechanics and bipedalism in early hominids. In Y Coppens and B Senut (eds.): Origine(s) de la Bipédie chez les Hominidés, Cah. Paléoanthrop. Paris: Editions du CNRS, pp. 159–167.

Latimer B (1991) Locomotor adaptations in *Australopithecus afarensis*: The issue of arboreality. In Y Coppens and B Senut (eds.): Origine(s) de la Bipédie chez les Hominidés, Cah. Paléoanthrop. Paris: Editions du CNRS, pp. 169–176.

Latimer BM, and Lovejoy CO (1989) The calcaneus of *Australopithecus afarensis* and its implications for the evolution of bipedality. Am. J. Phys. Anthropol. 78:369–386.

Latimer BM, and Lovejoy CO (1990a) Hallucal tarsometatarsal joint in *Australopithecus afarensis*. Am. J. Phys. Anthropol. 82:125–133.

Latimer BM, and Lovejoy CO (1990b) Metatarsophalangeal joints of *Australopithecus afarensis*. Am. J. Phys. Anthropol. 83:13–23.

Latimer BM, Lovejoy CO, Johanson DC, and Coppens Y (1982) Hominid tarsal, metatarsal, and phalangeal bones recovered from the Hadar Formation: 1974–1977 collection. Am. J. Phys. Anthropol. 57:701–719.

Latimer BM, Ohman JC, and Lovejoy CO (1987) Talocrural joint in African hominids: Implications for *Australopithecus afarensis*. Am. J. Phys. Anthropol. 74:155–175.

Leakey MD (1987) The hominid footprints. Introduction. In MD Leakey and JM Harris (eds.): Laetoli: A Pliocene Site in Northern Tanzania. Oxford: Clarenden Press, pp. 490–496.

Leakey MD, and Hay RL (1979) Pliocene footprints in the Laetolil Beds, northern Tanzania. Nature 278:317–323.

Lisowski FP, Albrecht GH, and Oxnard CE (1976) African fossil tali: Further multivariate morphometric studies. Am. J. Phys. Anthropol. 45:5–18.

Lovejoy CO, Johanson DC, and Coppens Y (1982) Hominid lower limb bones recovered from the Hadar Formation: 1974–1977. Am. J. Phys. Anthropol. 57:679–700.

McHenry HM (1986) The first bipeds: A comparison of the *A. afarensis* and *A. africanus* postcranium and implications for the evolution of bipedalism. J. Hum. Evol. 15:177–191.

McHenry HM (1991) First steps? Analyses of the postcranium of early hominids. In Y Coppens and B Senut (eds.): Origine(s) de la Bipédie chez les Hominidés, Cah. Paléoanthrop. Paris: Editions du CNRS, pp. 133–141.

Meldrum DJ (1993) On plantigrady and quadrupedalism. Am. J. Phys. Anthropol. 91: 379–381.

Morton DJ (1935) The Human Foot. New York: Columbia University Press.

Musiba CM, Tuttle RH, Hallgrímsson B, and Webb DM (1997) Swift and sure-footed on the savanna: A study of Hadzabe gaits and feet in northern Tanzania. Am. J. Hum. Biol. 9:303–321.

Nicol AC, and Paul JP (1988) Biomechanics. In B Helal and D Wilson (eds.): The Foot, Vol. 1. Edinburgh: Churchill Livingstone, pp. 75–86.

Oxnard CE (1972) Some African fossil foot bones: A note on the interpolation of fossils into a matrix of extant species. Am. J. Phys. Anthropol. 37:3–12.

Oxnard CE (1973) Form and Pattern in Human Evolution. Chicago: University of Chicago Press.

Oxnard CE (1984) The Order of Man. New Haven: Yale University Press.

Oxnard CE, and Lisowski FP (1980) Functional articulation of some hominoid foot bones: Implications for the Olduvai (Hominid 8) foot. Am. J. Phys. Anthropol. 52:107–117.

Remis M (1995) Effects of body size and social context on the arboreal activities of lowland gorillas in the Central African Republic. Am. J. Phys. Anthropol. *97*:413–433.

Robbins LM (1987) Hominid footprints from Site G. In MD Leakey and JM Harris (eds.): Laetoli: A Pliocene Site in Northern Tanzania. Oxford: Clarenden Press, pp. 497–502.

Sarmiento EE (1983) The significance of the heel process in anthropoids. Int. J. Primatol. *4*:127–152.

Sarmiento EE (1994) Terrestrial traits in the hands and feet of gorillas. American Museum Novitates, no. 3091, 56 pp.

Schmitt D, and Larson SG (1995) Heel contact as a function of substrate type and speed in primates. Am. J. Phys. Anthropol. *96*:39–50.

Schultz AH (1963) Relations between the lengths of the main parts of the foot skeleton in primates. Folia Primatol. *1*:150–171.

Stein TA (1995) Who's in Charge: Observations of Social Behavior in a Captive Group of Western Lowland Gorillas. M.A. thesis. The University of Chicago.

Susman RL, Stern JT, Jr, and Jungers WL (1984) Arboreality and bipedality in the Hadar homionids. Folia Primatol. *43*:113–156.

Trinkaus E (1975) Squatting among the Neandertals: A problem in the behavioral interpretation of skeletal morphology. J. Arch. Science *2*:327–351.

Trinkaus E (1983) The Shanidar Neandertals. New York: Academic Press.

Tuttle RH (1970) Postural, propulsive, and prehensile capabilities in the cheiridia of chimpanzees and other great apes. In GH Bourne (ed.): The Chimpanzee, Vol. 2. Basel: Karger, pp. 167–253.

Tuttle RH (1972) Functional and evolutionary biology of hylobatid hands and feet. In DM Rumbaugh (ed.): Gibbon and Siamang, Vol. 1. Basel: Karger, pp. 136–206.

Tuttle RH (1981) Evolution of hominid bipedalism and prehensile capabilities. Phil. Trans.Royal Soc., Lond. *B-292*: 89–94.

Tuttle RH (1985) Ape footprints and Laetoli impressions: A response to the SUNY claims. In PV Tobias (ed.): Hominid Evolution: Past, Present and Future. New York: Liss, pp. 129–133.

Tuttle RH (1987) Kinesiological inferences and evolutionary impilications from Laetoli bipedal trails G-1, G-2/3, and A. In MD Leakey and JM Harris (eds.): Laetoli: A Pliocene Site in Northern Tanzania. Oxford: Clarenden Press, pp. 503–523.

Tuttle RH (1988) What's new in African paleoanthropology? Ann. Rev. Anthropol. *17*:391–426.

Tuttle RH (1994) Up from electromyography: Primate energetics and the evolution of human bipedalism. In RS Corruccini and RL Ciochon (eds.): Integrative Paths to the Past: Paleoanthropological Advances in Honor of F.C. Howell, New York: Prentice Hall, pp. 269–284.

Tuttle RH (1996) The Laetoli hominid G footprints. Where do they stand today? Kaupia *6*:97–102.

Tuttle RH (submitted). Animalia, *Homo*, and the Kingdom of God. In TL Gilbert (ed.): The Epic of Creation: Scientific and Religious Perspectives on our Origins.

Tuttle RH, and Cortright GW (1988) The positional behavior, adaptive complexes and evolution of *Pongo pygmaeus*. In JH Schwartz (ed.): Orang-utan Biology. Oxford: Oxford University Press, pp. 311–330.

Tuttle RH, and Watts DP (1985) The positional behavior and adaptive complexes of *Pan gorilla*. In S Kondo (ed.): Primate Morphophysiology, Locomotor Analyses and Human Bipedalism. Tokyo: Univeristy of Tokyo Press, pp. 261–28.

Tuttle RH, Webb DM, and Baksh M (1991) Laetoli toes and *Australopithecus afarensis*. Hum. Evol. *6*:193–222.

Tuttle RH, Webb DM, Weidl E, and Baksh M (1990) Further progress on the Laetoli trails. J. Arch. Science *17*:347–362.

Tuttle RH, Webb DM, Tuttle N I, and Baksh M (1992) Footprints and gaits of bipedal apes, bears and barefoot people: Perspectives on Pliocene tracks. In S Matano, RH Tuttle, H Ishida, and M Goodman (eds.): Topics in Primatology, Vol. 3: Evolutionary Biology, Reproductive Endocrinology and Virology. Tokyo: University of Tokyo Press, pp. 221–242.

Tuttle RH, Yant L, Hallgrímsson B, and Basmajian JV (in press) Hominoid heels. Proc.of the XVth Congress of the International Primatological Society, Kuta Bali, Indonesia, August 4, 1994.

Vancata V (1991) The roots of hominid bipedality. In Y Coppens and B Senut (eds.): Origine(s) de la Bipédie chez les Hominidés, Cah. Paléoanthrop. Paris: Editions du CNRS, pp. 143–158.

White TD (1980) Evolutionary implications of Pliocene hominid footprints. Science *208*:175–176.

Wood BA (1974a) Evidence on the locomotor pattern of *Homo* from early Pleistocene of Kenya. Nature *251*:135–136.

Wood BA (1974b) A *Homo* talus from East Rudolf, Kenya. J. Anat. *117*:203–204.

Wood BA (1976) Remains attributable to *Homo* in the East Rudolf succession. In Y Coppens, FC Howell, G Ll Isaac, and REF Leakey (eds.): Earliest Man and Environments in the Lake Rudolf Basin. Chicago: University of Chicago Press, pp. 490–506.

23

EVOLUTION OF THE HOMINID HIP

Christopher Ruff

Department of Cell Biology and Anatomy
Johns Hopkins University School of Medicine
725 N. Wolfe Street
Baltimore, Maryland 21205

1. INTRODUCTION

The morphology of the hip region, and its functional implications, have figured prominently in discussions of the origin and nature of hominid bipedality (Dart, 1949; Broom and Robinson, 1950; Washburn, 1950; Le Gros Clark, 1955; Mednick, 1955; Napier, 1964, 1967; Day, 1969, 1973; Robinson, 1972; Lovejoy et al., 1973; McHenry, 1975; Wood, 1976; McHenry and Corruccini, 1978; Stern and Susman, 1983, 1991; Susman et al., 1984; Lovejoy, 1988; Berge, 1991; Jungers, 1991). During most of human bipedal gait, the body is balanced over one lower limb (Inman et al., 1981), a biomechanical problem not faced by quadrupeds. The solution to this problem has involved major changes in the form of the human pelvis and proximal femur (as well as structures more distal in the lower limb) from that of our primate quadrupedal contemporaries, and presumably ancestors (Le Gros Clark, 1959).

While there is little disagreement over what distinguishes the modern human hip from that of monkeys and apes, there has been considerable debate over the extent to which early hominids (i.e., australopithecines) approached the modern human condition in this respect, in terms of both morphology and, especially, function (e.g., Napier, 1964; Lovejoy et al., 1973; Stern and Susman, 1983; Lovejoy, 1988; Jungers, 1991). Despite continuing differences in interpretation of the fossil evidence, however, two recurring themes have emerged on which there is general agreement: a) morphology must be considered within its biomechanical context, and b) an appropriate "size" parameter must be used in comparative analyses. In fact, as illustrated in the following example, these two issues are actually closely related.

Many investigators have noted the apparently long femoral neck and small femoral head characteristic of *Australopithecus* compared to modern humans (see McHenry and Corruccini, 1978 and references therein). The most common method of quantifying these differences has been to calculate ratios between head diameter or neck length and proxi-

Primate Locomotion, edited by Strasser *et al.*
Plenum Press, New York, 1998

mal femoral shaft breadth (Napier, 1964; Day, 1969; Robinson, 1972; Wood, 1976). As pointed out by Lovejoy (1975) and Wolpoff (1976), however, these kinds of ratios are problematic, since it is impossible to distinguish to what extent the resulting variability is due to differences in femoral shaft robusticity or to differences in head or neck dimensions. Consequently, both authors (and Walker, 1973) recommended evaluating such dimensions relative to femur length rather than shaft breadth. Use of femur length in such a role, however, is not without its own problems. We have previously argued that for weight-bearing skeletal elements, body mass or some derivative of body mass is the most appropriate measure of "size" against which to compare other dimensions (also see Jungers, 1985; Ruff et al., 1993). Due to systematic differences in body build, it is unlikely that femur length (or even a power of length) showed the same proportionality to body mass in australopithecines as it does in modern humans (Ruff, 1991; Aiello, 1992; Franciscus and Holliday, 1992). Differences in body shape must also be taken into account when evaluating diaphyseal robusticity, even within *Homo* (Ruff et al., 1993).

In fact, there is probably a direct functional relationship between certain proximal femoral dimensions and shaft robusticity or strengthening. I have shown elsewhere (Ruff, 1995) that within modern humans the ratio of mediolateral to anteroposterior bending strength of the proximal femoral shaft is significantly positively correlated with the ratio of femoral neck length to femoral length. That is, as the femoral neck increases in length, the proximal shaft becomes more buttressed in the M-L plane. This supported theoretical predictions of a biomechanical model in which an increase in femoral neck length increased M-L bending loads on the femoral shaft, particularly proximally. The same model was used to explore differences in morphology between modern humans and early *Homo*. Early *Homo* (*H. erectus* and "*erectus*-like") was shown to have a long femoral neck and greatly increased M-L bending strength of the femoral shaft relative to modern humans, as predicted by the model (biacetabular breadth of the pelvis was also predicted to be large in early *Homo*). Interestingly, in terms of femoral shaft cross-sectional morphology, australopithecines were shown to be intermediate between modern and earlier *Homo*.

In the present study this same biomechanical model of the hip is applied to the *Australopithecus afarensis* A.L. 228-1 ("Lucy"), around which much of the more recent controversy regarding hip functioning in australopithecines has centered (Stern and Susman, 1983; Lovejoy, 1988; Ruff, 1988; Berge, 1991; Jungers, 1991). In fact, using ostensibly the same general approach applied to the same specimen, different investigators have come to diametrically opposing conclusions regarding the ability of Lucy's hip musculature to function efficiently during gait (Lovejoy, 1988; Jungers, 1991)! It is shown here that while A.L. 228-1 certainly was adapted to bipedality, it is very likely that she (and perhaps at least early australopithecines in general) exhibited some subtle differences in gait pattern from that of modern humans, as well as earlier *Homo*.

2. BIOMECHANICAL MODEL

McLeish and Charnley published a comprehensive study of lower limb loadings during the single support phase of gait in 1970. They radiographed subjects standing on one leg with the pelvis at various angles of inclination to the horizontal. Using a force platform under the supporting foot and knowledge of the orientation of the abductor musculature from cadaver dissections, they were able to use basic Newtonian principles to calculate the force vectors acting about the hip (for details see Ruff, 1995). Figure 1 is a modified reproduction of one of their radiographic tracings of a subject standing with the

Modern Human

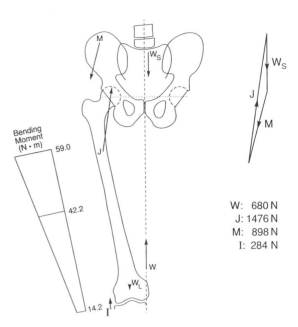

Figure 1. Force vectors and resulting femoral shaft mediolateral bending moments during one-legged stance in modern humans. N, Newtons; m, meters; W, body weight; W_L, weight of right lower limb; W_S: weight superimposed on right hip (W - W_L); M, gluteal abductor force; J, hip joint reaction force; I, iliotibial tract force about knee. All forces drawn to scale. Forces in vector triangle doubled to better show details. (Ruff, 1995, Figure 5; modified from Figure 8 in McLeish and Charnley, 1970.)

W: 680 N
J: 1476 N
M: 898 N
I: 284 N

pelvis at -0.5° inclination (down on the nonsupport side), which is similar to the average inclination during the midstance phase of gait (Eberhart et al., 1954). From the subject's body weight the actual magnitudes of the joint reaction force (J) and abductor force (M) have been calculated, along with the force necessary in the iliotibial tract to stabilize the knee (I). As shown previously (Ruff, 1995), the resulting magnitude of I is reasonable and acts as one check on the validity of the model.

To calculate the resulting mediolateral bending loads on the femur itself, the vectors shown in Figure 1, calculated relative to the pelvis, need to be reversed in direction. When this is done, M-L bending moments at any cross section in the femur may be derived using the formula shown in Figure 2. The resulting bending moments along the femoral shaft are shown on the left in Figure 1. These have also been shown to be reasonable in magnitude and distribution (Ruff, 1995).

What happens when we apply the same type of analysis to A.L. 228-1? Figure 3 is a tracing of A.L. 228-1's pelvis and femur drawn from a photographic slide of the skeletal reconstruction done by Dr. Peter Schmid (1983) and kindly made available to me by Dr. Robert Martin. When checked against casts of the pelvis and left femur (supplied courtesy of Dr. Bruce Latimer), all significant proportions (e.g., biacetabular breadth, biomechanical femoral neck length, femoral length) are virtually identical in the figure.[*] For the analysis the body center of gravity (W_s) was kept in the same position relative to biacetabular breadth, and the origin of the abductors (M) was maintained in the same posi-

---

[*] Reconstructions of the pelvis by Schmid (1983) and Tague and Lovejoy (1986)—the former the basis for the photograph and the latter for the cast measured here—are slightly different, but the major dimensions of interest for this study were found to be very similar in the two versions. It was not possible to use the recent new reconstruction of the pelvis by Häusler and Schmid (1995) since the published photographs are not in the correct orientation for the purposes of this study.

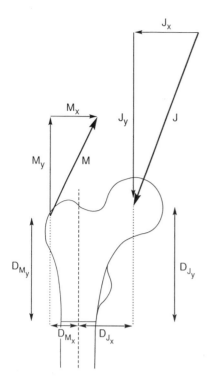

Bending Moment at Section of Interest:

$$J_y \cdot D_{J_x} - J_x \cdot D_{J_y} + M_y \cdot D_{M_x} + M_x \cdot D_{M_y}$$

Figure 2. Method of calculating femoral shaft bending moments from hip joint reaction and gluteal abductor forces. Clockwise bending moments are considered to be positive (Ruff, 1995).

tion relative to bi-iliac (total pelvic) and iliac blade breadth as in the modern human of Figure 1. (Note that the relative degree of sagittal or frontal orientation of the iliac blades in A.L. 228-1 should have little if any effect on the model, since for calculation of M-L moment arms all dimensions are projected into the coronal plane.) The precise position of the knee joint in one-legged stance was determined by first calculating the length of the tibia from the femur using a modern human African sample (Ruff, 1995) and assuming that the distal articular surface of the tibia was centered under the whole body midline (W). Because of significant distortion of the preserved proximal tibia (A.L. 228-1aq) and heavy reconstruction of the distal femur (A.L. 228-1ap) (Johanson et al., 1982a), casts of the better preserved knee joint of A.L. 129–1a,b (Johanson and Coppens, 1976) were used to establish the femoral-tibial angle at the knee. Slight differences in either tibia length or this angle make very little difference in the final results. All vectors were scaled to the same body weight (W) as in Figure 1, with the proportion of body weight superimposed over the hip (W_s) assumed to be the same. Bending moments in the femur were calculated as before.

The predicted change in magnitudes of hip joint reaction force (ΔJ), abductor force (ΔM), and bending moments at the subtrochanteric ($\Delta 80\%$ BM) and midshaft ($\Delta 50\%$ BM) femur of A.L. 228-1 relative to the modern human model are shown in Figure 3 along with the superimposed vector triangles of Figures 1 and 3 used to calculate these parameters. The abductor force in A.L. 228-1 is predicted to rise by 27%, while the hip joint reaction force rises by 12%. These increases are brought about by the very large increase in

A.L. 288-1

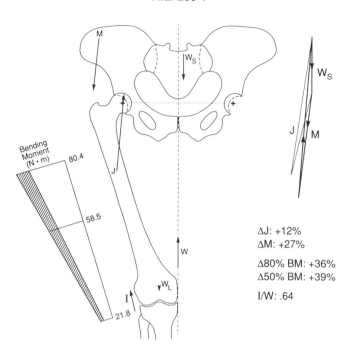

Figure 3. Force vectors and femoral shaft bending moments in A.L. 228-1 compared to modern human of Figure 1. All forces drawn to scale. Hatched area in bending moment diagram shows increase in A.L. 228-1; in vector triangle heavy lines are A.L. 228-1 and lighter lines are modern human (enlarged 2X). Numbers below triangle show increases in J, M, and bending moments at 80% (subtrochanteric) and 50% (midshaft) locations, and the ratio of I to W. All symbols as in Figure 1.

biacetabular distance, which is only partially offset by the wide iliac flare; thus in order to balance the pelvis, abductor force and joint reaction force must also increase (see Ruff, 1995 for further discussion). The predicted increases in mediolateral bending of the femoral shaft (relative to body weight) are substantial — about 35–40%. The relative magnitude of the iliotibial tract force necessary to stabilize the knee (I) is larger than in the modern human model, but still within reasonable limits (Ruff, 1995). [†]

So, if the modern human model applied to "Lucy", that is, if she walked the same way we do, one would expect evidence for the following in her skeleton: a) increased buttressing of the ilium against the increased force of the gluteal abductors; b) a somewhat larger femoral head to maintain hip joint stresses at about the same level under increased joint force; and, c) markedly increased M-L bending strength of the femoral shaft to counter much larger M-L bending moments. In order to properly evaluate these features, though, they must be considered relative to her body mass or to another appropriate parameter related to body mass. The following section describes how body mass was derived for A.L. 228-1 in this study.

[†] The force in the iliotibial tract would also probably reduce M-L bending of the femoral shaft (see Ruff, 1995), but since this would also affect bending of the modern (as well as fossil—see below) *Homo* femur, the effect on relative differences in M-L bending of the shaft should be minimal.

3. BODY MASS ESTIMATION

As discussed previously (Ruff et al., 1993), in analyses of the relative size of post-cranial skeletal features, it is difficult to avoid circular reasoning when estimating body mass. For example, the joint size of weight-bearing skeletal elements bears a close relationship to body mass (Ruff, 1988; Jungers, 1990) and so has been used in the past to estimate body mass in fossil hominids (Suzman, 1980; McHenry, 1991b, 1992; Ruff et al., 1997). However, it is obviously not appropriate to employ this methodology when evaluating the relative size of a hindlimb joint itself, e.g., the femoral head! The same caveat applies to all mechanically related features, including shaft cross-sectional dimensions. An alternative is to take a *non*-mechanical morphometric approach to the estimation of body mass, in which preserved skeletal elements are used to directly assess the size and shape of the body. We have used such an approach to estimate body mass in the early *Homo erectus* KNM-WT 15000 (Ruff and Walker, 1993), as well as in other Pleistocene *Homo* (Ruff, 1994; Ruff et al., 1997).

The morphometric approach is based on a "cylindrical model" of the human body, with stature as the height of the cylinder and bi-iliac breadth as the breadth (Ruff, 1991). The volume of a cylinder is equal to $\pi r^2 l$, where r is its radius (half of bi-iliac breadth) and l is its length or height (stature). Since body density can be assumed to be approximately constant and close to 1, mass should be equivalent to volume.

Multiple regressions using bi-iliac breadth and stature predict body mass with a very acceptable level of accuracy (standard errors of estimate of 3–6 kg) in a variety of modern human populations (Ruff and Walker, 1993; Ruff, 1994). Estimates using this technique are also concordant with those based on femoral head breadth regressions in Pleistocene *Homo* (Ruff et al., 1997). However, as noted previously (Ruff and Walker, 1993) the more elliptical shape of the pelvis, and presumably the trunk as a whole, in A.L. 228-1 (Tague and Lovejoy, 1986; Lovejoy, 1988) requires some adjustment in the method before it can be applied to her body mass estimation. Previously we attempted to do this using internal midplane anteroposterior breadth of the pelvis as an indicator of external A-P breadth of the body, since true external A-P breadths for modern samples were not available, but also realized its limitations as such an index (Ruff and Walker, 1993). Since then I have collected data for true external A-P breadth of the pelvis, together with bi-iliac breadth, in two osteological samples of modern humans, and so can apply a more appropriate correction factor for A.L. 228-1.

The two modern samples include an East African (Bantu) sample from the Kenya National Museums in Nairobi (N=30 adults) and a sample of US whites from the Terry Collection, Smithsonian Institution, Washington, DC (N=20 adults), each equally divided between males and females. The two hip bones and sacrum were first articulated and held together with several heavy rubber bands, and bi-iliac breadth measured using an osteometric board. Then, with the pelvis in anatomical position (anterior edge of pubic symphysis and anterior superior iliac spines in the same coronal plane), the A-P breadths from the anterior-most points to the most posterior edges of the sacrum and the ilium (posterior superior iliac spines) were measured, again using an osteometric board with the aid of some larger movable end pieces. Because A.L. 228-1's restored sacrum is probably slightly distorted (flattened) (Tague and Lovejoy, 1986), and because sacral curvature varies between modern males and females, only the A-P measurement to the posterior superior iliac spines is used here. The mean ratio of A-P to M-L (bi-iliac) pelvic breadths in the African sample is 0.5082 (±0.0313 SD), and in the US white sample 0.5073 (±0.0532 SD). The mean ratio of the pooled sample, rounded to 0.508 (±0.041 SD), is used here. A.L.

228-1's A-P/M-L ratio, measured in the same way, is 0.324, more than 4 standard deviations from the pooled modern sample.

The formula for volume (=mass) of a cylinder with an elliptical cross-section, modified from the one given above, is $\pi r_1 r_2 l$, where r_1 and r_2 refer to the two breadths of the cylinder and l is again its length. The modern reference sample used here to estimate A.L. 228-1's body mass here is a world-wide living sample of 56 sex/population means for matched body mass, stature, and body breadth, including those for African Pygmies (Ruff, 1994).[‡] Only bi-iliac breadth (not A-P breadth of the pelvis) was available as a body breadth measurement for this sample. Thus, using the elliptical cylindrical model, the following procedure was carried out in order to apply these data to A.L. 228-1:

$$
\begin{aligned}
\text{Body mass} \quad &: \quad \pi r_1 r_2 l \\
&= \quad 1/4\pi l D_1 D_2 \\
&= \quad 0.785 l D_1 D_2 \\
&= \quad 0.785 l D_1{}^2 (D_2/D_1)
\end{aligned}
$$

where D_1 and D_2 refer to the diameters in the M-L and A-P planes, respectively. For the modern sample, using the mean A-P/M-L ratio derived above, this body mass predictor becomes $0.785 \cdot$ stature $\cdot$ (bi-iliac breadth)$^2 \cdot 0.508$; for A.L. 228-1 it is $0.785 \cdot$ stature $\cdot$ (bi-iliac breadth)$^2 \cdot 0.324$. A.L. 228-1's stature is taken as 107 cm (Jungers, 1988a) and her bi-iliac breadth, adjusted for the addition of 0.5 cm soft tissue,[°] is 25.8 cm (Ruff, 1991), giving a value of 18115 cm^3 for this index.

Figure 4 is a plot of body mass against the morphometric body mass index in the modern reference sample, and the predicted body mass of A.L. 228-1. Because A.L. 228-1 lies well outside the range of modern values, including Pygmies, a Model II rather than least squares equation through the modern data is used to estimate her body mass (Olivier, 1976; Jungers, 1988a). For reasons given elsewhere (Aiello, 1992; Ruff et al., 1993), reduced major axis (RMA) analysis is used here.

The predicted body mass of A.L. 228-1 using this technique is 27.4 kg. This confirms her originally estimated body mass of 27–28 kg (Johanson and Edey, 1981; Latimer et al., 1987), and is also close to the midpoint of a range of previous body mass estimates—23 to 30.4 kg—derived using a variety of techniques (Johanson and Edey, 1981; McHenry, 1984; Latimer et al., 1987; Jungers, 1988b,c, 1991; Ruff and Walker, 1993; Porter, 1995). It is difficult to derive a true "confidence interval" for this estimate, since A.L. 228-1 lies well outside the range of modern values and the modern data points are them-

---

[‡] An error in one of the primary sources of data used for the list in Ruff, 1994 has since been discovered: the body mass for Aleut females reported by Laughlin in 1951 should have been 117 lbs. (53.4 kg) rather than 177 lbs. (80.5 kg) (Laughlin, pers. comm.). I am indebted to Dr. Steven Churchill for drawing my attention to this.

[°] We have recently suggested that for *Homo* an addition for soft tissue of about 1 cm rather than 0.5 cm may be more appropriate for individuals with bi-iliac breadths in this size range (Ruff et al., 1997). However, the sacroiliac and pubic symphyseal joint sizes of A.L. 228-1 are much smaller than in *Homo* specimens with comparable bi-iliac breadths, and it is likely that soft tissue thickness (cartilage, fibrocartilage) within these joints was also correspondingly thinner in A.L. 228-1. Thus, the original 0.5 cm correction may be more appropriate for this specimen. If 1 cm is added to A.L. 228-1's skeletal bi-iliac breadth, her body mass prediction index becomes 18824 cm^3, which gives an estimated body mass of 28.2 kg (see below), well within the error ranges plotted in subsequent figures. It should also be noted that another recent reconstruction of A.L. 228-1's pelvis (Häusler and Schmid, 1995) appears to produce a narrower skeletal bi-iliac breadth (about 1 cm less than in other reconstructions), which could result in a lower body mass estimate, although this would also depend upon the A-P dimension of the reconstruction, impossible to measure from published photographs.

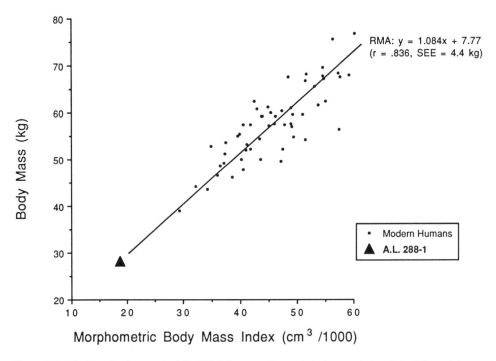

Figure 4. Prediction of body mass in A.L. 228-1 from morphometric body mass index derived from stature and body breadth (see text). Fifty-six modern human sex/population means derived from data in Ruff (1994). Reduced major axis line drawn through modern sample.

selves not individuals but sex/population means. The standard error of estimate of the regression equation through the moderns is 4.4 kg. In fact, all of the subsequent comparisons involve logarithmic transformations of body mass, which makes "confidence limits" of this kind nonsymmetrical. For ease in plotting, the average of the upper and lower $\log_e$-transformed differences of ±4.4 kg (±0.162) is given in these figures. This gives an actual range of (nontransformed) values on the plots of about 23–32 kg, encompassing the range of previous body mass estimates for A.L. 228-1 (see above). It seems very likely, therefore, that the range plotted actually includes Lucy's true body mass.

4. TESTING PREDICTIONS OF THE MODEL IN A.L. 228-1

4.1. Relative Femoral Head Size

One of the predictions of the biomechanical model of A.L. 228-1 (Figure 3) was that she would exhibit a somewhat larger femoral head relative to body mass than in modern humans. Figure 5 is log-log plot of femoral head superoinferior breadth against body mass in a modern human sample and A.L. 228-1. The modern human sample consists of 46 East African and 31 Pecos Pueblo Amerindian skeletons for which body mass was derived from bi-iliac breadth and reconstructed stature using regression equations based on living populations (Ruff et al., 1997). (Stature in the Pecos sample was calculated using Genoves' (1967) formulas for Mesoamericans, and in the East Africans using the equa-

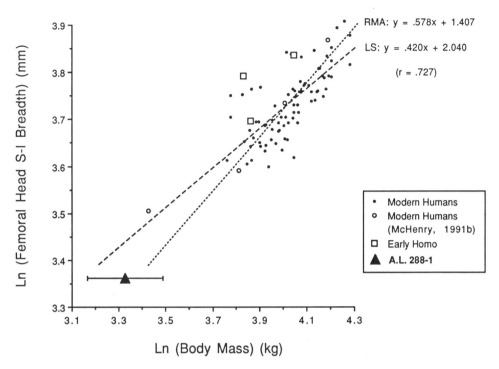

Figure 5. Log$_e$-transformed femoral head superoinferior breadth versus body mass in modern human individuals (Pecos Pueblo Amerindians and East Africans), A.L. 228-1, and early *Homo* (from smallest to largest body mass, KNM-ER 1481, 1472, and OH 28), and McHenry's (1991b) four modern human means, including Pygmies (smallest value). Both reduced major axis and least squares lines drawn through modern sample (McHenry's data not used in calculating regression). Error bars around A.L. 228-1 represent approximately ±4.4 kg. See text for details.

tions given by Feldesman and Lundy (1988) for South African blacks.) These two samples were used here in part because they have very different body builds, encompassing much of the range of variation in body breadth/height present among modern humans (Ruff, 1995). In addition, the four modern human means for femoral head breadth and body mass used by McHenry (McHenry, 1991b; 1992), which include data points for African Pygmies and the small-bodied Khoisan, are plotted for comparison. Finally, three early *Homo* (KNM-ER 1472, 1481 and OH 28) for which femoral head size can be measured or calculated (Ruff et al., 1993) and body mass estimated morphometrically (Ruff and Walker, 1993; Ruff et al., 1997) are included in the plot.[+] Both RMA and least squares regression lines through the modern East African-Pecos reference sample, extrapolated downwards to A.L. 228-1, are shown.

What is most apparent from Figure 5 is that A.L. 228-1 has a femoral head size that is about what would be predicted from her body mass, based on the scaling of these two

---

[+] In Ruff et al. (1997) body masses of several early *Homo* specimens with intact femoral heads were estimated using equations based on femoral head size to body mass in modern humans, and are larger than the morphometric estimates used here. The femoral head estimates are obviously inappropriate for the present analysis. The morphometric estimates used here were derived following general procedures described in Ruff et al. (1997).

parameters in modern humans. McHenry's (1991b) data points fall along the same general trajectory as that of my modern human sample. Indeed, the fact that he obtained a body mass estimate of 26.7 kg for A.L. 228-1 based on femoral head size and a modern human reference sample that is close to my morphometric estimate of 27.4 kg is indirect evidence that her femoral head scaled at about the same proportion to body mass as in modern humans. Thus, similar to my conclusion based on a different comparison (Ruff, 1988), but contrary to that of Jungers (Jungers, 1988c, 1991), Lucy did not have a relatively small femoral head. However, the prediction of the biomechanical model was that she should have had a relatively *large* femoral head, and there is no evidence from Figure 5 to support this. Interestingly, the three early *Homo* data points all lie in the upper range of the modern humans, i.e., they appear to have somewhat enlarged femoral heads, although this difference does not reach statistical significance (Ruff et al., 1993). I have argued elsewhere (Ruff, 1995) that hip joint reaction force in early *Homo* may have been slightly increased over that of modern humans, largely due to increased biacetabular breadth, which could explain this observation. This makes the *non*-deviation of femoral head size in A.L. 228-1, with her even more exaggerated biacetabular breadth, all the more striking.

It should also be noted that the analysis shown in Figure 5 intrinsically accounts for non-isometric scaling of femoral head size with body mass within hominids. Specifically, as shown previously (Ruff, 1988; McHenry, 1991b; Ruff et al., 1993), femoral head size is positively allometric relative to body mass in modern humans (and earlier *Homo*; see Figure 5 and Ruff et al., 1993), i.e., it increases in size faster than would be predicted to maintain geometric similarity with body mass. We have argued that reduced major axis analysis is the most appropriate method for evaluating such allometric relationships (Ruff et al., 1993); the RMA slope through the modern sample in Figure 5 is 0.578 (SE = 0.046), highly significantly different from the isometric slope of 0.333 (P<0.001). Even the lower least squares slope of 0.420 (same SE) suggests positive allometry (P<0.06). Lucy's femoral head size is consistent with this allometric scaling within hominids.

4.2. Cross-Sectional Femoral Shaft Dimensions

Another prediction of the biomechanical model was that A.L. 228-1 would show greatly increased mediolateral bending strength of the femoral diaphysis. Here not only body mass but also relative activity level and muscle strength in general must be considered, since it is known that the cortical bone in diaphyses is highly plastic in response to applied mechanical loads during life (Ruff and Runestad, 1992; Trinkaus et al., 1994). The A.L. 228-1ap femur has a break through it at approximately midshaft between its second and third segments (Johanson et al., 1982a). The endosteal boundary is clearly visible on the distal surface of a cast of the fused first and second segments (housed in the Kenya National Museums), which was photographed, digitized, and analyzed using SLICE (Nagurka and Hayes, 1980). Other dimensions of this cast were found to be quite close to those published for the original specimen (Johanson et al., 1982b).

Figure 6 is a log-log plot of midshaft femoral cortical area in modern humans (East Africans and Pecos Pueblo Amerindians), A.L. 228-1, and early *Homo* (KNM-ER 737, 1472, 1481, 1808, and OH 28) against estimated body weight. The modern and early *Homo* cross-sectional data were derived as described elsewhere (Ruff et al., 1993; Ruff, 1995); body weights for early *Homo* were derived as described above. Reduced major axis and least squares regression lines are again plotted through the modern sample.

It is clear from Figure 6 that A.L. 228-1 had a very robust femoral diaphysis relative to her body weight when compared to modern humans. Her midshaft cortical area is al-

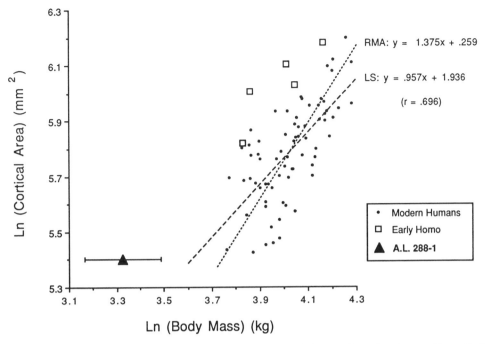

Figure 6. Log$_e$-transformed femoral midshaft cortical area versus body mass in modern humans (Pecos Pueblo Amerindians and East Africans), A.L. 228-1, and early *Homo* (from smallest to largest body mass, KNM-ER 1481, 1472, 737, OH 28, and KNM-ER 1808; data from Ruff et al., 1993, 1997). Reduced major axis and least squares lines drawn through modern sample. Error bars for A.L. 228-1 as in Figure 5.

most as large as that of the smallest modern humans in the sample, despite her body mass being some 35% lower than any of the modern humans. The early *Homo* specimens are also more robust than modern humans, as shown previously (Ruff et al., 1993). This deviation is best viewed as a reflection of overall increased muscularity and activity level in Plio-Pleistocene hominids (Ruff et al., 1993), and it is not surprising that an early australopithecine would exhibit as least as great an increase as later *Homo*. This means, however, that to properly evaluate mediolateral bending strength of the femoral shaft, this overall increased robusticity must be factored into the comparison. For this reason, ML bending strength is compared to AP bending strength at midshaft as in previous analyses of early *Homo* femora (Ruff, 1995). Anteroposterior bending strength should be a good index of overall mechanical loading of the femur that is not directly related to changes in M-L bending of the shaft. Measures of bending strength (section moduli) were derived from cross-sectional second moments of area as described previously (Ruff, 1995).

Figure 7 shows M-L versus A-P bending strength in the modern human sample, A.L. 228-1 and early *Homo*. Because there was no need to estimate body mass in this comparison (and thus no need to calculate stature and body breadth), both modern and earlier samples of *Homo* could be increased (see figure caption). A.L. 228-1 falls at or below RMA and least squares regression lines through the modern sample, indicating no increase in relative M-L bending strength of her femoral shaft. In fact, her midshaft cross section is almost circular, with nearly equal strengths in both planes. This is in stark contrast to the early *Homo* sections, which are strongly asymmetrical and buttressed against relatively high M-L bending loads. Again, as with relative femoral head size, the position of A.L.

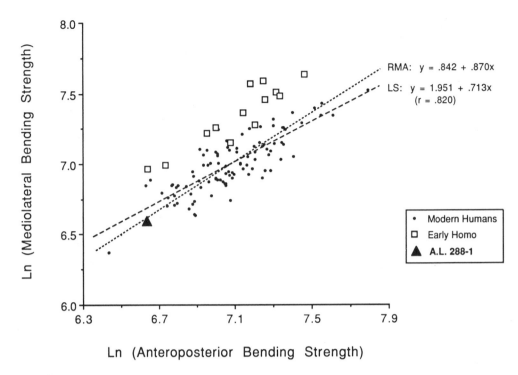

Figure 7. Log$_e$-transformed mediolateral versus anteroposterior bending strength of the femoral midshaft in modern humans (Pecos Pueblo Amerindians and East Africans), A.L.228-1, and early *Homo* (from smallest to largest A-P bending strength, Gesher Benot Ya'acov 1, KNM-ER 1481, Zhoukoudian 2, Zhoukoudian 5, KNM-ER 1472, Zhoukoudian 4, OH 28, Zhoukoudian 6, KNM-ER 737, KNM-ER 1808, Zhoukoudian 1, Aïn Maarouf 1, KNM-ER 803; data from Ruff, 1995). Reduced major axis and least squares lines drawn through modern sample.

228-1 can not be attributed to purely allometric effects: two of the smaller early *Homo* specimens with A-P bending strength close to that of A.L. 228-1 (KNM-ER 1481 and Gesher Benot Ya'acov 1) have almost 50% greater M-L bending strength at midshaft (see Figure 7).

I have previously noted that the femoral midshaft region is less reliable than the proximal femoral shaft for inferring changes in hip structure (Ruff, 1995). Thus, it would be desirable to carry out a similar analysis for a femoral subtrochanteric cross section. Unfortunately, no convenient natural break surface was available in this region for A.L. 228-1. Therefore, as an approximation, A-P and M-L external subtrochanteric breadths were used as general indices of A-P and M-L bending strengths of the proximal femoral shaft. Fossil data, including that for A.L. 228-1, were derived from McHenry (1988), Weidenreich (1941), Day (1969), and my own measurements, and modern human data were derived from a variety of sources (means for one African, six Amerindian, and four European archaeological samples). As traditionally measured, subtrochanteric breadths are more truly regarded as maximum and minimum breadths rather than M-L and A-P breadths, respectively, with the difference between the two types of measurements varying depending on femoral neck anteversion angle and other factors. These measurements also do not, of course, reflect variation in internal structure such as cortical thickness. However, they should give some indication of proximal shaft shape that can be used to evaluate general differences in morphology between groups (also see Ruff, 1987a).

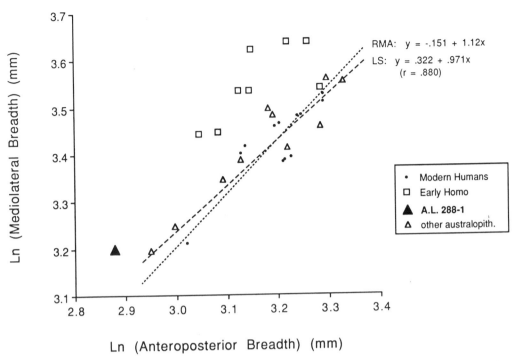

RMA: y = -.151 + 1.12x

LS: y = .322 + .971x

(r = .880)

- Modern Humans
- □ Early Homo
- ▲ A.L. 288-1
- △ other australopith.

Ln (Anteroposterior Breadth) (mm)

Figure 8. Log$_e$-transformed mediolateral versus anteroposterior femoral subtrochanteric external breadths in modern humans, A.L. 228-1 and other australopithecines, and early *Homo*. Early *Homo* data from Weidenreich, 1941; Day, 1971; McHenry, 1988; and my reconstruction for KNM-ER 1808 (see Ruff, 1995); from smallest to largest A-P breadth: KNM-ER 1481, 1472, Zhoukoudian 4, Zhoukoudian 1, KNM-ER 1808, 737, and 803. Australopithecine data from McHenry, 1988; in addition to A.L.228-1, from smallest to largest A-P breadth: A.L. 128–1, KNM-ER 1500, STS 14, OH 20, A.L.211-1, SK 97, SK 82, A.L.333-95, A.L.333W-40, A.L.333-3. Modern data points are population sample means; see text for general provenience (the smallest modern data point is for an African Bushman sample). Reduced major axis and least squares lines drawn though modern sample.

Figure 8 is a log-log plot of M-L versus A-P femoral subtrochanteric breadths in modern humans, early *Homo*, and australopithecines, including A.L 228-1. The other australopithecine data points, from McHenry's 1988 compilation, include both South and East African specimens. The australopithecines fall in the same general distribution as the modern human sample means. A.L. 228-1 is among the most platymeric (M-L expanded/A-P flattened) of them, with a positive deviation in M-L breadth from that predicted from A-P breadth of 8% relative to the least squares line through modern humans, and 14% relative to the RMA line through modern humans. Thus, her proximal shaft is moderately platymeric. However, the degree of platymeria present in the early *Homo* specimens is noticeably greater, with an average deviation in M-L breadth of about 18% from that predicted from A-P breadth using either method of line fitting through modern humans. One early *Homo* specimen (KNM-ER 803) falls close to the modern regression lines. However, its proximal shaft distal to the inferred position of the lesser trochanter (which is not preserved) is quite weathered with large chips missing; thus, its A-P and M-L breadths may not be reliable. Without this specimen, the average deviation of early *Homo* from the modern regression lines is about 20%, or about 1.5–2.5 times greater than that of A.L. 228-1.

These results reinforce previous cross-sectional analyses for a somewhat more limited fossil sample (Ruff, 1995) that indicated that australopithecines on average have moderately platymeric proximal femoral shafts, and early *Homo* have very markedly platymeric shafts. The predicted increase in M-L bending of the femoral shaft in early *Homo*, based on preserved femora and a range of possible pelvic morphologies, was actually less than that of A.L. 228-1 (Figure 3), yet early *Homo* shows much more evidence of shaft buttressing against such increased M-L loading. This strongly suggests that the relationship between morphology of the hip and mechanical loading of the femoral shaft was different in early *Homo* and A.L. 228-1.

4.3. Iliac Buttressing

The final prediction of the biomechanical model of A.L. 228-1 was that she would exhibit evidence for increased gluteal abductor force relative to modern humans. The development of an acetabulocristal buttress, or iliac pillar, is commonly viewed as a skeletal indicator of increased abductor force in hominids (e.g., Lovejoy et al., 1973). This feature is apparently developmentally plastic (Rader and Peters, 1993) and thus would be expected to reflect the actual level of force exerted by the gluteal abductors across the iliac blade.

The maximum breadth of the iliac crest, always occurring at the iliac tubercle, was used here as an indicator of the size of the acetabulocristal buttress. Because of its developmental plasticity, and evidence for overall increased muscularity and activity level in A.L. 228-1 (Figure 6), as with femoral shaft cross-sectional dimensions it is important to consider iliac buttressing relative to another skeletal feature that incorporates overall increased mechanical loading, but not specifically gluteal abductor loading. Here I chose the *minimum* iliac crest breadth, always occurring posterior to the iliac tubercle, as such an indicator.

Figure 9 shows iliac tubercle breadth plotted against minimum iliac crest breadth in a sample of modern East Africans (similar data were not available for the Pecos sample), A.L. 228-1, and two early *Homo* ilia (KNM-ER 3228 and Arago 44). A.L. 228-1 falls directly within the trajectory of the modern human data scatter (also note that both of her dimensions fall quite close to the lower end of this scatter, reinforcing the impression of her *general* skeletal robusticity relative to body mass). The two early *Homo* ilia, in contrast, fall above modern humans, especially KNM-ER 3228. Thus, while there is evidence for increased gluteal abductor force in early *Homo*, there is no such evidence in Lucy's pelvis.

5. DISCUSSION

A summary of biomechanical predictions and observed morphologies for A.L. 228-1 is given in Table 1. In each instance, the prediction based on the biomechanical model of the hip, assuming a gait pattern similar to that of modern humans (Figure 3), is not borne out by her actual morphology. Somewhat paradoxically, the foregoing analysis has demonstrated that A.L. 228-1 is in many respects very much like modern humans, with a femoral head size, iliac tubercle, and femoral shaft cross-sectional shape well within modern human limits, relative to appropriate size measures. However, the overall morphology of her hip region and its biomechanical consequences predict that she should *not* look like modern humans in these other respects. The fact that early *Homo*, who also appear to have departed from modern humans in some of the same aspects of pelvic/femoral morphology,

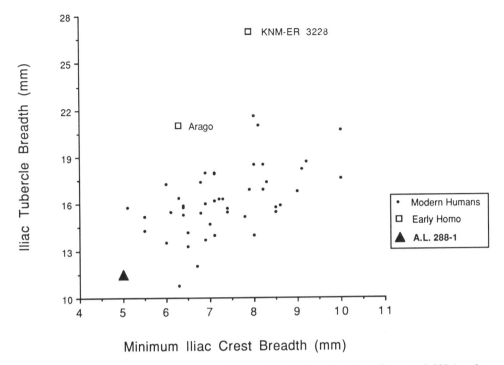

Figure 9. Iliac tubercle breadth versus minimum iliac crest breadth in modern East Africans, A.L.228-1, and two early *Homo* specimens.

including a wide interacetabular distance, long femoral neck, and possibly increased iliac flare, show evidence consistent with biomechanical predictions of the model (Ruff, 1995), strengthens the argument that A.L. 228-1, and possibly australopithecines as a whole, deviate from the model in some way. Stated in another way, if all else were equal, Lucy would not be expected to look exactly like modern humans in the morphology of her acetabulocristal buttress, femoral head, and femoral shaft shape, as is the case in early *Homo*. The fact that she does indicates that all else was *not* equal.

What could account for A.L. 228-1's deviation from the model predictions? One possibility is suggested by other results of McLeish and Charnley's original study. They found, not surprisingly, that a positive pelvic angle, that is, tilting of the pelvis up on the nonsupport side, greatly reduced abductor and hip joint reaction forces during single-legged stance. This is because a positive pelvic tilt also tilts the trunk over the support limb, moving the superimposed center of gravity of the body towards the support hip joint and greatly reducing the necessary balancing abductor moment (see Ruff, 1995: Figure 4).

Table 1. Predicted and observed morphology in A.L. 228-1 (relative to modern humans)

Biomechanical prediction	Observed morphology
Moderately increased hip joint reaction force	Little or no increase in relative hip joint size
Greatly increased gluteal abductor force	No increase in relative iliac tubercle size
Greatly increased M-L bending of femoral shaft	None to moderate increase in relative M-L bending strength of femoral shaft

Table 2. Effects of pelvic elevation on the nonsupport side in A.L. 228-1

| | A.L. 228-1 relative to modern | |
	Pelvis Level	Pelvis +4 Elevation
Hip joint reaction force	+12%	−4%
Gluteal abductor force	+27%	−5%
M-L bending proximal femur	+36%	+19%

In the same subject shown in my Figure 1, they found that a positive pelvic angle of 13° produced a decrease of more than 70% in abductor force and more than 40% in hip joint reaction force compared to that for a level pelvis.

I performed the same type of analysis for A.L. 228-1, except that I tilted her pelvis up on the nonsupport side by only 4° rather than 13°. Change in position of the body center of gravity relative to the hip joint was estimated from results given by McLeish and Charnley (1970) for three different positions of the pelvis in the same subject ("A", their figures 8–10). The relationship between pelvic angle and body weight moment arm was found to be virtually linear (R=0.999) over these three points, which ranged from -8.5° to 13° of pelvic tilt. Given the estimated position of the body center of gravity, the rest of the analysis was carried out as in Figure 3. Results are shown in Table 2. With this slight degree of pelvic elevation, predicted hip joint reaction force (relative to moderns) falls from +12% to -4%, abductor force falls from +27% to -5%, and mediolateral bending of the proximal femur falls from +36% to +19%. This in turn would predict approximately the same sized femoral head and degree of iliac buttressing, and a moderate increase in M-L buttressing of the femoral shaft relative to moderns, which is precisely what A.L. 228-1 exhibits.

The same mechanism of pelvic elevation on the nonsupport side was observed by Jenkins (1972) in young chimpanzees trained to walk bipedally. The amount of positive pelvic inclination in the chimpanzees was of approximately the same order as the 4° arbitrarily chosen here (my measurements from his figure 1). It is interesting that human patients with a painful hip demonstrate a similar phenomenon during weight-bearing on that side, apparently to reduce joint reaction force (Poss and Sledge, 1981). In both this and in so-called "gluteus medius limp" (resulting from weakness or paralysis of the gluteal abductors), lateral bending of the trunk over the support limb also occurs (Lehmann and De Lateur, 1990). This same lateral bending of the trunk is evident in McLeish and Charnley's subject with the elevated pelvis (1970: Figure 9) and was almost certainly also occurring in Jenkins' chimpanzee subjects, although this can not be determined from his published figure (which ends at the pelvis).

All this is not meant to suggest that A.L. 228-1 walked either like a modern bipedal chimpanzee or a modern human suffering from a gait abnormality. However, it does serve to illustrate that the same general biomechanical principles used in the present analysis do seem to operate in living animals that can be observed walking. It is possible that some lateral bending of the trunk and some pelvic elevation on the nonsupport side were both characteristic of A.L. 228-1's gait pattern. In any case, these motions, if they were present, must have begun during the late double-support phase of gait, i.e., prior to the assumption of single-limb support, through increased action of the abdominal and back musculature, as in modern humans with gluteal abductor weakness (Ducroquet et al., 1968: Plate 95). The fact that pelvic elevation occurs in bipedal chimpanzees, with their less well-developed hip abductor mechanism, illustrates that increased abductor force is not necessary to produce this gait pattern.

If the nonsupport hip was tilted slightly upward during gait in A.L. 228-1, this would also have had the effect of providing somewhat more ground clearance for her swing limb. Depending upon the particular reconstruction, A.L. 228-1 has been described as having a foot that was either at or slightly beyond the length predicted from her femur length in modern human adults (Jungers and Stern, 1983; Susman et al., 1984; White and Suwa, 1987). That is, relative to the probable length of her lower limb, she may have had a slightly long foot. Based on analogies with modern children, who also have relatively long feet, it has been proposed that she may have had to compensate for this during gait by increasing "vertical excursions, velocities and accelerations of the more proximal joints and limb segments" (Susman et al., 1984:147) in order for the foot to clear the ground, resulting in a less efficient gait pattern. However, only a few degrees of pelvic elevation on the swing side would have the same effect: 4° of elevation over A.L. 228-1's biacetabular breadth of 122 cm would result in an increase in height above the ground of the nonsupporting hip of 8.5 cm, while 2° of elevation would add 4.3 cm of "clearance". Thus, in addition to lowering abductor and hip joint reaction forces on the stance limb, a positive pelvic tilt could have facilitated movement of the swing limb during gait. This would have been especially critical if the stance limb also exhibited more hip and knee flexion, as suggested by some researchers (Stern and Susman, 1983; Preuschoft and Witte, 1991; Schmitt et al., 1996). A possible reduction in stride length in *Australopithecus* (Reynolds, 1987) could also be related to increased lateral trunk flexion during gait.

It is interesting in this regard that *Australopithecus* possessed six lumbar vertebrae, as opposed to the normal five found in modern humans (Robinson, 1972), since this would have increased the lateral flexibility of the trunk. However, this vertebral pattern is also found in early *Homo* (Latimer and Ward, 1993). Other mechanisms, including increased rotational mobility of the trunk during gait, have also been proposed to explain the retention of six lumbar vertebrae (Cartmill and Schmitt, 1996). The relationship between vertebral and hip morphology in hominid evolution is one that deserves further study (Ruff, 1995).

It is clear that A.L. 228-1, and australopithecines in general, were adapted for bipedality through a large number of alterations in skeletal morphology (see references at the beginning of this chapter). In addition to those features already described (McHenry, 1991a) may be added the presence of a relatively robust lower limb—the femoral diaphysis of A.L. 228-1 is very strong relative to her estimated body mass, much stronger than in modern humans and perhaps stronger even than in early *Homo* (Figure 6). Since great apes and modern humans fall very close to each other in lower limb cross-sectional diaphyseal size relative to body mass (Ruff, 1987b) this implies that Lucy was applying higher loads on her lower limb than great apes, i.e., that she was bipedal. (Modern humans have less robust diaphyses in general because of reduced muscularity and activity levels; Ruff, 1988; Ruff et al., 1993.) However, if she did walk with a slightly elevated pelvis on the swing side and/or laterally bent trunk over the support side, the additional muscular effort involved in tilting of the trunk over the stance limb would be expected to have made this gait pattern less energetically efficient than in modern humans. This, in combination with other features, such as relatively short lower limbs, may have limited her mobility on the ground to some degree. This is consistent with the view that while a facultative biped, she was likely not a long distance traveler (Rose, 1984; Hunt, 1994).

How far this interpretation can be extended to other australopithecines is difficult to determine without similar quantitative analyses, which are made difficult by the fragmentary nature of most other specimens. However, recent reconstructions of the pelvis of STS 14, an *Australopithecus africanus*, are similar to that of A.L. 228-1, including having a wide biacetabular breadth (Abitbol, 1995; Häusler and Schmid, 1995), and there are additional similarities in the proximal femur (Figure 8) and other areas of the lower limb (McHenry, 1986; 1994) between *A. afarensis* and other species of *Australopithecus*. Thus, it is possible that a slightly altered pattern of gait, and consequent restricted terrestrial mobility, characterized australopithecines in general, as compared to later *Homo*.

6. SUMMARY

To interpret variations in structure of the hominid hip it is first necessary to establish a valid biomechanical model based on observations of living humans. Here such a model is applied to the best preserved australopithecine pelvis and femur—A.L. 228-1 ("Lucy")—during the single-legged stance phase of gait, a critical component of bipedality. The model predicts several structural consequences, including a large femoral head and increased mediolateral buttressing of the femoral shaft and ilium, that are not borne out in her skeleton. In contrast, the same model works well for predicting skeletal structure in both modern and earlier *Homo* (Ruff, 1995). The most parsimonious explanation is that A.L. 228-1 did not walk exactly like *Homo*, thus violating the underlying assumptions of the model. One possibility is that instead of maintaining her pelvis level or slightly depressed on the nonsupport (swing) side during gait, she slightly elevated her nonsupport hip, reducing the abductor and hip joint reaction forces and mediolateral bending of the femoral shaft on the stance limb. This would also have had the effect of increasing ground clearance of the swing limb, which could have been advantageous if her feet were relatively long and/or her stance limb was more flexed than in modern humans.

This analysis demonstrates the value of considering morphology within a biomechanical context. It is also important, especially for small-bodied early hominids, to carefully consider the methods by which "size" is controlled in such analyses. A.L. 228-1 shows many of the hallmarks of bipedality, including a femoral shaft that is quite robust relative to her body mass, estimated here morphometrically at about 27–28 kg. However, a gait pattern characterized by increased lateral tilting of the trunk would probably have made walking less efficient, and could have reduced mobility on the ground. A slightly altered pattern of bipedal gait may have characterized australopithecines in general, with a completely modern pattern only established with the appearance of *Homo*.

ACKNOWLEDGMENTS

I would like to thank Henry McHenry and Elizabeth Strasser for inviting me to participate in this conference, Bob Martin for making available the slide of "Lucy's" reconstructed skeleton and Bruce Latimer for providing casts of her reconstructed pelvis and femur, Alan Walker and Erik Trinkaus for their collaboration in past and ongoing studies of structural variation within *Homo*, and three anonymous reviewers for their thoughtful comments. Data used in this study were collected in part through support from the National Science Foundation.

REFERENCES

Abitbol M (1995) Reconstruction of the STS 14 sacrum and pelvis: *Australopithecus africanus*. Am. J. Phys. Anthropol. *96* :143–158.

Aiello LC (1992) Allometry and the analysis of size and shape in human evolution. J. Hum. Evol. *22*:127–148.

Berge C (1991) Quelle est la significance fonctionelle du pelvis très large de *Australopithecus afarensis* (AL 228-1)? In Y Coppens and B Senut (eds.): Origine(s) de la Bipédie chez les Hominidés, Cah. Paléoanthrop. Paris: Editions du CNRS, pp. 113–119.

Broom R, and Robinson JT (1950) Notes on the pelves of the fossil ape-men. Am. J. Phys. Anthropol. *8*:489–494.

Cartmill M, and Schmitt D (1996) Pelvic rotation in human walking and running: Implications for early hominid bipedalism. Am. J. Phys. Anthropol. *22*:81.

Dart RA (1949) Innominate fragments of *Australopithecus prometheus*. Am. J. Phys. Anthropol. *7*:301–333.

Day MH (1969) Femoral fragment of a robust australopithecine from Olduvai Gorge, Tanzania. Nature *221*:230–233.

Day MH (1973) Locomotor features of the lower limb in hominids. Symp. Zool. Soc. Lond. *33*:29–51.

Ducroquet R, Ducroquet J, and Ducroquet P (1968) Walking and Limping. A Study of Normal and Pathological Walking. Philadelphia: Lippincott.

Eberhart HD, Inman VT, and Bresler B (1954) The principle elements in human locomotion. In PE Klopsteg and PD Wilson (eds.): Human Limbs and Their Substitutes. New York: McGraw-Hill, pp. 437–471.

Feldesman MR, and Lundy JK (1988) Stature estimates for some African Plio-Pleistocene fossil hominids. J. Hum. Evol. *17*:583–596.

Franciscus RG, and Holliday TW (1992) Hindlimb skeletal allometry in Plio-Pleistocene hominids with special reference to A.L. 228-1 ("Lucy"). Bull. et Mém. de la Société d'Anthropologie de Paris *n.s. 4*:5–20.

Genoves S (1967) Proportionality of the long bones and their relation to stature among Mesoamericans. Am. J. Phys. Anthropol. *26*:67–78.

Häusler M, and Schmid P (1995) Comparison of the pelves of Sts 14 and AL 228-1: Implications for birth and sexual dimorphism in australopithecines. J. Hum. Evol. *29*:363–383.

Hunt K (1994) The evolution of human bipedality: Ecology and functional morphology. J. Hum. Evol. *26*:183–202.

Inman VT, Ralston HJ, and Todd F (1981) Human Walking. Baltimore: Williams and Wilkins.

Jenkins FA (1972) Chimpanzee bipedalism: Cineradiographic analysis and implication for the evolution of gait. Science *178*:877–879.

Johanson DC, and Coppens Y (1976) A preliminary anatomical diagnosis of the first Plio/Pleistocene hominid discoveries in the Central Afar, Ethiopia. Am. J. Phys. Anthropol. *45*:217–234.

Johanson DC, and Edey MA (1981) Lucy, The Beginnings of Humankind. New York: Simon and Schuster.

Johanson DC, Lovejoy CO, Kimbel WH, White TD, Ward SC, Bush ME, Latimer BM, and Coppens Y (1982a) Morphology of the Pliocene partial hominid skeleton (A.L. 228-1) from the Hadar formation, Ethiopia. Am. J. Phys. Anthropol. *57*:403–451.

Johanson DC, Taieb M, and Coppens Y (1982b) Pliocene hominids from the Hadar Formation, Ethiopia (1973–1977): Stratigraphic, chronologic, and paleoenvironmental contexts, with notes on hominid morphology and systematics. Am. J. Phys. Anthropol. *57*:373–402.

Jungers WL (1985) Body size and scaling of limb proportions in primates. In WL Jungers (ed.): Size and Scaling in Primate Biology. New York: Plenum Press, pp. 345–381.

Jungers WL (1988a) Lucy's length: Stature reconstruction in *Australopithecus afarensis* (A.L. 228-1) with implications for other small-bodied hominids. Am. J. Phys. Anthropol. *76*:227–231.

Jungers WL (1988b) New estimates of body size in australopithecines. In FE Grine (ed.): Evolutionary History of the "Robust" Australopithecines. New York: Aldine de Gruyter, pp. 115–125.

Jungers WL (1988c) Relative joint size and hominid locomotor adaptations with implications for the evolution of hominid bipedalism. J. Hum. Evol. *17*:247–265.

Jungers WL (1990) Scaling of postcranial joint size in hominoid primates. In FK Jouffroy, MH Stack, and C Niemitz (eds.): Gravity, Posture and Locomotion in Primates. Florence: Il Sedicesimo, pp. 87–95.

Jungers WL (1991) A pygmy perspective on body size and shape in *Australopithcus afarensis* (AL 228-1, "Lucy"). In Y Coppens and B Senut (eds.): Origine(s) de la Bipédie chez les Hominidés, Cah. Paléoanthrop. Paris: Editions du CNRS, pp. 215–224.

Jungers WL, and Stern JT (1983) Body proportions, skeletal allometry and locomotion in the Hadar hominids: A reply to Wolpoff. J. Hum. Evol. *12*:673–684.

Latimer B, Ohman JC, and Lovejoy CO (1987) Talocrural joint in African hominids: Implications for *Australopithecus afarensis*. Am. J. Phys. Anthropol. *74*:155–175.

Latimer B, and Ward CV (1993) The thoracic and lumbar vertebrae. In A Walker and R Leakey (eds.): The Nariok-otome *Homo Erectus* Skeleton. Cambridge: Harvard Univ. Press, pp. 266–293.

Le Gros Clark WE (1955) The os innominatum of the recent Ponginae with special reference to that of the Australopithecinae. Am. J. Phys. Anthropol. *13*:19–27.

Le Gros Clark WE (1959) The Antecedents of Man. Edinburgh: Edinburgh University Press.

Lehmann JF, and De Lateur BJ (1990) Gait analysis: Diagnosis and management. In FJ Kottke and JF Lehmann (eds.): Krusen's Handbook of Physical Medicine and Rehabilitation. Philadelphia: Saunders, pp. 108–124.

Lovejoy CO (1975) Biomechanical perspectives on the lower limb of early hominids. In RH Tuttle (ed.): Primate Functional Morphology and Evolution. The Hague: Mouton, pp. 291–326.

Lovejoy CO (1988) Evolution of human walking. Sci. Am. *259*:118–125.

Lovejoy CO, Heiple KG, and Burstein AH (1973) The gait of *Australopithecus*. Am. J. Phys. Anthropol. *38*:757–780.

McHenry HM (1975) Biomechanical interpretation of the early hominid hip. J. Hum. Evol. *4*:343–355.

McHenry HM (1984) Relative cheek-tooth size in *Australopithecus*. Am. J. Phys. Anthropol. *64*:297–306.

McHenry HM (1986) The first bipeds: A comparison of the *A. afarensis* and *A. africanus* postcranium and implications for the evolution of bipedalism. J. Hum. Evol. *15*:177–191.

McHenry HM (1988) New estimates of body weight in early hominids and their significance to encephalization and megadontia in "robust" australopithecines. In FE Grine (ed.): Evolutionary History of the "Robust" Australopithecines. New York: Aldine de Gruyter, pp. 133–148.

McHenry HM (1991a) First steps? Analyses of the postcranium of early hominids. In Y Coppens and B Senut (eds.): Origine(s) de la Bipédie chez les Hominidés, Cah. Paléoanthrop. Paris: Editions du CNRS, pp. 133–141.

McHenry HM (1991b) Sexual dimorphism in *Australopithecus afarensis*. J. Hum. Evol. *20*:21–32.

McHenry HM (1992) Body size and proportions in early hominids. Am. J. Phys. Anthropol. *87*:407–431.

McHenry HM (1994) Early hominid postcrania. Phylogeny and function. In RS Corruccini and RL Ciochon (eds.): Integrative Paths to the Past (Advances in Human Evolution Series, Vol. 2). Englewood Cliffs: Prentice Hall, pp. 168–251.

McHenry HM, and Corruccini RS (1978) The femur in early human evolution. Am. J. Phys. Anthropol. *49*:473–488.

McLeish RD, and Charnley J (1970) Abduction forces in the one-legged stance. J. Biomech. *3*:191–209.

Mednick LW (1955) The evolution of the human ilium. Am. J. Phys. Anthropol. *13*:203–216.

Nagurka ML, and Hayes WC (1980) An interactive graphics package for calculating cross-sectional properties of complex shapes. J. Biomech. *13*:59–64.

Napier JR (1964) The evolution of bipedal walking in the hominids. Arch. Biol. (Leige) *75 (Suppl.)*:673–708.

Napier JR (1967) The antiquity of human walking. Sci. Am. *216*:56–66.

Olivier G (1976) The stature of *Australopithecus*. J. Hum. Evol. *5*:529–534.

Porter AMW (1995) The body weight of AL 228-1 ("Lucy"): A new approach using estimates of skeletal length and the body mass index. Int. J. Osteoarch. *5*:203–212.

Poss R, and Sledge CB (1981) Surgery of the hip in rheumatoid arthritis. In ED Harris, S Ruddy, and CB Sledge (eds.): Textbook of Rheumatology, Vol. II. Philadelphia: Saunders, pp. 1960–1972.

Preuschoft H, and Witte H (1991) Biomechanical reasons for the evolution of hominid body shape. In Y Coppens and B Senut (eds.): Origine(s) de la Bipédie chez les Hominidés, Cah. Paléoanthrop. Paris: Editions du CNRS, pp. 59–77.

Rader WT, and Peters CR (1993) Hypertrophy of the acetabulocristal buttress in *Homo sapiens*. Am. J. Phys. Anthropol. *92*:149–153.

Reynolds TR (1987) Stride length and its determinants in humans, early hominids, primates, and mammals. Am. J. Phys. Anthropol. *72*:101–115.

Robinson JT (1972) Early Hominid Posture and Locomotion. Chicago: Univ. Chicago Press.

Rose MD (1984) Food acquisition and the evolution of positional behavior: The case of bipedalism. In DJ Chivers, BA Wood, and A Bilsborough (eds.): Food Acquisition and Processing in Primates. New York: Plenum, pp. 509–523.

Ruff CB (1987a) Sexual dimorphism in human lower limb bone structure: Relationship to subsistence strategy and sexual division of labor. J. Hum. Evol. *16*:391–416.

Ruff CB (1987b) Structural allometry of the femur and tibia in Hominoidea and *Macaca*. Folia. Primatol. *48*:9–49.

Ruff CB (1988) Hindlimb articular surface allometry in Hominoidea and *Macaca*, with comparisons to diaphyseal scaling. J. Hum. Evol. *17*:687–714.

Ruff CB (1991) Climate, body size and body shape in hominid evolution. J. Hum. Evol. *21*:81–105.

Ruff CB (1994) Morphological adaptation to climate in modern and fossil hominids. Yrbk. Phys. Anthropol. *37*:65–107.

Ruff CB (1995) Biomechanics of the hip and birth in early *Homo*. Am. J. Phys. Anthropol. *98*:527–574.

Ruff CB, and Runestad JA (1992) Primate limb bone structural adaptations. Ann. Rev. Anthrop. *21*:407–433.

Ruff CB, Trinkaus E, and Holliday TW (1997) Body mass and encephalization in Pleistocene *Homo*. Nature *387*:173–176.

Ruff CB, Trinkaus E, Walker A, and Larsen CS (1993) Postcranial robusticity in *Homo*, I: Temporal trends and mechanical interpretation. Am. J. Phys. Anthropol. *91*:21–53.

Ruff CB, and Walker A (1993) Body size and body shape. In A Walker and R Leakey (eds.): The Nariokotome *Homo Erectus* Skeleton. Cambridge: Harvard Univ. Press, pp. 234–265.

Schmid P (1983) Eine rekonstruktion des skelettes von A.L. 228-1 (Hadar) und deren konsequenzen. Folia Primatol. *40*:283–306.

Schmitt D, Stern JR, and Larson SG (1996) Compliant gait in humans: Implications for substrate reaction forces during australopithecine bipedalism. Am. J. Phys. Anthropol. *Suppl. 22*:209.

Stern JT, and Susman RL (1983) The locomotor anatomy of *Australopithecus afarensis*. Am. J. Phys. Anthropol. *60*:279–317.

Stern JT, and Susman RL (1991) "Total morphological pattern" versus the "magic trait": Conflicting approaches to the study of early hominid bipedalism. In Y Coppens and B Senut (eds.): Origine(s) de la Bipédie chez les Hominidés, Cah. Paléoanthrop. Paris: Editions du CNRS, pp. 99–111.

Susman RL, Stern JT, Jr., and Jungers WL (1984) Arboreality and bipedality in the Hadar hominids. Folia Primatol. *43*:113–156.

Suzman IM (1980) A new estimate of body weight in South African australopithecines. In RE Leakey and BA Ogot (eds.): Proceedings of the 8th Panafrican Congress of Prehistory and Quaternary Studies Nairobi, 5 to 10 September 1977. Nairobi: The International Louis Leakey Memorial Institute for African Prehistory, pp. 175–179.

Tague RG, and Lovejoy CO (1986) The obstetric pelvis of A.L. 228-1 (Lucy). J. Hum. Evol. *15*:237–255.

Trinkaus E, Churchill SE, and Ruff CB (1994) Postcranial robusticity in *Homo*, II: Humeral bilateral asymmetry and bone plasticity. Am. J. Phys. Anthropol. *93*:1–34.

Walker A (1973) New *Australopithecus* femora from East Rudolf, Kenya. J. Hum. Evol. *2*:545–555.

Washburn SL (1950) The analysis of primate evolution with particular reference to the origin of man. Cold Spring Harbor Symp. Quant. Biol. *15*:67–78.

Weidenreich F (1941) The extremity bones of *Sinanthropus pekinensis*. Paleont. Sinica (N.S. D.) *5D*:1–150.

White TD, and Suwa G (1987) Hominid footprints at Laetoli: Facts and interpretations. Am. J. Phys. Anthropol. *72*:485–514.

Wolpoff MH (1976) Fossil hominid femora. Nature *264*:812–813.

Wood B (1976) Remains attributable to *Homo* in East Rudolf succession. In Y Coppens, FC Howell, GL Issac and REF Leakey (eds.): Earliest Man and Environments in the Lake Rudolf Basin. Chicago: Univ. of Chicago Press, pp. 490–506.

INDEX